AF846680

Soil Biology

Volume 49

Series Editor
Ajit Varma, Amity Institute of Microbial Technology,
Amity University Uttar Pradesh, Noida, UP, India

More information about this series at http://www.springer.com/series/5138

Muhammad Zaffar Hashmi • Vivek Kumar •
Ajit Varma
Editors

Xenobiotics in the Soil Environment

Monitoring, Toxicity and Management

Springer

Editors
Muhammad Zaffar Hashmi
Department of Meteorology
COMSATS Institute of Information Technology
Islamabad, Pakistan

Vivek Kumar
Amity Institute of Microbial Technology
Amity University Uttar Pradesh
Noida, India

Ajit Varma
Amity Institute of Microbial Technology
Amity University Uttar Pradesh
Noida, India

ISSN 1613-3382 ISSN 2196-4831 (electronic)
Soil Biology
ISBN 978-3-319-47743-5 ISBN 978-3-319-47744-2 (eBook)
DOI 10.1007/978-3-319-47744-2

Library of Congress Control Number: 2017931979

© Springer International Publishing AG 2017
This work is subject to copyright. All rights are reserved by the Publisher, whether the whole or part of the material is concerned, specifically the rights of translation, reprinting, reuse of illustrations, recitation, broadcasting, reproduction on microfilms or in any other physical way, and transmission or information storage and retrieval, electronic adaptation, computer software, or by similar or dissimilar methodology now known or hereafter developed.
The use of general descriptive names, registered names, trademarks, service marks, etc. in this publication does not imply, even in the absence of a specific statement, that such names are exempt from the relevant protective laws and regulations and therefore free for general use.
The publisher, the authors and the editors are safe to assume that the advice and information in this book are believed to be true and accurate at the date of publication. Neither the publisher nor the authors or the editors give a warranty, express or implied, with respect to the material contained herein or for any errors or omissions that may have been made. The publisher remains neutral with regard to jurisdictional claims in published maps and institutional affiliations.

Printed on acid-free paper

This Springer imprint is published by Springer Nature
The registered company is Springer International Publishing AG
The registered company address is: Gewerbestrasse 11, 6330 Cham, Switzerland

Foreword

Release of industrial and agricultural chemicals has resulted in widespread environmental pollution. The organic and inorganic compounds are released during the production, storage, transport, and use of these chemicals into the environment every year, as a result of various anthropogenic activities. While a few of them are naturally occurring but released in large quantities as a result of developmental activities, majority are foreign chemical substances that are not naturally produced in the biological milieu and have been given the name "xenobiotic" compounds. Xenobiotic compounds are mostly produced and released by (1) chemical and pharmaceutical industries producing a wide array of xenobiotics and synthetic polymers, (2) pulp and paper bleaching industry releasing natural and man-made halogenated aromatic compounds, (3) mining activities releasing heavy metals, (4) fossil fuels released through accidental spills, and (5) intensive agriculture releasing huge quantities of fertilizers, pesticides, and herbicides. The release of xenobiotic compounds has adversely affected the various components of the ecosystem including air, water, and soil and the ecosystem services they provide. This has resulted in a great concern to the inhabitants of this planet and prodded the global scientific community to develop ways and means of tackling this menace which impacts the ecosystem through both point and nonpoint sources.

Mercifully, a window of opportunity existed with the earliest inhabitants of this planet, the microorganisms, who had the repertoire of biochemical catalysts to decontaminate these xenobiotics. The large diversity of metabolic potential and the high genetic plasticity of microorganisms allow them to degrade almost all organic compounds of natural or anthropogenic (xenobiotics) origin and sequester or transform some of the heavy metals including those that are sources of environmental pollution. Thus, microorganisms are the major actors to eliminate or alleviate pollutions in the environment through the process of bioremediation. Such processes can be in situ involving cleanup of a contaminated site through in-place treatment or ex situ implementing treatment of soil or water that is removed from a contaminated site. The natural attenuation processes due to microbial activities as well as the possibilities of using microorganisms in the preventive treatments and bioremediation including biostimulation, bioaugmentation, rhizostimulation, bioleaching, and bioimmobilization are now being exploited albeit on a limited spatial scale. The main methods for microbial treatment of xenobiotics, the chemical structure and the origin of the major contaminants, as well as the mechanisms of degradation by microorganisms on the basis of physiological, biochemical, and genetic approaches now form the new vistas of microbiology.

This book, a collection of 24 scholarly presentations on xenobiotics, covering a wide range of topics from characterization of xenobiotic compounds to their environmental fate, is a worthy collection of knowledge on the discipline. I am sure this will act as a ready reckoner of information on xenobiotics and their biodegradation to the researchers and students alike. It should also enthuse the young innovators to develop technologies to implement the principles of xenobiotic handling on a large scale for maintaining the pristine condition of the Earth.

South Asia Nitrogen Centre
New Delhi, India

School of Biotechnology
KIIT University
Bhubaneswar, Odisha
India

T.K. Adhya

Preface

The expression "xenobiotic" is derived from the Greek word for xeno (foreign) and biotics (of or related to life). Xenobiotics are the compounds that are unimportant to an organism or in other words are not a part of its usual nutrition. The common examples of xenobiotics are the combinations that include drugs, food additives, synthetic chemicals, and environmental pollutants. These chemical agents are generally excluded from the body after metabolism to compounds that are excreted through the bile, kidney, lung, or dermis. Microbial enzymes that metabolize xenobiotics are very important for the pharmaceutical industry as they are responsible for the breakdown of drugs. Similarly, xenobiotic transporters also affect the duration that drugs are present in the body.

Man's use of xenobiotics dates from ancient times, but awareness about foreign compound metabolism dates from only the mid-nineteenth century when the understanding, methods, and techniques of organic chemistry were first applied to its study. For nearly a century thereafter, biotransformation was generally equated with "detoxification" or the elimination of a compound's biological activity.

Bioremediation can be defined as any process that uses microorganisms or their enzymes to return the environment altered by contaminants to its original condition. It can also be described as "a treatability technology that uses biological activity to reduce or lessen the concentration or toxicity of a pollutant. The process of bioremediation involves detoxification and mineralization, where toxic waste is converted into inorganic compounds such as carbon dioxide, water, and methane. When xenobiotic compounds are persistent in the environment, their biodegradation often proceeds through manifold steps utilizing diverse enzyme systems or dissimilar microbial populations.

Microbial biodegradation of xenobiotics is one of the significant and important ways to remove or detoxify the environmentally harmful compounds. The role of potential microorganisms to metabolize the xenobiotic compounds has been recognized as an effective means of toxic and hazardous waste removal. Contaminated wastewater, ground or surface waters, soils, sediments, and air where there has been either accidental or intentional release of pollutants or chemicals are the sites where bioremediation is employed.

Cleaning up the contamination and dealing rationally with wastes is, of course, in everybody's best interests. Considering the number of problems in the field of xenobiotics, the role of prospective microorganisms for environmental cleanup is highlighted in this work.

Twenty-four relevant articles are chosen to illustrate the problem with solution of the main areas of xenobiotics such as types and mode of action, atmospheric pollutants of Himalayan region, HCH and DDT residues in Indian soil, antibiotic resistance genes in soil, biomonitoring of humans, earthworms in biomonitoring, metagenomic strategies, techniques for quantification of soil-bound xenobiotic compounds, xenobiotics in food chain, biophasic dose–response phenomenon, agrochemicals and soil microbes, soil microbial and enzymatic diversity, in situ remediation assessment, transgenic approaches, bioavailability of xenobiotics, biodegradation, and many more. The distinct role of microbes in remediation of xenobiotics in future is emphasized considering the opportunities to contribute new approaches and directions in remediation of a contaminated environment, minimizing waste releases, and developing pollution prevention alternatives using the end-of-pipe technology. To take advantage of these opportunities, new strategies are also analyzed and produced. These methods would improve the understanding of existing biological processes in order to increase their efficiency, productivity, flexibility, and repeatability.

An enormous amount of natural and xenobiotic compounds are added to the environment every day. By exploring and employing the untapped potential of microbes and their products, there are possibilities of not only removing toxic compounds from the environment but also the conversion and production of useful end products. Basic methodologies and processes are highlighted in this book which will help in satisfying the expectations of different level of users/readers.

This book has been designed to serve as a comprehensive as well as a wide-ranging reference book. The authors thank all those who have contributed significantly in understanding the different aspects of the book and submitted their reviews and at the same time hope that it will prove of equally high value to advanced undergraduate, graduate students, research scholars, scientists, academicians, and designers of water, wastewater, and reclaimed soil. It is trusted that the enthusiasm and noteworthy opportunities presented in this work about our recent understanding of the challenges and relationships that brings about learning xenobiotic compounds and microbial synergistic approach will inspire readers to push the field forward to new frontiers.

Islamabad, Pakistan — Muhammad Zaffar Hashmi
Noida, India — Vivek Kumar
Noida, India — Ajit Varma

Contents

List of Contributors

Ghulam Hassan Abassi Department of soil Science, University College of Agriculture and Environmental Sciences, The Islamia University of Bahwalpur, Bahwalpur, Pakistan

Ali Abdel-Salam Department of soils and water, Faculty of Agriculture, Benha University, Banha, Egypt

Mohamed A. Abdel-Salam Department of Crop Sciences, Faculty of Agriculture, Minufiya University, Shibin El-Kom, Egypt

Maqshoof Ahmad Department of soil Science, University College of Agriculture and Environmental Sciences, The Islamia University of Bahwalpur, Bahwalpur, Pakistan

Niaz Ahmad Department of Soil Science, Faculty of Agricultural Sciences and Technology, Bahauddin Zakariya University, Multan, Pakistan

Zulkifl Ahmed Department of Building and Architecture Engineering, Bahauddin Zakariya University, Multan, Pakistan

Farjana Akter Sher-e-Bangla Agricultural University, Dhaka, Bangladesh

Hossain M. Anawar Department of Environmental Sciences, Faculty of Science and Engineering, Macquarie University, Sydney, NSW, Australia

Saeed Ahmad Asad Center for Climate Research and Development (CCRD), COMSATS University, Islamabad, Pakistan

Biljana Balabanova Faculty of Agriculture, University of Goce Delčev, Štip, Republic of Macedonia

Vishwas A. Bapat Department of Biotechnology, Shivaji University Vidyanagar, Kolhapur, India

Safdar Bashir Institute of Soil and Environmental Sciences, University of Agriculture Faisalabad, Faisalabad, Pakistan

Rupesh Kumar Basniwal Amity Institute for Advanced Research and Studies, Amity University, Noida, Uttar Pradesh, India

Irshad Bibi Institute of Soil and Environmental Sciences, University of Agriculture Faisalabad, Faisalabad, Pakistan

Paromita Chakraborty SRM Research Institute, SRM University, Kattankulathur, Tamil Nadu, India

Department of Civil Engineering, SRM University, Kattankulathur, Tamil Nadu, India

Muhammad Dawood Department of Environmental Sciences, Bahauddin Zakariya University, Multan, Pakistan

Abida Farooqi Department of Environmental Sciences, Faculty of Biological Sciences, Quaid-i-Azam University, Islamabad, Pakistan

Liliana Gianfreda Department of Agriculture, University of Naples Federico II, Naples, Italy

Pankaj Goyal Amity Institute of Microbial Technology, Amity University, Noida, Uttar Pradesh, India

Muhammad Zaffar Hashmi Department of Meteorology, COMSATS Institute of Information Technology, Islamabad, Pakistan

Ahmad Hasnain Department of Geography, Bahauddin Zakariya University, Multan, Pakistan

Waseem Hassan Department of Soil Science, Muhammad Nawaz Shareef University of Agriculture, Multan, Pakistan

Azhar Hussain Department of soil Science, University College of Agriculture and Environmental Sciences, The Islamia University of Bahwalpur, Bahwalpur, Pakistan

Umesh B. Jagtap Department of Botany, Government Vidarbha Institute of Science and Humanities, Amravati, Maharashtra, India

Moazzam Jamil Department of soil Science, University College of Agriculture and Environmental Sciences, The Islamia University of Bahwalpur, Bahwalpur, Pakistan

Md Abdul Kader Department of Soil Science, Bangladesh Agricultural University, Mymensingh, Bangladesh

Atif Kamal Department of Environmental Sciences, PMAS Arid Agriculture University, Rawalpindi, Pakistan

Environmental Biology and Ecotoxicology Laboratory, Department of Environmental Sciences, Faculty of Biological Sciences, Quaid-I-Azam University, Islamabad, Pakistan

Srujana Kathi Pondicherry University, Kalapet, India

Dambaru Ballab Kattel Institute of Tibetan Plateau Research, Chinese Academy of Sciences, Chaoyang District, Beijing, China

Department of Meteorology, COMSATS Institute of Information Technology, Islamabad, Pakistan

Fariha Khan Department of Biosciences, COMSATS Institute of Information Technology, Islamabad, Pakistan

Sohaib Khan Department of Anthropology, Bahauddin Zakariya University, Multan, Pakistan

Sanjenbam Nirmala Khuman Department of Civil Engineering, SRM University, Kattankulathur, Tamil Nadu, India

Bhupandar Kumar National Reference Trace Organics Laboratory, Central Pollution Control Board, Delhi, India

Manoj Kumar Amity Institute of Microbial Technology, Amity University, Noida, Uttar Pradesh, India

Vivek Kumar Amity Institute of Microbial Technology, Amity University, Noida, Uttar Pradesh, India

Bommanna Loganathan Department of Chemistry and Watershed Studies Institute, Murray State University, Murray, KY, USA

Adeel Mahmood Department of Environmental Sciences, Government College Women University, Sialkot, Pakistan

Riffat Naseem Malik Environmental Biology and Ecotoxicology Laboratory, Department of Environmental Sciences, Faculty of Biological Sciences, Quaid-I-Azam University, Islamabad, Pakistan

Zaffar Malik University College of Agriculture and Environmental Sciences, The Islamia University of Bahawalpur, Bahawalpur, Pakistan

Ashutosh Mohanty Truman Graduate School of Public Affair, Ulaanbaatar, Mongolia

Sidra Mukhtar Department of Environmental Sciences, Bahauddin Zakariya University, Multan, Pakistan

Nabeel Khan Niazi Institute of Soil and Environmental Sciences, University of Agriculture Faisalabad, Faisalabad, Pakistan

Anna Piotrowska-Długosz Department of Soil Science and Soil Protection, Division of Biochemistry, Faculty of Agriculture and Biotechnology, UTP University of Science and Technology, Bydgoszcz, Poland

Ram Prasad Amity Institute of Microbial Technology, Amity University, Noida, Uttar Pradesh, India

Abdul Qadir College of Earth and Environmental Science, University of the Punjab, Lahore, Pakistan

Maria A. Rao Department of Agriculture, University of Naples Federico II, Naples, Italy

Haythum M. Salem Department of Soils and Water, Faculty of Agriculture, Benha University, Banha, Egypt

Mahmoud F. Seleiman Department of Crop Sciences, Faculty of Agriculture, Minufiya University, Shibin El-kom, Egypt

Department of Agricultural Sciences, University of Helsinki, Helsinki, Finland

Chaofeng Shen College of Environmental and Natural Resource Sciences, Zhejiang University, Hangzhou, Zhejiang, People's Republic of China

Joginder Singh School of Biotechnology, Lovely Professional University, Phagwara, Punjab, India

Nim J. Singh School of Biotechnology, Gautam Buddha University, Noida, Uttar Pradesh, India

Zakaria M. Solaiman Soil Biology and Molecular Ecology Group, School of Earth and Environment (M087), and The UWA Institute of Agriculture, The University of Western Australia, Crawley, WA, Australia

Vladimir Strezov Department of Environmental Sciences, Faculty of Science and Engineering, Macquarie University, Sydney, NSW, Australia

Priyanku Teotia Department of Botany, CCS University, Meerut, India

Lekhendra Tripathee State Key Laboratory of Cryospheric Sciences, Cold and Arid Regions Environmental and Engineering Research Institute, Chinese Academy of Sciences, Lanzhou, China

Narendra Tuteja Amity Institute of Microbial Technology, Amity University, Noida, Uttar Pradesh, India

Kalim Ullah Department of Meteorology, COMSATS Institute of Information Technology, CIIT, Islamabad, Pakistan

Ajit Varma Amity Institute of Microbial Technology, Amity University, Noida, Uttar Pradesh, India

Abdul Wahid Department of Environmental Sciences, Bahauddin Zakariya University, Multan, Pakistan

Tandong Yao Institute of Tibetan Plateau Research, Chinese Academy of Sciences, Beijing, China

CAS Center for Excellence in Tibetan Plateau Earth Sciences, Beijing, China

Mazhar Iqbal Zafar Department of Environmental Sciences, Faculty of Biological Sciences, Quaid-i-Azam University, Islamabad, Pakistan

Zeshan Institute of Environmental Sciences and Engineering (IESE), School of Civil and Environmental Engineering (SCEE), National University of Sciences and Technology (NUST), Islamabad, Pakistan

Chapter 1
Xenobiotics, Types, and Mode of Action

Abdul Qadir, Muhammad Zaffar Hashmi, and Adeel Mahmood

1.1 Introduction

"Foreign chemical substance present in an organism, which is not normally/naturally expected to be present in that organism, is known as xenobiotic. All the substances that exist in unusually higher levels are also sheltered by xenobiotics. Unambiguously, drugs such as antibiotics are xenobiotic in human body because they are neither part of human food nor the body has the capability to self-produce.

Natural compounds can behave like xenobiotics, if they are transferred from one organism to another, i.e., uptake of the human hormone (natural) by fish existing downstream of sewage treatment plant, or may be the chemical defense against predator produced by some organisms (Afify et al. 1997).

The term xenobiotics as well as their impacts on the biota, however, is rarely used in the context of chemical pollutants (dioxins and polychlorinated biphenyls) because they exist as foreign substance in whole biological system, i.e., man-made substances, which did not exist in nature before its synthesis by the humans.

A. Qadir
College of Earth and Environmental Sciences, University of the Punjab, Lahore, Pakistan

M.Z. Hashmi
Department of Meteorology, COMSATS Institute of Information Technology, Islamabad 45550, Pakistan

A. Mahmood (✉)
Department of Environmental Sciences, Government College Women University, Sialkot 51310, Pakistan
e-mail: adilqau5@gmail.com

© Springer International Publishing AG 2017

M.Z. Hashmi et al. (eds.), *Xenobiotics in the Soil Environment*, Soil Biology 49,
DOI 10.1007/978-3-319-47744-2_1

1.1.1 Types of Xenobiotics

Environmental pollutants, hydrocarbons, food additives, oil mixtures, pesticides, otherxenobiotics, synthetic polymers, carcinogens, drugs, and antioxidants are the major groups of xenobiotics.

1.1.1.1 Drugs

These are the pharmaceutical medicines which are recommended to humans either for enhancing the work efficiency of body organs or because of any disorder.

1.1.1.2 Environmental Pollutants

Any compound that can change the normal ecosystem's functioning and produced naturally or intentionally might be included in this category. Environmental xenobitics mostly exist as unnatural/synthetic compounds.

1.1.1.3 Food Additives

Any food commodity having hazardous/unwanted effects on human and animal body is recognized as food additives.

1.1.1.4 Hydrocarbons

They are the derivatives of carbon and hydrogen as shown by the name. Hydrocarbons have toxic and carcinogenic impacts on the human body. These are the chemicals that are synthesized as combustion by-products.

1.1.1.5 Pesticides

These are chemicals used as fungicides, herbicides, and insecticides as well as associated organic pollutants may originate from two broad categories.

a. Produced unnaturally for various commitments.
b. Produced by industrial or anthropogenic activities as unintentional or accidental by-products.

Moreover, traces of pesticides may be produced from natural processes (Breivik et al. 2004). POPs that are produced intentionally may be classified into many subgroups. These chemicals correlate with many families of chlorinated and brominated aromatic, comprising PCBs (polychlorinated biphenyls), PBDEs (polybrominateddiphenyl ethers), PCDD/Fs (polychlorinated dibenzo-p-dioxins/ furans), PCNs (polychlorinated naphthalenes), and OCPs (organochlorines pesticides) including DDTs and its metabolites (toxaphene, chlordane, etc.). Few chemicals are produced from various sources, i.e., HCB (hexachlorobenzene) that is produced as unintentional by-product and also as industrial chemical (Bailey 2001). Another example includes PCBs, produced from both sources (Brown et al. 1995; Lohmann et al. 2000). While the comparative significance of by-product formation is ambivalent, an initial assessment accepts a little importance with respect to the PCBs' historical mass balance throughout the globe" (Breivik et al. 2002). The"Stockholm Convention" on deliberately produced POPs (HCBs and PCBs), the unwanted by-products, aims at the identification and quantification of their sources as well as the establishment of release inventories of unintended/ accidental production (Breivik et al. 2004).

1.1.1.6 Synthetic Polymers

Synthetic polymers are known as plastic polymers, including polystyrene, polyvinyl chloride, polyethylene, etc. Nylon is a synthetic polymer which is most common and extensively used in wrapping, cooked food's transportation packaging, and garments.

1.1.1.7 Oil Mixtures

Oil is a natural product. It can be subject to microbial degradation and has various rates of degradation for various compounds in it. Biodegradation has the potential to handle little oil seepages. But for large spills, pollution problem becomes very acute. Oil is recalcitrant in nature mostly because of its insolubility in water and the toxic nature of some of its constituents.

1.1.2 Other Xenobiotic Compounds

Most of the chemical compounds have cyclic rings and aliphatic structures processing exchange of the nitro-sulfonate, amino, methoxy, and carbonyl groups along with the occurrence of halogens.

1.1.3 Hazards from Xenobiotics

There are a number of potential hazardous impacts on humans and environment produced by xenobiotic compounds, which are briefly discussed below.

1.1.3.1 Toxicity

Xenobiotics compounds (halogenated aromatic hydrocarbons) are potentially toxic to biota significantly for the lower eukaryotes and even for humans.

1.1.3.2 Carcinogenicity

Pesticides that are xenobiotics are suspected carcinogens and also cause increased incidence of diabetes, endometriosis, and neurobehavioral impairment along with learning sickness and mental weakness. Persistent organic pollutants (POPs) are recognized as potential risk factor for human breast cancer by some authors (Safe 1994; Ross et al. 1995). Scientific findings from environmental impact assessment studies have shown that POPs cause immune and reproductive dysfunction, neurobehavioral disorder, endocrine disruption, and even cancer (Kelce et al. 1995). Reduced immunity in infants/children, infection, impairment, developmental abnormalities, neurobehavioral disorder, and tumor induction are a result of POP pollution. Children at developing stages are more at risk from pollutants. Growth and development of cells is very sensitive to environmental contamination as well as can easily be affected by POP exposure. Brain is of greatest concern, because POP-exposed children scored least on intelligence assessment during infancy (Bouwman 2003). Therefore, collective actions of POPs are discussed in the risk assessment procedures at screening level (Bailey 2001).

1.1.3.3 Progressive Buildup in the Environment

Many xenobiotics have persistent and recalcitrant nature in the environment, and this leads to their accumulation with time. They are manufactured and used in enough quantities, which enhance their accumulation in nature.

1.1.3.4 Bioaccumulation

Xenobiotic compounds like pesticides can increase and bioaccumulate in the food chain; apprehension exists for their impacts on the top predator species, for example, humans. Concerns about POP pollution are increasing in recent years because many compounds or metabolites are recognized as hormone disrupter as well as may alter the proper functioning of the reproductive and endocrine system of wildlife and humans. These types of pollutants can stay in fatty tissues for many years and result in chronic problems, i.e., birth defects, stunted growth, reduced ability, recurring diseases, long-lasting impairment of brain function, learning disabilities, cancer, respiratory problems including asthma, and neurological, behavioral, immunological, and reproductive deficiencies in animals and also in humans (Harrison et al. 1995).

1.1.4 Role in Biological System/Mode of Action

1.1.4.1 Xenobiotics–Protein Interaction

Xenobiotic interaction with protein (how xenobiotic and living cell organisms react to each other) defines the positive and negative impacts of xenobiotic compounds. For such purpose, xenobiotics must have

a. Penetration to targeted organism
b. Transported to the action site
c. Disrupt or alter vital function.

1.1.4.2 Transport Protein

Protein involved in the movement of small molecules, ions, and macromolecules such as another protein across a biological membrane is known as the membrane transport protein or transporter. These proteins are essential part of membrane proteins and have the capability to ease the movement/transportation of compounds as well as assist the active transport and protein diffusion; this mechanism is recognized as carrier-mediated transport. The carrier-mediated transport exists in two forms: (a) active transport (b) facilitated diffusion (Crompton 1999).

Movement of chemicals via membrane is boosted by facilitated diffusion protein, in the absence of energy input; ultimately, the related chemical can move only down a concentration gradient. This is most probably due to the high-specificity pore/channel formation, which then extends to the membrane. These "polar holes" through membrane are covered by specific residues of amino acids that lower the energy barrier to the movement of polar molecules (Fig. 1.1).

1.1.5 Conclusion

Foreign chemical substance that exists in an organism and not expected to be present normally or naturally in that organism falls into the category of xenobiotic. These chemicals include environmental pollutants, hydrocarbons, food additives,

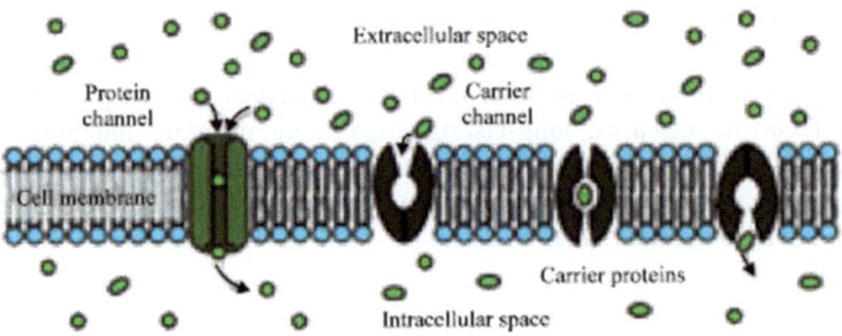

Fig. 1.1 Transportation of xenobiotics by facilitated diffusion protein (Afify 2010)

oil mixtures, pesticides, other xenobiotics, synthetic polymers, carcinogens, drugs, and antioxidants, which are the major groups of xenobiotics. Xenobiotic compounds show a variety of possible hazardous impacts on the environment and human beings like toxicity, progressive buildup, carcinogenesis, and bioaccumulation in the environment. Xenobiotic–protein interactions (how xenobiotic and living cells interaction takes place) define the beneficial as well as harmful impacts of xenobiotic compounds. For such reasons, xenobiotic compounds must show penetration into targeted organism, be transported to the action site, and should alter/disturb the vital function.

References

Afify AMR (2010) Biological function of xenobiotics through protein binding and transportation in living cells. Int J Agric Res 5:562–575

Afify AMR, Ragab AA, El-Baroty GS, Abo-Zeid M, Saleh MA (1997) Protein profile of humans milk of selected Egyptian pollutions. Egypt J Nutr 12:1–17

Bailey RE (2001) Global hexachlorobenzene emissions. Chemosphere 43:167–182

Bouwman H (2003) POPs in Southern Africa. In: Fiedler H (ed) The hand book of environmental chemistry, persistent organic pollutants, 3rd edn. Springer, Berlin

Breivik K, Sweetman A, Pacnya JM, Jones KC (2002) Towards a global historical inventory for selected PCB congeners a mass balance approach 2. Emissions. Sci Total Environ 290: 199–224

Breivik K, Alcock R, Li YF, Bailey RE, Fiedler H, Pacyna JM (2004) Primary sources of selected POPs: regional and global scale emission inventories. Environ Pollut 128:3–16

Brown JF, Frame GM, Olson DR, Webb JL (1995) The sources of the coplanar PCBs. Organohalogen Compd 26:427–430

Crompton M (1999) The mitochondrial permeability transition pore and its role in cell death. Biochem J 341:233–249

Harrison PTC, Humfrey CDN, Litcheld M, Peakall D, Shuker LK (1995) Environmental oestrogens: consequences to human health and wildlife. IEH-Medical Research Council, Leicester

Kelce WR, Stone CR, Laws SC, Gray LE, Kemppainen JA, Wilson EM (1995) Persistent DDT metabolite p,p-DDE is a potent androgen receptor antagonist. Nature 375:581–585

Lohmann R, Northcott GL, Jones KC (2000) Assessing the contribution of diffuse domestic burning as a source of PCDD/Fs, PCBs, and PAHs to the UK atmosphere. Environ Sci Technol 34: 2892–2899

Ross PS, De Swart RL, Reijnders PJH, Van Loveren H, Vos JG, Osterhaus ADME (1995) Contaminant related suppression of delayed-type hypersensitivity and antibody responses in harbor seals fed herring from the Baltic Sea. Environ Health Perspect 103:162–167

Safe SH (1994) Polychlorinated biphenyls (PCBs)-environmental impact, biochemical and toxic responses, and implications for risk assessment. Crit Rev Toxicol 24:87–149

Chapter 2
Atmospheric Pollutants and Its Transport Mechanisms in Soil Along the Himalayas, Tibetan Plateau, and Its Surroundings: A Brief Note

Dambaru Ballab Kattel, Kalim Ullah, Muhammad Zaffar Hashmi, Lekhendra Tripathee, Ashutosh Mohanty, and Tandong Yao

2.1 Introduction

The Himalayas, Tibetan Plateau, and its surroundings (HTPs) is one of the important and complex mountainous region, which plays a vital role in driving the regional, Asian, or even global climate system (Yao et al. 2012; Kattel et al. 2015). This region is also important for the flora, fauna, and cultural diversity, as well as livelihoods. In addition, soils in this regions are considered as the sink or the natural reservoir for different types of pollutants (e.g., potentially toxic trace metals) and are extremely persistent in the environment and nonbiodegradable causing threats to natural ecosystems (Tripathee et al. 2016). The HTPs is

D.B. Kattel (✉)
Institute of Tibetan Plateau Research, Chinese Academy of Sciences, No. 16 Lincui Road, Chaoyang District, Beijing 100101, China

Department of Meteorology, COMSATS Institute of Information Technology, CIIT, Islamabad, Pakistan
e-mail: katteldb@gmail.com; katteldb@itpcas.ac.cn; katteldb@comsats.edu.pk

K. Ullah • M.Z. Hashmi
Department of Meteorology, COMSATS Institute of Information Technology, CIIT, Islamabad, Pakistan

L. Tripathee
State Key Laboratory of Cryospheric Sciences, Cold and Arid Regions Environmental and Engineering Research Institute, Chinese Academy of Sciences, Lanzhou 730000, China

A. Mohanty
Truman Graduate School of Public Affair, Ulaanbaatar, Mongolia

T. Yao
Institute of Tibetan Plateau Research, Chinese Academy of Sciences, No. 16 Lincui Road, Chaoyang District, Beijing 100101, China

CAS Center for Excellence in Tibetan Plateau Earth Sciences, Beijing, China

© Springer International Publishing AG 2017

M.Z. Hashmi et al. (eds.), *Xenobiotics in the Soil Environment*, Soil Biology 49,
DOI 10.1007/978-3-319-47744-2_2

considered one of the most pristine terrestrial regions and hence can be regarded as an ideal place to monitor the natural background level of organic and inorganic pollutants (Cong et al. 2015a; Tripathee et al. 2014a). In recent days, many investigations have reported that the persistent organic pollutants (POPs) and polycyclic aromatic hydrocarbons (PAHs) are the major organic pollutants in this complex mountainous region (Gong et al. 2011, 2014; Wang et al. 2014; Devi et al. 2015). Furthermore, many studies have suggested that the existence of the organic, carbonaceous particles and inorganic pollutants in remote and high mountainous region is associated with the long-range atmospheric transport mechanism (Wang et al. 2010; Gong et al. 2014; Tripathee et al. 2014a, b; Cong et al. 2015a, b). Recent studies have reported that the regions which are located in lower latitudes are the considerable sources for the pollutants' distribution to the global, regional, and local scale. Numerous studies, which were carried out in the past, have highlighted the tropical, subtropical, arid, and alpine regions in the Asian continent as important source region, to understand the existence and distribution of pollutants in the environment (e.g., Shunthirasingham et al. 2010; Iwata et al. 1994; Jaward et al. 2005; Wang et al. 2010; Gong et al. 2011, 2014; Baek et al. 2013; Nasir et al. 2014; Devi et al. 2015). Majority of aforesaid investigations in different regions have concluded that the distributions of pollutants particularly in the Arctic, the Pacific, and the HTPs are associated with the characteristics of pollutants and atmospheric circulation systems. Thus, this chapter highlights a brief scenario of atmospheric pollutants and its existence in the Himalayas, Tibetan Plateau, and its surrounding regions based on the climatological point of view.

2.2 Atmospheric Pollutants and Its Sources

2.2.1 PAHs

Polycyclic aromatic hydrocarbons (PAHs), the majority of them being toxic in nature, are semi-volatile organic pollutants, composed of two or more fused aromatic rings (Devi et al. 2015). In general, aliphatic (*n*-alkanes) and PAHs are found in ambient air in both gas phase and particulate phase (Gong et al. 2011), which are generally generated from anthropogenic emissions, such as domestic heating, wood, fossil fuel and grass combustion, biomass burning, forest fires, road traffic, and industrial burning (Gong et al. 2011, 2014; Luo et al. 2016; Bi et al. 2016). In addition to this, some natural processes, such as Savanah burning volcanic eruptions and digenesis, also release substantial quantities of PAHs (Okuda et al. 2010; Ma et al. 2010; Tobiszewski and Namieśnik 2012). Strong incomplete combustion, incense burning, and monsoon climate may lead to different emissions

Table 2.1 High-molecular-weight PAHs compounds

No.	Name of compound	Abbreviation
1	Benzo (a)pyrene	BaP
2	Benzo (a) anthracene	BaA
3	Benzo (b)fluoranthene	BbF
4	Benzo(k)fluoranthene	BkF
5	Chrysene	CHR
6	Indeno (1,2,3-cd) pyrene	IcdP

of combustion pollutants in the Himalayas, Tibetan Plateau, and its surroundings (Gong et al. 2011, 2014).

Due to high and direct carcinogenetic and mutagenic properties of PAHs, health risk of it has been intensively focused and attracted widespread attention (DeMartinis et al. 2002; Devi et al. 2015). Once these PAHs compounds are released into the atmosphere, they may get redistributed into gas and particle phases and subsequently deposited on the soil media through wet and dry deposition mechanisms (Mackay 2001). Due to tendency of PAHs to deposit and accumulate in soil, sediments, and water, their release in soil poses a potential threat to human health, water, and environment (Menzie et al. 1992; Malik et al. 2011). Generally, high-molecular-weight PAH (HMW-PAHs) compounds, presented in Table 2.1, are carcinogenic and have the tendency to bioaccumulate in human body (USEPA 1993; Wang et al. 2007; Maliszewska-kordybach et al. 2009). Study of Devi et al. (2015) shows that the concentration of eight carcinogenic PAHs (BaA, CHR, BbF, BkF, BaP, DahA, IcdP, and BghiP) was high along the northern high mountainous region of the Indian subcontinent. They suggested that the source of high PAH contamination in Indian Himalayan region is due to mixture of incomplete combustion of coal, firewood, and traffic emission. Furthermore, recent studies on PAHs in soils from northern (Tibetan Plateau, China) and southern side (Nepal Himalayas) have suggested that the soils from southern side of the Himalayas have higher mean concentrations of PAHs than in Northern side (Luo et al. 2016; Bi et al. 2016). These results suggest that the pollutants deposit more on southern side of Himalayas than in Northern side. Recent study of precipitation chemistry between the northern and southern sides of the Himalayas by Tripathee et al. (2014a, b) confirms this result.

2.2.2 *POPs*

Persistent organic pollutants (POPs) are semi-volatile organic compounds. Numerous studies have reported that the agricultural activities are the main sources of POPs in a local level. Due to moderate vapor pressure characteristics, POPs are distributed globally (Hoff et al. 1992; Gong et al. 2014) and depend on the past local and regional characteristics and atmospheric transporting mechanisms (Wang et al. 2007).

Organo chlorine pesticides (OCPs) and polychlorinated biphenyls (PCBs) have been found in the environment to be toxic in nature (Hoff et al. 1992). Thus, these compounds are considered as hazardous compounds that pose high risk to the ecosystems and human health (Colborn and Smolen 1996; UNEP 2012). Nasir et al. (2014) noted from different literature (e.g., Iwata et al. 1994; Jaward et al. 2005; Baek et al. 2013) that, in comparison with developed nations, high concentrations of POPs generally observed in developing countries were suggestive of extensive past of ongoing usage for agricultural and vector-borne diseases as well as inability to enforce regulation to restrict indiscriminant use and disposal of industrial/agrochemicals. This result is due to the crop protection measures which are based on the intensive use of pesticides. Khan et al. (2002) reported that, overall, 83 % pesticides were used for the cotton, rice, and sugar crops. As an example, in Pakistan, extensive use of agrochemicals has been practiced since 1954 with 254 metric t reaching to 78,132 t per year annum in 2014 (Nasir et al. 2014). Nepal is also not far behind in the excessive use or deposition of DDT in soil and leaf from agricultural activities (e.g., Gong et al. 2014).

The concentrations and deposition fluxes of DDTs have been found in the organic/litter layer of the Tibetan forest soils (Wang et al. 2014). Occurrences of this compound in this region are not locally induced, mostly coming from the outside of the region, via atmospheric transport mechanisms (discussed in the following sections). A similar mechanism is also associated for depositions of POPs at higher elevation region along the southern slopes of the central Himalayas, Nepal. Furthermore, recent studies have revealed that forests play an important role in the global cycling of POPs (Wang et al. 2014). Generally, forest can increase the net atmospheric deposition of POPs to the terrestrial environment and can act as a pump that transports the pollutants from the atmosphere to the forest soil (Horstmann and McLachlan 1998; Su et al. 2006). Recent study of Gong et al. (2014) in Nepal further supports this finding. Wang et al. (2014) noted from different contemporary literature that forest soil is basically regarded as a final sink/strong reservoir of POPs due to their high content of organic matter, which has strong ability to absorb chemicals. Some studies, carried out in the past in Nepal, Pakistan, India, and Tibet (China), reported significant amount of residual PCBs, DDTs, and HCH in urban soils, coastal sediments, river streams, lakes, vegetation, leafs (e.g., Nasir et al. 2014; Gong et al. 2014; Devi et al. 2015) indicating that the region is becoming the hot spots for POPs.

2.2.3 Inorganic Pollutants

Mostly the inorganic pollutants (major ions such as Cl^-, NO_3^-, SO_4^{2-}, Na^+, NH_4^+, K^+, Mg^{2+}, Ca^{2+}, and HCO_3^-) in the Himalayas are derived from three distinct sources such as regional crustal dusts, anthropogenic emissions, and long-range transported from sea salts (Tripathee et al. 2014a). Moreover, trace elements are increasing in the environment as pollutants and many of these elements are extremely persistent and nonbiodegradable which can accumulate to toxic levels causing threats to natural ecosystems (e.g., air, water, soil, forests, etc.) and health. Previous studies have suggested that trace element concentrations are mostly derived from the crustal natural sources (e.g., Al, Mn, Fe, etc.) and anthropogenic sources (Cr, Co, Ni, Cu, Zn, Cd, and Pb) in the HTPs (Tripathee et al. 2014a; Guo et al. 2015). These studies have also illustrated that the concentrations of inorganic pollutants are higher during non-monsoon period and lower during monsoon period due to dilution effect. Nevertheless, these pollutants have decreasing trends from lower to the high altitudes and even lower concentrations in the northern sides of the Himalayas due to less population and no anthropogenic sources. In addition, recent study conducted by Tripathee et al. (2016) has found minimal elemental pollution in the central Himalayan soils; however, they have stated that the regions are at risk from increasing atmospheric pollution and long-range transport to the pristine Himalayas.

2.3 Driving Mechanisms and Temperature Sensitivity

Atmospheric pollutants derived from the urbanized areas and agricultural farm lands from South Asia can be transported to the HTPs and affect their fragile ecosystem and health (Gong et al. 2014; Tripathee et al. 2014a; Cong et al. 2015a, b). Generally, both types of pollutants such as organic (PAHs and POPs) and inorganic (major ions such as Cl^-, NO_3^-, SO_4^{2-}, Na^+, NH_4^+, K^+, Mg^{2+}, Ca^{2+}, and HCO_3^-) pollutants can undergo long-range transport and reach remote areas (e.g., Arctic, Antarctic, and the Tibetan Plateau) via atmospheric circulation system (Bustnes et al. 2009; Chiuchiolo et al. 2004; Wang et al. 2010; Tripathee et al. 2014a; Cong et al. 2015a, b).

In some literature (e.g., Wania and Mackay 1996), an interesting terminology such as "global fractionation and condensation" has been adopted to describe the discharge of POPs in tropical/subtropical environment for dissipation to polar and environmentally pristine regions through the atmospheric transport mechanisms. It is noted that the fate and transport of organic pollutants are strongly influenced by the climatic conditions and their gas/particulate partitioning (Gong et al. 2014). In general, the concentrations of PAHs are mostly sensitive with temperature. Some studies (e.g., Halsall et al. 2000; Kaupp and McLachlan 1999; Mader and Pankow 2002) have suggested that the gas-particulate partitioning depends mainly on the compounds' volatility and particulate characteristics. The colder environment such as high altitude mountains ceases the evaporation of volatile pollutants (Wilcke et al. 2002). Wang et al. (2014) reported that the low temperature, low pH, and microbial activity are more evident in the Tibetan forest soils than in the other situation. Therefore, they confirmed that due to cold climate of the plateau and the higher SOC content, the Tibetan forest soils are the final sink for DDTs, HCBs, and PCBs.

Evidence confirms that the atmosphere plays a major role in the subsequent transport or distribution of PAHs, POPs, and other inorganic pollutants in the global environment (Wania et al. 1998; Mackay, 2001; Wang et al. 2010; Tripathee et al. 2014a; Cong et al. 2015a), because it reflects ongoing emissions and responds rapidly to efforts at source reduction (Wang et al. 2010). The meteorological factors such as rainfall and wind patterns are potential driving forces for periodically changing concentrations of organic and inorganic pollutants in the environment (Wania et al. 1998; Gong et al. 2014; Tripathee et al. 2014b) and deposition of these compounds on soil media or reservoir (Devi et al. 2015). Numerous studies have been carried on the transport of pollutants in the mountainous regions in the Himalayas, Tibetan Plateau, and its surroundings (Sheng et al. 2013; Wang et al. 2014; Gong. et al. 2014, Nasir et al. 2014; Devi et al. 2015; Cong et al. 2015b). Monitoring data on pollutants in soil, grass, and ice samples from the Tibetan Plateau and its surroundings have suggested that it may be impacted by air masses derived from the Indian subcontinent as well as more local sources (Wang et al. 2010; Gong et al. 2014; Tripathee et al. 2014a; Devi et al. 2015; Cong et al. 2015b).

The climate over the Tibetan Plateau and its surroundings is mainly dominated by continental air (westerly) from the Mediterranean Sea and Eurasian continent during winter and maritime air from the Indian Ocean during summer (Kattel et al. 2013, 2015). A general pattern of atmospheric circulation over the region is presented in Fig. 2.1. Based on air mass trajectory analysis and ice core

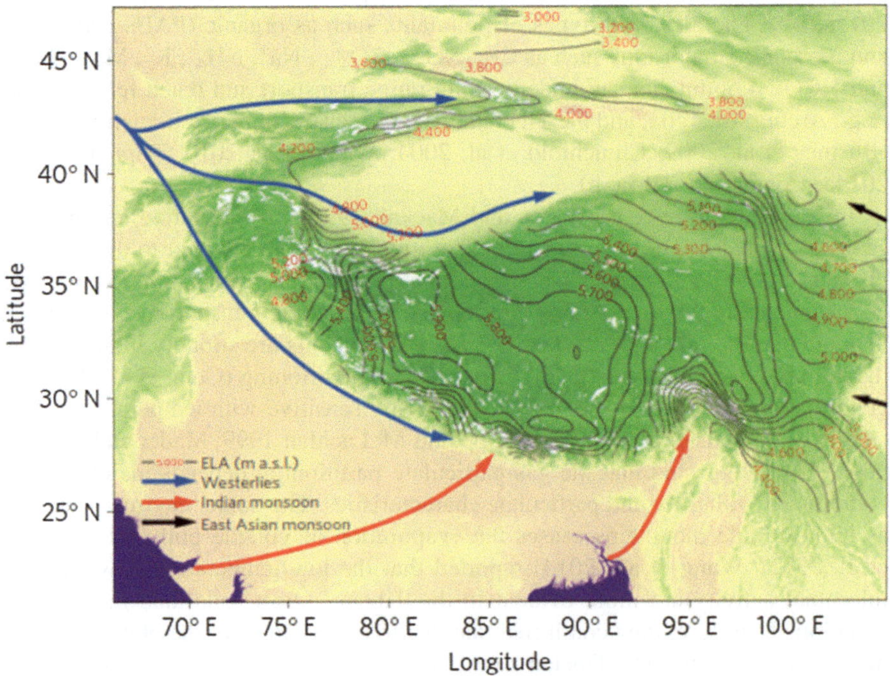

Fig. 2.1 Distribution of elevation and general patterns of the Indian monsoon and westerlies along the Himalayas, Tibetan Plateau, and its surroundings (adopted from Yao et al. 2012)

records of sulfate and black carbon, good correlations have been found between levels of pollutants in ice strata and trajectory originating from south Asia (Ming et al. 2008). Many studies (e.g., Sheng et al. 2013; Wang et al. 2014; Gong. et al. 2014; Nasir et al. 2014; Tripathee et al. 2014a; Devi et al. 2015; Guo et al. 2015) have reported that the largest deposition of pollutants was found during the monsoon seasons (May–September) in the region compared with the non-monsoon seasons (November–March), confirming the association of atmospheric circulation for the distribution of pollutants in the Himalayas, Tibetan Plateau, and surrounding regions. The Indian subcontinent has experienced heavy use of organochlorine pesticides (OCPs), such as HCHs and DDTs, while India continues the use of DDT to combat vector-borne diseases which are prevalent in the monsoon season (Zhang et al. 2008) and transported toward the Tibetan Plateau.

The monsoonal moist wind could transport these chemicals (organic and inorganic) to HTPs and its surrounding mountains, and the forest could further pump down these compounds from the atmosphere to the forest soil by litter deposition, and DDTs and HCHs were therefore strongly accumulated by the soil organic/litter horizons (long-term storage compartment) (Wang et al. 2014; Gong et al. 2014). Schematic diagram of the atmospheric circulation pattern along the Himalayas and over the Tibetan Plateau presented in Fig. 2.2a, b illustrates the transport mechanisms of pollutants in the high mountainous region in HTPs. In addition, DDTs are highly stable in the soil profiles, undergoing little translocation to the mineral layers and limited degradation. The continuous emission of DDTs in the Indian subcontinent and the sustained accumulation of DDTs by the Tibetan forest soil can occur following emission-deposition events (Wang et al. 2014). Predictions from model experiments by Fu et al. (2006) clearly show the Tibetan anticyclone could "trap" anthropogenic emissions lifted from South Asia, further supporting the argument.

The study of Sheng et al. (2013), Gong et al. (2014), Wang et al. (2014), Tripathee et al. (2014a), Cong et al. (2015b), and Guo et al. (2015) suggested that the monsoon-driven transport of pollutants from South Asia to the Tibetan Plateau and its surroundings is not a singular event, but a continuous and sustaining phenomena. The transporting of OCPs, such as HCHs and DDTs, as well as inorganic pollutants is continuous in the monsoon-dominated HTPs and its surrounding mountainous regions. It is noted that a recently published study on black carbon also demonstrated that the glaciers on the southern plateau can receive black soot, both from the south via Indian monsoon and from the west via winter westerlies (Xu et al. 2009).

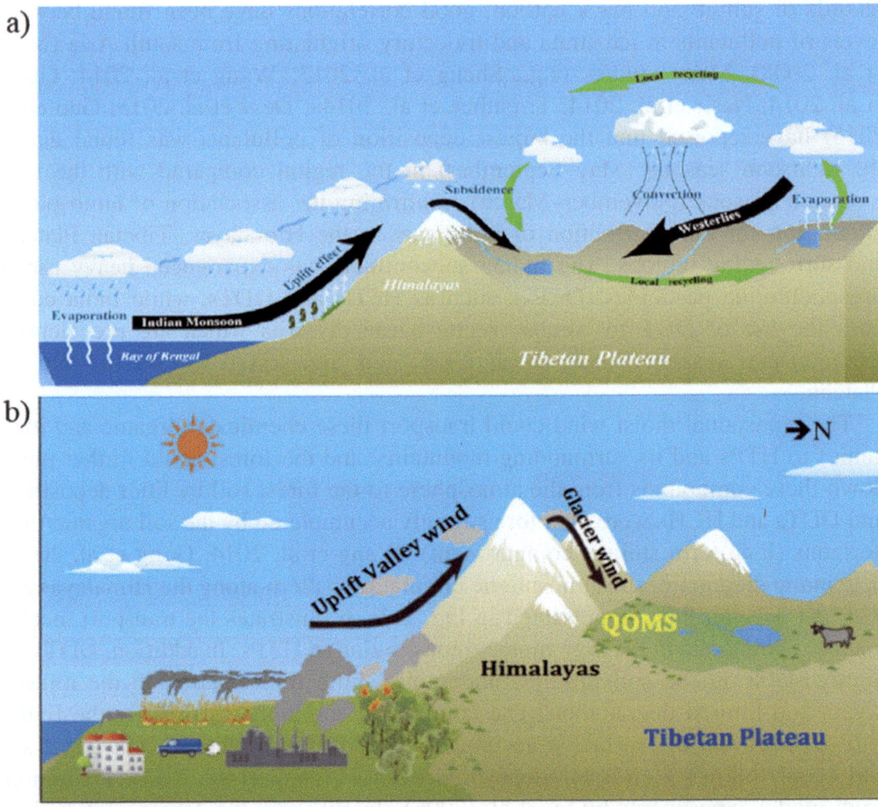

Fig. 2.2 (**a**) Schematic diagram of the atmospheric circulation along the Himalayas and over the Tibetan Plateau (adopted from Yao et al. 2013) and (**b**) aerosol transport mechanism from South Asia across the Himalayas (adopted from Cong et al. 2015b).

2.4 Conclusions

This chapter highlights a brief scenario of atmospheric pollutants and its existence in the Himalayas, Tibetan Plateau, and its surrounding regions based on the contemporary evidence. Observed organic pollutants in the region are due to the excessive use of pesticides in the agricultural activities and anthropogenic emissions, such as domestic heating, wood, fossil fuel and grasse combustion, road traffic, and industrial burning. In addition, inorganic ions and elemental pollutants are mostly derived from crustal (natural), anthropogenic (fossil fuel, biomass burning, agricultural activities, etc.), and long-range transported sea salt from ocean during monsoon season in the Himalayas. Existence of pollutants in the high mountainous region in HTPs and its surroundings is associated with long-range transport from urbanized areas and deposited via atmospheric circulation mechanisms.

Acknowledgments This work was supported by the CAS President's International Fellowship Initiative (PIFI) for Visiting Scientist and postdoctoral researcher (Grant Nos. 2016VEB013 and 2016PE007). The authors would like to thank the editors for their suggestions to improve this chapter.

References

Baek S-Y, Jurng J, Chang Y-S (2013) Spatial distribution of polychlorinated biphenyls, organochlorine pesticides, and dechlorane plus in. Northeast Asia Atmos Environ 64:40–46

Bi X, Luo W, Gao J, Xu L, Guo J, Zhang Q, deBoer J (2016) Polycyclic aromatic hydrocarbons in soils from the Central-Himalaya region: distribution, sources, and risks to humans and wildlife. Sci Total Environ 556:12–22

Bustnes JO, Gabrielsen GW, Verreault J (2009) Climate variability and temporal trends of persistent organic pollutants in the arctic: a study of glaucous gulls. Environ Sci Technol 44: 3155–3161

Chiuchiolo AL, Dickhut RM, Cochran MA, Ducklow HW (2004) Persistent organic pollutants at the base of the Antarctic marine food web. Environ Sci Technol 38:3551–3557

Colborn T, Samolen ME (1996) Epidemiological analysis of persistent organochlorine contaminants in cetaceans. Rev Environ Contam 146:91–172

Cong Z, Kang S, Kawamura K, Liu B, Wan X, Wang Z, Fu P (2015a) Carbonaceous aerosols on the south edge of the Tibetan Plateau: concentrations, seasonality and sources. Atmos Chem and Phys 15(3):1573–1584

Cong Z, Kawamura K, Kang S, Fu P (2015b) Penetration of biomass-burning emissions from South Asia through the Himalayas: new insights from atmospheric organic acids. Sci Rep 5: 9580

DeMartinis BS, Okamoto RA, Kado NY, Gundel LA, Carvalho LRF (2002) Polycyclic aromatic hydrocarbons in a bioassay-fractionated extract of PM10 collected in St. Paulo, Brazil. Atmos Environ 36:307–314

Devi NL, Yadav IC, Raha P, Shihua Q, Yang D (2015) Spatial distribution, source apointment and ecological risk assessment of residual organochlorine pesticides (OCPs) in the Himalayas. Environ Sci Pollut Res. http://dx.doi.org/10.1007/s11356-015-5237-5

Fu R, Hu Y, Wright JS, Jiang JH, Dickinson RE, Chen M, Filipiak M, Read WG, Waters JW, Wu DL (2006) Short circuit of water vapor and polluted air to the global stratosphere by convective transport over the Tibetan Plateau. Proc Natl Acad Sci USA 103:5664–5669

Gong P, Xiaoping X, Yao T (2011) Ambient distribution of particulate- and gas-phase n-alkanes and polycyclic aromatic hydrocarbons in the Tibetan Plateau. Environ Earth Sci 64:1703–1711

Gong P, Wang X, Li S, Yu W, Kattel DB, Wang W, Devkota LP, Yao T, Joswiak DR (2014) Atmospheric transport and accumulation of organochlorine compounds on the southern slopes of the Himalayas, Nepal. Environ Pollut 192:44–51

Guo J, Kang S, Huang J, Zhang Q, Tripathee L, Sillanpää M (2015) Seasonal variations of trace elements in precipitation at the largest city in Tibet, Lhasa. Atmos Res 153:87–97

Halsall CJ, Sweetman AJ, Barrie LA, Jones KC (2000) Modelling the behaviour of PAHs during atmospheric transport from the UK to Arctic. Atmos Environ 35:255–267

Hoff RM, Muir DCG, Grift NP (1992) Annual cycle of polychlorinated biphenyls and organohalogen pesticides in air in southern Ontario. Air concentration data. Environ Sci Technol 2: 266–275

Horstmann M, McLachlan MS (1998) Atmospheric deposition of semivolatile organic compounds to two forest canopies. Atmos Environ 32:1799–1809

Iwata H, Tanabe S, Sakai N, Nishimura A, Tatsukawa R (1994) Geographical distribution of persistent organochlorines in air, water and sediments from Asia and Oceania, and their implications for global re-distribution from lower latitudes. Environ Pollut 85(1):15–33

Jaward FM, DiGuardo A, Nizzetto L, Cassani C, Raffaele F, Ferretti R, Jones KC (2005) PCB and selected organochlorine compounds in Italian mountain air: the influence of altitude and forest ecosystem type. Environ Sci Technol 39(10):3455–3463

Kattel DB, Yao T, Yang K, Tian L, Yang G (2013) Temperature lapse rate in complex mountain terrain on the southern slopes of the central Himalayas. Theor Appl Climatol 113:671–671. doi:10.1007/s00704-012-0816-6

Kattel DB, Yao T, Yang W, Gao Y, Tian L (2015) Comparision of temperature lapse rates from the northern to the southern slopes of the Himalayas. Int J Climatol 35(15):4431–4443. doi:10.1002/joc.4297

Kaupp H, McLachlan MS (1999) Gas-partical partitioning of PCDD/Fs, PCBs, PCNs and PAHs. Chemosphere 38:3411–3421

Khan MA, Iqbal M, Ahmad I, Soomro MH, Chaudhary MA (2002) Economic evaluation of pesticides use externalities in the cotton zones of Punjab, Pakistan. Pak Dev Rev:683–698

Luo W, Gao J, Bi X, Xu L, Guo J, Zhang Q, Kang S (2016) Identification of sources of polycyclic aromatic hydrocarbons based on concentrations in soils from two sides of the Himalayas between China and Nepal. Environ Pollut 212:424–432

Ma WL, Li YF, Qi H, Sun DZ, Liu LY, Wang DG (2010) Seasonal variations of sources of polycyclic aromatic hydrocarbons (PAHs) to a northeastern urban city, China. Chemosphere 79:441–447

Mackay D (2001) Multimedia environmental models: the fugacity approach. Lewis/CRC, Boca Raton

Mader BT, Pankow JF (2002) Study of the effects of particle-phase carbon on the gas/particle partitioning of semivolatile organic compounds in the atmosphere using controlled field experiments. Environ Sci Technol 36:5218–5228

Malik A, Verma P, Singh AK, Singh KP (2011) Distribution of polycyclic aromatic hydrocarbons in water and bed sediments of the Gomti River, India. Environ Monit Assess 172:529–545

Maliszewska-kordybach B, Smreczak B, Pawlas AK (2009) Concentrations, sources, and spatial distribution of individual polycyclic aromatic hydrocarbons (PAHs) in agricultural soils in the Eastern part of the EU: Poland as a case study. Sci Total Environ 407:3746–3753

Menzie CA, Potocki BB, Santodonato J (1992) Exposure to carcinogenic PAHs in the environment. Environ Sci Technol 26:1278–1284

Ming J, Cachier H, Xiao C, Qin D, Kang S, Hou S, Xu J (2008) Black carbon record based on a shallow Himalayan ice core and its climatic implications. Atmos Chem Phys 8:1343–1352

Nasir J, Wang X, Xu B, Wang C, Joswiak DR, Rehman S, Lodhi A, Shafiq S, Jiklani R (2014) Selected organochlorine pesticides and polychlorinated biphenyls in urban atmosphere of Pakistan: concentration, spatial variation and sources. Environ Sci Technol 48:2610–2618

Okuda T, Okamoto K, Tanaka S, Shen ZX, Han YM, Huo ZQ (2010) Measurement and source identification of polycyclic aromatic hydrocarbons (PAHs) in the aerosol in Xi'an, China, by using automated column chromatography and applying positive matrix factorization (PMF). Sci Total Environ 408:1909–1914

Sheng J, Wang X, Gong P, Joswiak DR, Tian L, Yao T, Jones KC (2013) Monsoon-driven transport of organochlorine pesticides and polychlorinated biphenyls to the Tibetan Plateau: three year atmospheric monitoring study. Environ Sci Technol 47:3199–3208

Shunthirasingham C, Mmereki BT, Masamba W, Oyiliagu CE, Lei YD, Wania F (2010) Fate of pesticides in the arid subtropics, Botswana, Southern Africa. Environ Sci Technol 44(21): 8082–8088

Su Y, Wania F, Harner T, Lei YD (2006) Deposition of polybrominated diphenyl ethers, polychlorinated biphenyls, and polycyclic aromatic hydrocarbons to a boreal deciduous forest. Environ Sci Technol 41:534–540

Tobiszewski M, Namieśnik J (2012) PAH diagnostic ratios for the identification of pollution emission sources. Environ Pollut 162:110–119

Tripathee L, Kang S, Huang J, Sillanpää M, Sharma CM, Lüthi ZL, Guo J, Paudyal R (2014a) Ionic composition of wet precipitation over the southern slope of central Himalayas, Nepal. Environ Sci Pollut Res 21(4):2677–2687

Tripathee L, Kang S, Huang J, Sharma CM, Sillanpää M, Guo J, Paudyal R (2014b) Concentrations of trace elements in wet deposition over the central Himalayas, Nepal. Atmos Environ 95: 231–238

Tripathee L, Kang S, Rupakheti D, Zhang QG, Sharma CM, Huang J, Bajracharya RM, Gyawali A, Paudyal R, Sillanpää M (2016) Distribution and risk assessment of potentially toxic trace elements and rare earth elements in soils of the Langtang Himalayas Nepal. Environ Earth Sci 75: 1332

UNEP (2012) Global Environment Outlook-5. Environment for the future we want. United Nations Environmental Program (UNEP), Valletta

USEPA (1993) Provisional guidance for quantitative risk assessment of polycyclic aromatic hydrocarbons. EPA/600/R-93/089

Wang XP, Yao T, Cong ZY, Yan XL, Kang SC, Zhang Y (2007) Distribution of persistant organic pollutants in soil and grasses around Mt. Qomolangma, China. Arch Environ Contam Toxicol 52:153–162

Wang X, Gong P, Yao T, Jones KC (2010) Passive air sampling of organochlorine pesticides, polychlorinated biphenyls, and polybrominated diphenyl ethers across the Tibetan Plateau. Environ Sci Technol 44:2988–2993

Wang X, Xue Y, Gong P, Yao T (2014) Organochlorine pesticides and polychlorinated biphenyls in Tibetan forest soil: profile distribution and process. Environ Sci Pollut Res 21:1897–1904

Wania F, Mackay D (1996) Peer reviewed: tracking the distribution of persistent organic pollutants. Environ Sci Technol 30(9):390A–396A

Wania F, Haugen JE, Lei YD, Mackay S (1998) Temperature dependence of atmospheric concentrations of semivolatile organic compounds. Environ Sci Technol 32(8):1013–1021

Wilcke W, Krauss M, Amelung W (2002) Carbon isotop signature of polycyclic aromatic hydrocarbons (PAHs): evidence for different sources in tropical and temperate environments? Environ Sci Technol 36:3530–3535

Xu B, Cao J, Hansen J, Yao T, Joswia RD, Wang N, Wu G, Wang M, Zhao H, Yang W, Liu X, He J (2009) Black soot and the survival of Tibetan glaciers. Proc Natl Acad Sci USA 106: 22114–22118

Yao T, Thompson L, Yang W, Yu W, Gao Y, Guo X, Yang X, Duan K, Zhao H, Xu B, Pu J, Lu A, Xiang Y, Kattel DB, Joswiak D (2012) Different glacier status with atmospheric circulations in Tibetan Plateau and surroundings. Nat Clim Change 2:663–667. doi:10.1038/nclimate1580

Yao, TD, Masson-Delmotte V, Gao J, Yu WS, Yang XX, Risi C, Sturm C, Werner M, Zhao HB, He Y, Ren W, Tian LD, Shi CM, Hou SG (2013) A review of climatic controls on $\delta^{18}O$ in precipitation over the Tibetan Plateau: observations and simulations. Rev Geophys 51. http://dx.doi.org/10.1002/ rog.20023.

Zhang C, Chakraborty P, Li J, Sampathkumar P, Balasubramanian T, Kathiresan K, Takahashi S, Subramanian A, Tanabe S, Jones KC (2008) Passive atmospheric sampling of organochlorine pesticides, polychlorinated biphenyls, and polybrominated diphenyl ethers in urban, rural, and wetland sites along the coastal length of India. Environ Sci Technol 42:8218–8223

Chapter 3
HCH and DDT Residues in Indian Soil: Atmospheric Input and Risk Assessment

Paromita Chakraborty, Sanjenbam Nirmala Khuman, Bhupandar Kumar, and Bommanna Loganathan

3.1 Introduction

India is an agricultural country located in diverse climatic zones. Agriculture, with its allied sectors, is the largest source of livelihood in India, particularly in the vast rural areas contributing significantly to the Gross Domestic Product (GDP). Organochlorine pesticides (OCPs) such as dichlorodiphenyltricholoroethane (DDT) and hexachlorocyclohexane (HCH) have been extensively used in India for agricultural and public health purposes for more than five decades. DDT, HCH, and malathion (organophosphorous compound) constitute 70 % of the annual pesticide consumption (85,000 t) (Gupta 2004). OCPs were banned for agricultural practices in the late 1990s, but a substantial amount of these insecticides are still being used for exterminating insects that spread diseases such as malaria, kala-azar (black fever), etc. Due to widespread use, these pesticides continue to contaminate different environmental compartments because of their semi-volatile nature and long environmental lifetimes in soil and water (Kurt-Karakus et al. 2005).

P. Chakraborty (✉)
SRM Research Institute, SRM University, Kattankulathur, Tamil Nadu, India

Department of Civil Engineering, SRM University, Kattankulathur, Tamil Nadu, India
e-mail: paromita.c@res.srmuniv.ac.in

S.N. Khuman
SRM Research Institute, SRM University, Kattankulathur, Tamil Nadu, India

B. Kumar
National Reference Trace Organics Laboratory, Central Pollution Control Board, East Arjun Nagar, 110032 Delhi, India

B. Loganathan
Department of Chemistry and Watershed Studies Institute, Murray State University, Murray, KY, USA

© Springer International Publishing AG 2017

M.Z. Hashmi et al. (eds.), *Xenobiotics in the Soil Environment*, Soil Biology 49,
DOI 10.1007/978-3-319-47744-2_3

Owing to the hydrophobic and lipophilic properties and affinity towards particles, DDT and HCH accumulate in the organic matter of soil for longer period (Ockenden et al. 2003). Soil, therefore not only acts as the sink for these pollutants, but also acts as a secondary source by re-emitting these compounds into atmosphere (Harner et al. 2001; Wild and Jones 1995). Several studies (Babu et al. 2003; Rajendran and Subramanian 1999; Ramesh et al. 1991; Senthilkumar et al. 2001) have emphasized that HCHs and other OCPs with similar physicochemical properties are dissipated from soil under the tropical/subtropical conditions leading to their widespread distribution (Chakraborty et al. 2015; Agoramoorthy 2008; Rekha and Prasad 2006). In addition, our earlier studies have revealed the occurrence and distribution of persistent organic pollutants (POPs) in Indian atmosphere (Chakraborty and Zhang 2011). This chapter provides an overview of HCH and DDT residues in soil across northern, eastern, north-eastern, western, central, and southern part of India, their atmospheric input and the associated risk for human health.

3.2 Methodology

Concentrations of HCH isomers and DDT isomers and its metabolites in soil and air from different parts of India were compiled using recent literature. The data were used to elucidate the current status of contamination and human health risk estimation.

3.2.1 Sample Collection, Column Cleanup, and Instrumental Analysis

Surface soil samples were collected from the national capital city, New Delhi, and states, viz., Assam, Chattisgarh, Goa, Haryana, Karnataka, Maharashtra, Manipur, Tamilnadu, Tripura, Uttarpradesh, Uttarakhand, and West Bengal, and seven major Indian cities based on urban–suburban and rural transect. The details on the soil samples from different states have been given in Table 3.1. In most of the studies, Soxhlet was used for extraction of OCPs from the soil samples (Abhilash and Singh 2008; Chakraborty et al. 2015; Devi et al. 2013, 2015; Kumar et al. 2012; Singh et al. 2007). Some of the samples were processed by other extraction methods like shaker (Minh et al. 2006; Prakash et al. 2004; Ramesh et al. 1991), ultrasonication (Kumar et al. 2014) etc. After extraction, the sample extracts were subjected to column chromatography packed with alumina, silica gel, and sodium sulfate for cleanup. The sample extracts were further subjected to instrumental analysis in either Gas Chromatography (GC) equipped with electron capture detector (ECD) (Abhilash and Singh 2008; Agnihotri et al. 1996; Devi et al. 2013; IOG 2002; Jayashree and Vasudevan 2006; Kumar et al. 2012, 2014; Kumari et al. 2008; Mishra et al. 2003; Prakash et al. 2004; Ramesh et al. 1991; Singh et al. 2007) or Gas Chromatography interfaced with Mass Spectrometry techniques (Chakraborty et al. 2015).

Table 3.1 HCHs and DDTs residues in soil from various states of India

OCPs	Concentration (ng/g)	Soil type	Place	Year	References
HCHs	180–1586	Paddy fields	Dibrugarh	2009–2010	Mishra et al. (2003)
HCHs	345–1844	Paddy fields	Nagaon	2009–2010	
HCHs	75–2259	Tea gardens	Dibrugarh	2009–2010	
HCHs	223–1639	Tea gardens	Nagaon	2009–2010	
HCHs	178–1701	Others	Dibrugarh	2009–2010	
HCHs	98–1945	Others	Nagaon	2009–2010	
DDTs	75–2296	Paddy fields	Dibrugarh	2009–2010	
DDTs	166–2288	Paddy fields	Nagaon	2009–2010	
DDTs	218–2129	Tea gardens	Dibrugarh	2009–2010	
DDTs	351–1981	Tea gardens	Nagaon	2009–2010	
DDTs	172–1833	Others	Dibrugarh	2009–2010	
DDTs	181–1811	Others	Nagaon	2009–2010	
HCHs	122–638	Paddy fields	Dehradun	NA	Babu et al. (2003)
DDTs	13–238	Paddy fields	Dehradun	NA	
HCHs	5.83–85.083	Agricultural	Thiruvallur	2004–2005	Jayashree and Vasudevan (2006)
DDTs	1–10.5	Agricultural	Thiruvallur	2004–2005	
HCHs	89.40 (mean)	Agricultural	Aligarh	1998–1999	Nawab et al. (2003)
DDTs	34 (mean)	Agricultural	Aligarh	1998–1999	
Aldrin	1.46 (mean)	Agricultural	Aligarh	1998–1999	
HCHs	0.017–0.121	Forest	Assam	2006–2009	Devi et al. (2013)
DDTs	0.101–0.626	Forest	Assam	2006–2009	
Endos	0.161–0.463	Forest	Assam	2006–2009	
HCHs	0.015–0.097	Wildlife	Tripura	2006–2009	
DDTs	0.110–0.626	Wildlife	Tripura	2006–2009	
HCHs	0.018–0.149	Forest	Tripura	2006–2009	
DDTs	0.048–0.364	Forest	Tripura	2006–2009	
HCHs	0.006–0.140	Tea estate	Tripura	2006–2009	
DDTs	0.049–0.749	Tea estate	Tripura	2006–2009	
HCHs	0.029–0.234	Grassland	Tripura	2006–2009	
DDTs	0.096–0.549	Grassland	Tripura	2006–2009	
HCHs	0.029–0.234	Roadside	Manipur	2006–2009	
DDTs	0.096–0.549	Roadside	Manipur	2006–2009	
HCHs	0.080–2.950	Forest	Manipur	2006–2009	
DDTs	0.241–3.870	Forest	Manipur	2006–2009	
HCHs	0.079–1.642	Wetland	Manipur	2006–2009	
DDTs	0.328–5.208	Wetland	Manipur	2006–2009	
HCHs	7.1	Watershed	Vellar	1988–1989	Ramesh et al. (1991)
DDTs	1.5	Watershed	Vellar	1988–1989	
HCHs	0.003–0.33	Estuary	Hughli	1998–2000	Bhattacharya et al. (2003)
DDTs	0.003–0.119	Estuary	Hughli	1998–2000	

(continued)

Table 3.1 (continued)

OCPs	Concentration (ng/g)	Soil type	Place	Year	References
HCHs	6.4–212.2	Surface soil	New Delhi	2002	Prakash et al. (2004)
HCHs	0.2–212.2	Surface soil	Haryana	2002	
HCHs	6.91–637	Surface soil	Lucknow	2002	
HCHs	12.31–118.64	Sub surface soil	New Delhi	2002	
HCHs	10.97–382.97	Sub surface soil	Haryana	2002	
HCHs	0.08–7.25	Alluvial soil	Unnao, Gangetic plain	2003	Singh et al. (2007)
DDTs	BDL–74.06	Alluvial soil	Unnao, Gangetic plain	2003	
HCHs	53–99	Industrial	Lucknow	NA	Abhilash and Singh (2008)
HCHs	BDL-9	Dumpsite	Perungudi	1999–2001	Minh et al. (2006)
DDTs	BDL-63	Dumpsite	Perungudi	1999–2001	
HCHs	2–51	Agricultural	Haryana	NA	Kumari et al. (2008)
DDTs	1–66	Agricultural	Haryana	NA	
HCHs	14–158	Surface Alluvial soil	Farukabad	1991–1992	Agnihotri et al. (1996)
DDTs	27–337	Surface Alluvial soil	Farukabad	1991–1992	
HCHs	12–67	Subsurface Alluvial soil	Farukabad	1991–1992	
DDTs	28–295	Subsurface Alluvial soil	Farukabad	1991–1992	
HCHs	BDL-2.79	Surface soil	Itanagar, Guwahati, Tezpur, Dibrugarh	2012	Devi et al. (2015)
DDTs	0.28–2127	Surface soil	Itanagar, Guwahati, Tezpur, Dibrugarh	2012	
HCHs	0.24–59.8	Surface soil	Bangalore	2006–2007	Chakraborty et al. (2015)
DDTs	0.34–78	Surface soil	Bangalore	2006–2007	
HCHs	0.231–16.8	Surface soil	Chennai	2006–2007	
DDTs	0.35–10	Surface soil	Chennai	2006–2007	
HCHs	0.04–7.6	Surface soil	Mumbai	2006–2007	
DDTs	0.81–9.2	Surface soil	Mumbai	2006–2007	
HCHs	1.88–15.7	Surface soil	Goa	2006–2007	
DDTs	5.55–124.8	Surface soil	Goa	2006–2007	
HCHs	1.6–13.2	Surface soil	Agra	2006–2007	
DDTs	3.1–20.5	Surface soil	Agra	2006–2007	
HCHs	0.027–33.8	Surface soil	New Delhi	2006–2007	
DDTs	0.15–42	Surface soil	New Delhi	2006–2007	
HCHs	0.23–21.2	Surface soil	Kolkata	2006–2007	
DDTs	0.41–124	Surface soil	Kolkata	2006–2007	

3.2.2 *Risk Assessment*

Human exposure and consequent health risk to soil borne HCH and DDT residues in seven major Indian cities covering northern, eastern, western, and southern parts of India (Chakraborty et al. 2015) were estimated using the procedure described by the United States Environmental Protection Agency (USEPA). Incremental Lifetime Cancer Risk (ILCR) for each site was estimated (ATSDR 2005; USEPA 1989). ILCR for human was assessed from the estimated Lifetime Average Daily Dose (LADD) of DDTs and HCHs as per guidelines given by USEPA and Agency for Toxic Substances and Disease Registry (ATSDR) (ATSDR 2005; USEPA 1989). The equations used for estimating LADD and ILCR were as follows:

$$\mathrm{LADD}\left(\mathrm{mg\ kg^{-1}day^{-1}}\right) = (\mathrm{Cs} \times \mathrm{IR} \times \mathrm{F} \times \mathrm{EF} \times \mathrm{ED})/(\mathrm{BW} \times \mathrm{AT}), \quad (3.1)$$

where

Cs	Pollutant concentration in soil (mg kg^{-1}),
IR	Soil ingestion rate (100 mg day^{-1} for adult and 200 mg day^{-1} for children),
F	Unit conversion factor,
EF	Exposure frequency (365 days $year^{-1}$),
ED	Lifetime exposure duration (adults—70 years; children—12 years),
BW	Body weight (adults—70 kg; children—27 kg),
AT	Averaging time for carcinogens (EF × ED).

$$\mathrm{ILCR} = \mathrm{LADD} \times \mathrm{CSF} \quad (3.2)$$

where CSF is cancer slope factor for a particular compound intake (mg kg^{-1} day^{-1}).

3.3 Results and Discussion

3.3.1 *HCH and DDT: Production and Usage*

Unregulated use of synthetic pesticides started in India during 1948–1949 with the use of DDT for malaria control and HCH (also known as BHC) for locust control. Indian pesticides production started with the setting up of a BHC technical plant at Rishra near Kolkata in 1952. Hindustan Insecticides Ltd. set up two more units to manufacture DDT. Details of production and consumption of HCH and DDT in India have been given in Table 3.2.

India is the world's third largest consumer of technical HCH. The production, consumption, export, and import of HCH have been given in Table 3.2. About 10,43,000 t of HCH was produced between 1948 and 1997, and the total consumption of technical HCH in India during 1948–2000 was 10,57,000 t (IOG 2002; Wei

Table 3.2 Production, consumption, import, export, commencement, and ban of OCPs in India

Pesticide	Production (t)	Consumption (t)	Import/export (t)	No. of plants		Start year	Ban year	References
				Past	Now			
DDT	222721.82 1955–2009 5000 t/year 2011	85672 t (1988–2008)	Exp. 966 (1991–1999) Exp. 1.31 (2006–2010)	3	2	1955	1989	NIP (2011)
HCH	1042612 t (1948–1997)	1057000 t (1948–2000)	–	2	0	1954	1997	IOG (2002), Wei et al. (2007)
Lindane	6687 t (1990–2004)	2411 t (1995–2000)	Imp. 61 (2003–2005) Exp. 802.22 (1996–2008)	2	2	1995	2013	DGFT (2008); MCF (2001–2007), PPQS (2013))

et al. 2007). Cumulative consumption of the HCHs in India until 1985 was 5,75,000 t, and since then, about 45,000 t of HCH was used annually until it was banned in the year 1997.

During 2005-2010, DDT was used for malaria control in various states of India (Fig. 3.1) (NVBDCP 2010). It is very clear that DDT was extensively consumed by the north-eastern states of India particularly Assam. Since 1987, the production of DDT has decreased in India. DDT production was ceased in a plant at New Delhi in 1998. China is the largest consumer, importer, and exporter of DDT. The United States of America produced significant quantity of DDT, although production ceased in 1972. India and China are the only countries currently manufacturing and exporting DDT to other countries, where these insecticides are exclusively used for public health purposes. India is the only country where more than 1,00,000 t of DDT was applied since its inception, mainly in agricultural and vector control programs until it was banned for agricultural use in 1989. Among various states, Chhattisgarh has a major malaria problem and the state contributed about 13 % of the total malaria cases reported in the country. OCPs and synthetic pyrethroids have been used for national malaria control program in the past (Kumar et al. 2014).

Lindane is a gamma isomer of HCH, mainly used as insecticide. In India, lindane formulations are registered for usage in pharmaceutical products (Gupta 2004). In the year 2000, the production of lindane was 1107 t, and subsequently, India signed treaties against the usage of lindane. The production further declined in 2008 to 75 t (DGFT 2008; PPQS 2013). Being the largest exporter of lindane in the year 1998, India supplied 207 t to other countries. During 1990s with 1000 t of lindane production, India was reported to be the highest producer of lindane (DGFT 2008; MCF 2007; PPQS 2013). During 1990-2004, consumption of lindane in India was 6840 t, i.e., only 2 % of the global usage (Vijgen et al. 2006). Finally in 2013, Lindane has been banned in India.

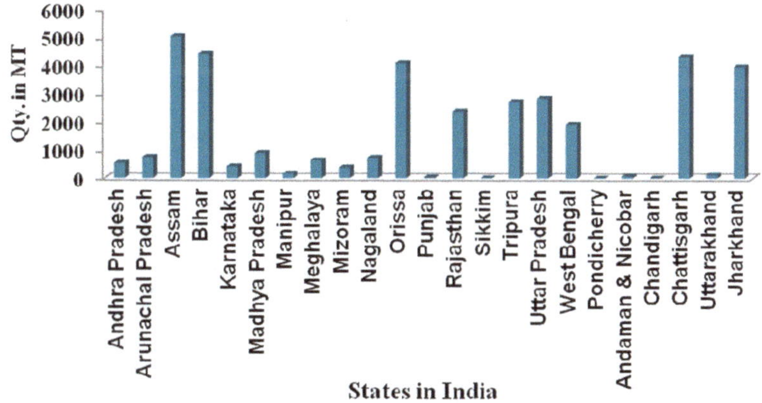

Fig. 3.1 DDT usage for malaria control in various states of India during 2005–2010 (NVBDCP 2010)

3.3.2 Region-Specific Distribution and Atmospheric Input

Maximum HCH and DDT residues observed in soil from different states of India between 2005-2015 have been given in Fig. 3.2.

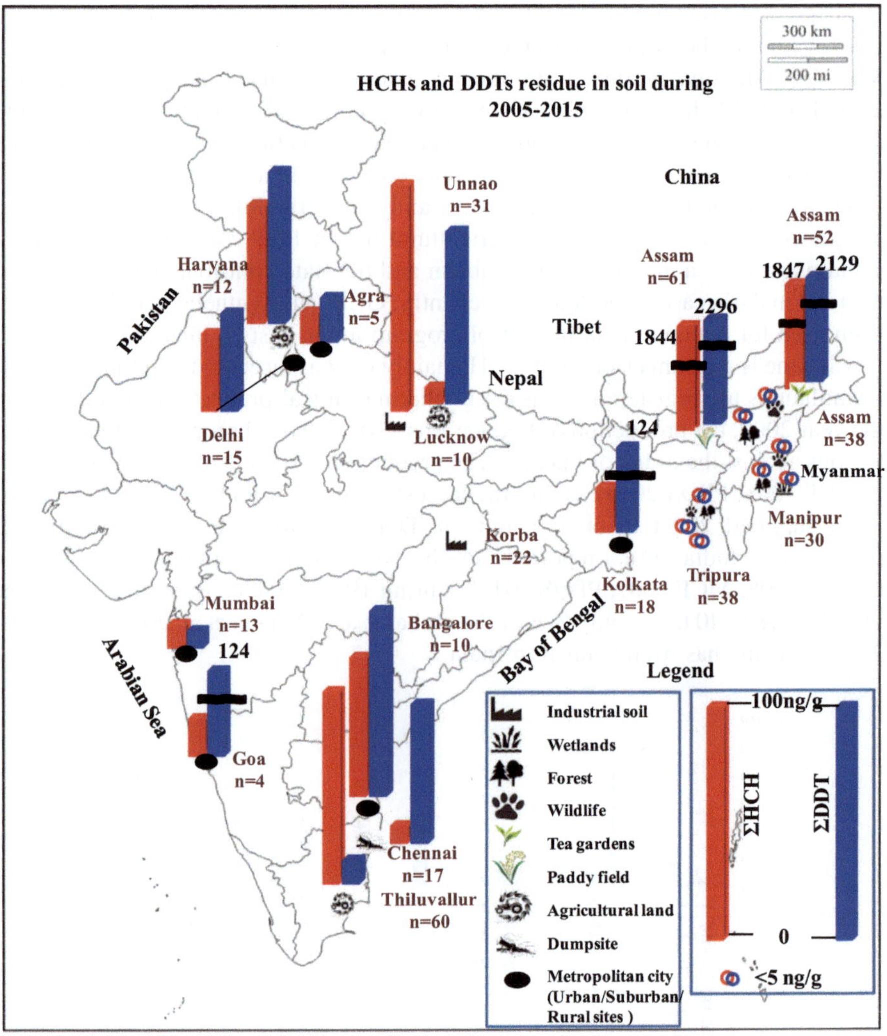

Fig. 3.2 Maximum HCH and DDT residues in surface soil from different states of India. Concentrations presented in this figure have been obtained from various studies in India (Abhilash and Singh 2008; Chakraborty et al. 2015; Devi et al. 2013; Jayashree and Vasudevan 2006; Kumar et al. 2014; Kumari et al. 1996; Mishra et al. 2003; Singh et al. 2007)

3.3.2.1 Northern India

HCH Elevated levels of HCH were observed in alluvial plains of Farukabad (Gupta 2004). In the capital city, New Delhi, and a nearby city, Agra, all the HCH isomers were prevalent in soil and ΣHCHs showed significant positive correlation with soil organic carbon (Chakraborty et al. 2015). Isomeric composition of soilborne HCH isomers in New Delhi ($\alpha = 4.4$ %, $\beta = 51.3$ %, $\gamma = 29.4$ %, $\delta = 15$ %) and Agra ($\alpha = 4.4$ %, $\beta = 51.3$ %, $\gamma = 29.4$ %, $\delta = 15$ %) showed elevated δ-HCH for sites in New Delhi close to Uttar Pradesh border as well as in all the sites of Agra located within Uttar Pradesh (Chakraborty et al. 2015). HCH isomers detected in the agricultural soil of New Delhi were attributed to the runoffs from the dump sites of the adjoining states (Prakash et al. 2004). In Haryana, a north Indian state in the western part of New Delhi, agricultural soil from paddy-wheat, cotton-wheat, and sugarcane fields were detected predominantly with γ-HCH (Kumari et al. 2008). Similarly at Kurukshetra in Haryana, all the HCH isomers, viz., α-HCH (33 %), β-HCH (35 %), γ-HCH (29 %), and δ-HCH (4 %), were observed (Kumar et al. 2012). Lindane usage was evident in all the studies. Chakraborty et al. (2010) reported that the wide range of fugacity fractions for δ-HCH showed deposition due to site-specific contamination especially in New Delhi and Agra possibly due to contamination from the nearby lindane manufacturing units (Prakash et al. 2004). Around Lindane producing factory, γ-HCH (lindane) was detected in all the soil samples, but a decreasing trend in the concentration of HCH was observed as the sampling sites extended from the center of lindane production to the outskirts of the industrial area in Lucknow (Abhilash and Singh 2008). Hence, the dumped waste from the HCH manufacturing unit in Haryana (Prakash et al. 2004) and the by-products released from lindane manufacturing unit in Uttar Pradesh (CAPE 2005) were attributed as the prime reasons for soilborne HCH isomers.

Higher atmospheric α-, β-, and δ-isomers of HCHs in New Delhi particularly from the suburban site, Gagan Vihar, and in the rural sites bordering the state of Uttar Pradesh have been evidenced in back trajectory analysis of air parcels during a passive air sampling (PAS) study (Chakraborty et al. 2010). Air parcels ending in New Delhi irrespective of the site of origin traversed across the lindane manufacturing unit located in Uttar Pradesh (Chakraborty et al. 2010). High atmospheric HCH isomers in Agra were also affected by a major cluster originating from north of Agra located more close to the lindane manufacturing unit (Chakraborty et al. 2010).

DDT Concentrations of DDDs and DDEs were on the higher side in soil from New Delhi, indicating historical usage of DDT. In addition, elevated *o,p′*-DDT levels and the highest *o,p′*-DDT/*p,p′*-DDT ratio were observed at Agra (average 7) followed by New Delhi (average 5), indicating ongoing usage of DDT (Chakraborty et al. 2010). The observed ratio of DDT/(DDD + DDE) for this study ranged between 0.09 and 2.39 with an average value of 0.75, indicating that DDT input in this area is both due to past and present usage (Chakraborty et al. 2015). In Kurukshetra city, *p,p′*-DDT, *o,p′*-DDT, *p,p′*-DDE, and *p,p′*-DDD occupied 12 %, 26 %, 30 %, and 33 % of total DDT respectively (Kumar et al. 2012). Among DDT analogues, *p, p′* DDE was found to be the dominant in Haryana, indicating past usage (Kumari et al. 2008). Mostly DDT isomers showed deposition at specified locations of New Delhi (Chakraborty et al. 2015). Very high concentration of DDT has been observed in Farukabad and somewhat lower concentration in Lucknow (Gupta 2004).

3.3.3 Eastern and North-Eastern India

HCH In most locations of Kolkata, a major metropolitan city in eastern India, α-HCH was found to be one- to two folds higher than γ-HCH, indicating ongoing use of technical HCH apart from Lindane (Chakraborty et al. 2015). Maximum soil-borne HCHs have been observed in Assam, a northeastern state of India with dominance of β-isomer particularly in paddy fields, and γ-HCH was dominant in tea garden (Mishra et al. 2003). Mean value of α/γ HCH in Dibrugarh (2.78) and Nagaon (2.51) suggests potential usage of technical HCH in Assam (Mishra et al. 2003). Soil from paddy fields contain substantially and significantly ($p < 0.05$) higher amount of HCHs compared to tea gardens, other agricultural fields, and fallow land (Mishra et al. 2003). More than 50 % of the soil samples taken from the forest cover of Manipur were detected with HCH with higher prevalence of β-HCH and δ-HCH concentration in Tripura (Devi et al. 2013). Therefore HCHs in the eastern and north-eastern states of India can be attributed to both lindane and technical HCH usage. High atmospheric HCH concentration was observed in Kolkata (Chakraborty et al. 2010). Most of the sampling sites demonstrated the usage and deposition of lindane in the background soil from the north-eastern states of India (Devi et al. 2013). Back trajectory analysis showed major air mass clusters originating from northern and eastern parts of India and traversed through Kolkata. Transboundary movement from Bangladesh before ending at Manipur, indicated potential long-range atmospheric transport of these pollutants from the source regions (Devi et al. 2011).

DDT An urban tilt of DDT has been observed in Kolkata. Apart from agricultural use (Guzzella et al. 2005), dicofol was used as an effective acaricide for tea cultivation in the north-eastern part of India (Saha et al. 2004). Soilborne DDTs from paddy fields showed substantially and significantly ($p < 0.05$) higher amount of DDT compared to tea gardens, other agricultural fields, and fallow land (Mishra et al. 2003). High average *o,p′*-DDT/*p,p′*-DDT ratios were observed in rural sites with highest DDT in soil from agricultural sites of Barasat, therefore suggesting the use of fresh DDT (Chakraborty et al. 2015). Dominance of *p,p′*-DDT has been observed in Assam with mean *p,p′*-DDT/((*p,p′*-DDE + *p,p′*-DDD) ratio of 1.25 and 1.82 in Dibrugarh and Nagaon respectively (Mishra et al. 2003). Higher *p,p′*-DDT for tea garden (having high organic carbon and high acidic soil) and more *p,p′*-DDE in paddy soil (having comparatively low organic carbon and high clay content) suggest more use of technical DDT to control malaria vectors or for intense paddy cultivation in the past and ongoing usage in tea plantation (Mishra et al. 2003). Low pesticide concentration was observed where there are less agricultural activities and vegetation cover. Higher concentration with the maximum load of soilborne DDT was found in wetland soil of Manipur (Devi et al. 2013). Wildlife sanctuary of Tripura contributed maximum amount of DDT load with dominance of metabolites, i.e., *p,p′*-DDE and *p,p′*-DDD (Devi et al. 2013). In Dibrugarh and Nagaon districts, 81,553 ha (24.09 %) and 160,035 ha (38.94 %) area is under paddy cultivation (summer, winter, and autumn paddy) with 66,309 ha (49.12 %) and 234,633 ha (61.25 %) total sown area (NIC 2004–2005) were subjected to high pesticide application, which may be a possible explanation for elevated pesticidal residues in these districts (Mishra et al. 2003).

3.3.4 Southern India

HCH Agricultural soil in Thiruvallur district of Tamil Nadu was highly contaminated with pesticidal residues particularly with higher concentration of γ-HCH, and the residual levels of α- and δ-HCH were lower than those of γ- and β-HCH (Jayashree and Vasudevan 2006). Tropical climate in southern India facilitates the post-application volatilization of 90 % HCHs from soil to atmosphere in the paddy fields (Takeoka et al. 1991), leading to the highest level of atmospheric gaseous phase HCHs in Chennai and very low HCH concentration in the particulate phase (average 0.55 %) (Chakraborty et al. 2010). Highest level of ΣHCHs was observed in Bangalore and γ-HCH was predominant, but a fair amount of α- and β-isomers were also present (Chakraborty et al. 2015). Elevated levels of γ-HCH in the urban sites are possibly due to its use in healthcare programs (Subramanian and Tanabe 2007).

DDT Over 60 % soilborne DDT in Chennai was comprised of DDT isomers (*o,p′*-DDT and *p,p′*-DDT), indicating ongoing technical DDT usage (Chakraborty et al. 2015). This reflects net volatilization of OCPs from soil to air particularly for the cities with higher ambient temperature under tropical climate (Chakraborty et al. 2015). In Bangalore, high DDT was found in a site which was a dumpsite in the past, whereas in other sites, the DDEs and DDDs were more prevalent (Chakraborty et al. 2015).

3.3.5 Central and Western India

HCH Residues of HCH isomers in soil of Korba, an industrial area in Chattisgarh, showed both technical and lindane usage (Kumar et al. 2014). Soilborne β-HCH dominated the concentrations of HCH isomers in Mumbai and Goa possibly due to the ongoing usage of technical HCH mainly for cotton cultivation practiced in the western and central parts of India (Chakraborty et al. 2015, 2010). Major contributions of atmospheric β-HCH were from Goa and Mumbai (Chakraborty et al. 2010). India was found to be a major source of global β-HCH emissions in 2000 (Li et al. 2003). Atmospheric models have shown that the contaminated air masses originating from the western and central parts of India were transported to Mumbai city via atmospheric transport (Chakraborty et al. 2010).

DDT DDT usage indicated past and ongoing application of technical DDT with higher concentration of soilborne DDE than parent isomers from Korba, attributed to the aerobic degradation of DDT coupled with the long-range atmospheric transport (LRAT) under tropical climatic conditions (Kumar et al. 2014). Soilborne DDT was high in Mumbai and Goa. Higher concentration of DDT in the urban centers particularly *p,p′*-DDT was found in coastal sites of Mumbai (Chakraborty et al. 2010) due to DDT application for vector control (Pandit et al. 2006) (Fig. 3.2).

3.4 Ecological Risk Assessment

Environmental risk assessment is expressed as the comparison of the estimated environmental concentration with guideline concentrations. Environmental quality guidelines such as soil quality guidelines (SQG) are usually based on toxicokinetics data of pollutants on plants and invertebrates from soil contact of different land uses. The soil quality guidelines for land uses are based on models designed to protect primary, secondary, and tertiary consumers from ingestion of contaminated soil and food. For all land uses, the soil contact values of contaminants are also called threshold effect concentration (TEC), above which adverse effects are not expected or rarely occur on the microorganisms and soils are considered to be clean or less polluted. Environmental guidelines for HCHs and DDTs in soil and sediment are not available in India. Therefore, recommended soil quality guidelines from National Oceanography and Atmospheric Administration (NOAA) USA and Canada government were applied in this study for the evaluation of ecotoxicological effects of HCHs and DDTs. The guideline concentration of 700 $\mu g\ kg^{-1}$ (agricultural and residential/parkland use) and 12,000 $\mu g\ kg^{-1}$ (commercial and industrial land use) for $\sum$DDT was established by Canadian Government. NOAA recommended 99.4–9940 $\mu g\ kg^{-1}$ for HCH in soil for mammals. Excluding the maximum values from the northeastern state, Assam, the concentrations of $\sum$DDT and $\sum$HCH in India were lower than the aforementioned guidelines.

Incremental Lifetime Cancer RiskAssessment in Indian Cities ILCRs for different exposure routes increased in the following order: inhalation < dermal contact < direct ingestion. The range and average of total calculated cancer risk for male child, female child, adult male, and adult female in all the seven Indian cities have been given in Tables 3.3, 3.4, 3.5, and 3.6.

Table 3.3 Male Child Cancer Risk (ILCR) for seven major Indian cities

	Bangalore	Chennai	Agra	New Delhi	Goa	Mumbai	Kolkata
Compound	Range (Avg)	Range (Avg)	Range (Avg)	Range (Avg)	Range (Avg)	Range (Avg±SD)	Range (Avg±SD)
α-HCH	3×10^{-10}–7×10^{-6} (1×10^{-6})	1×10^{-7}–5×10^{-6} (1×10^{-6})	4×10^{-7}–4×10^{-6} (2×10^{-6})	3×10^{-10}–3×10^{-6} (6×10^{-7})	1×10^{-7}–3×10^{-7} (2×10^{-7})	3×10^{-10}–2×10^{-6} (5×10^{-7})	4×10^{-10}–1×10^{-5} (2×10^{-6})
β-HCH	1×10^{-7}–1×10^{-5} (2×10^{-6})	1×10^{-7}–5×10^{-6} (1×10^{-6})	7×10^{-7}–8×10^{-6} (3×10^{-6})	4×10^{-11}–2×10^{-5} (2×10^{-6})	1×10^{-6}–1×10^{-5} (8×10^{-6})	1×10^{-10}–5×10^{-6} (2×10^{-6})	1×10^{-10}–1×10^{-5} (2×10^{-6})
γ-HCH	7×10^{-8}–4×10^{-5} (6×10^{-6})	1×10^{-7}–4×10^{-6} (1×10^{-6})	8×10^{-7}–3×10^{-6} (2×10^{-6})	2×10^{-11}–1×10^{-5} (1×10^{-6})	8×10^{-7}–1×10^{-6} (1×10^{-6})	1×10^{-10}–1×10^{-6} (7×10^{-7})	2×10^{-7}–8×10^{-6} (1×10^{-6})
δ-HCH	7×10^{-11}–2×10^{-6} (4×10^{-7})	1×10^{-10}–7×10^{-6} (7×10^{-7})	1×10^{-7}–1×10^{-6} (1×10^{-6})	BDL–4×10^{-6} (5×10^{-7})	6×10^{-8}–1×10^{-7} (1×10^{-7})	8×10^{-11}–4×10^{-7} (5×10^{-8})	7×10^{-11}–2×10^{-6} (4×10^{-7})
p,p'-DDE	4×10^{-9}–1×10^{-6} (2×10^{-7})	8×10^{-9}–6×10^{-8} (2×10^{-8})	4×10^{-8}–3×10^{-7} (2×10^{-7})	7×10^{-13}–6×10^{-7} (1×10^{-7})	1×10^{-7}–8×10^{-7} (4×10^{-7})	1×10^{-8}–1×10^{-7} (4×10^{-8})	4×10^{-9}–2×10^{-6} (2×10^{-7})
p,p'-DDD	1×10^{-9}–3×10^{-7} (5×10^{-8})	5×10^{-12}–8×10^{-9} (4×10^{-9})	1×10^{-8}–7×10^{-8} (4×10^{-8})	1×10^{-12}–8×10^{-8} (2×10^{-8})	4×10^{-11}–2×10^{-7} (9×10^{-8})	4×10^{-9}–4×10^{-8} (1×10^{-8})	3×10^{-9}–7×10^{-7} (8×10^{-8})
p,p'-DDT	1×10^{-8}–1×10^{-8} (1×10^{-8})	1×10^{-8}–9×10^{-7} (2×10^{-7})	9×10^{-11}–6×10^{-7} (2×10^{-7})	1×10^{-11}–4×10^{-7} (1×10^{-7})	4×10^{-10}–1×10^{-5} (3×10^{-6})	2×10^{-8}–2×10^{-7} (6×10^{-8})	2×10^{-11}–1×10^{-6} (2×10^{-7})

Table 3.4 Female Child Cancer Risk (ILCR) for seven major Indian cities

	Bangalore	Chennai	Agra	New Delhi	Goa	Mumbai	Kolkata
Compound	Range (Avg)	Range (Avg)	Range (Avg)	Range (Avg)	Range (Avg)	Range (Avg)	Range (Avg)
α-HCH	3×10^{-10}–7×10^{-6} (1×10^{-6})	1×10^{-7}–5×10^{-6} (1×10^{-6})	5×10^{-7}–4×10^{-6} (2×10^{-6})	3×10^{-10}–3×10^{-6} (6×10^{-7})	1×10^{-7}–3×10^{-7} (2×10^{-7})	4×10^{-10}–2×10^{-6} (5×10^{-7})	4×10^{-10}–1×10^{-5} (3×10^{-6})
β-HCH	1×10^{-7}–1×10^{-5} (2×10^{-6})	1×10^{-7}–5×10^{-6} (1×10^{-6})	7×10^{-7}–8×10^{-6} (3×10^{-6})	4×10^{-11}–2×10^{-5} (2×10^{-6})	1×10^{-6}–1×10^{-5} (8×10^{-6})	1×10^{-10}–5×10^{-5} (9×10^{-6})	1×10^{-10}–1×10^{-5} (2×10^{-6})
γ-HCH	5×10^{-8}–3×10^{-5} (4×10^{-6})	1×10^{-7}–3×10^{-6} (1×10^{-6})	6×10^{-7}–2×10^{-6} (2×10^{-6})	1×10^{-11}–9×10^{-6} (9×10^{-7})	6×10^{-7}–9×10^{-7} (7×10^{-7})	9×10^{-11}–1×10^{-6} (5×10^{-7})	1×10^{-7}–6×10^{-6} (1×10^{-6})
δ-HCH	7×10^{-11}–2×10^{-6} (4×10^{-7})	1×10^{-10}–8×10^{-6} (7×10^{-7})	2×10^{-7}–1×10^{-6} (1×10^{-6})	BDL–4×10^{-6} (6×10^{-7})	6×10^{-8}–1×10^{-7} (1×10^{-7})	9×10^{-11}–4×10^{-7} (5×10^{-8})	7×10^{-11}–2×10^{-6} (4×10^{-7})
p,p'-DDE	3×10^{-8}–1×10^{-5} (1×10^{-6})	6×10^{-8}–4×10^{-7} (1×10^{-7})	3×10^{-7}–2×10^{-6} (1×10^{-6})	5×10^{-12}–5×10^{-6} (1×10^{-6})	1×10^{-6}–6×10^{-6} (3×10^{-6})	8×10^{-8}–1×10^{-6} (2×10^{-7})	3×10^{-8}–1×10^{-5} (2×10^{-6})
p,p'-DDD	1×10^{-8}–2×10^{-6} (3×10^{-7})	3×10^{-11}–6×10^{-8} (3×10^{-8})	1×10^{-7}–5×10^{-7} (3×10^{-7})	1×10^{-11}–6×10^{-7} (1×10^{-7})	3×10^{-10}–1×10^{-6} (7×10^{-7})	3×10^{-8}–3×10^{-7} (8×10^{-8})	2×10^{-8}–5×10^{-6} (6×10^{-7})
p,p'-DDT	1×10^{-8}–9×10^{-7} (1×10^{-7})	1×10^{-8}–9×10^{-7} (2×10^{-7})	1×10^{-10}–7×10^{-7} (2×10^{-7})	1×10^{-11}–4×10^{-7} (1×10^{-7})	4×10^{-10}–1×10^{-5} (3×10^{-6})	3×10^{-8}–2×10^{-7} (6×10^{-8})	2×10^{-11}–1×10^{-6} (2×10^{-7})

Table 3.5 Male Adult Cancer Risk (ILCR) for seven major Indian cities

	Bangalore	Chennai	Agra	New Delhi	Goa	Mumbai	Kolkata
Compound	Range (Avg)	Range (Avg)	Range (Avg)	Range (Avg)	Range (Avg)	Range (Avg)	Range (Avg)
α-HCH	1×10^{-10}–2×10^{-6} (4×10^{-7})	5×10^{-8}–2×10^{-6} (5×10^{-7})	2×10^{-7}–1×10^{-6} (9×10^{-7})	1×10^{-10}–1×10^{-6} (2×10^{-7})	6×10^{-8}–1×10^{-7} (1×10^{-7})	1×10^{-10}–1×10^{-6} (2×10^{-7})	1×10^{-10}–5×10^{-6} (1×10^{-6})
β-HCH	4×10^{-8}–8×10^{-6} (1×10^{-6})	5×10^{-8}–2×10^{-6} (5×10^{-7})	3×10^{-7}–3×10^{-6} (1×10^{-6})	1×10^{-11}–8×10^{-6} (9×10^{-7})	5×10^{-7}–7×10^{-6} (3×10^{-6})	5×10^{-11}–2×10^{-5} (3×10^{-6})	5×10^{-11}–5×10^{-6} (1×10^{-6})
γ-HCH	2×10^{-8}–1×10^{-5} (2×10^{-6})	7×10^{-8}–1×10^{-6} (6×10^{-7})	3×10^{-7}–1×10^{-6} (1×10^{-6})	1×10^{-11}–5×10^{-6} (5×10^{-7})	3×10^{-7}–4×10^{-7} (4×10^{-7})	5×10^{-11}–7×10^{-7} (3×10^{-7})	8×10^{-8}–3×10^{-6} (6×10^{-7})
δ-HCH	3×10^{-11}–1×10^{-6} (1×10^{-7})	4×10^{-11}–3×10^{-6} (3×10^{-7})	8×10^{-8}–7×10^{-7} (4×10^{-7})	BDL–2×10^{-6} (2×10^{-7})	2×10^{-8}–7×10^{-8} (5×10^{-8})	3×10^{-11}–1×10^{-7} (2×10^{-8})	3×10^{-11}–9×10^{-7} (1×10^{-7})
p,p'-DDE	1×10^{-9}–6×10^{-7} (1×10^{-7})	3×10^{-9}–2×10^{-8} (1×10^{-8})	1×10^{-8}–1×10^{-7} (9×10^{-8})	5×10^{-12}–5×10^{-6} (1×10^{-6})	7×10^{-8}–3×10^{-7} (1×10^{-7})	6×10^{-9}–6×10^{-8} (1×10^{-8})	2×10^{-9}–1×10^{-6} (1×10^{-7})
p,p'-DDD	6×10^{-10}–1×10^{-7} (2×10^{-8})	2×10^{-12}–3×10^{-9} (1×10^{-9})	7×10^{-9}–2×10^{-8} (1×10^{-8})	1×10^{-11}–7×10^{-7} (1×10^{-7})	1×10^{-11}–8×10^{-8} (3×10^{-8})	1×10^{-9}–1×10^{-8} (4×10^{-9})	1×10^{-9}–2×10^{-7} (3×10^{-8})
p,p'-DDT	6×10^{-9}–6×10^{-9} (6×10^{-9})	5×10^{-9}–3×10^{-7} (8×10^{-8})	4×10^{-11}–2×10^{-7} (9×10^{-8})	1×10^{-10}–3×10^{-6} (1×10^{-6})	1×10^{-10}–4×10^{-6} (1×10^{-6})	9×10^{-9}–8×10^{-8} (2×10^{-8})	1×10^{-11}–5×10^{-7} (1×10^{-7})

Table 3.6 Female Adult Cancer Risk (ILCR) for seven major Indian cities

	Bangalore	Chennai	Agra	New Delhi	Goa	Mumbai	Kolkata
Compound	Range (Avg)	Range (Avg)	Range (Avg)	Range (Avg)	Range (Avg)	Range (Avg)	Range (Avg)
α-HCH	1×10^{-10}–3×10^{-6} (4×10^{-7})	6×10^{-8}–2×10^{-6} (6×10^{-7})	2×10^{-7}–2×10^{-6} (1×10^{-6})	1×10^{-10}–1×10^{-6} (3×10^{-7})	1×10^{-8}–3×10^{-7} (2×10^{-7})	1×10^{-10}–1×10^{-6} (2×10^{-7})	2×10^{-10}–6×10^{-6} (1×10^{-6})
β-HCH	5×10^{-8}–9×10^{-6} (1×10^{-6})	5×10^{-8}–2×10^{-6} (5×10^{-7})	3×10^{-7}–4×10^{-6} (1×10^{-6})	2×10^{-11}–9×10^{-6} (1×10^{-6})	6×10^{-7}–8×10^{-6} (3×10^{-6})	6×10^{-11}–2×10^{-5} (4×10^{-6})	6×10^{-9}–6×10^{-4} (1×10^{-5})
γ-HCH	3×10^{-8}–2×10^{-5} (2×10^{-6})	8×10^{-8}–2×10^{-6} (6×10^{-7})	3×10^{-7}–1×10^{-6} (1×10^{-6})	1×10^{-11}–5×10^{-6} (5×10^{-7})	3×10^{-7}–5×10^{-7} (4×10^{-7})	6×10^{-11}–8×10^{-7} (3×10^{-7})	9×10^{-8}–4×10^{-6} (7×10^{-7})
δ-HCH	3×10^{-11}–1×10^{-6} (1×10^{-7})	5×10^{-11}–3×10^{-6} (3×10^{-7})	9×10^{-8}–8×10^{-7} (5×10^{-7})	BDL–2×10^{-6} (2×10^{-7})	2×10^{-8}–8×10^{-8} (6×10^{-8})	3×10^{-11}–2×10^{-7} (2×10^{-8})	3×10^{-11}–1×10^{-6} (2×10^{-7})
p,p′-DDE	1×10^{-9}–7×10^{-7} (1×10^{-7})	3×10^{-9}–3×10^{-8} (1×10^{-8})	2×10^{-8}–1×10^{-7} (1×10^{-7})	3×10^{-13}–3×10^{-7} (7×10^{-8})	8×10^{-8}–3×10^{-7} (1×10^{-7})	5×10^{-9}–7×10^{-8} (1×10^{-8})	2×10^{-9}–1×10^{-6} (1×10^{-7})
p,p′-DDD	7×10^{-10}–1×10^{-7} (2×10^{-8})	2×10^{-12}–3×10^{-9} (1×10^{-9})	8×10^{-9}–3×10^{-8} (2×10^{-8})	8×10^{-13}–4×10^{-8} (1×10^{-8})	1×10^{-11}–9×10^{-8} (4×10^{-8})	1×10^{-9}–1×10^{-8} (5×10^{-8})	1×10^{-9}–3×10^{-7} (3×10^{-8})
p,p′-DDT	7×10^{-9}–4×10^{-7} (6×10^{-8})	6×10^{-9}–4×10^{-7} (9×10^{-8})	4×10^{-11}–3×10^{-7} (1×10^{-7})	8×10^{-12}–2×10^{-7} (7×10^{-8})	2×10^{-10}–5×10^{-6} (1×10^{-6})	1×10^{-8}–9×10^{-8} (2×10^{-8})	1×10^{-11}–6×10^{-7} (1×10^{-7})

ILCR varied only slightly in terms of gender differences. The highest and lowest average ILCR for male varied between 1×10^{-6} and 1×10^{-8} and female between 1×10^{-4} and 3×10^{-8}. Soil ingestion showed predominant risk for all the cities mostly ranging between 10^{-7} and 10^{-4}. Isomers of HCH have been found to be important source for ingestion risk in male and female children in Bangalore, Chennai, Agra, and Delhi (Tables 3.3 and 3.4). ILCR level of γ-HCH showed risk for all the Indian cities, predominantly Goa and Mumbai; this may be due to higher levels of γ-HCH found in their atmosphere (Chakraborty et al. 2010). Excluding Agra, the predominance of β-HCH showed high risk for all the Indian cities. In Mumbai, only one rural site was found at risk. However, β-HCH and γ-HCH showed predominant ILCR in all the cities due to their higher K_{OA} and relatively greater deposition in soil (Xiao et al. 2004) thereby leading to higher risk due to soil ingestion. Interestingly in Bangalore, only one rural site was found to have potential risk due to soilborne HCH. For children typically toddlers, soil ingestion pathway is higher in child leading to higher cancer risk due to exposure to soilborne pesticides from the infancy stage.

ILCR due to dermal contact exceeded 10^{-6} except for few sites in Goa, New Delhi, and Kolkata. Individual HCH isomers have an average ILCR range varying between 10^{-6} and 10^{-7} for all the age groups. ILCR for DDT in soil from seven major Indian cities were closely associated with the HCH residues. Lindane manufacturing units present in Agra, near New Delhi could be a reason for high risk in those areas. ILCRs caused by inhalation of soil particles were less among the three pathways ranging between 10^{-17} and 10^{-10}, indicating that the cancer risk caused by the inhalation of soil particles was negligible.

3.5 Conclusions

Indian climate varies from tropical region in the south to subtropical region in the north and temperate climate in the far north along the Himalayan Range. Such a diverse climatic variation can play a major role in the long-range atmospheric transport of POPs in India. Further monsoonal events can also cause significant impact on the movement of these compounds to the aquatic environment. Residues of HCH and DDT were due to extensive usage and large-scale production of DDT, technical HCH, and lindane for agricultural use in rural areas and vector (largely for mosquito) control in urban areas in last five decades. Soil is acting as sink for these compounds due to their tendency to bind with soil organic matter. Apart from local usage that is completely site specific in northern India, atmospheric transport can play an important role particularly in the movement of HCH isomers away from the source. Lower average ambient temperature during winter due to the subtropical climate in northern India was speculated to be the possible cause for deposition of HCH and DDT. All the cities showed potential risk due to exposure to soilborne HCH residues. We suggest that the soil ingestion pathway may be a potential cause for cancer risk due to chronic exposure from infant stage.

References

Abhilash P, Singh N (2008) Distribution of hexachlorocyclohexane isomers in soil samples from a small scale industrial area of Lucknow, North India, associated with lindane production. Chemosphere 73:1011–1015

Agnihotri N, Kulshrestha G, Gajbhiye V, Mohapatra S, Singh S (1996) Organochlorine insecticide residues in agricultural soils of the Indo-Gangetic plain. Environ Monit Assess 40:279–288

Agoramoorthy G (2008) Can India meet the increasing food demand by 2020? Futures 40:503–506

ATSDR (2005) Toxicology profile for polyaromatic hydrocarbons. ATSDR's toxicological profiles on CD-ROM. CRC Press, Boca Raton

Babu GS, Farooq M, Ray R, Joshi P, Viswanathan P, Hans R (2003) DDT and HCH residues in Basmati rice (*Oryza sativa*) cultivated in Dehradun (India). Water Air Soil Pollut 144:149–157

Bhattacharya B, Sarkar SK, Mukherjee N (2003) Organochlorine pesticide residues in sediments of a tropical mangrove estuary, India: implications for monitoring. Environ Int 29:587–592

CAPE (2005) Factsheet Lindane's Dirty Secret: Indian facilities dump toxic waste 2005 (Community Action for Pesticide Elimination)

Chakraborty P and Zhang G (2011) Organochlorine pesticides (OCPs), polychlorinated biphenyls (PCBs) and polybrominated diphenyl ethers (PBDEs) in the Indian atmosphere. In the chapter, contamination profiles of POPs in India in the book, global contamination trends of persistent organic chemicals. Taylor and Francis Books, Boca Raton

Chakraborty P, Zhang G, Li J, Xu Y, Liu X, Tanabe S et al (2010) Selected organochlorine pesticides in the atmosphere of major Indian cities: levels, regional versus local variations, and sources. Environ Sci Technol 44:8038–8043

Chakraborty P, Zhang G, Li J, Sivakumar A, Jones KC (2015) Occurrence and sources of selected organochlorine pesticides in the soil of seven major Indian cities: assessment of air–soil exchange. Environ Pollut 204:74–80

Devi NL, Qi S, Chakraborty P, Zhang G, Yadav IC (2011) Passive air sampling of organochlorine pesticides in a northeastern state of India, Manipur. J Environ Sci 23:808–815

Devi NL, Chakraborty P, Shihua Q, Zhang G (2013) Selected organochlorine pesticides (OCPs) in surface soils from three major states from the northeastern part of India. Environ Monit Assess 185:6667–6676

Devi NL, Yadav IC, Raha P, Shihua Q, Dan Y (2015) Spatial distribution, source apportionment and ecological risk assessment of residual organochlorine pesticides (OCPs) in the Himalayas. Environ Sci Pollu Res:1–13

DGFT (2008) Department of chemicals and petrochemicals, ministry of chemicals and director general of foreign trade, New Delhi. Government I. Eleventh five-year plan: 2008–2012, Planning Commision of India

Gupta PK (2004) Pesticide exposure—Indian scene. Toxicology 198:83–90

Guzzella L, Roscioli C, Vigano L, Saha M, Sarkar S, Bhattacharya A (2005) Evaluation of the concentration of HCH, DDT, HCB, PCB and PAH in the sediments along the lower stretch of Hugli estuary, West Bengal, northeast India. Environ Int 31:523–534

Harner T, Bidleman TF, Jantunen LMM, Mackay D (2001) Soil-air exchange model of persistent pesticides in the United States cotton belt. Environ Toxicol Chem 20:1612–1621

IOG (2002) Regionally based assessment of persistent toxic substances—Indian Ocean Regional Report global environment facility. United Nations Environmental Programme, Châtelaine

Jayashree R, Vasudevan N (2006) Residues of organochlorine pesticides in agricultural soils of Thiruvallur district, India. J Food Agric Environ 4:313–317

Kumar B, Mishra M, Verma VK, Kumar S, Sharma CS (2012) Distribution of dichlorodiphenyltrichloroethane and hexachlorocyclohexane in urban soils and risk assessment. J Xenobiot 3:1–9

Kumar B, Verma V, Mishra M, Gaur R, Kumar S, Sharma C (2014) DDT and HCH (organochlorine pesticides) in residential soils and health assessment for human populations in Korba, India. Hum Ecol Risk Assess 20:1538–1549

Kumari B, Singh R, Madan V, Kumar R, Kathpal T (1996) DDT and HCH compounds in soils, ponds, and drinking water of Haryana, India. Bull Environ Contam Toxicol 57:787–793
Kumari B, Madan V, Kathpal T (2008) Status of insecticide contamination of soil and water in Haryana, India. Environ Monitor Assess 136:239–244
Kurt-Karakus PB, Bidleman TF, Jones KC (2005) Chiral organochlorine pesticide signatures in global background soils. Environ Sci Technol 39:8671–8677
Li Y-F, Scholtz MT, Van Heyst BJ (2003) Global gridded emission inventories of β-hexachlorocyclohexane. Environ Sci Technol 37:3493–3498
MCF (2007) Performance of chemical & petrochemical industry at a glance Available via http://chemicals.nic.in/stat0107.pdf
Minh N, Minh T, Kajiwara N, Kunisue T, Subramanian A, Iwata H et al (2006) Contamination by persistent organic pollutants in dumping sites of Asian developing countries: implication of emerging pollution sources. Arch Environ Contam Toxicol 50:474–481
Mishra K, Sharma RC, Kumar S (2003) Contamination levels and spatial distribution of organochlorine pesticides in soils from India. Ecotoxicol Environ Saf 76:215–225
Nawab A, Aleem A, Malik A (2003) Determination of organochlorine pesticides in agricultural soil with special reference to γ-HCH degradation by Pseudomonas strains. Bioresour Technol 88:41–46
NIC (2004–2005) Agriculture contingency plan for district Nagaon Available via http://agricoop.nic.in/Admin_Agricoop/Uploaded_File/ASSAM16-NAGAON-26.7.2012.pdf
NIP (2011) National implementation plan, Stockholm convention on persistent organic Pollutants. Government of India, Pretoria
NVBDCP (2010) National vector borne disease control programme. National Rural Health Mission, Bhopal
Ockenden WA, Breivik K, Meijer SN, Steinnes E, Sweetman AJ, Jones KC (2003) The global re-cycling of persistent organic pollutants is strongly retarded by soils. Environ Pollut 121:75–80
Pandit G, Sahu S, Sharma S, Puranik V (2006) Distribution and fate of persistent organochlorine pesticides in coastal marine environment of Mumbai. Environ Int 32:240–243
PPQS (2013) Pesticide and documentation unit, Directorate of Plant Protection, Quarantine & Storage. Available via http://ppqs.gov.in/IpmPesticides_Cont.htm
Prakash O, Suar M, Raina V, Dogra C, Pal R, Lal R (2004) Residues of hexachlorocyclohexane isomers in soil and water samples from Delhi and adjoining areas. Curr Sci 87:73–77
Rajendran RB, Subramanian A (1999) Chlorinated pesticide residues in surface sediments from the River Kaveri, South India. J Environ Sci Health Part B 34:269–288
Ramesh A, Tanabe S, Murase H, Subramanian A, Tatsukawa R (1991) Distribution and behaviour of persistent organochlorine insecticides in paddy soil and sediments in the tropical environment: a case study in South India. Environ Pollut 74:293–307
Rekha NSN, Prasad R (2006) Pesticide residue in organic and conventional food-risk analysis. J Chem Health Saf 13:12–19
Saha K, Saha T, Banerjee H, Bhattacharyya A, Chowdhury A, Somchoudhury A (2004) Persistence of dicofol residue on tea under North-East Indian climatic conditions. Bull Environ Contam Toxicol 73:347–350
Senthilkumar K, Kannan K, Subramanian A, Tanabe S (2001) Accumulation of organochlorine pesticides and polychlorinated biphenyls in sediments, aquatic organisms, birds, bird eggs and bat collected from south India. Environ Sci Pollut Res 8:35–47
Singh KP, Malik A, Sinha S (2007) Persistent organochlorine pesticide residues in soil and surface water of northern Indo-Gangetic alluvial plains. Environ Monitor Assess 125:147–155
Subramanian A, Tanabe S (2007) Persistent toxic substances in India. Dev Environ Sci 7:433–485
Takeoka H, Ramesh A, Iwata H, Tanabe S, Subramanian A, Mohan D et al (1991) Fate of the insecticide HCH in the tropical coastal area of South India. Marine Pollut Bull 22:290–297
USEPA (1989) Risk assessment guidance for superfund. Human health evaluation manual. volume I. EPA/540/1-89/002

Vijgen J, Yi LF, Forter M, Lal R, Weber R (2006) The legacy of lindane and technical HCH production. Organohalog Compd 68:899–904

Wei D, Kameya T, Urano K (2007) Environmental management of pesticidal POPs in China: past, present and future. Environ Int 33:894–902

Wild SR, Jones KC (1995) Polynuclear aromatic hydrocarbons in the United Kingdom environment: a preliminary source inventory and budget. Environ Pollut 88:91–108

Xiao H, Li N, Wania F (2004) Compilation, evaluation, and selection of physical-chemical property data for α-, β-, and γ-hexachlorocyclohexane. J Chem Eng Data 49:173–185

Chapter 4
Antibiotics and Antibiotic Resistance Genes (ARGs) in Soil: Occurrence, Fate, and Effects

Muhammad Zaffar Hashmi, Adeel Mahmood, Dambaru Ballab Kattel, Sohaib Khan, Ahmad Hasnain, and Zulkifl Ahmed

4.1 Introduction

The era of anthropogenic antibiotics began after the discovery of penicillin in 1928 by Alexander Fleming (Diggins 1999). Natural soil is the reservoir of autochthonous soil microorganisms which biosynthesize antimicrobial secondary metabolites (Martinez-Fleites et al. 2006; Raaijmakers et al. 1997). Nevertheless, since the first introduction of antibiotics, new chemical entities of antimicrobials have been discovered and were synthesized on an industrial scale (Kumar and Gayen 2011). Antibiosis is a natural chemical regulation mechanism among organisms, especially

M.Z. Hashmi (✉)
Department of Meteorology, COMSATS Institute of Information Technology, Park Road, Islamabad, Pakistan
e-mail: hashmi_qau@yahoo.com

A. Mahmood
Department of Environmental Sciences, Government College Women University, Sialkot, 51310, Pakistan

D.B. Kattel
Department of Meteorology, COMSATS Institute of Information Technology, Park Road, Islamabad, Pakistan

Institute of Tibetan Plateau Research, Chinese Academy of Sciences, No. 16 Lincui Road, Beijing, Chaoyang District, 100101, PR China

S. Khan
Department of Anthropology, Bahauddin Zakariya University, Multan, Pakistan

A. Hasnain
Department of Geography, Bahauddin Zakariya University, Multan, Pakistan

Z. Ahmed
Department of Building and Architecture Engineering, Bahauddin Zakariya University, Multan, Pakistan

© Springer International Publishing AG 2017

M.Z. Hashmi et al. (eds.), *Xenobiotics in the Soil Environment*, Soil Biology 49,
DOI 10.1007/978-3-319-47744-2_4

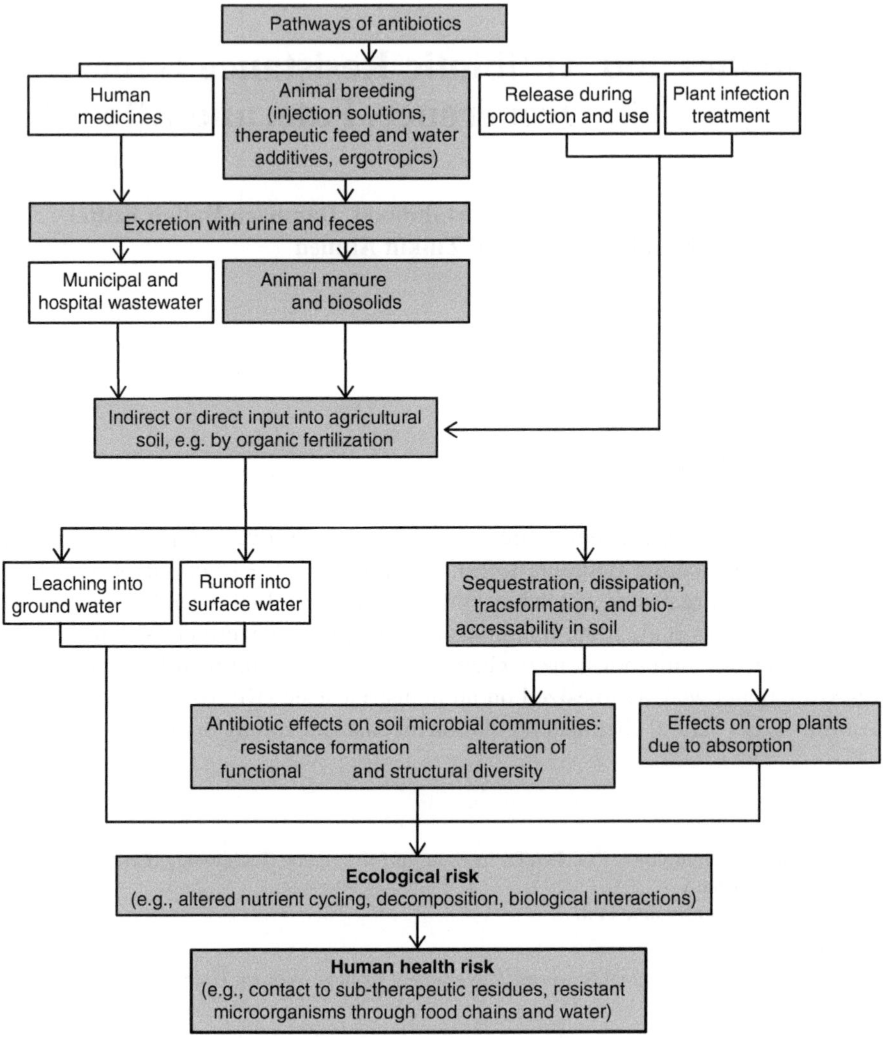

Fig. 4.1 Pathways of antibiotics (Adapted from Du and Liu 2012)

microorganisms. Hence, antibiotics also biosynthesize in soils (Gottlieb 1976). Today, to treat and control many bacterial infections, synthetic and naturally antibiotics have been widely used in veterinary and human medicines, but misuse or overuse of antibiotics contributes to the emergence and spread of antibiotic resistance genes (ARGs) in the environmental compartments (Tang et al. 2015). Antibiotic resistance genes (ARGs) are emerging environmental contaminants and may pose a threat to public health (Chen and Zhang 2013). There are some typical pathways of antibiotics by which they are introduced into the environment: (1) inadvertent release during production and use, (2) their application against plant diseases, (3) sewage sludge with antibiotic residues of human medicine, and (4) medicated animals' manure which received therapeutic feed/water additives, injection solutions, or ergotropics (see below; Fig. 4.1; adapted from Du and Liu

2012). As a result, antibiotics have been detected in different environmental compartments such as groundwater of farms and in aquatic and soil environments (Martinez 2009). Forty different tetracycline-resistance (*tet*) genes with three specific mechanisms (i.e., target modification with ribosomal protection protein, antibiotic efflux pumps, and antibiotic inactivation) have been characterized to date (Roberts 2005). Four sulfonamide-resistance (*sul*) gene types, including *sul1*, *sul2*, *sul3*, and *sulA*, have also been studied (Pei et al. 2006). In both animals and humans, a significant amount of antibiotics (up to 75 %) can be excreted in an unaltered state (Elmund et al. 1971) which could pose threat to soil environment in general and particularly to humans. So this chapter covers the antibiotics and ARGs' general information, their fate in soil environment, pollution caused by the ARGs, and their impact on human health.

4.2 Physical Properties of Antibiotics

The compounds which are synthesized through secondary metabolism of living organisms are known as antibiotics, with exceptions for semi- or completely synthetic substances. Antibiotics are heterogeneous compounds with different fields of usage such as antiinfectives, antimycotics, and anthelmintics, have different structural classes (tetracyclines, nucleosides), and reveal different molecular structures and diverse physical and chemical properties (Garrod and O'Grady 1968). Depending on the pH of the medium, most antibiotics tend to ionize; pk_a values are related with the diverse functional groups of the compounds. Some important antibiotic compounds' physicochemical properties ranges are listed in Table 4.1. However, a brief description of physicochemical properties of selected antibiotics is given below:

4.2.1 Tetracyclines

Tetracyclines (TCs) are polyketides, and structure is like that of a naphthacene ring (Fig. 4.2). The TCs have three pk_a values and are amphoteric. They exhibit stability in acids, but not in bases, and form salts in both media (Podojil et al. 1984). Most TCs are sparingly soluble in water, while the solubility of the corresponding hydrochlorides is much higher. The TCs strongly absorb light and, thus, are susceptible to photodegradation (Chen et al. 2008).

4.2.2 Sulfonamides

Sulfonamides (SAs) are relatively insoluble in water and characterized by two pk_a values (Fig. 4.3). SAs indicate protonation of the amino group at a pH of 2–3 and deprotonation of the R_1SO_2 NHR_2 moiety at a pH of 5–11 (Ingerslev and Halling-

Table 4.1 Physical properties of antibiotics (Osol and Remington 1980)

Compound class	Molar mass (g/mol)	Water solubility (mil/l)	log K_{ow}	pk_a	Henry's constant (Pa l/mol)
Tetracyclines chlortetracycline, oxytetracycline, tetracycline	444.5–527.6	230–52,000	−1.3 to 0.05	3.3/7.7/9.3	1.7×10^{-23} –4.8×10^{-22}
Sulfonamides sulfanilamide, sulfadiazine, sulfadimidine, sulfadimethoxine, sulfapyridine, sulfamethoxazole	172.2–300.3	7.5–1500	−0.1 to 1.7	2–3/4.5–10.6	1.3×10^{-12} –1.8×10^{-8}
Aminoglycosides kanamycin, neomycin, streptomycin	332.4–615.6	10–500[a]	−8.1 to −0.8	6.9–8.5	8.5×10^{-12} –4.1×10^{-8}
β-Lactams penicillins: ampicillin, meropenem, penicillin G; cephalosporins: ceftiofur, cefotiam	334.4–470.3	22–10,100	0.9 to 2.9	2.7	2.5×10^{-19} –1.2×10^{-12}
Macrolides erythromycin, oleandomycin, tylosin	687.9–916.1	0.45–15	1.6 to 3.1	7.7–8.9	7.8×10^{-36} –2.0×10^{-26}
Flouroquinolones ciprofloxacin, enrofloxacin, flumequin, sarafloxacin, oxolinic acid	229.5–417.6	3.2–17,790	−1.0 to 1.6	8.6	5.2×10^{-17} –3.2×10^{-8}
Imidazoles fenbendazole, metronidazole, oxfendazole	171.5–315.3	6.3–407	−0.02 to 3.9	2.4	2.3×10^{-13} –2.7×10^{-10}
Polypeptides avermectin, bacitracin, ivermactin, virginiamycin	499.6–1038	Not completely	−1.0 to 3.2		Negligible–2.8×10^{-23}
Polyethers monensin, salinomycin	670.9–751.0	2.2×10^{-6} –3.1×10^{-3}	5.4 to 8.5	6.4	2.1×10^{-18} –1.5×10^{-18}
Glycopeptides vancomycin	1450.7	>1000	Not soluble in octanol	5.0	Negligible
Quinoxaline derivatives olaquindox	263.3	1.0×10^{-6}	−2.2	10	1.1×10^{-18}

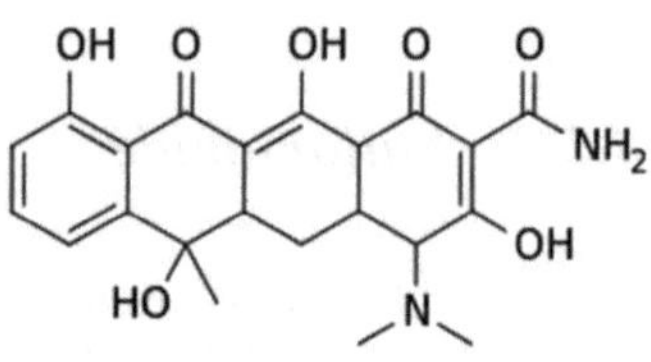

Fig. 4.2 The four rings of the basic tetracycline structure

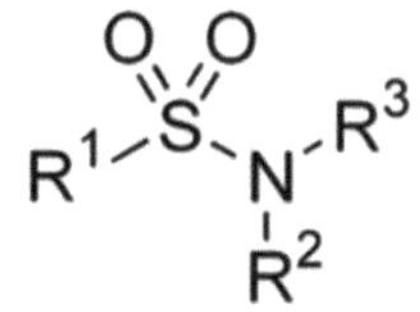

Fig. 4.3 The structure of the sulfonamide group

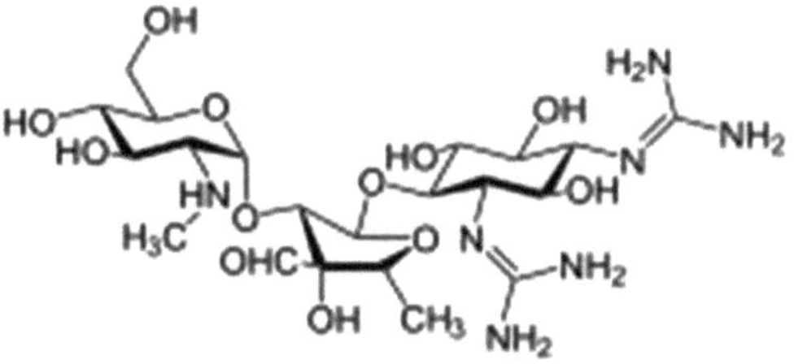

Fig. 4.4 The structure of the aminoglycosides group

Sørensen 2000). In general, the atmospheric SAs behave as weak acids and form salts in strongly acidic or basic solutions.

4.2.3 *Aminoglycosides*

Aminoglycosides like antibiotics are basic and strongly polar cationic compounds (Fig. 4.4). Their molecular structure is comprised of two or more amino sugars that are glycosidically bound to aminocyclitol. They are susceptible to photodegradation, mostly hydrophilic, and water soluble.

4.2.4 *Fluoroquinolones*

Fluoroquinolones (FQs) showed large chemical stability and are known as quinolones (Fig. 4.5). They are degraded by UV light, insensitive to hydrolysis and increased temperature. Their antibiotic potency depends mostly on the aromatic fluorine substituent at the C-6 position (Sukul and Spiteller 2007).

4.3 Natural Antibiotics Occurrence in Soil

New pharmaceuticals were discovered due to secondary metabolites of natural resources and often served as blueprints for the development of synthetic antibiotics (Sarmah et al. 2006). Soil microbial communities and plant roots serve as the reservoir of natural antimicrobial production (Gottlieb 1976; Mavrodi et al. 2012; Raaijmakers and Mazzola 2012), often comprising fungal, pseudomonad, and actinomycete species (Butler and Buss 2006; Raaijmakers et al. 2002). It has been reported that the production of antibiotic in natural microbial communities

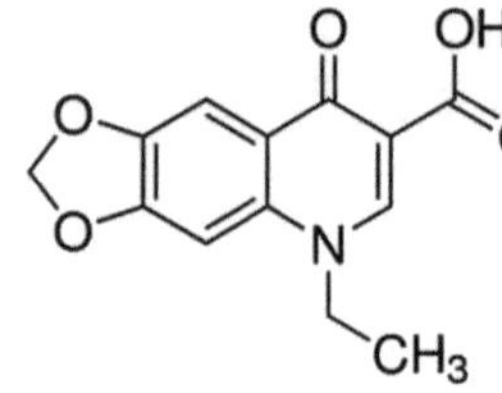

Fig. 4.5 Quinolone antibiotic

is suggested to promote microbial fitness, defense, signaling, gene regulation, and competiveness (Mavrodi et al. 2012). Hence, antibiotics are considered as a part of the disease regulation in soil. The detection and quantification of natural antimicrobials, in nutrient-poor soil, are often prohibited by fast degradation, strong sorption to the soil matrix, and concentrations near the detection limit (Mavrodi et al. 2012; Thomashow et al. 1997). However, the natural broad-spectrum antibiotic phenazine-1-carboxylic acid was successfully quantified at concentrations up to 1.6 μg/g^{-1} root fresh weight in nutrient-rich rhizosphere soil of wheat plants (Mavrodi et al. 2012). Rhizosphere samples of plants exhibited natural antibiotic concentrations in the range 0.02–5.0 μg/g^{-1} (Mavrodi et al. 2012; Thiele-Bruhn 2003). The antibiotic resistance evolution in natural microbial communities is thought due to the exposure to autochthonous antimicrobial compounds. Such exposure might also alter the response to synthetic antibiotics from anthropogenic sources (Aminov and Mackie 2007). Soil microbial communities are probably composed of more and less antibiotic-susceptible strains, suggesting also coadaptations to some synthetic antimicrobials.

4.4 Anthropogenic Antibiotics Occurrence in Soil

Anthropogenic synthesized antibiotics contamination of terrestrial soil and aquatic environments is a widespread problem, which begins through hospital waste and with the excretion by the animals (Boxall et al. 2004; Kümmerer 2008a). Most of antibiotics have been designed to be readily excreted after medication with rather short half-lives, for example, 0.1–26 h for SA antibiotics and 1.5–16 h for FQ antibiotics (Picó and Andreu 2007). Of the applied drug, 44 % SDZ (SA) was excreted unchanged as parent compound, with acetyl conjugates (26 %) and hydroxylated compounds (19 %) as major metabolites (Lamshöft et al. 2010). Of the applied DIF (FQ), approximately 96 % were excreted as parent compound, with sarafloxacin as the major metabolite (Lamshöft et al. 2010). The concentrations of SA in contaminated manure typically ranged from 1 to 10 mg kg^{-1}, occasionally up to 235 mg kg^{-1} fresh weight (Hamscher and Mohring 2012; Kumar et al. 2005). Monitoring of manure samples from Austria exhibited SAs, tetracyclines, and FQ concentrations up to 91 mg kg^{-1}, 46 mg kg^{-1}, and 8 mg kg^{-1}, respectively (Martínez-Carballo et al. 2007). Feedlot pig manure from China contained tetracycline and FQ concentrations up to 60 mg kg^{-1} and 47 mg kg^{-1}, respectively (Zhao et al. 2010). Consequently, agriculture soils are enriched with large quantities of antibiotic compounds with contaminated animal manure. Under long-term conventional farming practice, a tetracycline concentration of 200 μg kg^{-1} was extractable from soil (Hamscher et al. 2003). Farmland soils of southern China had FQ and SA concentrations of even 1537 μg kg^{-1} and 321 μg kg^{-1}, respectively (Li and Zhang 2011). SDZ concentrations up to 90 μg kg^{-1} were reported in the soil of wheat-planted and manure-fertilized agricultural landscapes (Grote et al. 2007). Antibiotic concentration in field soil increased due to mixtures of different antibiotics such as SAs, FQs, and tetracyclines, as indicated in the northern Marmara region of Turkey (Karcı and Balcıoğlu 2009). However, due to low extraction efficiencies of many compounds, antibiotic concentrations in soil are probably underestimated.

4.5 Antibiotics Fate in the Soil Environment

When the antibiotics are added to the soil solid phase, they are liable to microbial transformation. Since the 1940s, there are some compounds like sulfonamides which have been extensively used in the animal husbandry and pig production (Kümmerer 2008b; Zhou et al. 2012). The changes occur because of the abovementioned biotransformation results in the retransformation of metabolites into the parent compound similarly as in fertilizers (Giang et al. 2015; Jechalke et al. 2014). For many antibiotics, mineralization accounts for less than 2 % of the added compounds (Forster et al. 2009; Junge et al. 2012). Because of poor light penetration, the photodegenration in the soils as an alternative pathway of pharmaceutical degradation is limited (Ozaki et al. 2011). Final pathways to be considered are potential transfers of the antibiotics from soil into the atmosphere, hydrosphere, and biosphere (Fig. 4.6). Low vapor pressure is responsible for the fate of the

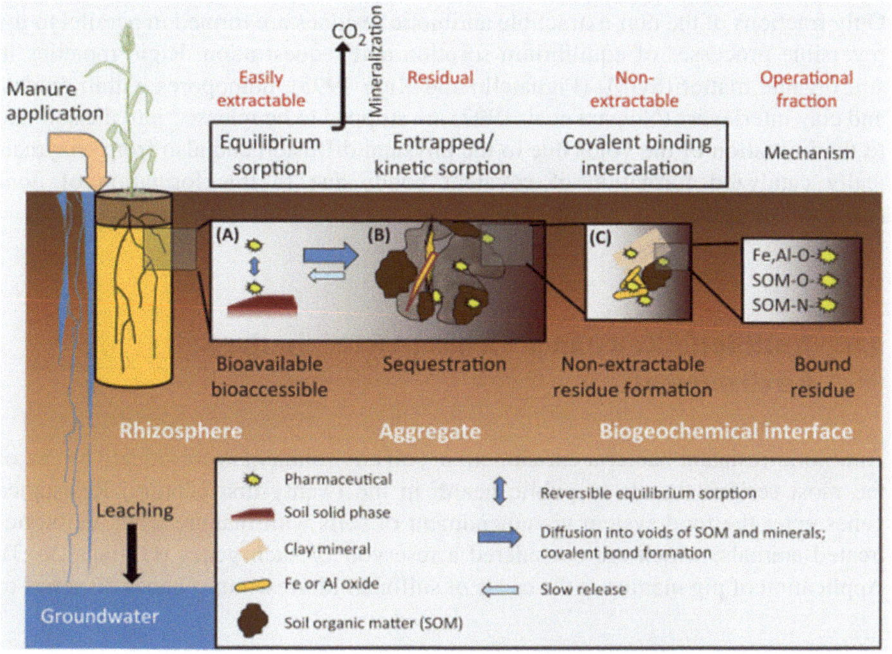

Fig. 4.6 Antibiotics fate in soil and its compartments (Adapted from Jechalke et al. 2014)

pharmaceuticals in soil and here volatilization is not relevant. All antibiotics may disperse because of the surface runoff and particle-facilitated transport in the environment (Burch et al. 2014; Joy et al. 2013, 2014). The vertical percolation or leaching that occurs in the groundwater is restricted to a few antibiotics such as sulfanomides, which mainly occurs in preferential flow path. So that is the reason that the largest fraction of most antibiotics applied to soils with manure is usually returned in the surface soil (Ostermann et al. 2013; Srinivasan and Sarmah 2014). For the removal of antibiotics from the soil, the plant uptake by several pharmaceuticals is done (Felizeter et al. 2012; Sabourin et al. 2012), but there concentrations are commonly so small in the plant tissues that there plant uptake might not represent a major pathway (Engelhardt et al. 2015; Fang et al. 2014; Rosendahl et al. 2012). Sorption and desorption reactions are responsible for the interaction of the veterinary antibiotics with the soil solid phase. Biotransformation, biological effects, their mobility, and uptake by plants are controlled by the sorption and desorption processes. Sorption is not only a function of polarity and water solubility for most antibiotics, and its effect on the chemical specifications and charge of the compounds is particularly controlled by the pH value (e.g., Schaffer et al. 2012; Yang et al. 2012). The history of fertilization and other solutes can also affect sorption. Sorption of a Sulfonamide antibiotic (sulfamethoxazole) was reduced in soils fertilized and irrigated with untreated wastewater for 14–100 years, most likely as a result of a saturation of high-affinity sorption sites, sorption competition with other solutes, and changes of soil organic matter properties, whereas sorption of ciprofloxacin was not affected (Dalkmann et al. 2014).

Resulting in hysteresis and a subsequent kinetic sorption and desorption with slow release rates, the primary sorption reaction of antibiotics is often governed by the reversible equilibrium process (Doretto and Rath 2013; Kasteel et al. 2010). Only fractions of the non-extractible antibiotic residues are formed in parallel to the reversible processes of equilibrium sorption and sequestration. Rigid moieties in soil organic matter (SOM) (Pignatello and Xing 1995), nomopores within oxides, and clay interlayers (Nowara et al. 1997) are stopped to be released into the soil due to the formation of the voids due to the physical diffusion and also from enzymatically catalyzed formation of covalent bonds due to this formation of non-extractible residues, e.g., Bialk et al. (2007) and Gulkowska et al. (2013).

4.6 Antibiotic Resistance Genes (ARGs) in the Soil Environment

Antibiotic-resistant bacteria enrichment in soil environment is considered as one of the most serious threats to public health in the twenty-first century. Resistance genes enter the food system via amendment of soils with manure from antibiotic-treated animals, which are considered a reservoir of such genes (Gillings 2013). Application of pig manure is the cause of sulfonamide-resistance genes dispersal to soil bacteria. Further, it has been found that dairy cow manure amendment enhanced the proliferation of genes encoding β-lactamases and resident antibiotic-resistant bacteria in soil even though the cows from which the manure was derived had not been treated with antibiotics.

Manure amendment and resistance exhibit a correlation which has been proved by a number of studies in previous few decades. For example, the prevalence of sulfonamide-resistant isolates was increased in a field soil after 2 years of slurry application of antibiotic-containing manure, compared to preapplication soil (Byrne-Bailey et al. 2009). In further microcosm experiments, the accumulation of resistance genes by the agricultural practice of repeated manure amendment was shown (Heuer et al. 2011). Sulfonamide concentrations as low as 0.1 mg kg^{-1} of soil could have a selective effect on resistant populations in soil (Heuer et al. 2008).

However, there are fewer reports available on the long-term effects of antibiotics pollution with antibiotics and resistant bacteria. A recent study by Tang et al. (2015) revealed the accumulation of antibiotic resistance genes in paddy soils from four field experiments in south of China over decades of increasing use of antibiotics.

4.7 Environmental and Public Health Effects of Antibiotic Resistance Genes Pollution

Antibiotic resistance genes' pollution could increase the human pathogens by obtaining resistance. When the human microbia containing the residues is released in the environment in which the bacteria-enriched resistance elements are present, then the determinants by human-linked bacteria increase the possibility of acquiring novel resistance. So that is why it is stressed that residuals from hospitals in which human commensal and infective bacteria and also antibiotics are released should be minimized at all costs to avoid interchange of genetic material (Sirés and Brillas 2012). As stated by Baquero et al., the possibility of the variation of genetics and the novel mechanisms resistance's possible emergence that are reintroduced in the human environment is because of the other types of microbiota in different ecosystems (from soil sediments of groundwater to animal microbiota) coming in contact with the human microbiota (Baquero et al. 2008). The physiology of natural microbial populations and their dynamics is challenged by the spread of resistance genes that currently exist in the human- or animal-associated microbial which are found without the antibiotic pollution in different environments, as indicated by several reports. Bacteria that is found in humans and animals has the genes on several occasions. The above fact has not been focused or studied in detail, but it is said that the environmental populations are responsible for the spread of those genes. Now the question arises here that in the recipient organisms these antibiotic resistance genes may produce relevant changes or not. The case has been proven by some studies. For instance, resistance to either glycopeptides or beta-lactam antibiotics strongly modifies the structure of the peptidoglycan in Gram-positive bacteria (Mainardi et al. 2008; Martinez et al. 2009), and it has been described that antibiotic resistance of small colony variants of *S. aureus* is associated with the rewiring of bacterial metabolism (Heinemann et al. 2005). For the bacterial metabolism, there is an indication that acquisition could have incalculable consequences. This could also lead to the evolution of the environmental biosphere. Up till now, the studies that have been made so far in which the antibiotic resistance has an effect of antibiotic pollution are different regarding the antibiotic genes that come with pollution. Firstly, this results in the outburst of the resistant bacteria that exist in the humans or animals. When the antibiotic exposure is ceased, the microbial community outcome is largely settled. If a gene transfer unit that already contains antibiotic resistance genes in the environment also exists as contaminants, then their diffusion will be favored by the occurrence of the antibiotics, and when the discharge of the antibiotics is reduced or ceased, then the modification of the antibiotic resistance will be lower.

4.8 Conclusion

This chapter provides a comprehensive overview of antibiotics and ARGs in soil, potentially providing knowledge for managing antibiotic resistance emanating from agricultural activities. Although there are numerous recent studies on antibiotics and antibioticresistance genes, many questions still need to be addressed more systematically. Information on the effects of reductions in antibiotic use on the overall level of resistance is needed in order to evaluate management options. Along the article, we review the impact that pollution by antibiotics or by antibiotic resistance genes may have for both human health and the evolution of environmental microbial populations.

Acknowledgement The Research was funded by TWAS-COMSTECH Research award_15-384 RG/ENG/AS_C and HEC Start Up Research Grant 21-700/SRGP/R&D/HEC/2016.

References

Aminov RI, Mackie RI (2007) Evolution and ecology of antibiotic resistance genes. FEMS Micro Lett 271:147–161

Baquero F, Negri M, Morosini M, Blazquez J (2008) The antibiotic selective process: concentration-specific amplification of low-level resistant populations. Antibiotic resistance: origins, evolution, selection and spread. Ciba Found Symp 787:93

Bialk HM, Hedman C, Castillo A, Pedersen JA (2007) Laccase-mediated Michael addition of 15N-sulfapyridine to a model humic constituent. Environ Sci Tech 41:3593–3600

Boxall A, Fogg L, Blackwell P, Blackwell P, Kay P, Pemberton E, Croxford A (2004) Veterinary medicines in the environment. Rev Environ Contam Toxicol 180:1–91

Burch TR, Sadowsky MJ, LaPara TM (2014) Fate of antibiotic resistance genes and class 1 integrons in soil microcosms following the application of treated residual municipal wastewater solids. Environ Sci Tech 48:5620–5627

Butler MS, Buss AD (2006) Natural products—the future scaffolds for novel antibiotics? Biochem Pharm 71:919–929

Byrne-Bailey K, Gaze WH, Kay P, Boxall A, Hawkey PM, Wellington EM (2009) Prevalence of sulfonamide resistance genes in bacterial isolates from manured agricultural soils and pig slurry in the United Kingdom. Antimicrob Agents Chemother 53:696–702

Chen H, Zhang M (2013) Occurrence and removal of antibiotic resistance genes in municipal wastewater and rural domestic sewage treatment systems in eastern China. Environ Intern 55:9–14

Chen Y, Hu C, Qu J, Yang M (2008) Photodegradation of tetracycline and formation of reactive oxygen species in aqueous tetracycline solution under simulated sunlight irradiation. J Photochem Photobiol A Chem 197:81–87

Dalkmann P, Willaschek E, Schiedung H, Bornemann L, Siebe C, Siemens J (2014) Long-term wastewater irrigation reduces sulfamethoxazole sorption, but not ciprofloxacin binding, in Mexican soils. J Environ Qual 43:964–970

Diggins FW (1999) The true history of the discovery of penicillin, with refutation of the misinformation in the literature. Br J Biomed Sci 56:83

Doretto KM, Rath S (2013) Sorption of sulfadiazine on Brazilian soils. Chemosphere 90:2027–2034

Du L, Liu W (2012) Occurrence, fate, and ecotoxicity of antibiotics in agro-ecosystems. A review. Agron Sustain Dev 32:309–327

Elmund GK, Morrison S, Grant D, Nevins M (1971) Role of excreted chlortetracycline in modifying the decomposition process in feedlot waste. Bull Environ Contam Toxicol 6:129–132

Engelhardt I, Sittig S, Šimůnek J, Groeneweg J, Pütz T, Vereecken H (2015) Fate of the antibiotic sulfadiazine in natural soils: experimental and numerical investigations. J Contam Hydrol 177:30–42

Fang H, Han Y, Yin Y, Pan X, Yu Y (2014) Variations in dissipation rate, microbial function and antibiotic resistance due to repeated introductions of manure containing sulfadiazine and chlortetracycline to soil. Chemosphere 96:51–56

Felizeter S, McLachlan MS, de Voogt P (2012) Uptake of perfluorinated alkyl acids by hydroponically grown lettuce (*Lactuca sativa*). Environ Sci Tech 46:11735–11743

Forster M, Laabs V, Lamshöft M, Groeneweg J, Zühlke S, Spiteller M, Krauss M, Kaupenjohann M, Amelung M (2009) Sequestration of manure-applied sulfadiazine residues in soils. Environ Sci Tech 43:1824–1830

Garrod LP, O'grady F (1968) Antibiotics & chemotherapy. Edinburgh University Press, Livingstone

Giang CND, Sebesvari Z, Renaud F, Rosendahl I, Minh QH, Amelung W (2015) Occurrence and dissipation of the antibiotics sulfamethoxazole, sulfadiazine, trimethoprim, and enrofloxacin in the Mekong Delta, Vietnam. PLoS One 10:e0131855

Gillings MR (2013) Evolutionary consequences of antibiotic use for the resistome, mobilome and microbial pangenome. Front Microbiol 4

Gottlieb D (1976) The production and role of antibiotics in soil. J Antibio 29:987–1000

Grote M, Schwake-Anduschus C, Michel R, Stevens H, Heyser W, Lan G, Betsche T, Freitag M (2007) Incorporation of veterinary antibiotics into crops from manured soil. Landbauforschung Volkenrode 57:25–32

Gulkowska A, Sander M, Hollender J, Krauss M (2013) Covalent binding of sulfamethazine to natural and synthetic humic acids: assessing laccase catalysis and covalent bond stability. Environ Sci Technol 47:6916–6924

Hamscher G, Mohring SA (2012) Veterinary drugs in soil and in the aquatic environment. Chem Ing Tech 84:1052–1061

Hamscher G, Pawelzick HT, Sczesny S, Nau H, Hartung J (2003) Antibiotics in dust originating from a pig-fattening farm: a new source of health hazard for farmers? Environ Health Perspect 111:1590

Heinemann M, Kümmel A, Ruinatscha R, Panke S (2005) In silico genome-scale reconstruction and validation of the Staphylococcus aureus metabolic network. Biotechnol Bioeng 92:850–864

Heuer H, Focks A, Lamshöft M, Smalla K, Matthies M, Spiteller M (2008) Fate of sulfadiazine administered to pigs and its quantitative effect on the dynamics of bacterial resistance genes in manure and manured soil. Soil Biol Biochem 40:1892–1900

Heuer H, Solehati Q, Zimmerling U, Kleineidam K, Schloter M, Müller T, Focks A, Thiele-Bruhn S, Smalla K (2011) Accumulation of sulfonamide resistance genes in arable soils due to repeated application of manure containing sulfadiazine. Appl Environ Microbiol 77:2527–2530

Ingerslev F, Halling-Sørensen B (2000) Biodegradability properties of sulfonamides in activated sludge. Environ Toxicol Chem 19:2467–2473

Jechalke S, Heuer H, Siemens J, Amelung W, Smalla K (2014) Fate and effects of veterinary antibiotics in soil. Trends Microbiol 22:536–545

Joy SR, Bartelt-Hunt SL, Snow DD, Gilley JE, Woodbury BL, Parker DB, Marx DB, Li X (2013) Fate and transport of antimicrobials and antimicrobial resistance genes in soil and runoff following land application of swine manure slurry. Environ Sci Technol 47:12081–12088

Joy SR, Li X, Snow DD, Gilley JE, Woodbury B, Bartelt-Hunt SL (2014) Fate of antimicrobials and antimicrobial resistance genes in simulated swine manure storage. Sci Total Environ 481:69–74

Junge T, Claßen N, Schäffer A, Schmidt B (2012) Fate of the veterinary antibiotic 14C-difloxacin in soil including simultaneous amendment of pig manure with the focus on non-extractable residues. J Environ Sci Health Part B 47:858–868

Karcı A, Balcıoğlu IA (2009) Investigation of the tetracycline, sulfonamide, and fluoroquinolone antimicrobial compounds in animal manure and agricultural soils in Turkey. Sci Total Environ 407:4652–4664

Kasteel R, Mboh CM, Unold M, Groeneweg J, Vanderborght J, Vereecken H (2010) Transformation and sorption of the veterinary antibiotic sulfadiazine in two soils: a short-term batch study. Environ Sci Technol 44:4651–4657

Kumar M, Gayen K (2011) Developments in biobutanol production: new insights. Appl Energy 88:1999–2012

Kumar K, Gupta SC, Chander Y, Singh AK (2005) Antibiotic use in agriculture and its impact on the terrestrial environment. Adv Agron 87:1–54

Kümmerer K (2008a) Antibiotics in the environment. In: Kümmerer K (ed) Pharmaceuticals in the environment. Springer, Heidelberg, pp 75–93

Kümmerer K (2008b) Pharmaceuticals in the environment – a brief summary. In: Kümmerer K (ed) Pharmaceuticals in the environment. Springer, Heidelberg, pp 3–21

Lamshöft M, Sukul P, Zühlke S, Spiteller M (2010) Behaviour of 14 C-sulfadiazine and 14 -C-difloxacin during manure storage. Sci Total Environ 408:1563–1568

Li B, Zhang T (2011) Mass flows and removal of antibiotics in two municipal wastewater treatment plants. Chemosphere 83:1284–1289

Mainardi J-L, Villet R, Bugg TD, Mayer C, Arthur M (2008) Evolution of peptidoglycan biosynthesis under the selective pressure of antibiotics in Gram-positive bacteria. FEMS Microbiol Rev 32:386–408

Martinez JL (2009) Environmental pollution by antibiotics and by antibiotic resistance determinants. Environ Pollut 157:2893–2902

Martinez JL, Fajardo A, Garmendia L, Hernandez A, Linares JF, Martínez-Solano L, Sánchez MB (2009) A global view of antibiotic resistance. FEMS Microbiol Rev 33:44–65

Martínez-Carballo E, González-Barreiro C, Scharf S, Gans O (2007) Environmental monitoring study of selected veterinary antibiotics in animal manure and soils in Austria. Environ Pollut 148:570–579

Martinez-Fleites C, Proctor M, Roberts S, Bolam DN, Gilbert HJ, Davies GJ (2006) Insights into the synthesis of lipopolysaccharide and antibiotics through the structures of two retaining glycosyltransferases from family GT4. Chem Biol 13:1143–1152
Mavrodi OV, Mavrodi DV, Parejko JA, Thomashow LS, Weller DM (2012) Irrigation differentially impacts populations of indigenous antibiotic-producing Pseudomonas spp. in the rhizosphere of wheat. Appl Environ Microbiol 78:3214–3220
Nowara A, Burhenne J, Spiteller M (1997) Binding of fluoroquinolone carboxylic acid derivatives to clay minerals. J Agric Food Chem 45:1459–1463
Osol A, Remington JP (1980) Remington's pharmaceutical sciences. Mack, Easton
Ostermann A, Siemens J, Welp G, Xue Q, Lin X, Liu X, Amelung W (2013) Leaching of veterinary antibiotics in calcareous Chinese croplands. Chemosphere 91:928–934
Ozaki N, Bester K, Moldrup P, Henriksen K, Komatsu T (2011) Photodegradation of the synthetic fragrance OTNE and the bactericide triclosan adsorbed on dried loamy sand – results from models and experiments. Chemosphere 83:1475–1479
Pei R, Kim S-C, Carlson KH, Pruden A (2006) Effect of river landscape on the sediment concentrations of antibiotics and corresponding antibiotic resistance genes (ARG). Water Res 40:2427–2435
Picó Y, Andreu V (2007) Fluoroquinolones in soil—risks and challenges. Anal Bioanal Chem 387:1287–1299
Pignatello JJ, Xing B (1995) Mechanisms of slow sorption of organic chemicals to natural particles. Environ Sci Technol 30:1–11
Podojil M, Blumauerova M, Vanek Z, Culik K (1984) The tetracyclines: properties, biosynthesis, and fermentation. In: Vandamme EJ (ed) Drugs and the pharmaceutical sciences. Marcel Dekker, New York
Raaijmakers JM, Mazzola M (2012) Diversity and natural functions of antibiotics produced by beneficial and plant pathogenic bacteria. Annu Rev Phytopathol 50:403–424
Raaijmakers JM, Weller DM, Thomashow LS (1997) Frequency of antibiotic-producing Pseudomonas spp. in natural environments. Appl Environ Microbiol 63:881–887
Raaijmakers JM, Vlami M, De Souza JT (2002) Antibiotic production by bacterial biocontrol agents. Antonie van Leeuwenhoek 81:537–547
Roberts MC (2005) Update on acquired tetracycline resistance genes. FEMS Microbiol Lett 245:195–203
Rosendahl I, Siemens J, Kindler R, Groeneweg J, Zimmermann J, Czerwinski S, Lamshöft M, Laabs V, Wilke BM, Vereecken H, Amelung W (2012) Persistence of the fluoroquinolone antibiotic difloxacin in soil and lacking effects on nitrogen turnover. J Environ Qual 41:1275–1283
Sabourin L, Duenk P, Bonte-Gelok S, Payne M, Lapen DR, Topp E (2012) Uptake of pharmaceuticals, hormones and parabens into vegetables grown in soil fertilized with municipal biosolids. Sci Total Environ 431:233–236
Sarmah AK, Meyer MT, Boxall AB (2006) A global perspective on the use, sales, exposure pathways, occurrence, fate and effects of veterinary antibiotics (VAs) in the environment. Chemosphere 65:725–759
Schaffer M, Boxberger N, Börnick H, Licha T, Worch E (2012) Sorption influenced transport of ionizable pharmaceuticals onto a natural sandy aquifer sediment at different pH. Chemosphere 87:513–520
Sirés I, Brillas E (2012) Remediation of water pollution caused by pharmaceutical residues based on electrochemical separation and degradation technologies: a review. Environ Int 40:212–229
Srinivasan P, Sarmah AK (2014) Assessing the sorption and leaching behaviour of three sulfonamides in pasture soils through batch and column studies. Sci Total Environ 493:535–543
Sukul P, Spiteller M (2007) Fluoroquinolone antibiotics in the environment. In: Sukul P (ed) Reviews of environmental contamination and toxicology. Springer, Berlin, pp 131–162
Tang X, Lou C, Wang S, Lu Y, Liu M, Hashmi MZ, Liang X, Li Z, Liao Y, Qin W, Fan F, Xu J, Brookes PC (2015) Effects of long-term manure applications on the occurrence of antibiotics

and antibiotic resistance genes (ARGs) in paddy soils: evidence from four field experiments in south of China. Soil Biol Biochem 90:179–187

Thiele-Bruhn S (2003) Pharmaceutical antibiotic compounds in soils – a review. J Plant Nutr Soil Sci 166:145–167

Thomashow LS, Bonsall RF, Weller DM (1997) Antibiotic production by soil and rhizosphere microbes in situ manual of environmental microbiology. ASM Press, Washington, DC, pp. 493–499

Yang W, Lu Y, Zheng F, Xue X, Li N, Liu D (2012) Adsorption behavior and mechanisms of norfloxacin onto porous resins and carbon nanotube. Chem Eng J 179:112–118

Zhao L, Dong YH, Wang H (2010) Residues of veterinary antibiotics in manures from feedlot livestock in eight provinces of China. Sci Total Environ 408:1069–1075

Zhou LJ, Ying GG, Liu S, Zhao JL, Chen F, Zhang RQ, Peng FQ, Zhang QQ (2012) Simultaneous determination of human and veterinary antibiotics in various environmental matrices by rapid resolution liquid chromatography – electrospray ionization tandem mass spectrometry. J Chroma A 1244:123–138

Chapter 5
Human Being as Biomonitor of Soil Xenobiotics

Atif Kamal and Riffat Naseem Malik

5.1 Introduction

5.1.1 Soil/Dust in Populated Zones: Natural Sink of Xenobiotics

Exposure to the soil is one of the most significant environmental risk factor for human health. Soil, which is also a part of the indoor dust in rural and urban environments, can be considered as a major sink and reservoir of myriad pollutants and widespread xenobiotics. In addition to the provision of agricultural food services, the dependence of human on soil is ever increasing. Additionally, dust is another form of mixture of particles, which largely contain the soil particles and associated xenobiotics. One of the properties of dust is that it can travel longer distance due to small particles (Goudie 2014). The indoor dust is one of the major sink for a large number of xenobiotics, especially those which are particulate bound (Kamal et al. 2015a, b). In fact, the house dust is a heterogeneous mixture, in its composition; it may contain soil particles, human skin, particulate matter, spores of fungi, and human hair, etc. (Yu et al. 2012; Maertens et al. 2004). The residence and retention of xenobiotics in the soil can be for a longer duration in a variety of situations (Dong et al. 2009). The chemistry of soil/dust can be very complex; it

A. Kamal
Department of Environmental Sciences, PMAS Arid Agriculture University, Rawalpindi, Pakistan

Environmental Biology and Ecotoxicology Laboratory, Department of Environmental Sciences, Faculty of Biological Sciences, Quaid-I-Azam University, Islamabad 45320, Pakistan

R.N. Malik (✉)
Environmental Biology and Ecotoxicology Laboratory, Department of Environmental Sciences, Faculty of Biological Sciences, Quaid-I-Azam University, Islamabad 45320, Pakistan
e-mail: r_n_malik2000@yahoo.co.uk

© Springer International Publishing AG 2017

M.Z. Hashmi et al. (eds.), *Xenobiotics in the Soil Environment*, Soil Biology 49,
DOI 10.1007/978-3-319-47744-2_5

can be relocated and resettled; dust can even comprise coal dust, fine dust, and indoor aerosol; therefore, it poses considerable hazard to human health. In some regions of the world, dust is largely composed of soil particles (such as in Pakistan, India, and Bangladesh); the composition of dust varies therefore among different regions of the world. In fact, the degree of anthropogenic activities directly relates with the variation in the composition of dust and the concentration of xenobiotics in the soil (Man et al. 2013).

The "xenobiotics" is a term which refers to foreign chemicals; it is a combination of the Greek word xenos (meaning foreigner or strangers) and bios (meaning life). Xenobiotics thus refer to the chemicals which may be carcinogens, drugs, food additives, hydrocarbons, pesticides, and many other forms of environmental pollutants (Wikipedia 2015). Many of these xenobiotic chemicals whether introduced into the air or directly to the crop field end up in the soil and water bodies. However, the accumulation of xenobiotics in the soil is an important aspect that directly or indirectly influences the sustainability of natural ecosystem. In general, the xenobiotics are not even recognized by the biochemical mechanisms of plants or microbial community; thus, they are resistant to degradation (i.e., a process which transforms the complex molecules into simple ones) to a greater extent. Moreover, they also influence the chemical stability of soil microenvironment. The chemical reactions in soil necessarily should facilitate conversion of xenobiotics to simpler compounds (mineralization) or sometimes alternatively xenobiotics undergo activation (conversion into toxic molecule). These two conditions determine the fate of xenobiotics in the soil, whether it will complete, be degraded (mineralized), or will persist for a longer time (stabilization). Changes in the structure of parent xenobiotic compound can make physiochemically different compounds, which are sometimes more toxic than the parent compounds. The process of mineralization converts chemical compounds into small inorganic molecules (e.g., H_2O, NH_3, H_2S, CO_2, etc.), facilitated by the complex interaction of biotic and abiotic processes. Some of the biotic processes can also convert the xenobiotics into different derivatives; such process includes oxidation. However, a clear distinction between the products of biotic and abiotic reactions is not possible, because most of the time, degradation products could be a result of both. The sulfur-containing pesticides, cyclodienes, anilines, aromatic compounds, and phenoxy alkanoic acids can undergo oxidation; mineralization (as discussed earlier) is an example of oxidation, and almost all the organic compound can be mineralized under this process (Mansour et al. 1992). Other processes include the reduction of carbon, nitrogen, sulfur contents in alkanes, nitro compounds, disulfides, DDT, cyclodiene, and halogenated propane. Hydrolysis (such as ester hydrolysis, epoxide hydrolysis, hydrolytic dehydrogenation) of many insecticides, halogenated compounds, chlorinated phenols, etc., and synthetic processes of phenols and anilines are examples of abiotic processes (Scheunert 1992).

In biomonitoring, human subjects can serve as a bioindicator of eco-health status. Since the exposure to soil/dust (bound contaminants) among human varies with the differences in the sociodemographic conditions of the communities. The exposure to house dust via accidental ingestion only accounts up to 24 % B2 carcinogens, in low-income households (Chuang et al. 2010). Many xenobiotics have poor solubility/volatility, and therefore most of them are bound to dust/particulates in the atmosphere, e.g., PAHs (Kamal et al. 2015a, b). It is therefore a valuable area of research to investigate the dust-bound contaminant as a non-dietary route of exposure among human communities (Dirtu et al. 2010).

5.1.2 *Organic and Inorganic Xenobiotics in Soil/Dust have a link with Human Activities*

The binding sites of Xenobiotic on soil matrices are numerous, due to which a wide variety of such compounds can be found adsorbed to soil/dust and sediments (Fig. 5.1). Many soil/dust samples could have attached oxides such as Al_2O_3, Fe_2O_3, CaO, SiO_2, etc. (Zaady et al. 2001). The dust storms can also be source of pathogens, heavy metals, dioxins, herbi- and pesticides, and radioactive materials, which in turn may be in the form of aerosol, cesium (Ogorodnikov 2011), uranium (Csavina et al. 2012), or adsorbed materials (Goudie 2014). Storms of dust can bring in particulate matters (PM), arsenic, and other salts (Morman and Plumlee 2013).

In the urban and rural environments, soil (in the outdoor environment) and dust (indoor environment) serve as a sink of a large number of pollutants. All the organic and inorganic forms of xenobiotics ultimately end up in soil/dust sink, some of which are retained for a long period of time. Others may undergo biochemical degradation or may be adsorbed to the dust/soil particles and become inert. In the urban environment (in particular), the traffic pollution and industrial exhaust release large quantities of air pollutants, whereas in suburban and rural environments, the use of agricultural chemicals and household utilization of crude biomass fuels are sometimes major source of pollution. The motor vehicles are the most significant source of contaminants in the urban environment, even in developing countries. One of the major pollutants released from the combustion exhaust (of biomass or fossil fuel or plastic material) are the PMs, which are carcinogenic

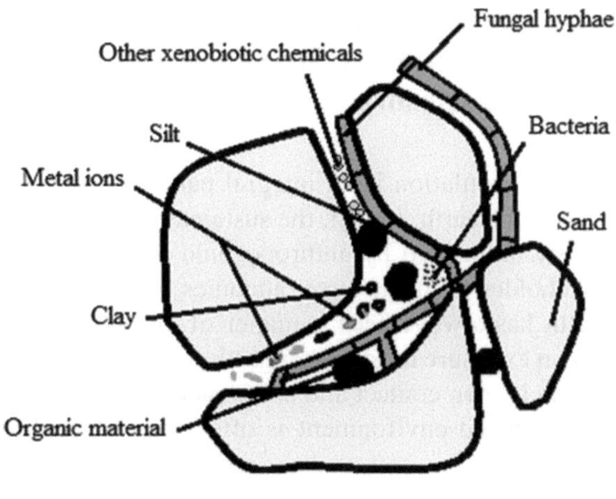

Fig. 5.1 Typical soil particle with biological and chemical constituents

to human (Hsieh et al. 2011). Among organic contaminates, the polychlorinated dibenzo-*p*-dioxins and dibenzofurans (PCDD/Fs) and many novel forms of fire retardant materials have become dominant in urban environment. These pollutants are similar to polycyclic aromatic hydrocarbons (PAHs) another group of potentially hazardous compounds, polychlorinated biphenyls (PCBs), polybrominated dibenzo-*p*-dioxins and dibenzofurans (PBDD/Fs), and polybrominated diphenyl ethers (PBDEs). They are semi-volatile and can exist both in gas and particle phases (Wang et al. 2008; Wang et al. 2010). Due to such unique pollution pattern in the urban environment, the humans more frequently exposed. Humans from diverse occupations are known to have high serum concentrations of these organic compounds and their isomers; these results are observed in both genders (Zamir et al. 2008). Some of the congeners of PBDEs, especially BDE-209, also have tendency to bioaccumulate in humans (Qu et al. 2007). Nowadays, the presence of organic halogenated compounds (OHCs) in the urban environment is one of the great concern for human health; these compounds can enter human body via all routes of exposure (Dirtu et al. 2010). The human being, highly (and chronically) exposed to environmental conditions bearing particular xenobiotics contamination, can serve as bioindicator of that particular environment. Several recent studies on paired human serum and dust samples show that indoor dust significantly enhances exposure to organic contaminant, traces of which can be found in the human serum with sound correlation (Kamal et al. 2011, 2014b, 2015b). In some case, serum and dust samples revealed residues of banned pesticides in which a possible use of these pesticides (such as DDT) in the exposure sites or its vicinities was shown. Consequently, these chemicals are gradually accumulated, in the soil, directly or indirectly, and translocated (as a fraction of fugitive dust/particles) into various compartments of ecosystem depending upon the land use pattern and climatic condition. Human beings in contact with such soil/dust gradually uptake the contaminants which ultimately results in bioaccumulation.

5.2 Human Exposure to Soil/Dust-Bound Xenobiotics via Skin Contact, Ingestion and Inhalation Routes of Exposure

Human population is an integral part of rural and urban ecosystem, in populated zone of the earth. In fact, the sustainability of eco-health in urban and rural areas is largely influenced by anthropogenic activities, and human being is itself the key stakeholder of the adverse outcomes. The connection between the soil and human health has always been a matter of discussion among scientific community. The human exposure to soil via ingestion, inhalation, and dermal routes is the key factor behind human contact and subsequent health implications. Human exposure in the contaminated environment is often inevitable; thus, human being can serve as a bioindicator of ambient environmental contamination. The communities of the humans are influenced by the land use activities, nature of occupation, welfare facilities, and socioeconomic barriers among populations, and therefore, differences of exposure and biomarkers in human biological matrices differ significantly across these lines of sociodemographic divisions. Since bioaccumulation often takes place after a prolonged exposure to xenobiotics, the biomarkers in human biological matrices can be an indirect measure of eco-health status; however, the biomarkers can also be influenced by the biochemistry and physical activities of individuals.

5.2.1 Skin Contact and Dermal Intake of Xenobiotics

Human skin is comprised of *stratum corneum* (the outmost layer of epidermis having dead cells), epidermis, dermis (0.2–0.3 cm thick), and hypodermis (also known as *superficial fascia*) (Hole 1992); there are many toxic compounds (and possible carcinogens) in the soil which can penetrate skin due to their lipophilic nature. According to Pannatier et al. (1978), skin is 5 % of the adult human body, and it covers an area approximately equal to 1.8 m^2. Skin absorption is largely facilitated by the *stratum corneum* lipids; the route is in fact an intercellular pathway (Michaels et al. 1975; Elias 1983). The particles with a size ranging between 100 and 200 μm are easily retained by the human skin (Lewis et al. 1999). The bioavailability of a contaminant present in the soil and via dermal exposure pathway (necessary for the skin absorption) depends upon the transfer of the compound from carrier (such as from soil vehicle into the *stratum corneum*) and subsequent penetration of compound through the *stratum coneum* to the cells (Pugh and Chilcott 2008). Since the physical and chemical properties of the compounds' penetration into the skin can greatly influence the dermal bioavailability (Brain and Chilcott 2008). The dermal bioavailability of organic compound is the amount of that compound in soil that after exposure can be absorbed from the epidermis and reaches target organic and/or tissue via blood circulation USEPA (1992a, b, 1996). The dermal bioavailability can be described in terms of equation (Roy et al. 1998a, b).

$$\mathrm{Abs}(d) = \mathrm{Qsoil}/\mathrm{Tsoil},$$

where Qsoil = amount of chemical compound absorbed into the skin
Tsoil = total concentration of that compound in the soil, and

According to Ziegler and Simonart (2001), the soil particles which are rich in Fe contents can cause the suppression of the immune system, possibly more frequently via epidermis of feet and hand regions. The petroleum products, PAHs, dioxins, and pesticides are examples of xenobiotics which have greater penetrability via skin absorption. Absorption of dust-bound PAHs and possibly other contaminant can be influenced by the binding of contaminate to the soil/dust particle. Benzo(a)pyrene is for instance present in dust/particulate-bound or petroleum products (Kamal et al. 2015a). The skin surface area exposed to the environment is an important determinant of exposure, which in the case of outdoor workers (construction workers in particular) can be exposed to soil/dust-bound xenobiotics, depending upon the varying use of personnel protective equipments (PPEs). In developing countries, no such precautions are practiced by construction industry and outdoor works, which is why the exposure is always high among them (Kamal et al. 2012, 2014a, b). In the urban environment, the soil contaminated with petrochemical fraction of chemicals contaminates can be a significant risk factor for workers within the workplaces (Kamal et al. 2011; Ud-Din et al. 2013). Many studies have shown that the (excessive lifetimes) carcinogenic risk is mostly higher via dermal contact pathway than other routes of exposure, especially higher than the inhalation exposure (Wang et al. 2010; Kamal et al. 2014a, b, 2015a, b). The importance of this pathway of exposure can be evaluated by the risk assessment conducted by the USEPA which according to Johnson and Kissel (1996), 37 out of 235 such assessments have shown that the excessive lifetime carcinogenic risk was high (i.e. $> 10^{-4}$).

5.2.2 *Ingestion (Oral Intake)*

The ingestion is one of the important routes of exposure; however, the ingestion of soil/dust adsorbed on food or skin is not voluntary and occurs accidently or unintentionally (Maertens et al. 2004). Deliberate ingestion of soil is often observed in children in many countries, and this is more frequently observed in the children of the households with low socioeconomic status. The pica/dirt eating among children (Hawley 1985; Roberts and Dickey 1995) is known throughout the world and is one of the reckless behaviors in children. On a population scale, the unintentional intake of soil is a common incidence (Ferguson and Marsh 1993) accounting for around 10 mg/day/adult ingestion of dust among adult members of the population (Stanek et al. 1997). According to some estimates, the hands of children and adult can absorb 28 mg and 51 mg dust, respectively (Hawley 1985). Literature studies have no consensus on the soil ingestion rate among population and different age groups of exposed population; thus, there are varying rates of ingestion available, so there exist uncertainties in this regard (Man et al. 2013). However, there are numerous rates of soil ingestion available on the USEPA websites, for different exposure and age groups, and are being used in the risk assessment studies. The vulnerable children population has been studied the most in the context of soil ingestion, for having more hand-to-mouth behavior than adults (Clausing et al. 1987). The soil ingestion is reported in the literature in varying rates of intake; therefore, these values are not often reliable (Calabrese et al. 1991). Accidental ingestion (and other non-dietary intake) is one of the routes of human exposure to xenobiotics. The contaminant is mobilized from the dust, inside the gastrointestinal tract, thereby making it bioaccessible, by which the contaminant is transported to the circulatory system across the epithelium of the intestine. There is limited data available in the case of human beings regarding the bioaccessibility of contaminants, much of which comes from the experimental studies (Yu et al. 2012 and reference therein).

The soil ingestion pathway renders exposure to heavy metals, minerals, and pathogens. In general, the concentration of Pb, Cd, and Zn remains high in many urban dust samples from traffic pollution and petrochemical and paint products (Mielke et al. 1999). Nowadays, the fire retardants and PAHs' contamination is also a burning topic of debate, because these contaminants pose significant health risk to the population exposed to workplace soil and/or house dust in both urban and rural environments. The sources of former are consumer products and electronic equipments and of latter are traffic, biomass consumption, and industrial activities.

5.2.3 Inhalation Route of Exposure

The lungs are connected with outer environment via bronchioles, bronchi, and nostrils. The alveoli are small sacs present in the lungs where the exchange of the oxygen and carbon dioxide takes place though capillaries. The bronchioles are linked with tiny cilia, which are hairlike structures and play an important role in protecting from the dust (Hole 1992). The cilia can remove dust particles which are non-respirable via a process of ciliary escalator. The lungs can inhale little fractions of the dust particles; some of the soluble dust can ultimately be absorbed into the blood, while insoluble dust can be retained permanently, which can cause the scarring of bronchial tissues (pneumoconiosis) and can be a cause of chronic morbidities among humans and can cause asthma, allergies, and nasal cancers (Hughes 2007).

Many contaminants are bound to PM, which are easily inhalable (especially those having ≤ 5 µm diameters) Harvey (1997). The fugitive dust is the clay-sized particles, which emerges as a result of soil erosion (Brady and Weil 1999). In addition to this, anthropogenic activities such as cultivation of soil can also introduce considerable amount of the mineral dust into the atmosphere. Top soil is responsible for bringing the radon (Rn) into the homes, for which the one-meter layer of the soil top is most important. The daughter isotopes of radioactive materials usually adhere to the particles of dust and thus can be inhaled by humans causing damage to the cells in the bronchial tissues where dust particles are retained. However, the concentration of such radioactive fraction also varies among different regions of the world (Baird 1998).

5.3 Soil-Bound Xenobiotics: Health Risk Factor for General Population

A number of respiratory disorders among humans could be linked with the dust particles attributable to the pathogenic agents present in the soil/dust reservoir. The biological agents (viruses, fungi, bacterial sports, and pollen grains) and other toxic elements can affect the epithelium of respiratory system (Goudie 2014; Leski et al. 2011). In older humans, cardiopulmonary disorders, asthma in younger children, could be traced in some countries linked with the same phenomenon (Kanatani et al. 2010). The exposed populations are susceptible to the short-term effects of suspended particulates (Jiménez et al. 2010; Yu et al. 2012; Prospero et al. 2008; Chien et al. 2012; Miri et al. 2007). Asbestos mineral dust is notorious for causing lung cancer in humans, especially in the occupationally exposed population. The inhalation of gaseous form of radioactive radon particles is also a cause of lung cancer (Henshaw et al. 1990). Other studies have reported the occurrence of allergic rhinitis and pneumonia on exposure to dust storms from different parts of the countries (Cheng et al. 2008; Kang et al. 2012; Chang et al. 2006; Rutherford et al. 1999) lung cancer (Giannadaki et al. 2013), *Meningococcal meningitis*, (Thomson et al. 2009; Deroubaix et al. 2013; Martigny and Chiapello 2013), and conjunctivitis (Goudie 2014; McMoli et al. 1984; Yang 2006). Several different toxic elements can result in skin problems, such as irritation by nickle (Otani et al. 2012). Ingestion of organic contaminants can result in subsequent accumulation of soil in the human alimentary canal which can be the cause of abdominal discomfort, obstruction of colon, colic pain, and many related diseases (Bateson and Lebroy 1978; Solien 1954). The soil indirectly impacts humans and their environment by contributing to the greenhouse effect which was brought into the notice of scientific community in 1989 (Abrahams 2002); since then the soil has gained an importance as a factor behind greenhouse effect. Soil is also a sink for CH_4, N_2O, and CO_2 gases (Bridges and Batjes 1996). Global warming can impact overall environment of the earth, resulting in shift of vector-borne disease pattern, global crop production, and earth's energy budget (Abrahams 2002).

5.3.1 Construction Workers (and Child Labor) in Developing Countries Face Increased Carcinogenic Risk Associated with Soil/Dust Exposure

Construction works pose varied degrees of hazards to the workers worldwide. The facilities for self-protection and workplace safety (for workers welfare) are well maintained and provisions are sufficient in most of the developed countries. However, the situation is worse in a number of underdeveloped or developing countries, especially in south Asian region. Construction works (such as excavation, tilling, dozing, beck making, etc.), agriculture (driving heavy vehicle on soil), exposure in some disposal sites, and mining activities generate the particulate matter/fugitive dust; and a considerable exposure of workers to the dust occurs in proximities of the roadside (in particular, the unpaved one) (EPA 1996, 1992a, b). Additionally, the outdoor working conditions in these regions are even worse, involving many different working tasks and workplace ambience. For instance, the brick-making profession is widely distributed in many developing countries and is one of the hazardous occupations which pose serious health risks to the workers. Additionally, the construction workers in the urban and suburban areas spend an extended period of time in building homes, commercial areas, etc., on a daily basis. Therefore, these areas of occupational sector render serious health risks to the workers. The risk factors are supported by the fact that in the carcinogenic risk evaluation, the consideration of exposure among construction and outdoor workers is very high (Kamal et al. 2014a, b, 2015b).

Younger workers in this field are more vulnerable, because they lack experience and hazard awareness. Metals, PAHs, and various pesticides are among key contaminants that a worker may encounter in construction works. Incidental ingestion and skin contacts are the main routes by which workers in construction and outdoor activities are exposed more frequently. In developing countries, more workers are exposed in the brick-making industry, and carcinogenic risk from PAHs' exposure among these workers is very high (Kamal et al. 2015a, b).

In developing countries, child labor is another risk factor for youngsters, who due to financial constrains are forced to work at very early ages. Children are naturally vulnerable to foreign chemicals because of their sensitive immune system and high frequency of exposure than adults. Thus, nutritional, physiological, and behavioral difference among child and adult workers demands separate risk assessment and poses different levels of risk to both groups. In addition to child labor, non-occupationally exposed children in the vicinity of such workplace are at risk of exposure alike. The hand-to-mouth behavior of children is a reason behind high oral intake of dust/soil contamination. Children are not decision-makers, they are usually dependent upon adults for provision and decision regarding safety and self-protective measure. Naturally, the lower body weight of children (versus adults) and high intake of dust/soil contents is a reason behind the uptake of higher dose of toxic substance per unit body weight. Another important aspect is the growth stages of the young children because of earlier stages of development of the organ, the damage to their organs can become permanent in later stages of life.

5.3.2 Current and Emerging Sources of Xenobiotics

Industrial revolution and urban span (and mismanage expansion) has resulted in complex pollution patterns and shift of natural balance of ecosystem in many of the world's populated regions. Organic xenobiotics are introduced into the environment from both intentional anthropogenic activities (e.g., application of herbicides and pesticides in the crop field, burning of waste materials for energy purposes) or without intention. There are many occasions where chlorinated solvents, pharmaceutical drugs, and aromatic hydrocarbons are released into the environment as a by-product of various processes. Application of pesticide for crop protection introduces huge amounts of persistent organic compounds into the environment, which directly interact with the soil (Kastner et al. 2014).

The combustion processes and road traffic are also contributors of a wide range of pollutants in the urban environment. The pollution from particular matter (PM) is one of the key nuisances in the environment as like dust particles they also bear numerous adsorbed metals and oxides (SOx and NOx) and silicates, and these can be deposited in building materials to varying degrees. In addition to this, PAHs and chlorinated compounds such as PCBs are also deposited in the building material in the industrial zones. These compounds are lipophilic and hazardous to human health. PCBs and other forms of fire retardants are widely used nowadays in the insulation of electronic wiring, thermoplastics (Prieto-Taboada et al. 2013 and reference therein), and also in the textile industries such as mattresses and related products. Sources which are known for the emission of PCDD/Fs include incinerators of waste material (Wang et al. 2010), wood burning (McNamara et al. 2011), rice straw burning (Lin et al. 2007), iron ore sintering (Kuo et al. 2011), and the boilers, etc. Apart from these stationary sources of PCDD/F (Chuang et al. 2010), the emission of PCDD/F from mobile source such as vehicle is very significant in many countries because vehicles are very important sources of $PM_{2.5}$. According to Wenger et al. (2008a, b), exhaust from the heavy duty diesel engine contains PAHs (nine prominent fractions) and 2,3,7,8-PCDD/Fs. Studies on the incinerators and metallurgical processes have suggested that PBDEs are either produced or survived from the PBDE-contaminated feeding materials in the combustion system (Wang et al. 2010). The diesel engine exhaust contains gas phase total PAHs (85.8–98.4 %), total PCDD/Fs (53.4–61.4 %), total PCBs (61.1–81.9 %), and total PBDEs (5–8.7 %). PBDEs which are emitted from vehicle exhaust are formed during the process of combustion and may be present due to bromine in the fuel. Both the PBDEs and PCDD/Fs form by combustion from the same process (Wang et al. 2010; Artha et al. 2011). This combustion system is the main contributor of brominated PBDEs in urban environment, and vehicle exhaust is the prominent source (Wang et al. 2010).

Combustion of biomass fuel (and organic/synthetic substance) can release numerous chemical compounds into the atmosphere. During complete combustion, some heavy metals, HCl, CO_2 gases, like SOx and NOx, are released. However, the incomplete combustion of fuel can result in the emission of volatile organic compound, ammonia, PAHs, PM, and PCDD/F. Metals such as copper, chromium, mercury, nickel, cadmium, arsenic, lead, zinc, etc., are constituents of the basic composition of the biomass fuel, while the sulfur and nitrogen contents in fuel wood-related biomass are low as compared to the fossil fuel. In addition to the release of these metals in the combustion of biomass and fossil fuels, the dioxin and furans are also formed during the combustion of the organic materials containing carbon, oxygen, and chlorine contents (Villeneuve et al. 2012 and reference therein).

Nowadays, the introduction and massive use of electronics equipment in commercial and households have given rise to pollutants of novel categories (such as novel brominated fire retardants) and commonly known PCBs and PBDEs and polycyclic aromatic hydrocarbons (PAHs). PAHs are released largely by anthropogenic activities, mainly during incomplete combustion of organic material. The PAH sources' transport and fate is well discussed in many previous studies (Kamal et al. 2015a, b). PAHs are more ubiquitous because in urban environment vehicular traffic (mobile source) and in rural areas the burning of biomass fuel (fuel wood/coal, crop residues) and various industries using fuel wood/coal or rubber in their combustion process can be source of atmospheric PAHs. In addition to this, numerous petroleum products, especially used engine oil, contain high concentration of PAHs. Usually high-molecular-weight PAHs are adsorbed to soil/dust particles and particulate matter in the environment (Kamal et al. 2015a, b). Another form of PAHs present in the environment is halogenated polycyclic aromatic hydrocarbons (HPAHs). HPAHs are halogenated products of PAHs, which contain the halogen attachments to the aromatic rings. Limited studies have investigated the soil/dust constituents of HPAHs; nevertheless, these were detected in soil/dust samples taken from various industrial sites, including e-waste recycling sites and agricultural and industrial zones. Comparing two sites, more HPAHs were observed in the industrial soil/dust samples, which shows the influence of land use types on their emission (Sun et al. 2011 and reference therein).

In the contemporary world, the new emerging organic compounds include PFCs (especially perfluorooctanesulfonic acid PFOS), which are used as surface-acting agents, for coating, or in foam used in firefighting. The phthalates, perfluorinated compounds (PFCs), and brominated flame retardants (BFRs) as discussed earlier, phenols (bisphenol A (BPA), benzophenones, and parabens). BFRs are intentionally incorporated in carpets, paint materials, furniture, building material, plastics, and electronic equipment to prevent them from catching fire (Casasa et al. 2012; Hooper and McDonald 2000; Segev et al. 2009). Another form of industrial additives is phthalate, which is used in the plasticizers, packaging of food materials, containers of food, and plastic toys (Shea 2003). BPAs are also used in consumer products including resin (epoxy), cans, and containers of food made of plastic, medical device, toys, baby feeders, etc. Benzophenone is present in cosmetic products (Casasa et al. 2012 and reference therein).

E-waste recycling by open-burning and dismantling materials generates a huge concentration of heavy metals. The heavy metals from the e-waste include chromium (Cr), arsenic, lead, and cadmium. Moreover, e-waste also contains persistent substances in large quantities, including PBDES and PCBs, which pose significant exposure risk to adults and children alike. Most of these contaminants can be traced to outdoor soil and indoor dust. The circuit boards release numerous metals and fire retardants (brominated and chlorinated compounds). In addition to this, the household and commercial wiring and thermoplastic materials are also coated with fire retardants. The e-waste recycling or dealing workers are highly exposed to such compounds; being lipophilic, these compounds can easily gain entry into human body via all three routes of exposure (Wang et al. 2010). In addition to these sources, industrialization and traffic swarm in industrial cities also release numerous xenobiotics into environment and can be traced in outdoor to indoor soil/dust and ambient air in varying concentration. In general, the mega cities and urban areas are polluted due to traffic and industrial exhaust emission.

Automobile and motorcycles are a significant source of air pollution in urban areas of developing countries. The atmospheric pollutants ultimately settle down in soil and dust particles and therefore can be traced in a wide range of residential and commercial settings. While in rural areas, various industries such as small-scale brick manufacturing, utilizing cheap fuel in the form of scrap rubbers, tires in addition to fuel wood, small-scale steel industries, and tanneries are some of the sources that introduce numerous chemicals into the environment, most of which as discussed earlier are persistent and capable of long-range transport (e.g., PAHs, PBDEs, etc., Kamal et al. 2015a, b).

5.4 Concluding Remarks

The eco-health in the industrialized world is greatly influenced by anthropogenic activities; in turn, human beings are significantly influenced by the resulting environmental deterioration and the imbalance of natural cycles. Nowadays, a lot of xenobiotics chemicals can be traced in the soil and indoor dust, sources of which range from specific to general, limited to widespread, and natural to anthropogenic. The emerging xenobiotics are risk factor for human population in both urban and rural settings. Human biomonitoring nowadays has provided a lot of evidence that these xenobiotics not only accumulate in human body but also pose significant health risk to a greater extent, especially if exposure occurs in young age. The exposure to xenobiotics varies among human population due to difference in the land use pattern, physiochemical difference, and personnel activities. Despite these differences, the exposure to soil and indoor dust is common all over the world; therefore, exposed human population has a potential to serve as a bioindicator of eco-health status.

References

Abrahams PW (2002) Soils: their implications to human health. Sci Total Environ 291:1–32
Artha AA, Wu EMY, Wang LC, Chen CH, Chang-Chien GP (2011) Thermal formations of polybrominated diphenyl ethers in raw and waterwashed fly ash. Aerosol Air Qual Res 11:393–400
Baird C (1998) Environmental chemistry. Freeman, New York
Bateson EM, Lebroy T (1978) Clay eating by aboriginals of the northern territory. Med J Aust 10:1–3
Brady NC, Weil RR (1999) The nature and properties of soils, 12th edn. Prentice Hall, Upper Saddle River, p. 881
Brain KR, Chilcott RP (2008) Physicochemical properties affecting skin absorption. In: Chilcot RP, Price S (eds) Principles and practice of skin toxicology. Wiley, London
Bridges EM, Batjes NH (1996) Soil gaseous emissions and global climatic change. Geography 81:155–169
Calabrese EJ, Stanek EJ, Gilbert CE (1991) Evidence of soil-pica behavior and quantification of soil ingested. Hum Exp Toxicol 10(4):245–249
Casas L, Fernandez MF, Llop S, Guxens M, Ballester F, Olea N (2012) Urinary concentrations of phthalates and phenols in a population of Spanish pregnant women and children. Environ Int 37:858–866
Chang CC, Lee IM, Tsai SS, Yang CY (2006) Correlation of Asian dust storm events with daily clinic visits for allergic rhinitis in Taipei, Taiwan. J Toxicol Environ Health A 69:229–235
Cheng MF, Ho SC, Chiu HF, Wu TN, Chen PS, Yang CY (2008) Consequences of exposure to Asian dust events on daily pneumonia hospital admissions in Taipeh, Taiwan. J Toxicol Environ Health 71:1295–1299
Chien LC, CH Y, HL Y (2012) Estimated effects of Asian dust storms on spatiotemporal distributions of clinic visits for respiratory diseases in Taipei children (Taiwan). Environ Health Perspect 120:1215–1220
Chuang SC, Chen SJ, Huang KL, Wu EMY, Chang-Chien GP, Wang LC (2010) Gas/particle partitioning of dioxins in exhaust gases from automobiles. Aerosol Air Qual Res 10:489–496
Clausing P, Brunekreef B, Van Wijnen JH (1987) A method for estimating soil ingestion by children. Int Arch Occup Environ Health 59:73–82
Csavina J, Field J, Taylor MP, Gao S, Landázuri A, Betterton EA, Sáez AE (2012) A review on the importance of metals and metalloids in atmospheric dust and aerosol from mining operations. Sci Total Environ 433C:58–73 http://doi.org/10.1016/j.scitotenv.2012.06.013
Deroubaix A, Martiny N, Chiapello I, Marticoréna B (2013) Suitability of OMI aerosol index to reflect mineral dust surface conditions: preliminary application for studying the link with meningitis epidemics in the Sahel. Remote Sens Environ 133:116–127
Dirtu AC, Ali N, Van den Eede N, Neels H, Covaci A (2010) Country specific comparison for profile of chlorinated, brominated and phosphate organic contaminants in indoor dust. Case study for Eastern Romania. Environ Int 49:1–8
Dong W, Hu C, Chen S, Zhang Y (2009) Tillage and residue management effects on soil carbon and CO_2 emission in a wheat–maize doublecropping system. Nutr Cycl Agroecosyst 83:27–37. doi:10.1007/s10705-008-9195-x
Elias PM (1983) Epidermal lipids, barrier function, and desquamation. J Invest Dermatol 80:44–49
EPA (U.S. Environmental Protection Agency) (1992a) Workbook of screening techniques for assessing impacts of toxic air pollutants (Revised) EPA-454/R-92-024. Office of air quality planning and standards. US. Environmental Protection Agency, Research Triangle Park
EPA (U.S. Environmental Protection Agency) (1992b) Fugitive dust background document and technical information document for best available control measures, EPA-450/2-92-004.

Office of air quality planning and standards. U.S. Environmental Protection Agency, Research Triangle Park
EPA (U.S. Environmental Protection Agency) (1996) Soil screening guidance: user's guide, vol 9355, 2nd edn. Office of Emergency and Remedial Response, U.S. Environmental Protection Agency, Washington, DC, pp. 4–23
Ferguson CC, Marsh JA (1993) Assessing human health risks from ingestion of contaminated soil. Land Contam Reclam 1:177–185
Giannadaki D, Pozzer A, Lelieveld J (2013) Modeled global effects of airborne desert dust on air quality and premature mortality. Atmos Chem Phys Disc 13:957–968
Goudie AS (2014) Desert dust and human health disorders. Environ Int 63:101–113
Harvey RG (1997) Polycyclic aromatic hydrocarbons. Wiley, New York
Hawley JK (1985) Assessment of health risk from exposure to contaminated soil. Risk Anal 5:289–302
Henshaw DL, Eatough JP, Richardson RB (1990) Radon as a causative factor in induction of myeloid leukaemia and other cancers. Lancet 335:1008–1012
Hole John W Jr (1992) Essential of human anatomy and physiology, 4th edn. William C Brown Publishers, Hardcover
Hooper K, McDonald TA (2000) The PBDEs: an emerging environmental challenge and another reason for breast-milk monitoring programs. Environ Health Perspect 108:387–392
Hsieh LT, Wu EMY, Wang LC, Chang-Chien GP, Yeh YF (2011) Reduction of toxic pollutants emitted from heavy-duty diesel vehicles by deploying diesel particulate filters. Aerosol Air Qual Res 11:709–715
Hughes F (2007) Introduction to health and safety at work, 3rd edn. The handbook for students on NEBOSH and other introductory H&S courses. Butterworth-Heinemann, Jordan Hill, Oxford OX2 8DP 30 Corporate Road, Burlington, MA 01803
Jiménez E, Linares C, Martínez D, Díaz J (2010) Role of Saharan dust in the relationship between particulate matter and short-term daily mortality among the elderly in Madrid (Spain). Sci Total Environ 408:5729–5736
Johnson JE, Kissel JC (1996) Prevalence of dermal pathway dominance in risk assessment of contaminated soils: a survey of superfund risk assessments, 1989–1992. Hum Ecol Risk Assess 2:356–365
Kamal A, Qayyum M, Cheema IU, Rashid A (2011) Biological monitoring of blood naphthalene level as a marker of occupational exposure to PAHs among car-mechanics and spray painters in Rawalpindi. BMC Public Health 11:467. doi:10.1186/1471-2458-11-467
Kamal A, Malik RN, Fatima N, Rashid A (2012) Chemical exposure in occupational settings and related health risks, a neglected area of research in Pakistan. Environ Toxicol Pharmacol 34 (1):46–58
Kamal A, Malik RN, Martellini T, Cincinelli A (2014a) PAHs exposure biomarkers are associated with clinico-chemical changes in the brick kiln workers in Pakistan. Sci Tot Environ 490:521–527
Kamal A, Malik RN, Martellini T, Cincinelli A (2014b) Cancer risk evaluation of brick kiln workers exposed to dust bound PAHs in Punjab province, Pakistan. Sci Total Environ 493:562–570
Kamal A, Cincinelli A, Martellini T, Malik RN (2015a) A review of PAH exposure from the combustion of biomass fuel and their less surveyed effect on the blood parameters. Environ Sci Pollut Res. doi:10.1007/s11356-014-3748-0
Kamal A, Cincinelli A, Martellini T, Malik RN (2015b) Source, profile and carcinogenic risk assessment for cohorts occupationally exposed to dust-bound PAHs in Lahore and Rawalpindi cities (Punjab province, Pakistan). Environ Sci Pollut Res Int 14:10580–10591. doi:10.1007/s11356-015-4215-2
Kanatani KT et al (2010) Desert dust exposure is associated with increased risk of asthma hospitalization in children. Am J Respir Crit Care Med 182:1475–1481
Kang JH, Keller JJ, Chen SC, Lin HC (2012) Asian dust storm events are associated with an acute increase in pneumonia hospitalization. Ann Epidemiol 22:257–263

Kastner J, Knorre DV, Himanshu H, Erb M, Baldwin IT, Meldau S (2014) Salicylic acid, a plant defense hormone, is specifically secreted by a molluscan herbivore. PLoS One 99(1):e86500. doi:10.1371/journal.pone.0086500

Kuo YC, Chen YC, Yang CW, Mou JL, Shih TS, Tsai PJ (2011) Identification the content of the Windbox dust related to the formation of PCDD/Fs during the iron ore sintering process. Aerosol Air Qual Res 11:351–359

Leski TA, Malanoski AP, Gregory MJ, Lin B, Stenger DA (2011) Application of a broad-range resequencing array for detection of pathogens kin desert dust samples from Kuwait and Iraq. Appl Environ Microbiol 77:4285–4292

Lewis RG, Fortune CR, Willis RD, Camann DE, Antley JT (1999) Distribution of pesticides and polycyclic aromatic hydrocarbons in house dust as a function of particle size. Environ Health Perspect 107:721–726

Lin LF, Lee WJ, Li HW, Wang MS, Chang-Chien GP (2007) Characterization and inventory of PCDD/F emissions from coal-fired power plants and other sources in Taiwan. Chemosphere 68:1642–1649

Maertens RM, Bailey J, White PA (2004) The mutagenic hazards of settled house dust: a review. Mutat Res 567:401–425

Man YB, Kang Y, Wang HS, Lau W, Li H, Sun XL, Giesy JP, Chow KL, Wong MH (2013) Cancer risk assessments of Hong Kong soils contaminated by polycyclic aromatic hydrocarbons. J Hazard Mater 261:770–776

Mansour M, Scheunert I, Korte F (1992) Fate of persistend organic compounds in soil and water. In: Petruzzelli D, Freidreich G (eds) Migration and fate of pollutants in soils and suboils, vol 32. Springer, Heidelberg

Martigny N, Chiapello I (2013) Assessments for the impact of mineral dust on the meningitis incidence in West Africa. Atmos Environ 70:245–253

McMoli TE, Bordoh AN, Munube GMK, Bell EJ (1984) Epidemic acute haemorrhagic conjunctivitis in Lagos, Nigeria. Br J Ophthalmol 68:401–404

McNamara ML, Noonan CW, Ward TJ (2011) Correction factor for continuous monitoring of wood smoke fine particulate matter. Aerosol Air Qual Res 11:315–322

Michaels AS, Chandrasekaran SK, Shaw JE (1975) Drug permeation through human skin: theory and in vitro experimental measurement. Aiche J 21:985–996

Mielke HW, Gonzales CR, Smith MK, Mielke PW (1999) The urban environment and children's health: soils as an integrator of lead, zinc, and cadmium in New Orleans, Louisiana, U.S.A. Environ Res 81(2):117–129

Miri A, Ahmadi H, Ghanbari A, Moghaddamnia A (2007) Dust storms impacts on air pollution and public health under hot and dry climate. Int J Energy Environ 1:101–105

Morman A, Plumlee GS (2013) The role of airborne mineral dusts in human disease Suzette. Aeolian Res 9:203–212

Ogorodnikov BI (2011) A dust storm over the Ukraine and Belarus territory contaminated by radionuclides after the Chernobyl accident. Russ Meteorol Hydrol 36:613–623

Otani S, Onishi K, Mu H, Yokoyama Y, Hosoda T, Okamoto M et al (2012) The relationship between skin symptoms and allergic reactions to Asian dust. Int J Environ Res Public Health 9:4606–4614

Prieto-Taboada N, Ibarrondo I, Gomez-Laserna OI, Martinez-Arkarazo MA, Olazabal JM, Madariaga (2013) Buildings as repositories of hazardous pollutants of anthropogenic origin. J Hazard Mater 248–249:451– 460

Prospero JM, Blades E, Naidu R, Mathison G, Thani H, Lavoie MC (2008) Relationship between African dust carried in the Atlantic trade winds and surges in pediatric asthma attendances in the Caribbean. Int J Biometeorol 52:823–832

Pugh WJ, Chilcott RP (2008) Principles of diffusion and thermodynamics. In: Chilcott RP, Price S (eds) Principles and practice of skin toxicology. Wiley, London

Qu W, Bi X, Gy S, Sy L, Jm F, Li L (2007) Exposure to polybrominated diphenyl ethers among workers at an electronic waste dismantling region in Guangdong, China. Environ Int 33:1029–1034

Roberts JW, Dickey P (1995) Exposure of children to pollutants in house dust and indoor air. Rev Environ Contam T 143:59–78

Roy TA, Krueger AJ, Mackerer CR, Neil W, Arroyo AM, Yang JJ (1998a) SAR models for estimating the percutaneous absorption of polynuclear aromatic hydrocarbons. SAR QSAR Environ Res 9:171–185

Roy TA, Krueger AJ, Taylor BB, Mauro DM, Goldstein LS (1998b) Studies estimating the dermal bioavailability of polynuclear aromatic hydrocarbons from manufactured gas plant tar-contaminated soils. Environ Sci Technol 32:3113–3117

Rutherford S, Clark E, McTainsh GH, Simpson R, Mitchell C (1999) Characteristics of rural dust events shown to impact on asthma severity in Brisbane, Australia. Int J Biometeorol 42:217–225

Scheunert I (1992) Transformation and degradation of pesticides in soil. In: Ebing W (ed) Chemistry of plant protection, vol 8. Springer, Heidelberg

Segev O, Kushmaro A, Brenner A (2009) Environmental impact of flame retardants (persistence and biodegradability). Int J Environ Res Public Health 6:478–491

Shea KM (2003) Pediatric exposure and potential toxicity of phthalate plasticizers. Pediatrics 111:1467–1474

Solien NL (1954) A cultural explanation of geophagy. Fla Anthrop 7:1–9

Stanek EJ, Calabrese EJ, Barnes R, Pekow P (1997) Soil ingestion in adults—results of a second pilot study. Ecotoxicol Environ Saf 36:249–257

Sun YB, Zhou QX, Xu YM, Wang L, Liang XF (2011) Phytoremediation for 7 co–contaminated soils of benzo[a]pyrene (B[a]P) and heavy metals using ornamental 8 plant *Tagetes patula*. J Hazard Mater 186:2075–2082

Thomson MC, Jeanne I, Djingarey M (2009) Dust and epidemic meningitis in the Sahel: a public health and operational research perspective. IOP Conf Ser Earth Environ Sci 7. doi:10.1088/1755-1307/7/1/012017

Ud-Din I, Rashid A, Mahmood T, Khalid A (2013) Effect of land use activities on PAH contamination in urban soils of Rawalpindi and Islamabad, Pakistan. Environ Monit Assess. doi:10.1007/s10661-013-3204-5

Villeneuve J, Palacios JH, Savoie P, Godbout S (2012) Critical review of emission standards and regulations regarding biomass combustion in small scale units (<3 MW). Bioresour Technol 111:1–11

Wang LC, Chiou TW, Chang-Chien GP (2008) Abating the emissions of polychlorinated dibenzo-pdioxins and dibenzofurans from municipal solid waste incinerators by spraying bio-solution in feeding. Waste J Environ Eng Manage 18:123–128

Wang LC, Hsi HC, Wang YF, Lin SL, Chang-Chien GP (2010) Distribution of polybrominated diphenyl ethers (PBDEs) and polybrominated dibenzop- dioxins and dibenzofurans (PBDD/Fs) in municipal solid waste incinerators. Environ Pollut 158:1595–1602

Wenger D, Gerecke AC, Heeb NV, Naegeli H, Zenobi R (2008a) Catalytic diesel particulate filters reduce the in vitro estrogenic activity of diesel exhaust. Anal Bioanal Chem 390:2021–2029

Wenger D, Gerecke AC, Heeb NV, Zennegg M, Kohler M, Naegeli H, Zenobi R (2008b) Secondary effects of catalytic diesel particulate filters, reduced aryl hydrocarbon receptor-mediated activity of the exhaust. Environ Sci Technol 42:2992–2998

Wikipedia (2015) www.wikipedia.com

Yang CY (2006) Effects of Asian dust storm events on daily clinical visits for conjunctivitis in Taipei, Taiwan. J Toxicol Environ Health A 69:1673–1680

Yu HL, Chien LC, Yang CH (2012) Asian dust storm elevates children's respiratory health risks: a spatiotemporal analysis of children's clinic visits across Taipei (Taiwan). PLoS One 7(7): e41317. doi:10.1371/journal.pone.0041317

Zaady E, Offer ZY, Shachak M (2001) The content and contribution of deposited aeolian organic matter in a dry land ecosystem of the Negev Desert, Israel. Atmos Environ 35:769–776

Zamir R, Athanasiadou M, Nahar N, Mamun MIR, Mosihuzzaman M, Bergman A (2008) Persistent organohalogen contaminants in plasma from groups of humans with different occupations in Bangladesh. Chemosphere 74:453–459

Ziegler JL, Simonart T (2001) Snoeck R.Kaposi's sarcoma, oncogenic viruses and iron. J Clin Virol 20:127–130

Chapter 6
Use of Earthworms in Biomonitoring of Soil Xenobiotics

Muhammad Dawood, Abdul Wahid, Muhammad Zaffar Hashmi, Sidra Mukhtar, and Zaffar Malik

6.1 Introduction

Soil, being a complex mixture of mineral constituents, organic matter and soil fauna is constantly losing its health as a result of rapid industrialization and other anthropogenic activities. Intensive use of biocides and fertilizers in agriculture, industrial activities, urban waste, and atmospheric deposition has led to addition of a number of xenobiotics into the soil substrate, thus posing severe threats to soil quality and productivity. Xenobiotic-led soil pollution is causing decrease in soil fertility, alteration of soil structure, disturbance of the balance between flora and fauna residing in the soil, contamination of the crops, and contamination of groundwater, constituting a threat for living organisms. The most diffusive chemicals occurring in soil are heavy metals, pesticides, petroleum hydrocarbons, polychlorobiphenyl (PCBs), and dibenzopdibenzo-p-dioxins/dibenzofurans (PCDD/Fs). Although the main types of xenobiotics of water and soil media are almost the same, their fate and effect on living species may vary considerably. The very small size (<2 μm) of soil colloids, i.e., clay (mineral particles) and humus (organic matter), with larger surface area to volume ratios makes them unique from water in possessing high binding capacity to organic and inorganic chemicals. The

M. Dawood (✉) • A. Wahid • S. Mukhtar
Department of Environmental Sciences, Bahauddin Zakariya University, Multan 60800, Pakistan
e-mail: dawood_mian25@hotmail.com

M.Z. Hashmi
Atmospheric Chemistry Division, Department of Meteorology, COMSATS Institute of Information Technology Islamabad, Islamabad, Pakistan

Z. Malik
University College of Agriculture and Environmental Sciences, The Islamia University of Bahawalpur, Bahawalpur, Pakistan

© Springer International Publishing AG 2017
M.Z. Hashmi et al. (eds.), *Xenobiotics in the Soil Environment*, Soil Biology 49,
DOI 10.1007/978-3-319-47744-2_6

binding of pollutants to soil colloids reduces their mobility and their bioavailability and modulates their biological effects. In addition, soil aging and weathering are important factors influencing the available fraction of pollutants in soils and decreasing their environmental bioavailability for uptake by organisms and for exerting toxic effects (Spurgeon and Hopkin 1995; Peijnenburg et al. 1997; Peijnenburg and Vrijver 2009; Lionetto et al. 2012). Consequently, xenobiotic toxicity toward terrestrial species is different in nature and cannot be extrapolated from aquatic species (Vasseur and Bonnard 2014).

The soil environments are also the sources of food and habitat for different organisms, particularly the invertebrates. These organisms living at different trophic levels may be all exposed even if only few members of the food web are affected by exposure to xenobiotics. Direct exposure of xenobiotics to lower trophic level organisms will greatly enhance the chances of contamination, bioaccumulation, and biomagnification of the contaminant to the higher levels in a food web through various food chain routes. Apparently, it seems better to assess the exposure and intensity of xenobiotic in a particular environment by using lower trophic level organisms as instant and early warning systems.

For these reasons, biomonitoring approaches, such as the use of living organisms (as bioindicators) or response of xenobiotics to cellular and biochemical processes (as biomarkers), becomes imperative to assess the soil quality and health and to visualize the pollution status of this important medium of the environment (Kammenga et al. 2000). Soil invertebrates have advantage to be used as bioindicator of pollutants over the vertebrates in a sense that invertebrates are usually in direct contact with the soil capillary water or feeding sources; while the latter are indirectly linked with the food chain and are seldom used in bioindication programs (Kammenga et al. 2000). Earthworms are considered as one of the best soil invertebrate organisms known for their soil formation and organic matter breakdown characteristics in terrestrial environmental conditions. Owing to their peculiar and meticulous interactions with soil medium, earthworms are significantly affected by xenobiotic exposure arising from intensive use of agricultural chemicals, industrial emissions, and atmospheric deposition. Earthworms are part of a very complex soil food web and occupy a lower position in the food chain. Due to their specific morphological structures and ecological relevance, earthworms have been used as bioindicators and, as they are also biosensors of sublethal concentrations, they could serve as a warning sign for the early effects of soil contamination (Lanno et al. 2004). Earthworms are often suggested as bioindicators of soil quality because they are an important part of the soil system and also because they are frequent, easy to collect, and rather simple to identify (Buckerfield et al. 1997; Paoletti 1999; USDA-NRCS 2009). The occurrence and the effects of earthworms are generally associated with good soil quality. Markert et al. (2003) defined bioindication as qualitative indication of environmental properties and biomonitoring as quantified bioindication in order to detect trends in time and space. Thus, earthworms are equally used as bioindicators as well as biosensor processes for early warning and effective biomonitoring programs to assess the xenobiotics exposure and status into soil environments.

6.2 Earthworms as Bioindicator of Soil Pollutants

Earthworms belonging to phylum Annelida, class Chaetopoda, and order Oligochaeta make up the greatest proportion (>80 %) of biomass of terrestrial invertebrates and form major soil macrofauna. Being the first group of multicellular, eucoelomate invertebrates succeeding to inhabit terrestrial environment, the so-called earthworms occupy a unique position in animal kingdom. Species richness, abundance, and distribution pattern of earthworms specifies their ecological importance by contributing as edaphic and climatic factors of the geographical zone (Kale and Karmegam 2010).

Through horizontal and vertical stratification, they play an important role in pedogenesis and soil profiling, thus affecting the physical, chemical, and microbiological properties of soil (Barlett et al. 2010). Earthworms' abundance is generally considered as valuable indicator of improved soil fertility by conferring enhanced mineralization and humification of organic matter by food consumption, respiration, and gut passage (Lavelle and Spain 2001). They may also enhance microbial activity and total mass, and by increasing the surface area of organic particles by their casting activity, they can foster the mobilization of nutrients within the medium (Emmerling and Paulsch 2001). Apart from these, earthworms are good at improving water infiltration and soil aeration through their unique burrowing activities (Lionetto et al. 2012).

Organic matter decomposition and nutrients cycling are two important features of earthworms, which makes them ideal candidate to be used as indicator organisms for the assessment of biological impact of xenobiotics (Spurgeon et al. 2003). Comparatively simpler manipulation in earthworms makes it easier to measure various life-cycle parameters to be used in monitoring programs. These parameters include growth and reproduction of the organism, accumulation and excretion of pollutants, and biochemical responses. Thus, earthworms are suitable organisms for soil ecotoxicological research.

6.3 Exposure and Uptake of Xenobiotics

Earthworms are significantly affected by soil xenobiotics due to their direct and particular interaction with soil substrate. In case of earthworms, exposure to xenobiotics can be either through ingestion of pore water having dissolved pollutants or through alimentary tract intake of soil-carrying contaminants adsorbed to solid particles. Firstly, being habitant of soil medium, they are in direct contact with soil pore water as well as with the dissolved xenobiotics in it. The skin of earthworm is highly permeable to water (Wallwork 1983) that makes it the main route for contaminant uptake (Jajer et al. 2003; Vijver et al. 2003). Pollutant uptake via the dermal route can thus be directly related to their chemical concentration in pore/capillary water. Secondly, being soil eaters, earthworms ingest larger amounts of soil as their intrinsic nature and are continuously exposed to contaminants adsorbed to solid particles. This type of pollutant intake is carried through their alimentary tract (Morgan et al. 2004). According to Vijver et al. (2003), the dermal route is the main route for heavy metals uptake, while pore water uptake via ingestion does not significantly contribute to metal accumulation in the earthworm.

6.4 Bioaccumulation of Xenobiotics

Earthworms have the potential to uptake and accumulate various organic and inorganic pollutants of the soil (Morrison et al. 2000). In case of metal pollutants, they have got efficient sequestration mechanisms which help them tolerate relatively higher tissue metal concentrations, particularly the trace metals such as Cu and Zn (Peijnenburg 2002; Andre et al. 2009). Post alimentary canal is the primary site for heavy metals accumulation. This portion is chiefly comprised of intestine and the related chloragogenous tissue that segregates the absorptive epithelia from the coelomic cavity (Andre et al. 2009). The chloragogenous tissue is composed of pedunculated cells, and its main functions are synthesis of hemoglobin, homeostasis of cation composition in the blood and coelomic fluid, maintenance of a balanced pH level, storage of nutrients and waste, and uptake and detoxification of toxic cations (Jamieson 1992). Therefore, chloragogenous tissue represents a major metal sink (Morgan et al. 2002).

Bioaccumulation of xenobiotics in earthworms is highly variable and depends upon a number of factors. These may be species of earthworm, physicochemical properties of xenobiotics, metal speciation and concentration, soil type and its physicochemical characteristics, temperature, and exposure time. Bioaccumulation of xenobiotics in earthworms is normally expressed as biota to soil accumulation factor (BSAF), which can be calculated as:

$$\text{BSAF} = \text{Concentration in Earthworm}/\text{Concentration in Soil}$$

As earthworms are a major source of food for other animals like amphibians, reptiles, birds, and mammals, the bioaccumulation of xenobiotics by earthworms is characterized by the risk of the transfer of pollutants to higher trophic levels (Marino et al. 1992).

6.5 Suitability of Earthworms as Accumulation Indicators

For a number of reasons, earthworms are well suited to serve as accumulation indicators for the presence of bioavailable chemicals in the soil:

1. Earthworms are soil dwellers and are in constant contact with some portion of the soil.
2. Earthworms reside in contaminated sites, allowing field validation of chemical bioavailability.
3. Earthworms are found in a wide variety of soil types and horizons.
4. Their specific morphological features allow them to uptake contaminants directly from the soil. The exterior epidermal surface of the earthworm is vascularized with no cuticle, allowing the uptake of contaminants directly from the soil.
5. Earthworms ingest soil or specific fractions of soil, providing a means for the dietary uptake of contaminants.
6. Earthworms have a large mass, so contaminant concentrations can be determined in individual organisms.
7. There is a low level of mixed-function oxidase (MFO) activity, allowing greater potential for the accumulation of organic compounds that would normally be metabolized in other organisms.
8. We have an understanding of their physiology and metabolism of metals (Lanno et al. 2004).

Environmental chemicals in earthworms may also be a relevant issue for wildlife protection since earthworms are an important food of many vertebrate and invertebrate species (Beyer and Stafford 1993). There are two important parameters when earthworms are used as accumulation indicators: The body concentration in the earthworm indicates the risk of secondary poisoning for predators feeding on earthworms. The BSAF or simply the bioaccumulation factor (BAF = concentration of chemical in worm/concentration of chemical in soil) indicates the bioavailability of the contaminant in the soil.

6.6 Xenobiotic/Chemicals Accumulating in Earthworms

A variety of xenobiotics of organic and inorganic nature have been found to be accumulated by earthworms. Heavy metals have been reported by different researchers to be extensively uptaken and accumulated by earthworms. Cd, Hg, and Zn are greatly bioconcentrated in the earthworm with a BAF > 1 (Neuhauser et al. 1995; Rahtkens and von der Trenck 2006; Ernst et al. 2008; Nahmani et al. 2009; Tischer 2009). Bioconcentrated heavy metals are distributed in the alimentary tract of earthworms which is subdivided into a foregut, midgut, and hindgut. Most of the metabolic activities are done in midgut that has direct or indirect role in the uptake, transport, storage, and excretion of metals (Hopkin 1989). Surplus metals are mostly stored as metal-bound proteins and/or metal-containing granules in the chloragogenous tissue surrounding the lumen of the gut.

Different species of earthworm possess different potential to deal with and accumulate heavy metals in different ways. *Lumbricus terrestris* rich in active calciferous glands accumulates less Pb, whereas *Aporrectodea longa* accumulates relatively higher Pb due to production of waste.

Cadmium is mainly stored in the gut wall (Andersen and Laurensen 1982) and nephridia (Prinsloo et al. 1990) in the form of metallothioneines (metal-binding proteins).

As far as xenobiotics of organic nature are concerned, the so-called equilibrium partition theory can provide estimation of their accumulation in earthworms (Jager 1998). Basis of this theory is that the substances/chemical present only in soil water can get accumulated by earthworms. Henson-Ramsey et al. (2009) proposed a pharmacokinetic model to estimate bioconcentration of organic xenobiotics in earthworms. A relatively higher accumulation for persistant organic pollutants (POPs) has been reported with a BAF for DDT = 5, Dieldrin = 8, and heptachlor epoxide = 10 (Beyer and Gish 1980). Whereas, PCBs are usually less accumulated in earthworms (BAF = 1.8) (Beyer and Stafford 1993). PAHs are uptaken in little amounts with an average BAF of 0.1 (range 0.03–0.26) (Ma et al. 1998).

6.7 Factor Affecting Bioaccumulation of Xenobiotics by Earthworms?

High variation in the BAF of xenobiotics in earthworms is probably due to external factors along with specie-specific variability. Several abiotic and biotic factors govern the complex process of bioaccumulation of an environmental chemical into the earthworm. Various environmental factors including pH, electrical conductivity, organic carbon, nitrogen content, atmospheric temperature, soil temperature, soil moisture, humidity, and rainfall play major roles in uptake and bioaccumulation of soil xenobiotics into earthworms.

However, with the particular soil environments, the bioaccumulation chiefly depends on the concentration, chemical speciation (chemical bioavailability), and on the spatial distribution (physical bioavailability) of the chemical.

Concentration Pollutant concentration in the soil is a very important factor affecting bioaccumulation of the chemical into organism. While working with heavy bioaccumulation into earthworms, Neuhauser et al. (1995) investigated the bioconcentration of Cd, Cu, Ni, Zn, and Pb from sewage sludge-contaminated soils in *Aporrectodea tuberculata* and *L. rubellus* and compiled data from another 20 published studies. A significant correlation between the metal concentration in the soil and the metal concentration in the worm turned out for Cd, Zn, Pb, and Cu but not for Ni. The BAF declines at high concentrations in soil. This holds in particular for Cd and the essential metals Zn and Cu. Tischer (2009) analyzed earthworms without gut clearance from 84 soil-monitoring sites and found significant positive correlations between metal contents in earthworms and metal contents in soil for various metals.

Chemical Speciation Alberti et al. (1996) showed that lead that has been added to the soil as a salt is accumulated more than lead from an aged soil contamination. It is a general experience that aging reduces the bioavailability of contaminants. Soil properties like pH and redox potential are important determinants for the speciation/solubility of metals in the soil.

Distribution Between Microcompartments Morgan and Morgan (1999) showed experimentally that the Pb distribution in the soil profile affected the pattern of tissue concentration between the epigeic *L. rubellus* and the endogenic *Aporrectodea caliginosa*. Plant roots are microcompartments in the soil with enriched concentrations of heavy metals (Ernst et al. 2008).

Among the factors in the worm vitality is a prerequisite for the indication of bioaccumulation. Only worms can be analyzed from contaminated sites that have been able to survive the contamination. Often these worms have adapted to the contamination and therefore are not representative for the average population.

Consequently, accumulation monitoring with earthworms will probably be biased at highly contaminated sites. The concentration of a substance in the worm is regulated by the interplay of physiological mechanisms such as uptake, internal demand, internal sequestration, and excretion. These mechanisms are subject to adaptations as well as to the regulatory influence of environmental factors such as the pH or the Ca-ion concentration (Nahmani et al. 2009). Behavioral aspects include the feeding and habitat choice of the earthworm (ecological group). Ernst et al. (2008) showed in a PCA analysis that there were specific patterns in the Cd, Pb, and Hg content in earthworms of ecological groups in relation to the metal content in different compartments of the soil-litter system.

6.8 Biomarkers in Earthworms for Impact Assessment of Soil Xenobiotics

More recently the use of biomarkers to estimate either exposure or resultant effects of chemicals have received considerable attention. Biomarkers in theory offer two outstanding advantages in ecotoxicological investigations. First, by their very nature they are responsive only to the bio- or toxicoactive fraction of the accumulated body burden of one or more potential toxicants; to this extent they reflect, albeit indirectly, the analytically elusive bioavailability fraction of environmental chemicals within chosen test organisms. Secondly they integrate the interactive effects of several chemicals in complex mixture (Weeks 1995).

Biomarkers used as measures of toxic effects in organisms at the level below individual (molecular or cellular) level (Van Gestel and Van Brummelen 1996) represent early or initial responses to environmental perturbation and contamination.

In ecotoxicology of earthworm, there exist a range of biomarkers of toxic compounds, including biomarkers from the molecular to the organismal level like metallothioneins, stress proteins, cholinesterases, detoxification enzymes, parameters of oxidative stress, and others (Novais et al. 2011). Besides, the investigation of new, potent biomarkers in earthworms is in accelerating phase for the impact assessment of soil pollutants.

6.9 Metallothioneins

It is well-known fact that metallothioneins (MTs) are low-molecular-weight cysteine-rich metal-binding proteins that are produced internally as a defense mechanism to counter the heavy metal stress. MTs play their metal stress-alleviating role either by regulating the homeostasis of essential metals like Cu and Zn and/or by detoxifying nonessential metals such as Ag, Cd, and Hg (Amiard et al. 2006). In addition to their role as metal chelators, MTs are also considered as free radical scavengers (Min 2007). These metallothioneins are produced as a result of low exposure to nonessential trace metal. Induction of MTs on low exposure to metals has been observed in many organisms including earthworms. Various organisms including the earthworms have exhibited the induction of metallothioneins due to metal exposure (Strzenbaum et al. 2001). Significant induction of MTs has been reported by researchers on different species of earthworms, for example, *Lumbricus rubellus*, *Eisenia fetida*, and *Eisenia andrei* exposed to cadmium (Demuynck et al. 2006; Brulle et al. 2007; Ndayibagira et al. 2007; Calisi et al. 2009); in *Lumbricus terrestris* exposed to cadmium, copper, and mercury (Khan and Ali 2007; Calisi et al. 2011); and in *Lumbricus mauritii* exposed to Pb- and Zn-contaminated soil (Maity et al. 2011). It is well-known fact that earthworms have high degree of tolerance to heavy metal exposure, most probably due to induction of MTs. Metallothionein induction is therefore considered as the most widely used biomarkers in earthworms for the early detection of exposure to heavy metals in soil monitoring.

6.10 Acetylcholinesterase

Acetylcholinesterase (AChE) is a key enzyme in the nervous system, terminating nerve impulses by catalyzing the hydrolysis of neurotransmitter acetylcholine. AChE is the target site of inhibition by organophosphorus and carbamate pesticides. In particular, organophosphorus pesticides inhibit the enzyme activity by covalently phosphorylating the serine residue within the active site group. They irreversibly inhibit AChE, resulting in excessive accumulation of acetylcholine, leading to hyperactivities and consequently impairment of neural and muscle system.

The monitoring of acetylcholinesterase activity in the brains of fish and birds in the field has become a technique commonly used for diagnosing the exposure to cholinergic poisons (Zinkl et al. 1991). Acetylcholinesterase represents the main cholinesterase in earthworms (Rault et al. 2007).

AChE inhibition in earthworms is presently regarded as giving early warning of adverse effects of pesticides (Booth and O'Halloran 2001) and consistently included among the batteries of biomarkers employed for early assessments of pollutant impact on wildlife in terrestrial ecosystems. However, concerning AChE in earthworms, only a few pesticides in use have been tested against relatively few earthworm species both in laboratory tests and under field conditions (Rao et al. 2003; Calisi et al. 2009). As pointed out by Scott-Fordsmand and Weeks (2000), the potential use of AChE in earthworms as biomarker of pesticide exposure has not been sufficiently explored.

6.11 Hemoglobin Oxidation

Hematological changes are believed to be one of the early warning signals of the toxic effects of xenobiotics in vertebrate organisms (Dauwe et al. 2006). However, these are poorly explored in invertebrates including the earthworms. Blood of the earthworms contains hemoglobin (Hb), which is a large extracellular hemoprotein flowing in a closed circulatory system. This respiratory pigment possesses the fundamental role in earthworm physiology; however, its sensitivity to environmental pollutants is yet to be explored.

Exposure to certain heavy metals including cadmium, copper, and mercury significantly induces changes in either Hb concentration or its oxidation state as has been shown by Calisi et al. (2011) in the earthworm *Lumbricus terrestris*. Similarly, another study has shown enhancement of blood Hb concentration by earthworms as result of exposure to Cd, Hg, and Cu. In addition to changes in the Hb concentration, heavy metals showed a dramatic effect on the oxidation state of the respiratory pigment. Future studies will be addressed to evaluate if the observed response is specific for heavy metal exposure or represents a biomarker of general health of earthworms in polluted sites. In any case it demonstrated to be a suitable biomarker of exposure/effect to be included in a multibiomarker strategy in earthworm in soil-monitoring assessment.

6.12 Glutathione S-transferase (GST) and Antioxidant enzymes

Many agrochemicals such as OP insecticides are able to induce oxidative stress (Lukaszewicz-Hussain 2010), a situation in which the production of reactive oxygen species (ROS) overcomes the cellular antioxidant mechanisms (molecular and enzymatic), leading to the oxidative damage of biomolecules (e.g., lipids, proteins, or DNA). Glutathione level is one of the most used biomarker of prooxidant exposure in fish (van der Oost et al. 2003) and birds (Koivula and Eeva 2010). Glutathione transferases (GSTs) form a ubiquitous super family of multifunctional dimeric enzymes (w50 kDa) with roles in detoxification. Several studies have been concerned with changes in glutathione concentration and glutathione-dependent enzymes in terrestrial invertebrates, i.e., earthworm on metals and pesticide exposure (Aly and Schröder 2008; Maity et al. 2008). Biomarkers of oxidative stress have been mainly explored in earthworms exposed to, or inhabiting in, metal-polluted environments. For example, earthworm GST activity is a noteworthy detoxification system, which is induced in earthworms exposed to organochlorine pesticides (Hans et al. 1993). However, no effects on this enzyme activity were observed in earthworms exposed to the OP fenitrothio (Booth and O'Halloran 2001) or the CM carbaryl (Ribera et al. 2001). Herbicides also induce the GST activity of earthworms.

6.13 Biomarkers of Toxic Metal Exposure

Potential metal contaminants are produced from various industries such as mining, metal, pigment, and chemical manufacturing as well as in combustion of fossil fuels, including the exhaust emission of motor vehicles. In the field of toxicology, it is essential to be able to measure the exposure to a toxic agent, the extent of any toxic response, and also to predict the likely effects. Hence, integrating measures of different types of responses to toxic stress of exposed individuals and populations offers a powerful tool for documenting the extent of exposure and the effects of environmental metal contamination. Therefore, use of biomarkers for environmental monitoring of individuals and populations exposed to chemical pollution has gained much attention in the last decades, because it offers great opportunities for a fast and sensitive detection of chemical stresses within organisms (Handy et al. 2003).

6.14 Biomarkers of Pesticide Exposure

Nowadays it is widely accepted that current agricultural practices cause a loss of biodiversity. Moreover, the introduction of vast areas of monocultives (e.g., biofuel crops) contributes to increase the risk for crop loss by pest infestation (Ali et al. 2006). Plant protection products (PPPs) are still necessary for combating pests. The massive use of pesticides leads to a set of environmental hazards on nontarget organisms of ecological and agronomic concerns such as earthworms, pollinators, or natural enemies of pests. Moreover, the occurrence of pesticide residues in soil can change microbial communities and soil enzyme activities involved in nutrient cycles. These effects can lead, in turn, to a loss of soil quality. Biomarkers are often used to provide mechanistic understandings on the toxic effects observed at the whole-organism level. Classical biochemical biomarkers have been used in terrestrial invertebrates, mainly earthworms, to assess exposure to organophosphate (OP) and carbamate (CM) pesticides.

6.15 Cellular Biomarkers

Coelomic fluid of earthworm shows very interesting features from ecotoxicological point of view for the development of novel cellular biomarkers. It helps in transport of pollutant in the exposed organism (Engelmann et al. 2004). Five cell types were observed by Calisi et al. (2009) in *Eisenia fetida* celomic fluid, corresponding to the previously described coelomocyte cell types: leukocyte types I (basophilic) and II (acidophilic), granulocytes, neutrophils, and eleocytes. Immunoactive coelomocytes and their viability are one of the most promising surrogate assays to assess immunotoxic risks (Burch et al. 1999).

Several studies are concerned with the effects of environmental pollution including heavy metals on earthworm immune functions mediated by coelomocytes (Scott-Fordsmand and Weeks 2000).

Fugere et al. (1996) recorded the inhibition of phagocytic activity of coelomocytes exposed in vitro on heavy metal (Cd, Zn) solutions. Homa et al. (2007) studied the effects of heavy metal toxicity on the coelomocytes of the earthworm *A. chlorotica* and reported that the coelomocytes viability and activity was significantly reduced. This was explained as a result of accumulation of heavy metals in the body tissues of the earthworms including coelomocytes which in turn affect their and viability.

In earthworms, lysosomal fragility as a result of heavy metal toxicity has been developed as a promising biomarker by Weeks and Svendsen (1996). The method used is the neutral red retention assay (NRR-assay), a histochemical technique based on the principle that only healthy lysosomes can retain the acidotrophic, red dye after an initial uptake. The time lysosomes able to retain the dye is called neutral red retention time (NRR time), and the NRR time decreases when lysosomes are exposed to membrane disrupters such as heavy metals. The NRR assay has earlier been used to assess the effects of copper, cadmium, nickel, and zinc on earthworm coelomocytes (Weeks and Svendsen 1996; Scott-Fordsmand and Weeks 2000; Spurgeon et al. 2003).

6.16 Biomarkers of Genotoxicity

Coelomocytes of earthworms' coelomic fluid are also taken for monitoring the genotoxic effect of soil pollutants in earthworms. Many pollutants in soil like metals, metallooids and agricultural pollutants etc. may cause deleterious effect on the structure of DNA; thus, it is an important indicator for evaluation of dreadful effects of pollutants on earthworm health (Frenzilli et al. 2001; Reinecke and Reinecke 2004). Hence, the evaluation of metal sublethal toxicity should always include biomarkers of DNA damage because such damage may result in inappropriate gene expression and, subsequently, in more concerning genotoxic and mutagenic effects (Hartsock et al. 2007). The single cell gel electrophoresis (or comet assay) and micronucleus test are the two most extensively used methods in the detection of genotoxicity of chemicals in the environment. The single cell gel electrophoresis assay (also known as the comet assay) is a simple, rapid, and sensitive technique for analyzing and quantifying DNA damage in individual cells, as single- and double-strand breaks, alkali-labile sites, and oxidative DNA base damage (Cotelle and Ferard 1999). The comet assay is an effective method for determining DNA damage levels in the coelomocytes of earthworms exposed to genotoxic compounds, both in vivo and in vitro, in several studies (Reinecke and Reinecke 2004). Instead of comet assays, micronucleus assays have emerged as one of the preferred methods for assessing chromosome damage because they enable both chromosome loss and chromosome breakage to be measured reliably accumulated during lifespan of the cell in vertebrates and invertebrates. Sforzini et al. (2010) provided the first step of validation of this test on earthworm (*Eisenia andrei*) cells.

The abovementioned earthworm biomarkers have gained importance for the evaluation of xenobiotic assessment and contaminant effects on soil organisms. These organisms are considered suitable indicators of environmental change in soil environments. MT induction, which is mostly used to metal exposure, has already been extensively investigated. Other biomarkers (e.g., ChE), which are mostly used in vertebrate-based monitoring programs, have received little attention in earthworm studies. Additional studies, involving different species of earthworm, with different endpoints, temperature regimes, and soil types, are necessary to be carried out. Research should be extended to ecologically relevant species of earthworms and also to other soil fauna to get a comprehensive knowledge on the malfunction in the soil biological processes due to soil pollutants. So, there is a need to acquire more knowledge on the chemical nature, mode of action, and means of degradation of pollutants in soil, so that harm caused to soil fauna as well as to organisms higher up in the food chain can be minimized.

References

Alberti G, Hauk B, Kohler HR, Storch V (1996) Decomposition. Ecomed, Landsberg

Ali SA, Khan I, Ali AS (2006) Friendly earthworms. Sci Rep CSIR New Delhi 43(1):28–30

Aly MA, Schröder P (2008) Effect of herbicides on glutathione S-transferases in the earthworm, Eisenia fetida. Environ Sci Pollut Res Int 15:143–149

Amiard JC, Amiard-Triquet C, Barka S, Pellerin J, Rainbow PS (2006) Metallothioneins in aquatic invertebrates: their role in metal detoxification and their use as biomarkers. Aquat Toxicol 76 (2):160–202

Andersen C, Laurensen J (1982) Distribution of heavy metals in *Lumbricus terrestris. Aporrectodea longa* and *A. rosea* measured by atomic absorption and X-ray fluorescence spectrometry. Pedobiologia 24:347–356

Andre J, Charnock J, Sturzenbaum SR, Kille P, Morgan AJ, Hodson ME (2009) Metal speciation in field populations of earthworms with multi-generational exposure to metalliferous soils: cell fractionation and high energy synchrotron analysis. Environ Sci Technol 43(17):6822–6829

Barlett MD, Briones MJI, Neilson R, Schmidt O, Spurgeon D, Creamer RE (2010) A critical review of current methods in earthworm ecology: from individuals to populations. Eur J Soil Biol 46(2):67–73

Beyer WN, Gish CD (1980) Persistence in earthworms and potential hazards to birds of soil applied DDT, dieldrin and heptachlor. J Appl Ecol 17:295–307

Beyer WN, Stafford C (1993) Survey and evaluation of contaminants in earthworms and in soils derived from dredged material at confined disposal facilities in the Great Lakes Region. Environ Monit Assess 24:151–165

Booth LH, O'Halloran K (2001) A comparison of biomarker responses in the earthworm *Aporrectodea caliginosa* to the organophosphorus insecticides diazinon and chlorpyrifos. Environ Toxicol Chem 20(11):2494–2502

Brulle F, Mitta G, Leroux R, Lemière S, Leprêtre A, Vandenbulcke F (2007) The strong induction of metallothionein gene following cadmium exposure transiently affects the expression of many genes in *Eisenia fetida*: a trade-off mechanism? Comp Biochem Physiol C 144 (4):334–341

Buckerfield JC, Lee KE, Davoren CW, Hannay JN (1997) Earthworms as indicators of sustainable production in dryland cropping in southern Australia. Soil Biol Biochem 29:547–554

Burch SW, Fitzpatrick LC, Goven AJ, Venables BJ, Giggleman MA (1999) In vitro earthworm coelomocyte assay for use in terrestrial identification evaluation. Bull Environ Contam Toxicol 52:547–554

Calisi A, Lionetto MG, Schettino T (2009) Pollutant-induced alterations of granulocyte morphology in the earthworm *Eisenia foetida*. Ecotoxicol Environ Saf 72(5):1369–1377

Calisi A, Lionetto MG, Sanchez-Hernandez JC, Schettino T (2011) Effect of heavy metal exposure on blood haemoglobin concentration and methemoglobin percentage in *Lumbricus terrestris*. Ecotoxicology 20(4):847–854

Cotelle S, Ferard JF (1999) Comet assay in genetic ecotoxicology: a review. Environ Mol Mutagen 34(4):246–255

Dauwe T, Janssens E, Eens M (2006) Effects of heavy metal exposure on the condition and health of adult great tits (*Parus major*). Environ Pollut 140(1):71–78

Demuynck S, Grumiaux F, Mottier V, Schikorski D, Lemière S, Leprêtre A (2006) Metallothionein response following cadmium exposure in the oligochaete *Eisenia fetida*. Comp Biochem Physiol C 144:134–146

Emmerling C, Paulsch D (2001) Improvement of earthworm (Lumbricidae) community and activity in mine soils from open-cast coal mining by the application of different organic waste materials. Pedobiologia 45(5):396–407

Engelmann P, Molnar L, Palinkas L, Cooper EL, Nemeth P (2004) Earthworm leukocyte populations specifically harbor lysosomal enzymes that may respond to bacterial challenge. Cell Tissue Res 316:391–401

Ernst G, Zimmermann S, Christie P, Frey B (2008) Mercury, cadmium and lead concentrations in different ecophysiological groups of earthworms in forest soils. Environ Pollut 156:1304–1313

Frenzilli G, Nigro M, Scarcelli V, Gorbi S, Regoli F (2001) DNA integrity and total oxyradical scavenging capacity in the Mediterranean mussel, *Mytilus galloprovincialis*: a field study in a highly eutrophicated coastal lagoon. Aqua Toxicol 53:1932

Fugere N, Brousseau P, Krzystyniak K, Coderre D, Fourier M (1996) Heavy metal-specifi c inhibition of phagocytosis and different in vitro sensitivity of heterogeneous coelomocytes from *Lumbricus terrestris* (Oligochaeta). Toxicology 109:157–166

Handy R, Galloway T, Depledege M (2003) A proposal for the use of biomarkers for the assessment of chronic pollution and in regulatory toxicology. Ecotoxicology 12:331–343

Hans RK, Khan MA, Farooq M, Beg MU (1993) Glutathione-Stransferase activity in an earthworm (*Pheretima posthuma*) exposed to three insecticides. Soil Biol Biochem 25:509–511

Hartsock WJ, Cohen JD, Segal DJ (2007) Uranyl acetate as a direct inhibitor of DNA binding proteins. Chem Res Toxicol 20:784–789

Henson-Ramsey H, Levine J, Kennedy-Stoskopf S, Taylor SK, Shea D, Stoskopf MK (2009) Development of a dynamic pharmacokinetic model to estimate bioconcentration of xenobiotics in earthworms. Environ Model Assess 14:411–418

Homa J, Stürzenbaum SR, Morgan AJ, Plytycz B (2007) Disrupted homeostasis in coelomocytes of Eisenia fetida and *Allolobophora chlorotica* exposed dermally to heavy metals. Euro J Soil Biol 43:273–280

Hopkin SP (1989) Ecophysiology of metals in terrestrial invertebrates. Elsevier Applied Science, London

Jager T (1998) Mechanistic approach for estimating bioconcentration of organic chemicals in earthworms. Environ Toxicol Chem 17:2080–2090

Jajer T, Fleuren RHLJ, Hogendoorn EA, de Korte G (2003) Elucidating the routes of exposure for organic chemicals in the earthworm, *Eisenia andrei* (Oligochaeta). Environ Sci Technol 37 (15):3399–3404

Jamieson BGM, (1992) Oligochaeta. In: Harrison FW, Gardiner SL (eds) Microscopic anatomy of invertebrates, Wiley, New York, pp 217–322. ISBN 978-0471561170

Kale RD, Karmegam N (2010) The role of earthworms in tropics with emphasis on Indian ecosystems. Appl Environ Soil Sci 2010:16

Kammenga JE, Dallinger R, Donker MH, Köhler HR, Simonsen V, Triebskorn R, Weeks JM (2000) Biomarkers in terrestrial invertebrates for ecotoxicological soil risk assessment. Rev Environ Contam Toxicol 164:93–147

Khan I, SA A, Ali SA (2007) Biomass and behavioural responses of earthworm *L. terrestris* to Copper Chloride. Iran J Toxicol 2:64–71

Koivula MJ, Eeva T (2010) Metal-related oxidative stress in birds. Environ Pollut 158:2359–2370

Lanno R, Wells J, Conder J, Basta N (2004) The bioavailability of chemicals in soil for earthworms. Ecotoxicol Environ Saf 57:39–47

Lavelle P, Spain A (2001) Soil ecology. Kluwer Scientific, Amsterdam. ISBN 978-0-7923-7123-6

Lionetto MG, Calisi A, Schettino T (2012) Earthworms biomarkers as tools for soil pollution assessment. In: Hernandz-Soriano MC (ed) Soil health and land use management. InTech-Open Access Publisher in Science, Technology and Medicine, Rijeka, pp 305–332

Lukaszewicz-Hussain A (2010) Role of oxidative stress in organophosphate insecticide toxicity short review. Pestic Biochem Physiol 98:145–150

Ma WC, van Kleunen A, Immerzeel J, de Maagd PGJ (1998) Bioaccumulation of polycyclic aromatic hydrocarbons by earthworms: assessment of equilibrium partitioning theory in in-situ studies and water experiments. Environ Toxicol Chem 17:730–1737

Maity S, Roy S, Chaudhury S, Bhattacharya S (2008) Antioxidant responses of the earthworm *Lampito mauritii* exposed to Pb and Zn contaminated soil. Environ Pollut 151(1):17

Maity S, Roy S, Bhattacharya S, Chaudhury S (2011) Metallothionein responses in the earthworm *Lampito mauritii* (Kinberg) following lead and zinc exposure: a promising tool for monitoring metal contamination. Eur J Soil Biol 47:69–71

Marino F, Ligero A, Cosin DJD (1992) Heavy metals and earthworms on the border of a road next to Santiago. Soil Biol Biochem 24(12):705–1709

Markert BA, Breure AM, Zechmeister HG (2003) Definition, strategies and principles for bioindication/biomonitoring of the environment. In: BA M et al (eds) Bioindicators and biomonitors. Trace metals and other contaminants in the environment, vol 6. Elsevier, Amsterdam, pp 3–40

Min KS (2007) The physiological significance of metallothionein in oxidative stress. J Pharmaceut Soc Jpn 127(4):695–702

Morgan JE, Morgan AJ (1999) The accumulation of metals (Cd, Cu, Pb, Zn and Ca) by two ecologically contrasting earthworm species (*Lumbricus rubellus* and *Aporrectodea caliginosa*): implications for ecotoxicological testing. Appl Soil Ecol 13:9–20
Morgan AJ, Turner MP, Morgan JE (2002) Morphological plasticity in metal sequestering earthworm chloragocytes: morphometric electron microscopy provides a biomarker of exposure in field populations. Environ Toxicol Chem 21(3):610–618
Morgan AJ, Stürzenbaun SR, Winters C, Grime GW, Aziz NAA, Kille P (2004) Differential metallothionein expression in earthworm (*Lumbricus rubellus*) tissues. Ecotoxicol Environ Saf 57(1):11–19
Morrison DE, Robertson BK, Alexander M (2000) Bioavailability to earthworms of aged DDT, DDE, DDD, and dieldrin in soil. Environ Sci Technol 34(4):709–713
Nahmani J, Hodson ME, Devin S, Vijver MG (2009) Uptake kinetics of metals by the earthworm Eisenia fetida exposed to field-contaminated soils. Environ Pollut 157:2622–2628
Ndayibagira A, Sunahara GI, Robidoux PY (2007) Rapid isocratic HPLC quantification of metallothionein-like proteins as biomarkers for cadmium exposure in the earthworm *Eisenia andrei*. Soil Biol Biochem 39(1):194–201
Neuhauser EF, Cukic ZV, Malecki MR, Loehr RC, Durkin PR (1995) Bioconcentration and biokinetics of heavy metals in the earthworm. Environ Pollut 89:293–301
Novais SC, Gomes SIL, Gravato C, Guilhermino L, De Coen W, Soares AMVM, Amorim MJB (2011) Reproduction and biochemical responses in *Enchytraeus albidus* (Oligochaeta) to zinc or cadmium exposures. Environ Pollut 159(7):1836–1843
Paoletti MG (1999) The role of earthworms for assessment of sustainability and as bioindicators. Agric Ecosyst Environ 74:137–155
Peijnenburg WJGM (2002) Bioavailability of metals to soil invertebrates. In: HE Allen (ed) Bioavailability of metals in terrestrial ecosystems: importance of partitioning for bioavailability to invertebrates, microbes, and plants, Society of Environmental Toxicology and Chemistry (SETAC), Pensacola, pp 89–112. ISBN 978-1880611463
Peijnenburg WJGM, Vrijver MG (2009) Earthworms and their use in eco(toxico)logical modeling. In: Deviller J (ed) Ecotoxicology modeling. Springer, Heidelberg, pp 177–204
Peijnenburg WJGM, Posthuma L, Eijsackers HPJ, Allen HE (1997) A conceptual framework for implementation of bioavailability of metals for environmental management purposes. Ecotoxicol Environ Saf 37:163–172
Prinsloo MW, Reinecke SA, Przybylowicz WJ, Mesjas-Przybylowicz J, Reinecke AJ (1990) Micro-PIXE studies of Cd distribution in the nephridia of the earthworm *Eisenia fetida* (Oligochaeta). Nucl Instrum Methods Phys Res B 158:317–322
Rahtkens K, von der Trenck T (2006) Schwermetalle in Regenwurmern Baden-Wurttembergs. Umweltwissenschaften Schadstoff-Forschung 18(3):164–173
Rao JV, Pavan YS, Madhavendra SS (2003) Toxic effects of chlorpyrifos on morphology and acetylcholinesterase activity in the earthworm, *Eisenia foetida*. Ecotoxicol Environ Saf 54 (3):296–301
Rault M, Mazzia C, Capowiez Y (2007) Tissue distribution and characterization of cholinesterase activity in six earthworm species. Comp Biochem Physiol B Biochem Mol Biol 147:2340–2346
Reinecke SA, Reinecke AJ (2004) The comet assay as biomarker of heavy metal genotoxicity in earthworms. Arch Environ Contam Toxicol 46:208–215
Ribera D, Narbonne JF, Arnaud C, Saint-Denis M (2001) Biochemical responses of the earthworm *Eisenia foetida andrei* exposed to contaminated artificial soil, effects of carbaryl. Soil Biol Biochem 33:1123–1130
Scott-Fordsmand JJ, Weeks JM (2000) Biomarkers in earthworms. Rev Environ Contam Toxicol 165:117–159
Sforzini S, Saggese I, Oliveri L, Viarengo A, Bolognesi C (2010) Use of the comet and micronucleus assays for in vivo genotoxicity assessment in the coelomocytes of the earthworm *Eisenia andrei*. Comp Biochem Physiol Part A 157(1):S13

Spurgeon DJ, Hopkin SP (1995) Extrapolation of the laboratory based OECD earthworm toxicity test to metal-contaminated field sites. Ecotoxicology 4:190–205

Spurgeon DJ, Weeks JM, Van Gestel CAM (2003) A summary of eleven years progress in earthworm ecotoxicology: the 7th international symposium on earthworm ecology, Cardiff, Wales, 2002. Pedobiologia 47(5–6):588–606

Strzenbaum SR, Winters C, Galay M, Morgan AJ, Kille P (2001) Metal ion trafficking in earthworms. Identification of a cadmium-specific metallothionein. J Biol Chem 276 (36):34013–34018

Tischer S (2009) Earthworms (Lumbricidae) as bioindicators: the relationship between in-soil and in-tissue heavy metal content. Pol J Ecol 57:531–541

USDA-NRCS (2009) Earthworms. Soil quality indicator information sheet. http://soils.usda.gov/SQI/assessment/files/earthworms_sq_biological_indicator_sheet.pdf

van der Oost R, Beyer J, Vermeulen NPE (2003) Fish bioaccumulation and biomarkers in environmental risk assessment: a review. Environ Toxicol Pharmacol 13:57–149

Van Gestel CAM, Van Brummelen TC (1996) Incorporation of the biomarkers concept in ecotoxicology calls for a redefinition of terms. Ecotoxicology 5:217–225

Vasseur P, Bonnard M (2014) Ecogenotoxicology in earthworms: a review. Curr Zool 60:255–272

Vijver MG, Vink JPM, Miermans CJH, Van Gestel CAM (2003) Oral sealing using glue: a new method to distinguish between intestinal and dermal uptake of metals in earthworms. Soil Biol Biochem 35(1):125–132

Wallwork JA (1983) Annelids: the first coelomates. earthworms biology. Edward Arnold, London. ISBN 0713128844

Weeks JM (1995) The value of biomarkers for ecological risk assessement: academic toys of legislative tools? J Appl Ecol 2:215–216

Weeks JM, Svendsen C (1996) Neutral-red retention by lysosomes from earthworm coelomocytes: a simple biomarker for exposure of soil invertebrates. Environ Toxicol Chem 15:1801–1805

Zinkl JG, Lockhart WL, Kenny SA, Ward FJ (1991) The effects of cholinesterase inhibiting insecticides on fish. In: Mineau P (ed) Cholinesterase inhibiting insecticides: their impact on wildlife and the environment. Elsevier, Amsterdam, pp 233–254

Chapter 7
Emerging Metagenomic Strategies for Assessing Xenobiotic Contaminated Sites

Srujana Kathi

7.1 Introduction

Soil is a complex dynamic biological system where it is difficult to link specific taxa to metabolic processes (Burns et al. 2013). Soil metagenomics, which comprises isolation of soil DNA and the production and screening of clone libraries, can provide a cultivation-independent assessment of the largely untapped genetic reservoir of soil microbial communities (Daniel 2005). The term 'metagenome' was proposed by Handelsman et al. (1998) to describe the genomes of the total microbiota found in nature that can be understood as the whole collection of genomic information of all microorganisms in a given environment (Di Bella et al. 2013). The increasing availability of genes, genomes and metagenomes as well as the growing understanding of their functionalities are supported by considerations of microbial physiology, biochemistry, genetic regulation and engineering in pure strains or defined communities for implementation in bioremediation (Agathos and Boon 2015).

Metagenomics is a rapidly growing area of genome sciences that seeks to characterize the composition of microbial communities, their operations and their dynamically co-evolving relationships with the habitats even in unculturable environments (Franzosa et al. 2015; Taupp et al. 2011; Yong and Zhong 2010; Turnbaugh and Gordon 2008). Metagenomics involves sequencing the total DNA extracted from environmental samples (Thomas et al. 2015). Metagenomics offers the possibility to retrieve unknown sequences or functions from the environment, in contrast to methods relying on PCR amplification which are based on prior knowledge of gene sequences (Stenuit et al. 2008).

S. Kathi (✉)
Pondicherry University, Kalapet, India
e-mail: srujanaaramesh@gmail.com

© Springer International Publishing AG 2017

M.Z. Hashmi et al. (eds.), *Xenobiotics in the Soil Environment*, Soil Biology 49,
DOI 10.1007/978-3-319-47744-2_7

The primary objective of any metagenome sequencing project is the total characterization of a community, taxonomic breakdown and relative abundance of the various species and genic composition of each member of the community including number, functional capacity and intra-species/intra-population heterogeneity of the genes (Scholz et al. 2012). Metagenomics can be employed to identify the functional potential and the taxonomic identity of all organisms in an environmental sample, without leaving any information regarding the active members of the microbial community involved (El Amrani et al. 2015). In such a case, metagenomic techniques such as single isotope probing can be applicable for the identification of active members of the microbial community and associated genes essential for biodegradative processes (Uhlik et al. 2013).

Functional metagenomics includes screening of environmental-DNA libraries for enzymatic activities or metabolite synthesis (Tannieres et al. 2013). The application of metagenomics might aid in the isolation of novel catabolic pathways for degradation of xenobiotic compounds, indicating the functional genetic capacity for contaminant degradation and providing molecular tools useful for identification of the microbial taxa encoding the biodegradative gene (Kakirde et al. 2010). Metagenomic approaches enable to identify several novel genes encoding cellulolytic, pectinolytic, proteolytic and lipolytic enzymes and many new enzymes for screening and identification of unexplored microbial consortia involved in soil xenobiotic degradation (Bashir et al. 2014).

Xenobiotics are foreign compounds to living organisms whose molecules are not easily recognized by existing degradative enzymes and tend to accumulate in soil and water. Xenobiotics include polyaromatic, chlorinated and nitroaromatic compounds, known to be toxic, carcinogenic and mutagenic for living organisms (Eyers et al. 2004). The toxicity of these compounds for the environment and for biota results from their resistance to natural degradation owing to their structural complexity (Ufarte et al. 2015). During microbial degradation of xenobiotics, all changes in the chemical structure are due to the action of enzymes. These enzymes possess a broad range of specificity to accommodate several molecules of similar structure. If such enzymes are identified and isolated, they can be engineered by directed evolution to improve their efficiency with respect to a particular compound (Theerachat et al. 2012).

It is estimated that soil metagenome accommodates approximately 6000–10,000 *Escherichia coli* genomes in undisturbed organic soils and 350–1500 genomes in disturbed of which only 5 % has been cultured and studied in the laboratory (Desai and Madamwar 2007). Metagenomic analyses have enabled researchers to explore the previously uncultivable microorganisms and exploit their genetic potential in the bioremediation of contaminated soil (Martin et al. 2006; Malik et al. 2008; Simon and Daniel 2011). The metagenomic DNA of polluted environments is a potential genetic resource from which phylogenetic affiliation of uncultured bacterial species could be determined and their genetic potential can be tapped by identifying novel biocatalyst, xenobiotic and metal-detoxifying genes with utility in bioremediation processes (Meier et al. 2015, 2016; Desai and Madamwar 2007).

Predictive elative metabolic turnover (PRMT) converts metagenomic sequence data into relative metrics for the consumption or production of specific metabolites (Larsen et al. 2011).

Metagenomics lacks the tools to determine whether sufficient coverage is available for the type of analysis planned or whether one can interpret data of a certain depth for a community of a given complexity. Therefore, the standard low coverage in metagenomic studies generates a dataset that reflects a random subsampling of the genomic content of the individual community members (Desai et al. 2012).

7.2 Approaches to Metagenomics

There are different approaches to metagenomics: (1) shotgun metagenomics where all DNA is sheared and sequenced and functions and taxonomy are derived from homology search in databases, (2) activity-driven studies that are designed to search for specific microbial functions, (3) sequence-driven studies that link genome information with phylogenetic or functional marker genes of interest and (4) direct determination of the whole collection of genes within an environmental sample without constructing a metagenomic library (Suenaga 2012; Harismendy et al. 2009; Shendure and Ji 2008; Brulc et al. 2009). The basic steps involved in metagenomics of soil-bound xenobiotic compounds has been analysed by means of schematic process workflow (Fig. 7.1).

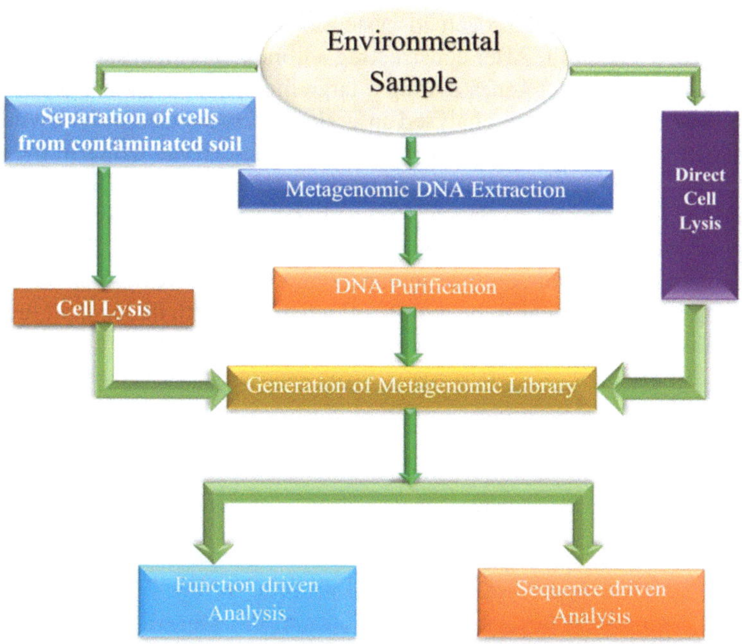

Fig. 7.1 A schematic workflow of steps involved in soil xenobiotic metagenomics

The four metagenomic approaches described above based on their random and directed sequencing strategies can be characterized as unselective (shotgun analysis and next-generation sequencing) and targeted (activity-driven and sequence-driven studies) metagenomics, respectively. Unselective metagenomics is a simple and cost-effective DNA sequencing option (Chen and Pachter 2005). The number of metagenomic projects has exploded in recent years, and hundreds of environmental samples have been unravelled by shotgun sequencing (Ivanova et al. 2010). Whole-metagenome shotgun sequencing and amplicon sequencing have been applied to study diverse microbiomes, ranging from natural environments to the built environment and the human body (Tyson et al. 2004). In addition to enrichment culture approaches, isolated environmental DNA can be subjected to whole genome amplification, that is, multiple displacement amplification (MDA) to provide sufficient genetic substrate for library production (Taupp et al. 2011).

A shotgun metagenomic approach relies on sequencing of total DNA extracted from a given sample, without prior cloning into a vector (Jansson 2015). The application of this approach involves the design of PCR primers or hybridization probes for the target genes that are derived from conserved regions of already known protein families, which a priori limits the chances for obtaining fundamentally new proteins (Ferrer et al. 2009). The activity-based approach involves construction of small to large insert expression libraries, especially those made in lambda phage, cosmid or copy-control fosmid vectors, which are further implemented for a direct activity screening (Lorenz and Eck 2005). Three different function-driven approaches have been used to recover novel biomolecules: phenotypical detection of the desired activity heterologous complementation of host strains or mutants and induced gene expression (Simon and Daniel 2011). The major limitation of this approach to systems microbiology is that metagenomic libraries have a size limit. After constructing a library, a critical step is to screen for clones that contain target genes among a large number of clones. Here, dozens of thousands of clones may be analysed in a single screen. Certainly, owing to the limitation of efficient expression of the metagenome-derived genes in the selected host, the numbers of positive clones will not be high. Furthermore, in activity-based screening, it is necessary to develop specialized screening systems to detect the activity of the products of the gene of interest (Ferrer et al. 2009). Targeted metagenomic studies that combine metagenomic library screening and subsequent sequencing analysis appear to be a more effective means to understand the content and composition of genes for key ecological processes in microbial communities (Suenaga 2012).

7.3 Metagenomics of Xenobiotics

7.3.1 PAHs

Polycyclic aromatic hydrocarbons (PAHs) are ubiquitous, persistent and toxic organic compounds in the environment (Cao et al. 2015; Srujana and Khan 2012). The advent of metagenomic approaches has revealed a higher degree of diversity in the degradation pathways and enzymes (Suenaga et al. 2007; Brennerova et al. 2009). Using a function-driven metagenomic approach, Sierra-Garcia et al. (2014) reported metagenomic fragments comprising of genes belonging to different pathways, showing novel gene arrangements. These results reinforce the potential of the metagenomic approach for the identification and elucidation of new genes and pathways in poorly studied environments and contribute to a broader perspective on the hydrocarbon.

Ring-hydroxylating dioxygenases/oxygenases (RHDs) play a crucial role in the biodegradation of a range of aromatic hydrocarbons found on polluted sites, including PAHs (Chemerys et al. 2014; Peng et al. 2010). RHDs are multicomponent metalloenzymes, which catalyse the first step in the bacterial degradation of various aromatic hydrocarbons (Jouanneau et al. 2011). Hydroxylation of an aromatic ring is the essential catalytic reaction for aromatic-ring degradation by bacteria in nature. Mostly, the hydroxylation is catalysed by an oxygenase family, Rieske oxygenase (RO). ROs catalyse a broad range of aromatic-ring compounds including mono- and polycyclic aromatic and hetero-aromatic compounds that are composed of terminal oxygenase and electron transfer components. The terminal oxygenase component has a Rieske cluster as a redox component that receives electrons from the electron transfer components and mononuclear iron as a catalytic site for dioxygen activation (Inoue and Nojiri 2014). The *nah* genes for PAH catabolism of *Pseudomonas* strains are highly homologous and usually organized in two operons: the upper *nah*1, which control initial oxidation of naphthalene and subsequent degradation to salicylate, and the *nah2* operon for salicylate oxidation. However, the location of both operons and their relative expression may vary. New variants of salicylate hydroxylase genes were found. Also isofunctional genes for salicylate oxidation could be often detected within one *Pseudomonas* strain. The unique genetic organization is described for *P. putida* AK5 which degrades PAH via salicylate and gentisate, combining 'classical' n*ahl* operon and newly described *sgp*-operon (Boronin et al. 2010).

In addition to the enzyme-encoding genes involved in aromatic compound degradation, metagenomic libraries were screened earlier for regulatory elements that sense aromatic compounds (Suenaga et al. 2009). The implementation of stable-isotope probing (SIP) to track PAH degraders led to the detection of novel bacteria with remarkable biodegradation potential. SIP approaches also exposed the affiliation of uncultured microorganisms with PAH-degrading bacteria identified in contaminated soils (Chemerys et al. 2014; Uhlik et al. 2012; Singleton et al. 2005). Fluorescence-based reporter assay system termed as substrate-induced gene expression (SIGEX) can be used for identification of transcriptional regulators that sense benzoate and naphthalene (Uchiyama and Miyazaki 2013).

Using a metagenomic approach, microbial communities were monitored in Alert biopiles over time to identify microorganisms and functional genes linked to the high hydrocarbon degradation rates in soils undergoing treatment using an unbiased, culture and PCR-independent method. *Pseudomonas* sp. expressing hydrocarbon-degrading genes were most abundant in diesel-contaminated Canadian High Arctic soils. After sequencing the metagenome of soil biopiles through a time course, the results were compared with uncontaminated soil and then quantified the expression and the abundance of key functional genes for abundant microorganisms identified in the metagenomic datasets (Yergeau et al. 2012). A culture-independent approach to assess the microbial aerobic catabolome for PAH degradation was used to study the microbial community of a PAH-contaminated soil subjected to 10 years of in situ bioremediation, basing on Illumina-based deep sequencing of amplicons targeting the V5–V6 region of 16S rRNA gene. A metagenomic library was prepared in pCCFos and 425,000 clones subjected to activity-based screening for key catabolic ring-cleavage activities using 2,3-dihydroxybiphenyl as a substrate. Since most of the genes encoding extradiol ring-cleavage enzymes on 672 fosmids could not be identified using primers based on currently available sequence information, 200 fosmid inserts were sequenced using the Illumina technology. Manually curated databases for catabolic key gene families involved in degradation of aromatics were developed named as AromaDeg to overcome the misannotations in databases. Sequence information of the fosmid inserts revealed not only the presence of novel extradiol dioxygenase genes but also additional key genes of aromatic metabolic pathways only distantly related to previously described variants (Duarte 2014).

An et al. (2013) reported that 160 microbial community compositions were compared in ten hydrocarbon resource environments (HREs) and sequenced 12 metagenomes to characterize their metabolic potential. In addition to common anaerobic communities, cores from oil sands and coal beds had unexpectedly high proportions of aerobic hydrocarbon-degrading bacteria. Likewise, most metagenomes had high proportions of genes for enzymes involved in aerobic hydrocarbon metabolism. Time-course analysis of microbial communities using a combination of metagenomics with metatranscriptomics and metaproteomics and stable-isotope probing technique will greatly contribute to the evaluation of the ecological functions of microbial genes at the community level (Muller et al. 2014; Kato et al. 2015). Loviso et al. (2015) investigated the potential to degrade PAHs of yet-to-be-cultured bacterial populations from chronically polluted intertidal sediments. They identified uncultured micro-organism having the potential to degrade aromatic hydrocarbons with various chemical structures thereby providing valuable information for the design of environmental molecular diagnostic tools for biotechnological application of RHO enzymes. When spatial and temporal variations of microbial communities and reconstructed metagenomes along the rice rhizosphere gradient during PAHs degradation were investigated, distance from root surface and PAH concentrations were found to affect the microbial communities and metagenomes in rice rhizosphere. The abundance of dioxygenase genes relating to PAH degradation in metagenomes mirrored the PAH degradation potential in rice rhizosphere (Ma et al. 2015).

7.3.2 *Organochlorinated Compounds*

Biphenyl dioxygenase (BphA) is a key enzyme in the aerobic catabolism of PCBs which carries out the initial attack on the inert aromatic nucleus. It belongs to class II of aryl-hydroxylating dioxygenases (ARHDOs) that typically hydroxylate substituted benzenes, like toluenes and biphenyls. This enzyme represents a catabolic bottleneck, as its substrate range is typically narrower than that of subsequent pathway enzymes. Metagenomic approaches can be applied to demonstrate the feasibility of the applied approach to functionally characterize dioxygenase activities of soil metagenomes via amplification of incomplete genes (Standfuß-Gabisch et al. 2012).

γ-Hexachlorocyclohexane also known as lindane (γ-HCH/γ-BHC) is a xenobiotic halogenated insecticide that was previously used worldwide, and this compound still remains in the environment and causes serious environmental concern (Vijgen et al. 2011). Activity-based screening techniques were applied to clone a gene-encoding γHCH dehydrochlorinase with its flanking regions from a cosmid-based library of DNA that was extracted from a γHCH-added suspension of HCH-contaminated soil. A total of 11 cosmid clones showing the γHCH dehydrochlorinase activity were obtained through the screenings. All the clones had a *linA* gene identical to known one, but its flanking regions showed some structural variations with known ones, suggesting high likelihood of genetic divergence in the *linA* flanking regions (Ito et al. 2012).

7.3.3 *Nitroaromatics*

Nitroaromatic compounds such as nitrobenzene or nitrotoluene are widely used as pesticides, dyes, polymers or explosives and are considered as priority pollutants (Kulkarni and Chaudhari 2007). The two main explosives, 2,4,6-trinitrotoluene (TNT) and hexahydro-1,3,5-trinitro-1,3,5-triazine [Royal Demolition Explosive (RDX)], are major nitroaromatic environmental pollutants and present distinct problems for bioremediation (Rylott et al. 2011). Rational design of enzymatic activity has been used to improve the degradation of nitroaromatic compounds. Nitrobenzene 1,2-dioxygenase catalyses the conversion of nitrobenzene to catechol and nitrite. The residues near the active site of this enzyme were modified for controlling substrate specificity. The substitution of amino acid at the position 293 (F293Q) expanded substrate specificity, resulting in 2.5-fold faster oxidization rate against 2,6-dinitrotoluene (Singh et al. 2008). Lee et al. (2005) reported that the residues of 2,6-dinitrotoluene near active sites chosen for site-directed mutagenesis and the replacement at the position 258 significantly changed the enantiospecificity.

Biodegradation of para-nitrophenol (PNP) proceeds via two distinct pathways in *Burkholderia sp.* strain SJ98, having 1,2,3-benzenetriol (BT) and hydroquinone (HQ) as their respective terminal aromatic intermediates. A ~41 kb fragment from the genomic library of *Burkholderia sp.* strain SJ98 has been sequenced and analysed. This DNA fragment was found to harbour all the lower pathway genes. Later the whole genome of strain SJ98 was sequenced and annotated and found two ORFs (viz. pnpA and pnpB) showing maximum identity at amino acid level with *p*-nitrophenol 4-monooxygenase (PnpM) and *p*-benzoquinone reductase. This is the first report for studying the genes for PNP degradation in strain SJ98 which are found to be arranged differentially in the form of non-contiguous gene clusters (Vikram et al. 2013).

7.4 Challenges and Future Prospects

Although metagenomics is revealing new information about phylogenetic and functional genes in some soils, it is not possible to adopt the information available to date to all soils (Jansson 2015). Owing to the complexity and heterogeneity of the biotic and abiotic components of soil ecosystems, the construction and screening of soil-based libraries is difficult and challenging (Daniel 2005). Soil metagenomics is susceptible to limitations that are common to all molecular techniques. Soil DNA extraction procedures are not fully efficient, where adsorption of cells and the adherence of DNA onto soil components cause losses of genetic information, and the DNA exploitation techniques currently in use provide access mainly to populations that dominate in soil (Lombard et al. 2011). Metagenomics provides little information on quantitative physiological characteristics such as maximum specific growth rate, saturation constant, pH, temperature for growth, susceptibility to predation and the speed of recovery after starvation. It is also difficult to draw meaningful information from correlations between the physicochemical characteristics of soil and metagenomic data (Prosser 2015).

Data analysis is the key limiting factor in metagenomic studies as increased data volumes are posing significant challenges to the existing analysis tools and indeed to the community providing analysis systems. This growth in dataset size, along with computational complexity of analysis, has left the metagenomics community in an unsustainable position, in terms of both financial cost and feasibility of analysis itself (Desai et al. 2012). While metagenome sequencing can provide useful estimates of the relative change in abundance of specific genes and taxa between environments or over time, this does not investigate the relative changes in the production or consumption of different metabolites (Larsen et al. 2011). Adapted sampling strategies and the combination of DNA extraction methods can help to recover these minority populations, which are normally masked by the dominant ones. Pyrosequencing and Illumina/Solexa technologies also offers a chance to access the rare biosphere, but is still concerned by the overriding effect of the dominant biota. Methods such as prior separation of particular minority cells via flow cytometry, or separation/fractionation of DNA by G+C % or in a SIP-based approach, will certainly help to tease out specific minority populations (Lombard et al. 2011).

Acknowledgements The author would like to thank Pondicherry University for providing Internet access to collect the information essential for writing this book chapter.

References

Agathos SN, Boon N (2015) Editorial overview: environmental biotechnology. Curr Opin Biotechnol 33:5–7

An D, Caffrey SM, Soh J, Agrawa A, Brown D, Budwill K, Dong X, Dunfield PF, Foght J, Gieg LM, Hallam SJ (2013) Metagenomics of hydrocarbon resource environments indicates aerobic taxa and genes to be unexpectedly common. Environ Sci Technol 47(18):10708–10717

Bashir Y, Singh P, Kumar KB (2014) Metagenomics: an application based perspective. Chin J Biotechnol 1–7. doi:10.1155/2014/146030

Boronin AM, Kosheleva IA, Sokolov SL, Izmalkova TY, Sazonova OI (2010) Variability of plasmids and PAH degradation genetic systems of pseudomonas. J Biotechnol 150:45–51

Brennerova MV, Josefiova J, Brenner V, Pieper DH, Junca H (2009) Metagenomics reveals diversity and abundance of meta-cleavage pathways in microbial communities from soil highly contaminated with jet fuel under air-sparging bioremediation. Environ Microbiol 11(9): 2216–2227

Brulc JM, Antonopoulos DA, Miller MEB, Wilson MK, Yannarell AC, Dinsdale EA, Edwards RE, Frank ED, Emerson JB, Wacklin P, Coutinho PM (2009) Gene-centric metagenomics of the fiber-adherent bovine rumen microbiome reveals forage specific glycoside hydrolases. Proc Natl Acad Sci 106(6):1948–1953

Burns RG, DeForest JL, Marxsen J, Sinsabaugh RL, Stromberger ME, Wallenstein MD, Weintraub MN, Zoppini A (2013) Soil enzymes in a changing environment: current knowledge and future directions. Soil Biol Biochem 58:216–234

Cao J, Lai Q, Yuan J, Shao Z (2015) Genomic and metabolic analysis of fluoranthene degradation pathway in *Celeribacter indicus* P73T. Sci Rep 5:67–71

Chemerys A, Pelletier E, Cruaud C, Martin F, Violet F, Jouanneau Y (2014) Characterization of novel polycyclic aromatic hydrocarbon dioxygenases from the bacterial metagenomic DNA of a contaminated soil. Appl Environ Microbiol 80(21):6591–6600

Chen K, Pachter L (2005) Bioinformatics for whole-genome shotgun sequencing of microbial communities. PLoS Comput Biol 1(2):e24

Daniel R (2005) The metagenomics of soil. Nat Rev Microbiol 3(6):470–478

Desai C, Madamwar D (2007) Extraction of inhibitor-free metagenomic DNA from polluted sediments, compatible with molecular diversity analysis using adsorption and ion-exchange treatments. Bioresour Technol 98(4):761–768

Desai N, Antonopoulos D, Gilbert JA, Glass EM, Meyer F (2012) From genomics to metagenomics. Curr Opin Biotechnol 23(1):72–76

Di Bella JM, Bao Y, Gloor GB, Burton JP, Reid G (2013) High throughput sequencing methods and analysis for microbiome research. J Microbiol Method 95(3):401–414

Duarte M (2014) Functional soil metagenomics: elucidation of polycyclic aromatic hydrocarbon degradation potential after 10 years of in situ bioremediation. In: Annual meeting and exhibition 2014, Simb, 20–24 July 2014

El Amrani A, Dumas AS, Wick LY, Yergeau E, Berthomé R (2015) "Omics" insights into PAH degradation toward improved green remediation biotechnologies. Environ Sci Technol 49(19): 11281–11291

Eyers L, George I, Schuler L, Stenuit B, Agathos SN, El Fantroussi S (2004) Environmental genomics: exploring the unmined richness of microbes to degrade xenobiotics. Appl Microbiol Biotechnol 66(2):123–130

Ferrer M, Beloqui A, Timmis KN, Golyshin PN (2009) Metagenomics for mining new genetic resources of microbial communities. J Mol Microbiol Biotechnol 16(1–2):109–123

Franzosa EA, Hsu T, Sirota-Madi A, Shafquat A, Abu-Ali G, Morgan XC, Huttenhower C (2015) Sequencing and beyond: integrating molecular 'omics' for microbial community profiling. Nat Rev Microbiol 13(6):360–372

Handelsman J, Rondon MR, Brady SF, Clardy J, Goodman RM (1998) Molecular biological access to the chemistry of unknown soil microbes: a new frontier for natural products. Chem Biol 5(10):R245–R249

Harismendy O, Ng PC, Strausberg RL, Wang X, Stockwell TB, Beeson KY, Schork NJ, Murray SS, Topol EJ, Levy S, Frazer KA (2009) Evaluation of next generation sequencing platforms for population targeted sequencing studies. Genome Biol 10(3):R32

Inoue K, Nojiri H (2014) Structure and function of aromatic-ring hydroxylating dioxygenase system. In: Biodegradative bacteria. Springer, Tokyo, pp 181–205

Ito M, Ono A, Ohtsubo Y, Tsuda M, Nagata Y (2012) Cloning of γ-hexachlorocyclohexane dehydrochlorinase gene with its flanking regions from soil by activity-based screening techniques. Eur J Soil Biol 52:16–19

Ivanova N, Tringe SG, Liolios K, Liu WT, Morrison N, Hugenholtz P, Kyrpides NC (2010) A call for standardized classification of metagenome projects. Environ Microbiol 12(7):1803–1805

Jansson J (2015) Soil metagenomics. Encyclopedia of metagenomics: environmental metagenomics. Springer, New York, pp. 600–609

Jouanneau Y, Martin F, Krivobok S, Willison JC (2011) Ring-hydroxylating dioxygenases involved in PAH biodegradation: structure, function, biodiversity. Microbial bioremediation of non-metals: current research. Caister Academic, Norflok, pp. 149–175

Kakirde KS, Parsley C, Liles MR (2010) Size does matter: application-driven approaches for soil metagenomics. Soil Biol Biochem 42(11):1911–1923

Kato H, Mori H, Maruyama F, Toyoda A, Oshima K, Endo R, Fuchu G, Miyakoshi M, Dozono A, Ohtsubo Y, Nagata Y (2015) Time-series metagenomic analysis reveals robustness of soil microbiome against chemical disturbance. DNA Res 22(6):413–424

Kulkarni M, Chaudhari A (2007) Microbial remediation of nitro-aromatic compounds: an overview. J Environ Manage 85(2):496–512

Larsen PE, Collart FR, Field D, Meyer F, Keegan KP, Henry CS, McGrath J, Quinn J, Gilbert JA (2011) Predicted relative metabolomic turnover (PRMT): determining metabolic turnover from a coastal marine metagenomic dataset. Microb Inf Exp 1(1):1–11

Lee KS, Parales JV, Friemann R, Parales RE (2005) Active site residues controlling substrate specificity in 2-nitrotoluene dioxygenase from *Acidovorax* sp. strain JS42. J Ind Microbiol Biotechnol 32(10):465–473

Lombard N, Prestat E, van Elsas JD, Simonet P (2011) Soil-specific limitations for access and analysis of soil microbial communities by metagenomics. FEMS Microbiol Ecol 78(1):31–49

Lorenz P, Eck J (2005) Metagenomics and industrial applications. Nat Rev Microbiol 3(6): 510–516

Loviso CL, Lozada M, Guibert LM, Musumeci MA, Sarango CS, Kuin RV, Marcos MS, Dionisi HM (2015) Metagenomics reveals the high polycyclic aromatic hydrocarbon-degradation potential of abundant uncultured bacteria from chronically polluted subantarctic and temperate coastal marine environments. J Appl Microbiol 119(2):411–424

Ma B, Lyu XF, Zha T, Gong J, He Y, Xu JM (2015) Reconstructed metagenomes reveal changes of microbial functional profiling during PAHs degradation along a rice (*Oryza sativa*) rhizosphere gradient. J Appl Microbiol 118(4):890–900

Malik S, Beer M, Megharaj M, Naidu R (2008) The use of molecular techniques to characterize the microbial communities in contaminated soil and water. Environ Int 34(2):265–276

Martin HG, Ivanova N, Kunin V, Warnecke F, Barry KW, McHardy AC, Yeates C, He S, Salamov AA, Szeto E, Dalin E (2006) Metagenomic analysis of two enhanced biological phosphorus removal (EBPR) sludge communities. Nat Biotechnol 24(10):1263–1269

Meier MJ, Paterson ES, Lambert IB (2015) Metagenomic analysis of an aromatic hydrocarbon contaminated soil using substrate-induced gene expression. Appl Environ Microbiol 82(3): 897–909

Meier MJ, Paterson ES, Lambert IB (2016) Use of substrate-induced gene expression in metagenomic analysis of an aromatic hydrocarbon-contaminated soil. Appl Environ Microbiol 82(3):897–909

Muller EE, Pinel N, Laczny CC, Hoopmann MR, Narayanasamy S, Lebrun LA, Roume H, Lin J, May P, Hicks ND, Heintz-Buschart A (2014) Community-integrated omics links dominance of a microbial generalist to fine-tuned resource usage. Nat Commun 5:5603

Peng RH, Xiong AS, Xue Y, Fu XY, Gao F, Zhao W, Tian YS, Yao QH (2010) A profile of ring-hydroxylating oxygenases that degrade aromatic pollutants. Rev Environ Contam Toxicol 206:65–94 (Springer: New York)

Prosser JI (2015) Dispersing misconceptions and identifying opportunities for the use of 'omics' in soil microbial ecology. Nat Rev Microbiol 13(7):439–446

Rylott EL, Lorenz A, Bruce NC (2011) Biodegradation and biotransformation of explosives. Curr Opin Biotechnol 22(3):434–440

Scholz MB, Lo CC, Chain PS (2012) Next generation sequencing and bioinformatic bottlenecks: the current state of metagenomic data analysis. Curr Opin Biotechnol 23(1):9–15

Shendure J, Ji H (2008) Next-generation DNA sequencing. Nat Biotechnol 26(10):1135–1145

Sierra-Garcia IN, Alvarez JC, de Vasconcellos SP, de Souza AP, dos Santos Neto EV, de Oliveira VM (2014) New hydrocarbon degradation pathways in the microbial metagenome from Brazilian petroleum reservoirs. PLoS One 9(2):e90087

Simon C, Daniel R (2011) Metagenomic analyses: past and future trends. Appl Environ Microbiol 77(4):1153–1161

Singh S, Kang SH, Mulchandani A, Chen W (2008) Bioremediation: environmental clean-up through pathway engineering. Curr Opin Biotechnol 19(5):437–444

Singleton DR, Powel SN, Sangaiah R, Gold A, Ball LM, Aitken MD (2005) Stable-isotope probing of bacteria capable of degrading salicylate, naphthalene, or phenanthrene in a bioreactor treating contaminated soil. Appl Environ Microbiol 71(3):1202–1209

Srujana K, Khan AB (2012) Isolation and characterisation of polycyclic aromatic hydrocarbon degrading soil microbes from automobile workshop sediments. J Environ Sci Technol 5(1): 74–83

Standfuß-Gabisch C, Al-Halbouni D, Hofer B (2012) Characterization of biphenyl dioxygenase sequences and activities encoded by the metagenomes of highly polychlorobiphenyl-contaminated soils. Appl Environ Microbiol 78(8):2706–2715

Stenuit B, Eyers L, Schuler L, Agathos SN, George I (2008) Emerging high-throughput approaches to analyze bioremediation of sites contaminated with hazardous and/or recalcitrant wastes. Biotechnol Adv 26(6):561–575

Suenaga H (2012) Targeted metagenomics: a high-resolution metagenomics approach for specific gene clusters in complex microbial communities. Environ Microbiol 14(1):13–22

Suenaga H, Ohnuki T, Miyazaki K (2007) Functional screening of a metagenomic library for genes involved in microbial degradation of aromatic compounds. Environ Microbiol 9(9): 2289–2297

Suenaga H, Mizuta S, Miyazaki K (2009) The molecular basis for adaptive evolution in novel extradiol dioxygenases retrieved from the metagenome. FEMS Microbiol Ecol 69(3):472–480

Tannieres M, Beury-Cirou A, Vigouroux A, Mondy S, Pellissier F, Dessaux Y, Faure D (2013) A metagenomic study highlights phylogenetic proximity of quorum-quenching and xenobiotic-degrading amidases of the AS-family. PLoS One 8(6):e65473

Taupp M, Mewis K, Hallam SJ (2011) The art and design of functional metagenomic screens. Curr Opin Biotechnol 22(3):465–472

Theerachat M, Emond S, Cambon E, Bordes F, Marty A, Nicaud JM, Chulalaksananukul W, Guieysse D, Remaud-Siméon M, Morel S (2012) Engineering and production of laccase from Trametes versicolor in the yeast *Yarrowia lipolytica*. Bioresour Technol 125:267–274

Thomas T, Gilbert J, Meyer F (2015) A 123 of metagenomics. In: Nelson KE (ed) Encyclopedia of metagenomics. Springer, New York, pp 1–9

Turnbaugh PJ, Gordon JI (2008) An invitation to the marriage of metagenomics and metabolomics. Cell 134(5):708–713

Tyson GW, Chapman J, Hugenholtz P, Allen EE, Ram RJ, Richardson PM, Solovyev VV, Rubin EM, Rokhsar DS, Banfield JF (2004) Community structure and metabolism through reconstruction of microbial genomes from the environment. Nature 428(6978):37–43

Uchiyama T, Miyazaki K (2013) Metagenomic screening for aromatic compound-responsive transcriptional regulators. PLoS One 8(9):e75795

Ufarte L, Laville E, Duquesne S, Potocki-Veronese G (2015) Metagenomics for the discovery of pollutant degrading enzymes. Biotechnol Adv 33(8):1845–1854

Uhlik O, Wald J, Strejcek M, Musilova L, Ridl J, Hroudova M, Vlcek C, Cardenas E, Mackova M, Macek T (2012) Identification of bacteria utilizing biphenyl, benzoate, and naphthalene in long-term contaminated soil. PLoS One 7(7):e40653

Uhlik O, Leewis MC, Strejcek M, Musilova L, Mackova M, Leigh MB, Macek T (2013) Stable isotope probing in the metagenomics era: a bridge towards improved bioremediation. Biotechnol Adv 31(2):154–165

Vijgen J, Abhilash PC, Li YF, Lal R, Forter M, Torres J, Singh N, Yunus M, Tian C, Schäffer A, Weber R (2011) Hexachlorocyclohexane (HCH) as new Stockholm convention POPs—a global perspective on the management of lindane and its waste isomers. Environ Sci Pollut Res 18(2):152–162

Vikram S, Pandey J, Kumar S, Raghava GPS (2013) Genes involved in degradation of para-nitrophenol are differentially arranged in form of non-contiguous gene clusters in *Burkholderia* sp. strain SJ98. PLoS One 8(12):e84766

Yergeau E, Sanschagrin S, Beaumier D, Greer CW (2012) Metagenomic analysis of the bioremediation of diesel-contaminated Canadian high arctic soils. PLoS One 7:e30058

Yong YC, Zhong JJ (2010) Recent advances in biodegradation in China: new microorganisms and pathways, biodegradation engineering, and bioenergy from pollutant biodegradation. Process Biochem 45(12):1937–1943

Chapter 8
An Overview of Extraction, Clean-up and Instrumentation Techniques for Quantification of Soil-Bound Xenobiotic Compounds

Srujana Kathi

8.1 Introduction

Soils and sediments are a major sink for organic pollutants (Riefer et al. 2011). Depending on physical and chemical characteristics of both the contaminants and the macromolecular geopolymers, there are different types of incorporation and binding modes. Adsorbed or reversibly linked xenobiotics by van der Waals forces and ionic bonds can mostly be separated by exhaustive solvent extraction or changes in the extracting conditions. Stronger incorporation occurs through sequestration and aging by forming covalent bonds (Kalathoor et al. 2015). The determination of xenobiotic chemicals in sediments and other environmental solid matrices requires efficient extraction and sensitive analytical methods owing to the concentrations ranging from a few ng/g to several μg/g (Pintado-Herrera et al. 2016). Rapid, reliable analytical methods are necessary to monitor and to control the widespread distribution of emerging contaminants in the environment. Sample preparation is a key aspect of analysis at trace levels, and it is often the most time-consuming and least sophisticated step of the analytical procedure (Albero et al. 2015).

Generally, sample pretreatment comprises solvent extraction, extract concentration, chromatography clean-up, eluent concentration and injection into an instrument (GC/LC) for analysis (Szulejko et al. 2014). Sample preparation of target compounds from biological, pharmaceutical, environmental and food matrices is one of the most time-consuming steps in analytical procedures along with multistep extraction followed by clean-up procedures. Extraction techniques are based on the processes running on the phase such as liquid–liquid and liquid–solid (Buszewski and Szultka 2012). Extraction from soil matrices is a process in which solutes

S. Kathi (✉)
Pondicherry University, Kalapet, India
e-mail: srujanaaramesh@gmail.com

© Springer International Publishing AG 2017
M.Z. Hashmi et al. (eds.), *Xenobiotics in the Soil Environment*, Soil Biology 49,
DOI 10.1007/978-3-319-47744-2_8

desorb from the sample matrix and then dissolve into a selected solvent. The analytical determination of organic pollutants in solid matrixes such as soils and sediments presents complex operations of sample preparation, mainly due to the difficulty of quantitatively leaching the analyte from the solid sample due to very strong interactions established between analytes and solid matrix (Richter et al. 2006). A number of modern solid–liquid extraction techniques have been described in recent years in an attempt to increase extraction efficiency, decrease the organic solvent consumption and increase sample throughput (Tadeo et al. 2010). Traditional sample preparation methods include Soxhlet extraction (USEPA 3540), automated Soxhlet extraction (USEPA 3541 ultrasonic extraction) (USEPA 3550C) and microwave-assisted extraction (MAE) (USEPA 3556) (Souza-Silva et al. 2015; Sanchez-Prado et al. (2010) and accelerated solvent extraction (ASE) and supercritical fluid extraction (SFE) (USEPA 3560, 3561, 3562) (Rouvière et al. 2012)). Recently MAE combined with solvent bar has emerged as green and effective method (Guo et al. 2013).

Quantification of contaminants in soil extracts can be achieved by employing gas chromatography–mass spectrometry (GC–MS), gas chromatography–tandem mass spectrometry (GC–MS/MS), comprehensive two-dimensional gas chromatography coupled to microelectron capture detection (GC × GC–μECD), liquid chromatography–mass spectrometry (LC–MS) or liquid chromatography–tandem mass spectrometry (LC–MS/MS) (Albero et al. 2012). LC–MS analysis can be performed using several ionisation modes and is used if the focus is on HMW PAH or if the sample is liquid and direct infusion is preferred to sample preparation. Among the available ionisation techniques, atmospheric pressure chemical ionisation (APCI, suited for semipolar compounds) has been used for PAH analysis where the limit of detection (LOD) using APCI is 200 pg for benzo[a]pyrene (Stader et al. 2013).

8.2 Sample Pretreatment

The quality of sample preparation is a key factor in determining the success of chemical analysis. The objective of this challenging and critical step is to transfer the analyte into a form that is pre-purified, concentrated and compatible with the analytical system (Saleh et al. 2009). An ideal sample preparation methodology should be accurate, fast and precise and should consume little solvent. Modern extraction methods should also meet other demands such as maintaining sample integrity, high throughput and compatibility with subsequent analysis. Additionally, they should be easily adapted for fieldwork requiring low-cost materials (Rodriguez-Mozaz et al. 2007). Sample pretreatment involves sample homogenisation followed by sample drying. The homogenisation step aids to increase the surface area of the sample, enhancing the accessibility of bound analytes in sample matrices to the solvents (Runnqvist et al. 2010). Sample drying can be achieved with air-drying, lyophilisation and mixing the sample with common dehydrating agents, such as anhydrous sodium sulphate and diatomaceous earth/cellulose (Cocco et al. 2011). The removal of moisture from homogenised samples prior to extraction enhances the extraction and helps to eliminate the extraction variability among samples with different moisture contents, especially when nonpolar extraction solvents are used (Subedi et al. 2015).

In the process of the development of efficient, economical and miniaturised sample preparation methods, solid-phase microextraction (SPME) (Xu et al. 2013) and liquid–liquid microextraction (LLME) have been developed. SPME is a solvent-free extraction technique that incorporates sample pretreatment, concentration and sample introduction all into a single procedure. LLME is a single-step extraction method that requires a very high sample to solvent ratio which leads to a higher enrichment factor of analytes. So, conventional LLME has been proposed in several USEPA methods as an efficient alternative to liquid–liquid extraction (LLE) (Saleh et al. 2009). LLE is among the oldest and more widespread techniques for the extraction of a wide range of organic pollutants. However, LLE is time-consuming, requires large amounts of organic solvents that are possibly toxic and is difficult to systematise (Regueiro et al. 2008).

8.3 Solvent Extraction

Majority of xenobiotic compounds are lipophilic in nature; their extraction methods are based on the isolation from the sample matrix, controlled both kinetically and thermodynamically. The most commonly used extraction methods are Soxhlet, pressurised liquid extraction (also known as accelerated solvent extraction (ASE)), supercritical fluid extraction (SFE), matrix solid-phase dispersion (MSPD), microwave-assisted extraction (MAE), ultrasonic-assisted extraction (UAE), liquid–liquid extraction (LLE), solid-phase extraction (SPE) and SPME (Guo and Kannan 2015) (Fig. 8.1).

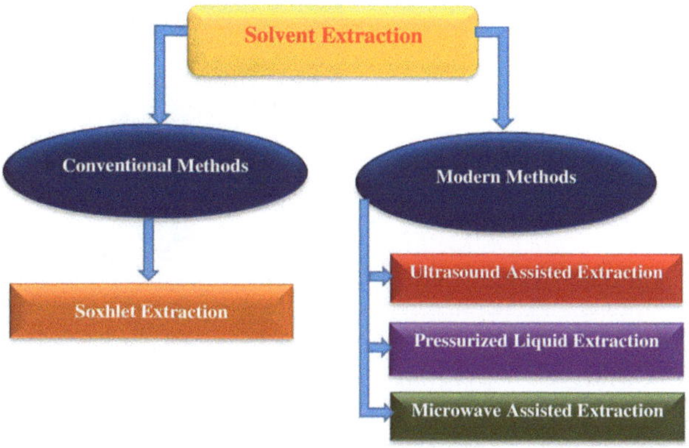

Fig. 8.1 An overview of solvent extraction methods

8.3.1 Soxhlet Extraction

In 1879, von Soxhlet developed a Soxhlet extractor which has for a long time been the most widely used leaching technique (De Castro and Priego-Capote 2010). Soxhlet extraction has been the traditional method used for extraction of xenobiotics from soils and sediments requiring large amounts of solvent and are often carried out for 20 h or more (Sporring et al. 2005). The fastidiousness of high-pressure Soxhlet extraction is that the extractants do not reach supercritical conditions. This can be achieved by placing the extractor in a cylindrical stainless-steel autoclave or by the use of either commercial or laboratory-made supercritical fluid–Soxhlet extractors (De Castro and Priego-Capote 2010) (Fig. 8.2).

Fig. 8.2 Soxhlet extraction unit

8.3.2 Ultrasound-Assisted Extraction

Ultrasound-assisted extraction (UAE) is an inexpensive and efficient alternative to classical extraction techniques. Another advantage of UAE over Soxhlet is that UAE extracts thermolabile analytes that may be altered in the working conditions of Soxhlet. Extraction of analytes mainly depends on the polarity of the solvent, the nature and the homogeneity of the sample, the ultrasound frequency and the sonication time (Albero et al. 2015). Ultrasound comprises mechanical waves (from 20 kHz to 10 MHz) that need an elastic medium to spread. For the classification of ultrasound applications, the key criterion is the amount of energy generated, characterised by sound power, sound intensity or sound energy density. The uses of ultrasound are broadly distinguished into two groups: high intensity and low intensity (Pico 2013). Recently, a novel microextraction technique, named ultrasound-assisted emulsification microextraction (USAEME), has been developed which is based on the emulsification of a micro volume of water-immiscible extraction solvent in the sample aqueous solution by ultrasound radiation (Regueiro et al. 2008). The application of ultrasonic radiation accelerates the mass transfer process of the analytes between the two immiscible phases, which, together with the large surface of contact between the two phases, leads to an increment in the extraction efficiency in a minimum amount of time. The analytes in the sample are extracted into the fine droplets. The two phases can be readily separated by centrifugation, and the enriched analytes in the sediment phase can be determined by either chromatographic or spectrometric methods (Wu et al. 2010).

8.3.3 Pressurised Liquid Extraction

Pressurised liquid extraction (PLE) is a technique operated by automation that makes use of elevated temperature and pressure to achieve thorough extraction from solid matrices. Hence, this technique reduces solvent consumption and enhances sample throughput when compared with traditional procedures. Therefore, it can be considered an environment-friendly technique, generating small volumes of waste and reducing costs and time (Vazquez-Roig and Pico 2015). García-Galán et al. 2013 developed, validated and practically applied a fully automated analytical PLE method for the determination of 22 sulphonamides, including 5 of their acetylated metabolites, in both sewage sludge and soils. The sensitivity achieved allowed the detection of the target sulphonamides at levels down to 0.03 ng g^{-1} in sewage sludge and 0.01 ng g^{-1} in soil samples. Modular pressurised liquid extraction (M-PLE) procedures were developed for simultaneous extraction, clean-up and fractionation of polychlorinated dioxins and furans (PCDD/Fs) in soil, sediment and sludge samples. The procedures utilise two coupled extraction cells including an upstream cell filled with the sample and layers of silica and acid- and base-modified silica and a downstream cell filled with activated carbon. The silica layers were added to remove polar or hydrolysable matrix components and carbon to perform planarity-based fractionation (Do et al. 2013).

8.3.4 Microwave-Assisted Extraction

Microwave-assisted extraction (MAE) is based on the nonionising radiation that causes molecular motion by migration of ions and rotation of dipoles, without changing the molecular structure (Luque-Garcıa and De Castro 2003). MAE requires small volumes of solvents for the extraction in minimum extraction time (Fountoulakis et al. 2005). MAE can be used for the extraction of pollutants such as polycyclic aromatic hydrocarbons (PAHs), polychlorinated biphenyl (PCB) pesticides, phenols and organometallic components from environmental matrices. Matrix effects due to strong adsorption of the solutes onto the matrix are particularly crucial for environmental matrices (Sanchez-Prado et al. 2010). Khan and Kathi (2014) evaluated heavy metal contamination of roadside surface soil adjacent to automobile workshops by digesting 0.5 g of soil aqua regia using microwave-assisted digestor in accordance with the USEPA method SW 3051.

8.4 Comparison of Extraction Methods

A comparison between Soxhlet extraction, PSE, SWE and SFE has been previously carried out for the extraction of PAHs from PAH-contaminated soils and for the extraction of alkanes and PAHs in urban air particulate matter (Hawthorne et al. 2000). With SWE, xenobiotics (PAHs) were efficiently extracted in urban particulate matter with little extraction of alkanes (Richter et al. 2006). Some studies have been made reporting comparative efficiencies of techniques used for the extraction of samples contaminated by PAHs (Itoh et al. 2008).

8.5 Clean-up Procedures

The detection capability of the instrumentation is correlated strongly with the efficacy of the clean-up. While in the 1970s only solvent extraction and homemade columns (e.g. Silicagel, AlO) were used for clean-up, solid-phase extraction (SPE; 1980s), immuno-affinity chromatography (IAC; 1990s) and molecular imprinted polymers (MIPs; end of 1990s) took over the job. HPLC fractionation results in several purified fractions each containing a limited number of analytes and matrix components. In that aspect comprehensive (two-dimensional) GC or LC prior to MS may be important in the future (De Brabander et al. 2009). Matrix effects and possible interferences in the analysis can be avoided by performing a clean-up step, which also results in a pre-concentration of the target compound (García-Galán et al. 2009). Lipids can significantly reduce the performance of GC–MS due to the accumulation in the injection port, column and ionisation source. Owing to the complexity of the biological matrices, the presence of interfering compounds in the extract requires an intensive clean-up before extracted samples can be submitted to the separation and determination step. Gel permeation chromatography (GPC) using a narrow and short GPC clean-up might reduce solvent usage until around 60 ml (Wang and Guo 2010). To remove matrix components in the clean-up step, modifications of the original dispersive SPE step by using graphitised carbon black and C_{18} sorbent, SPE in cartridge or Florisil cartridges have been used (Pinto et al. 2010). Quick, easy, cheap, effective, rugged and safe (QuEChERS), developed by Anastassiades et al. (2003) is a relatively quick procedure that makes use of acetonitrile salt-out extraction involving excess amounts of salts (anhydrous magnesium sulphate mixed with sodium chloride, sodium acetate or sodium citrate/disodium citrate sesquihydrate), followed by a solid-phase dispersive clean-up step involving the acetonitrile extract and a mixture of magnesium sulphate and primary secondary (PSA) sorbent. Additional clean-up treatments of the extract can also include other dispersive sorbents such as C_{18} and graphitised carbon, as well as using solid-phase extraction cartridges (Zhang et al. 2011).

8.6 Instrumental Analysis

8.6.1 Gas Chromatography

GC is the most used separation technique applied for the determination of PAHs and aliphatic hydrocarbons, PBDEs, pesticides, PBBs, PCDDs and organotin compounds (Zuloaga et al. 2012). The late 1950s involved hyphenation with a classical non-spectroscopic GC detection such as a thermal conductivity detector (TCD) for the analysis of pesticides. This preceded introduction of GC–MS for pesticide analysis. Other detectors coupled to GC for pesticide analysis reported in the 1960s and 1970s include micro-coulometric detector (halogen, sulphur; <0.1 ppm), electron capture detector (ECD; halogen, nitrogen; 10 ppb), thermionic specific detector (TSD; phosphorous, nitrogen; 0.01 μg), emission spectrometric detector (phosphorus, sulphur, nitrogen; 0.1 ng) and flame photometric detector (FPD; phosphorus, sulphur; 0.1, 1 ng). GC–ECD (μECD) is conventionally used in standard testing methods for pesticide analysis (Nolvachai et al. 2015). GC–MS is normally applied by using electron ionisation (EI) at vacuum, which allows compound identification with $>240{,}000$ spectra available in standard database libraries which is relatively independent of instrumental designs and MS conditions apart from the standard 70 eV ionisation energy (Pico et al. 2007). The basic parameter for the characterisation of mass analyser ability to resolve peaks in mass spectra is a

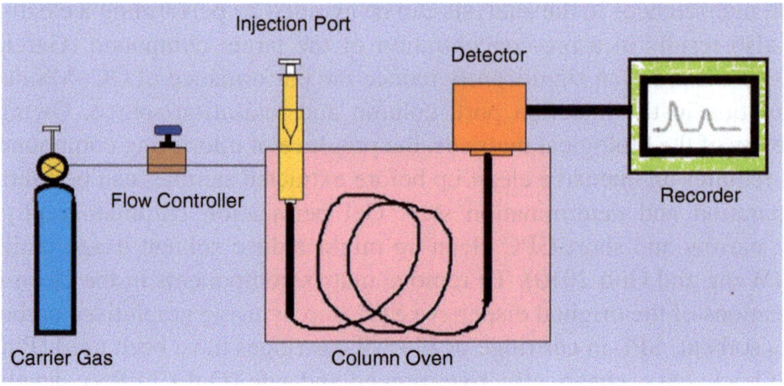

Fig. 8.3 Block diagram of a gas chromatography system

resolving power (RP), which is defined as the *m*/*z* value of particular peak divided by the peak full width at half maximum (FWHM): RP = $(m/z)/\Delta m/z$. The resolution is the inverse of RP expressed as $\Delta m/z$ for a given *m*/*z* value (Sparkman 2000) (Fig. 8.3).

Analytical multidimensional gas chromatography (MDGC) and the excellent separation efficiency it achieves serve advanced characterisation of complex volatile and semi-volatile samples, which is unlikely to be accomplished by single-dimensional chromatography (Marriott et al. 2012). Comprehensive 2D GC, with respect to one-dimensional GC, is characterised by (1) increased selectivity and peak capacity, (2) enhanced sensitivity and (3) increased identification power due to the formation of ordered 2D chromatogram patterns relative to homologous compounds (Tranchida et al. 2015). 2D GC includes heart-cut MDGC (GC–GC) and comprehensive two-dimensional GC (GC × GC) (Nolvachai et al. 2015). In multidimensional GC (MDGC) selected fractions of the GC elutes are online subjected to a second GC separation. This is a useful technique when only one, or a few, target analyte has to be determined. When wide-ranging screening has to be performed or the detection of unknowns is of particular interest, MDGC becomes extremely time-consuming and cannot really solve the analytical problems. Comprehensive two-dimensional GC (GC × GC) should then be used. Here, the entire sample is subjected to two independent separations, using a very rapid (2–8 s) second-dimension separation in order not to lose the resolution achieved on the first column. The total runtime is therefore essentially the same as with a conventional one-dimensional separation (Adahchour and Udo 2014). A series of experiments were conducted to compare GC–MS–MS pseudo-multiple reaction monitoring mode (PMRM) and GC–MS–MS classic multiple reaction monitoring (CMRM) detection modes for soil extract sample analysis. Results for the recovery of PAHs demonstrated the advantage of PMRM triple quad analysis over single quad SIM mode GC–MS. PMRM provided additional mass filtering due to dual quadrupole ion focusing and potential reduction or destruction of isobaric interferences in the collision cell, thus improving sensitivity for most of the targeted compounds (Shang et al. 2014).

Simultaneous multicomponent analysis of organic pollutants in complex environmental matrices can be accomplished using multidimensional gas chromatography (MDGC) (Muscalu et al. 2011). MDGC techniques, such as comprehensive two-dimensional gas chromatography coupled to time-of-flight mass spectrometry (GC × GC-TOF/MS), combine enhanced chromatographic and mass spectral resolution for the confident identification of individual compounds (López et al. 2013; Pena-Abaurrea et al. 2014; Jennerwein et al. 2014). This method is useful on complex matrices because the high resolution provided by GC × GC enables to reduce any overlap between target analytes and matrix interferences and to get lower LODs than with GC–TOF/MS (Engel et al. 2013). Although sensitivity and linearity range of GC × GC–TOF/MS are lower than that of GC–HRMS, it could be a relevant alternative to study the impact of technological or physiological transformation on the level of contaminants (Engel et al. 2013). Planche et al. (2015) reported that GC × GC–TOF/MS method assessed in their study allows for simultaneous analysis of 206 dioxin-like micropollutants by using column set Rtx-Dioxin2/BPX-50. While discussing the novel analytical methods suitable for flame retardants and plasticizers Ballesteros-Gómez et al. (2013) pointed out that GC × GC coupled to Atmospheric Pressure Chemical Ionization (APCI) and a High Resolution (HR)–Time-Of-Flight (TOF)–Mass Spectrometry (MS) appeared to be more sensitive compared to Liquid Chromatography (LC)–APCI/Atmospheric Pressure Photoionization (APPI)–HRTOF–MS with absolute detection limits in the range of 0.5–25 pg. The combination of quadrupole time-of-flight MS (QqTOF-MS), which consists of a quadrupole and a collision cell coupled with TOFMS, enables product ion scan modes with fast full-scan TOFMS analysis (~50 Hz) of the entire product ion profile from a selected precursor ion. The TOFMS sector normally provides accurate mass analysis. Therefore, structural elucidation may be provided from the accurate mass values of unknown analytes and their fragment ions; thereby, it is possible to achieve high-sensitivity full-spectral scanning (Botitsi et al. 2010).

The emerging xenobiotic compounds such as brominated flame retardant (BFR) compounds, hexabromobenzene, pentabromo ethylbenzene and decabromo diphenylethane might also be analysed based on gas chromatography coupled to mass spectrometry with negative chemical ionisation. Moreover, liquid chromatography–quadrupole linear ion trap mass spectrometry can be applied for the determination of hexabromocyclododecanes, tetrabromobisphenol A and their related compounds such as bisphenol A, monobromo bisphenol A, dibromo bisphenol A and tribromo bisphenol A (Gorga et al. 2013). SPME coupled with GC–MS has been demonstrated to be a simple, rapid, solvent-free and sensitive method for the determination of organochlorine pesticides in textile samples. The optimised method exhibited wide linearity, low detection limits and good repeatability (Cai et al. 2013).

Analysis of volatile organic compound (VOC) is usually performed using gas chromatography (GC) coupled with either mass spectrometry through electron impact ionisation (EI-MS) or flame ionisation detection (FID) (Rosell et al. 2006; Arambarri et al. 2004). GC–MS provides a higher degree of selectivity and sensitivity than flame ionisation detection and achieves lower limits of quantification. To measure VOCs at trace levels, a pre-concentration step is needed. Several techniques are used to measure VOCs in a variety of matrices, including purge and trap for measurement in water (Mesarchaki et al. 2014; Ueta et al. 2015; Aranda-Rodriguez et al. 2015), soil (Yao et al. 2013; Buszewski et al. 2016), human saliva (Huda et al. 2013), urine and blood (Cordell et al. 2013; Kakuta et al. 2015) or headspace solid-phase microextraction (HS-SPME) (Kermani et al. 2013) for measurement in water (Martínez et al. 2016; George et al. 2015), air (Barro et al. 2004), soil (Higashikawa et al. 2013), sewage sludge (Kotowska et al. 2012), bitumen (Tang and Isacsson 2006), urine and blood (Kester et al. 2016; Aggio et al. 2016) and faeces (Dixon et al. 2011). Although the purge-and-trap method is more sensitive, HS-SPME is usually preferred because no special equipment is required (Meyer-Monath et al. 2014).

In a recent review, the concept of low-pressure (LP) vacuum outlet gas chromatography (GC) and its theoretical applicability to fast analysis of GC-amenable chemicals was appreciated. LPGC is implemented by placing the outlet of a short, wide (10–15 m, 0.53 mm inner diameter) analytical column under vacuum conditions, which speeds the separation by reducing viscosity of the carrier gas, thereby leading to a higher optimal flow rate for the most separation efficiency. The analytical column is commonly coupled to a short, narrow uncoated restriction capillary that also acts as a guard column, to keep the inlet at normal operating pressures. In addition to higher sample throughput, LPGC provides other benefits, including lower detection limits, less chance of analyte degradation, reduced peak tailing, increased sample loadability and more ruggedness without overly narrow peaks that would necessitate excessively fast data acquisition rates (Sapozhnikova and Lehotay 2015).

8.6.2 Liquid Chromatography

Liquid chromatography (LC) is the laboratory-scale technique of choice for isolation and purification of materials that cannot be handled by crystallisation or distillation. The scale of the process varies from analytical scale, semi-preparative scale and preparative scale (Sciarrone et al. 2015). LC–MS records an enormous growth in recent years due to the application potential in analytical chemistry, biochemistry, pharmaceutical analysis, clinical analysis and many other fields, where the qualitative and quantitative characterisation of complex organic, bioorganic and organometallic mixtures is needed. Current trends in terms of mass analysers, ionisation techniques, fast LC–MS, LC–MALDI–MS, ion mobility spectrometry used in LC–MS, quantitation issues specific to MS and emerging mass spectrometric approaches complementary to LC–MS need to be understood (Holčapek et al. 2012). Major advances in column phase and technology have recently included perfusive packings, partially porous particles, inorganic–organic hybrids, monoliths, high-temperature columns, sub-2 mm particles and nanocolumn. High-purity silica and novel bonding chemistries have made LC columns more efficient and reliable, while commercial instrumentation capable of enhanced sensitivity, reliability, minimal band dispersion or ultrahigh-pressure operation has been developed, meeting the performance levels required for optimal operation. Besides the choice of stationary phases, selectivity of a separation can be further tuned by adjusting experimental parameters, such as mobile phase composition and pH value, use of ion-pairing agents, elution (either isocratic or gradient), flow rate and temperature (Donato et al. 2012). By combining new 1.7 μm reversed-phase packing material and a chromatographic system, operating at ca. 12,000 psi, ultra-performance liquid chromatography (UPLC) has enabled dramatic increases in chromatographic performance to be obtained for complex mixture separation. This increase in performance is manifested in improved peak resolution, together with increased speed and sensitivity. UPLC offers significant advantages over conventional reversed-phase HPLC amounting to a more than doubling of peak capacity, an almost tenfold increase in speed and a three- to fivefold increase in sensitivity compared to that generated with a conventional 3.5 μm stationary phase (Wilson et al. 2005) (Fig. 8.4).

LC–MS has secured a central analytical technique for metabolite identification with the continuous developments and improvements in LC and MS technologies. Recently, a wide range of experimental strategies and post-acquisition data processing and mining modes have emerged driven by the need to identify and characterise metabolites at ever-increasing sensitivity and in ever more complex samples. The classical and practical mass spectrometry-based techniques include low-resolution MS (quadruple, ion trap, linear ion trap, etc.), high-resolution MS (time-of-flight, hybrid time-of-flight instruments, Qrbitrap, Fourier transform ion

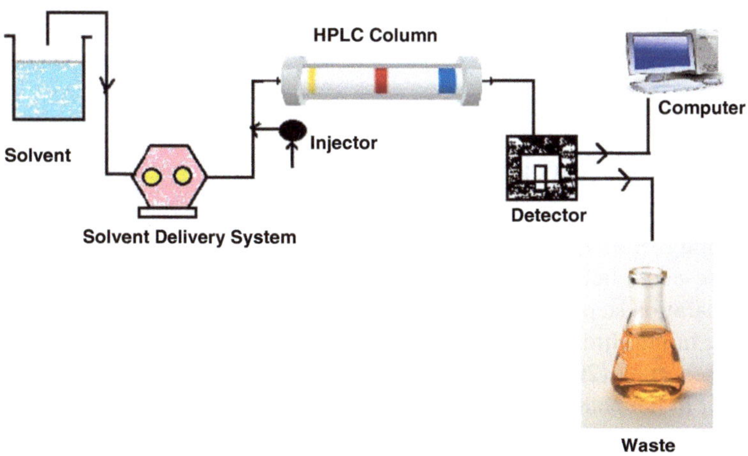

Fig. 8.4 Schematic overview of liquid chromatography

cyclotron resonance MS, etc.) and corresponding post-acquisition data processing and mining modes (precursor ion filtering, neutral loss filtering, mass defect filter, isotope pattern filtering, etc.) (Liang et al. 2011). LC–MS currently competes with GC–MS for the status of reference technique in the analysis of pesticides. The combination of different analyser designs of mass spectrometry to increase versatility and to allow multiple experiments to be carried out simultaneously using a single instrument appears to be a hot topic of present research. There are different mass analysers that enhance tandem mass spectrometry (MS–MS) capabilities, such as quadrupole ion trap (QIT), triple quadrupole (QqQ) and quadrupole time-of-flight (QqTOF), each of which has different features (Soler et al. 2007). LC–MS, in particular LC–triple quadrupole (QqQ) MS, has today become the technique of choice in the field of the analysis of antimicrobial residues (Bogialli and Corcia 2009). Liquid chromatography–tandem mass spectrometry (LC–MS–MS) can reduce analysis time and avoid derivatisation steps in the determination of biocides. Hence, biocides in different environmental matrices have been analysed by LC–MS–MS because of its simplicity of operation, high selectivity and excellent sensitivity (Chen et al. 2012).

Liquid chromatography–mass spectrometry (LC–MS) is as an excellent analytical tool in the determination of pesticides (Masia et al. 2014) by a single quadrupole LC–MS; however, there are still several analytical shortcomings to be overcome. The most important shortcoming derives from the characteristics of the mass spectrum of pesticides, which often gives only a molecule adduct or a weak fragment ion. A successful option to solve the above-mentioned analytical problems is the use of electron ionisation (EI) and a single quadrupole as the system of choice. EI may offer in LC–MS a unique fragmentation pathway and exceptional detection limits, and its use has been reported to determine a large number of pesticides and herbicides in water and in other matrices (Picó et al. 2004). High-resolution quadrupole time-of-flight mass spectrometry (QqTOF-MS) is particularly interesting for hyphenated LC–MS applications using ultrahigh-performance liquid chromatography (UHPLC), where sub-2 μm particle columns generate chromatography peaks with peak widths of only a few seconds (Picó et al. 2015).

8.7 Current Approaches

The coupling of electrochemistry (EC) to different MS techniques is a quickly growing research field in analytical chemistry. Depending on the analytical problem, a separation step can be further integrated according to the instrumental set-up and where liquid chromatography (LC) is preferred for this purpose (Jahn and Karst 2012). Chen et al. (2012) investigated the degradation of a variety of xenobiotics from polar to nonpolar and analysed their degradation products by online coupling of electrochemistry with mass spectrometry (EC–MS). Furthermore, they evaluated possible binding reactions with regard to the generation of non-extractable residues with some model substances such as catechol, phthalic acid, γ-l-glutamyl-lcysteinyl-glycine and l-histidine, deduced from a natural organic matter structure model and identified possible binding sites. EC–MS is a promising, fast and simple screening method to investigate the environmental behaviour of xenobiotics.

8.8 Conclusions

While the conventional sample extraction procedures continue to be applied for extracting soil-bound xenobiotic compounds, their limitations such as long analytical times, manual manipulation of the extracts, large consumption of sample and reagents and generation of large amounts of wastes have been overcome by modern extraction techniques. We have made an attempt to present recent advances in the detection by GC or LC coupled to various kinds of advanced detectors. Systematic research is needed to compare (GC × GC or LC × LC), UPLC, the use of different inlets in GC applications or the use of sampling techniques such as purge and trap, dynamic headspace and headspace required for the analysis of emerging xenobiotic compounds in soil. This article provides analytical characteristics of the procedures that have been applied recently in xenobiotic sample preparation methods and their applications in combination with chromatographic mass spectrometric analysis.

Acknowledgements The author would like to thank Pondicherry University for providing Internet access to collect the information essential for writing this book chapter.

References

Adahchour M, Udo AT (2014) Multidimensional and comprehensive two-dimensional gas chromatography. In: Dettmer-Wilde K, Engewald W (eds) Practical gas chromatography. Springer, Berlin, pp 461–502

Aggio RB, Mayor A, Coyle S, Reade S, Khalid T, Ratcliffe NM, Probert CS (2016) Freeze-drying: an alternative method for the analysis of volatile organic compounds in the headspace of urine samples using solid phase micro-extraction coupled to gas chromatography-mass spectrometry. Chem Cent J 10(1):1–10

Albero B, Sánchez-Brunete C, Miguel E, Pérez RA, Tadeo JL (2012) Determination of selected organic contaminants in soil by pressurized liquid extraction and gas chromatography tandem mass spectrometry with in situ derivatization. J Chromatogr A 1248:9–17

Albero B, Sanchez-Brunete C, Garcia-Valcarcel AI, Pérez RA, Tadeo JL (2015) Ultrasound-assisted extraction of emerging contaminants from environmental samples. TrAC Trends Anal Chem 7:110–118

Anastassiades M, Lehotay SJ, Štajnbaher D, Schenck FJ (2003) Fast and easy multiresidue method employing acetonitrile extraction/partitioning and "dispersive solid-phase extraction" for the determination of pesticide residues in produce. J AOAC Int 86(2):412–431

Arambarri I, Lasa M, Garcia R, Millán E (2004) Determination of fuel dialkyl ethers and BTEX in water using headspace solid-phase microextraction and gas chromatography–flame ionization detection. J Chromatogr A 1033(2):193–203

Aranda-Rodriguez R, Cabecinha A, Harvie J, Jin Z, Marchand A, Tardif R, Nong A, Haddad S (2015) A method for quantification of volatile organic compounds in blood by SPME-GC–MS/MS with broader application: from non-occupational exposure population to exposure studies. J Chromatogr B 992:76–85

Ballesteros-Gómez A, de Boer J, Leonards PE (2013) Novel analytical methods for flame retardants and plasticizers based on gas chromatography, comprehensive two-dimensional gas chromatography, and direct probe coupled to atmospheric pressure chemical ionization-high resolution time-of-flight-mass spectrometry. Anal Chem 85(20):9572–9580

Barro R, Ares S, Garcia-Jares C, Llompart M, Cela R (2004) Development of a sensitive methodology for the analysis of chlorobenzenes in air by combination of solid-phase extraction and headspace solid-phase microextraction. J Chromatogr B 1045(1):189–196

Bogialli S, Di Corcia A (2009) Recent applications of liquid chromatography–mass spectrometry to residue analysis of antimicrobials in food of animal origin. Anal Bioanal Chem 395 (4):947–966

Botitsi HV, Garbis SD, Economou A, Tsipi DF (2010) Current mass spectrometry strategies for the analysis of pesticides and their metabolites in food and water matrices. Mass Spectrom Rev 30:907–939

Buszewski B, Szultka M (2012) Past, present, and future of solid phase extraction: a review. CRC Crit Rev Anal Chem 42(3):98–213

Buszewski B, Ligor T, Ulanowska A (2016) Determination of volatile organic compounds: enrichment and analysis. In: Handbook of trace analysis. Springer, Heidelberg, pp 403–430

Cai JA, Zhu F, Ruan W, Liu L, Lai R, Zeng F, Ouyang G (2013) Determination of organochlorine pesticides in textiles using solid-phase microextraction with gas chromatography–mass spectrometry. Microchem J 110:280–284

Chen ZF, Ying GG, Lai HJ, Chen F, Su HC, Liu YS, Peng FQ, Zhao JL (2012) Determination of biocides in different environmental matrices by use of ultra-high-performance liquid chromatography–tandem mass spectrometry. Anal Bioanal Chem 404(10):3175–3188

Cocco E, Guignard C, Hoffmann L, Bohn T (2011) Rapid analysis of polychlorinated biphenyls in fish by pressurised liquid extraction with in-cell clean up and GC-MS. Int J Environ Anal Chem 91(4):333–347

Cordell RL, Pandya H, Hubbard M, Turne MA, Monks PS (2013) GC-MS analysis of ethanol and other volatile compounds in micro-volume blood samples-quantifying neonatal exposure. Anal Bioanal Chem 405(12):4139–4147

De Brabander HF, Noppe H, Verheyden K, Bussche JV, Wille K, Okerman L, Vanhaecke L, Reybroeck W, Ooghe S, Croubels S (2009) Residue analysis: future trends from a historical perspective. J Chromatogr A 1216(46):7964–7976

De Castro ML, Priego-Capote F (2010) Soxhlet extraction: past and present panacea. J Chromatogr A 1217(16):2383–2389

Dixon E, Clubb C, Pittman S, Ammann L, Rasheed Z, Kazmi N, Keshavarzian A, Gillevet P, Rangwala H, Couch RD (2011) Solid-phase microextraction and the human fecal VOC metabolome. PLoS One 6(4):e18471

Do L, Xuan TH, Lundstedt S, Haglund P (2013) Modular pressurized liquid extraction for simultaneous extraction, clean-up and fractionation of PCDD/Fs in soil, sediment and sludge samples. Anal Methods 5(5):1231–1237

Donato P, Cacciola F, Tranchida PQ, Dugo P, Mondello L (2012) Mass spectrometry detection in comprehensive liquid chromatography: basic concepts, instrumental aspects, applications and trends. Mass Spectrom Rev 31(5):523–559

Engel E, Ratel J, Blinet P, Chin ST, Rose G, Marriott PJ (2013) Benchmarking of candidate detectors for multiresidue analysis of pesticides by comprehensive two-dimensional gas chromatography. J Chromatogr A 1311:140–148

Fountoulakis M, Drillia P, Pakou C, Kampioti A, Stamatelatou K, Lyberatos G (2005) Analysis of nonylphenol and nonylphenol ethoxylates in sewage sludge by high performance liquid chromatography following microwave-assisted extraction. J Chromatogr A 1089(1):45–51

García-Galán MJ, Díaz-Cruz MS, Barceló D (2009) Combining chemical analysis and ecotoxicity to determine environmental exposure and to assess risk from sulfonamides. TrAC Trends Anal Chem 28(6):804–819

García-Galán MJ, Díaz-Cruz S, Barceló D (2013) Multiresidue trace analysis of sulfonamide antibiotics and their metabolites in soils and sewage sludge by pressurized liquid extraction followed by liquid chromatography–electrospray-quadrupole linear ion trap mass spectrometry. J Chromatogr A 1275:32–40

George MJ, Marjanovic L, Williams DBG (2015) Solvent-assisted headspace sampling using solid phase microextraction for the analysis of phenols in water. Anal Chem 87(19):9559–9562

Gorga M, Martínez E, Ginebreda A, Eljarrat E, Barceló D (2013) Determination of PBDEs, HBB, PBEB, DBDPE, HBCD, TBBPA and related compounds in sewage sludge from Catalonia (Spain). Sci Total Environ 444:51–59

Guo Y, Kannan K (2015) Analytical methods for the measurement of legacy and emerging persistent organic pollutants in complex sample matrices. In Zeng E (ed) Persistent Organic Pollutants (POPs): analytical techniques, environmental fate and biological effects, 1st edn. Comprehensive Analytical Chemistry, vol 67, pp 1–56

Guo L, Zhao ZY, Zhao X, Bai J, Han SY (2013) Preparation and performance evaluation of chitosan local medicine composite membrane material. In Advanced materials research, vol 668. Trans Tech Publications, Switzerland, pp 9–12

Hawthorne SB, Grabanski CB, Martin E, Miller DJ (2000) Comparisons of Soxhlet extraction, pressurized liquid extraction, supercritical fluid extraction and subcritical water extraction for environmental solids: recovery, selectivity and effects on sample matrix. J Chromatogr A 892 (1):421–433

Higashikawa FS, Cayuela ML, Roig A, Silva CA, Sánchez-Monedero MA (2013) Matrix effect on the performance of headspace solid phase microextraction method for the analysis of target volatile organic compounds (VOCs) in environmental samples. Chemosphere 93 (10):2311–2318

Holčapek M, Jirásko R, Lísa M (2012) Recent developments in liquid chromatography–mass spectrometry and related techniques. J Chromatogr A 1259:3–15

Huda AK, de Lacy CB, Ratcliffe N (2013) An investigation of volatile organic compounds from the saliva of healthy individuals using headspace-trap/GC-MS. J Breath Res 7(3):036004

Itoh N, Numata M, Aoyagi Y, Yarita T (2008) Comparison of low-level polycyclic aromatic hydrocarbons in sediment revealed by Soxhlet extraction, microwave-assisted extraction, and pressurized liquid extraction. Anal Chim Acta 612(1):44–52

Jahn S, Karst U (2012) Electrochemistry coupled to (liquid chromatography/) mass spectrometry-current state and future perspectives. J Chromatogr A 1259:16–49

Jennerwein MK, Eschner M, Gröger T, Wilharm T, Zimmermann R (2014) Complete group-type quantification of petroleum middle distillates based on comprehensive two-dimensional gas chromatography time-of-flight mass spectrometry (GC × GC-TOFMS) and visual basic scripting. Energy Fuel 28(9):5670–5681

Kakuta S, Nishiumi S, Yoshida M, Fukusaki E, Bamba T (2015) Profiling of volatile compounds in APC Min/+ mice blood by dynamic headspace extraction and gas chromatography/mass spectrometry. J Chromatogr B 1003:35–40

Kalathoor R, Zeiner M, Schmidt B, Schäffer A, Schwarzbauer J (2015) First evidence for covalent linkage of acidic metabolites of metalaxyl and DDT as non-extractable pesticide residues in soil and sediment. Environ Chem Lett 13(4):431–437

Kermani FR, Tugulea AM, Hnatiw J, Niri VH, Pawliszyn J (2013) Application of automated solid-phase microextraction to determine haloacetonitriles, haloketones, and chloropicrin in Canadian drinking water. Water Qual Res J Can 48(1):85–98

Kester ME, Freeman EW, Songsasen N, Huff TB (2016) Automated headspace solid-phase microextraction of urinary VOCs from eleven maned wolves (*Chrysocyon brachyurus*): a recursive workflow for GC–MS analysis. In Chemical signals in vertebrates 13, Springer, Switzerland, pp 477–498

Khan AB, Kathi S (2014) Evaluation of heavy metal and total petroleum hydrocarbon contamination of roadside surface soil. Int J Environ Sci Technol 11(8):2259–2270

Kotowska U, Żalikowski M, Isidorov VA (2012) HS-SPME/GC–MS analysis of volatile and semi-volatile organic compounds emitted from municipal sewage sludge. Environ Monit Assess 184 (5):2893–2907

Liang Y, Wang G, Xie L, Sheng L (2011) Recent development in liquid chromatography/mass spectrometry and emerging technologies for metabolite identification. Curr Drug Metab 12 (4):329–344

López SH, Gómez MJ, Hernando M, Fernández-Alba AR (2013) Automated dynamic headspace followed by a comprehensive two-dimensional gas chromatography full scan time-of-flight mass spectrometry method for screening of volatile organic compounds (VOCs) in water. Anal Methods 5(5):1165–1177

Luque-Garcıa JL, De Castro ML (2003) Ultrasound: a powerful tool for leaching. TrAC Trends Anal Chem 22(1):41–47

Marriott PJ, Chin ST, Maikhunthod B, Schmarr HG, Bieri S (2012) Multidimensional gas chromatography. TrAC Trends Anal Chem 34:1–21

Martínez C, Gómez V, Borrull F, Pocurull E (2016) Headspace-solid phase microextraction: useful technique to characterize volatile and semi-volatile organic compounds in water reuse applications. Desalin Water Treat 56:1–9

Masia A, Blasco C, Picó Y (2014) Last trends in pesticide residue determination by liquid chromatography–mass spectrometry. Trends Environ Anal Chem 2:11–24

Mesarchaki E, Yassaa N, Hein D, Lutterbeck HE, Zindler C, Williams J (2014) A novel method for the measurement of vocs in seawater using needle trap devices and gc–ms. Mar Chem 159:1–8

Meyer-Monath M, Beaumont J, Morel I, Rouget F, Tack K, Lestremau F (2014) Analysis of BTEX and chlorinated solvents in meconium by headspace-solid-phase microextraction gas chromatography coupled with mass spectrometry. Anal Bioanal Chem 406(18):4481–4490

Muscalu AM, Reiner EJ, Liss SN, Chen T, Ladwig G, Morse D (2011) A routine accredited method for the analysis of polychlorinated biphenyls, organochlorine pesticides, chlorobenzenes and screening of other halogenated organics in soil, sediment and sludge by GCxGC-μ ECD. Anal Bioanal Chem 401(8):2403–2413

Nolvachai Y, Kulsing C, Marriott PJ (2015) Pesticides analysis: advantages of increased dimensionality in gas chromatography and mass spectrometry. Crit Rev Environ Sci Technol 45 (19):2135–2173

Pena-Abaurrea M, Jobst KJ, Ruffolo R, Shen L, McCrindle R, Helm PA, Reiner EJ (2014) Identification of potential novel bioaccumulative and persistent chemicals in sediments from Ontario (Canada) using scripting approaches with GC × GC-TOF MS analysis. Environ Sci Technol 48(16):9591–9599

Pico Y (2013) Ultrasound-assisted extraction for food and environmental samples. TrAC Trends Anal Chem 43:84–99

Picó Y, Blasco C, Font G (2004) Environmental and food applications of LC–tandem mass spectrometry in pesticide-residue analysis: an overview. Mass Spectrom Rev 23:45–85. doi:10.1002/mas.10071

Pico Y, La FM, Soler C, Barcelo D (2007) Identification of unknown pesticides in fruits using ultra-performance liquid chromatography-quadrupole time-of-flight mass spectrometry. Imazalil as a case study of quantification. J Chromatogr A 1176:123–134

Picó Y, Farré M, Barceló D (2015) Quantitative profiling of perfluoroalkyl substances by ultrahigh-performance liquid chromatography and hybrid quadrupole time-of-flight mass spectrometry. Anal Bioanal Chem 407(15):4247–4259

Pintado-Herrera MG, González-Mazo E, Lara-Martín PA (2016) In-cell clean-up pressurized liquid extraction and gas chromatography–tandem mass spectrometry determination of hydrophobic persistent and emerging organic pollutants in coastal sediments. J Chromatogr A 1429:107–118

Pinto CG, Laespada MEF, Martín SH, Ferreira AMC, Pavón JLP, Cordero BM (2010) Simplified QuEChERS approach for the extraction of chlorinated compounds from soil samples. Talanta 81(1):385–391

Planche C, Ratel J, Mercier F, Blinet P, Debrauwer L, Engel E (2015) Assessment of comprehensive two-dimensional gas chromatography-time-of-flight mass spectrometry based methods for investigating 206 dioxin-like micropollutants in animal-derived food matrices. J Chromatogr A 1392:74–81

Regueiro J, Llompart M, Garcia-Jares C, Garcia-Monteagudo JC, Cela R (2008) Ultrasound-assisted emulsification–microextraction of emergent contaminants and pesticides in environmental waters. J Chromatogr A 1190(1):27–38

Richter P, Jiménez M, Salazar R, Maricán A (2006) Ultrasound-assisted pressurized solvent extraction for aliphatic and polycyclic aromatic hydrocarbons from soils. J Chromatogr A 1132(1):15–20

Riefer P, Klausmeyer T, Schwarzbauer J, Schäffer A, Schmidt B, Corvini PF (2011) Rapid incorporation and short-term distribution of a nonylphenol isomer and the herbicide MCPA in soil-derived organo-clay complexes. Environ Chem Lett 9(3):411–415

Rodriguez-Mozaz S, de Alda MJL, Barceló D (2007) Advantages and limitations of on-line solid phase extraction coupled to liquid chromatography–mass spectrometry technologies versus biosensors for monitoring of emerging contaminants in water. J Chromatogr A 1152(1):97–115

Rosell M, Lacorte S, Barceló D (2006) Simultaneous determination of methyl tert-butyl ether, its degradation products and other gasoline additives in soil samples by closed-system purge-and-trap gas chromatography–mass spectrometry. J Chromatogr A 1132(1):28–38

Rouvière F, Buleté A, Cren-Olivé C, Arnaudguilhem C (2012) Multiresidue analysis of aromatic organochlorines in soil by gas chromatography-mass spectrometry and QuEChERS extraction based on water/dichloromethane partitioning. Comparison with accelerated solvent extraction. Talanta 93:336–344

Runnqvist H, Bak SA, Hansen M, Styrishave B, Halling-Sørensen B, Björklund E (2010) Determination of pharmaceuticals in environmental and biological matrices using pressurised liquid extraction—are we developing sound extraction methods? J Chromatogr A 1217(16):2447–2470

Saleh A, Yamini Y, Faraji M, Rezaee M, Ghambarian M (2009) Ultrasound-assisted emulsification microextraction method based on applying low density organic solvents followed by gas chromatography analysis for the determination of polycyclic aromatic hydrocarbons in water samples. J Chromatogr A 1216(39):6673–6679

Sanchez-Prado L, Garcia-Jares C, Llompart M (2010) Microwave-assisted extraction: application to the determination of emerging pollutants in solid samples. J Chromatogr A 1217(16):2390–2414

Sapozhnikova Y, Lehotay SJ (2015) Review of recent developments and applications in low-pressure (vacuum outlet) gas chromatography. Anal Chim Acta 899:13–22

Sciarrone D, Pantò S, Ragonese C, Dugo P, Mondello L (2015) Evolution and status of preparative gas chromatography as a green sample-preparation technique. TrAC Trends Anal Chem 71:65–73

Shang D, Kim M, Haberl M (2014) Rapid and sensitive method for the determination of polycyclic aromatic hydrocarbons in soils using pseudo multiple reaction monitoring gas chromatography/tandem mass spectrometry. J Chromatogr A 1334:118–125

Soler C, James KJ, Picó Y (2007) Capabilities of different liquid chromatography tandem mass spectrometry systems in determining pesticide residues in food: application to estimate their daily intake. J Chromatogr A 1157(1):73–84

Souza-Silva ÉA, Jiang R, Rodríguez-Lafuente A, Gionfriddo E, Pawliszyn J (2015) A critical review of the state of the art of solid-phase microextraction of complex matrices I. Environmental analysis. TrAC Trends Anal Chem 71:224–235

Sparkman OD (2000) Mass spectrometry desk reference. Global View, Pittsburgh

Sporring S, Bøwadt S, Svensmark B, Björklund E (2005) Comprehensive comparison of classic Soxhlet extraction with Soxtec extraction, ultrasonication extraction, supercritical fluid extraction, microwave assisted extraction and accelerated solvent extraction for the determination of polychlorinated biphenyls in soil. J Chromatogr A 1090(1):1–9

Stader C, Beer FT, Achten C (2013) Environmental PAH analysis by gas chromatography–atmospheric pressure laser ionization–time-of-flight–mass spectrometry (GC-APLI-MS). Anal Bioanal Chem 405(22):7041–7052

Subedi B, Aguilar L, Robinson EM, Hageman KJ, Björklund E, Sheesley RJ, Usenko S (2015) Selective pressurized liquid extraction as a sample-preparation technique for persistent organic pollutants and contaminants of emerging concern. TrAC Trends Anal Chem 68:119–132

Szulejko JE, Kim KH, Brown RJ, Bae MS (2014) Review of progress in solvent-extraction techniques for the determination of polyaromatic hydrocarbons as airborne pollutants. TrAC Trends Anal Chem 61:40–48

Tadeo JL, Sánchez-Brunete C, Albero B, García-Valcárcel AI (2010) Application of ultrasound-assisted extraction to the determination of contaminants in food and soil samples. J Chromatogr A 1217(16):2415–2440

Tang B, Isacsson U (2006) Determination of aromatic hydrocarbons in bituminous emulsion sealants using headspace solid-phase microextraction and gas chromatography–mass spectrometry. J Chromatogr A 1137(1):15–21

Tranchida PQ, Franchina FA, Dugo P, Mondello L (2015) Comprehensive two-dimensional gas chromatography-mass spectrometry: recent evolution and current trends. Mass Spectrom Rev 35(4):524–534

Ueta I, Mitsumori T, Suzuki Y, Kawakubo S, Saito Y (2015) Determination of very volatile organic compounds in water samples by purge and trap analysis with a needle-type extraction device. J Chromatogr A 1397:27–31

Vazquez-Roig P, Pico Y (2015) Pressurized liquid extraction of organic contaminants in environmental and food samples. TrAC Trends Anal Chem 71:55–64

Wang JH, Guo C (2010) Ultrasonication extraction and gel permeation chromatography clean-up for the determination of polycyclic aromatic hydrocarbons in edible oil by an isotope dilution gas chromatography–mass spectrometry. J Chromatogr A 1217(28):4732–4737

Wilson ID, Nicholson JK, Castro-Perez J, Granger JH, Johnson KA, Smith BW, Plumb RS (2005) High resolution "ultra performance" liquid chromatography coupled to oa-TOF mass spectrometry as a tool for differential metabolic pathway profiling in functional genomic studies. J Proteome Res 4(2):591–598

Wu Q, Li Z, Wu C, Wang C, Wang Z (2010) Application of ultrasound-assisted emulsification microextraction for the determination of triazine herbicides in soil samples by high performance liquid chromatography. Microchim Acta 170(1–2):59–65

Xu J, Zheng J, Tian J, Zhu F, Zeng F, Su C, Ouyang G (2013) New materials in solid-phase microextraction. TrAC Trends Anal Chem 47:68–83

Yao CH, Jia LM, Wei QB, Fu D, Ni H, Heilongjiang PEMC (2013) Determination of volatile organic compounds in soil by liquid leaching and headspace and portable gas chromatograph mass spectrometer. Chem Eng 216(9):19–21

Zhang K, Wong JW, Yang P, Tech K, DiBenedetto AL, Lee NS, Hayward DG, Makovi CM, Krynitsky AJ, Banerjee K, Jao L (2011) Multiresidue pesticide analysis of agricultural commodities using acetonitrile salt-out extraction, dispersive solid-phase sample clean-up, and high-performance liquid chromatography–tandem mass spectrometry. J Agric Food Chem 59 (14):7636–7646

Zuloaga O, Navarro P, Bizkarguenaga E, Iparraguirre A, Vallejo A, Olivares M, Prieto A (2012) Overview of extraction, clean-up and detection techniques for the determination of organic pollutants in sewage sludge: a review. Anal Chim Acta 736:7–29

Chapter 9
Xenobiotics in the Food Chain: Quantitative Analysis, Toxic Impact, and Usage History

Adeel Mahmood, Muhammad Zaffar Hashmi, and Abdul Qadir

9.1 Introduction

Humans and animals are continuously exposed to a multitude of xenobiotic chemicals, and attention to this issue has increased over the past few years. The toxic nature and properties of such pollutants in the environment are a global issue because these chemicals persist in nature, have toxic properties, and bioaccumulate upwards in the food chain (Guo et al. 2008; Eqani et al. 2012). The buildup of these chemicals over time can produce effects that can be predicted based on the information available on even one of these compounds (Larsen et al. 2003). Continuous contact with these xenobiotics might cause cancer, reproductive disorders, neurological damage, and a reducti on in immunity (Kalyoncu et al. 2009).

9.2 Source Perspectives

The world's population increased during 1950–2000 from 2.5 billion to 6.1 billion, and by the year 2050 it is estimated to be 9.1 billion; thus the world's population doubled in the past 50 years, but perhaps will grow at a slower rate in the 50 years

A. Mahmood (✉)
Department of Environmental Sciences, Government College Women University, Sialkot, 51310, Pakistan
e-mail: adilqau5@gmail.com

M.Z. Hashmi
Department of Meteorology, COMSATS Institute of Information Technology, Islamabad, 45550, Pakistan

A. Qadir
College of Earth and Environmental Science, University of the Punjab, Lahore, Pakistan

© Springer International Publishing AG 2017

M.Z. Hashmi et al. (eds.), *Xenobiotics in the Soil Environment*, Soil Biology 49,
DOI 10.1007/978-3-319-47744-2_9

from 2000 to 2050. The world's population is increasing 1.2 % annually (77 million people per year), and six countries contribute half of this annual increase: China, India, Pakistan, Indonesia, Nigeria, and Bangladesh. The increasing population has led to an increase in food demand globally, and to meet this food demand, crop productivity needs to be increased. Xenobiotic chemical pesticides have been used extensively all over the world to eradicate insect-borne diseases for a better crop yield. These chemicals are capable of persisting in the environment for decades; migrating in air, soil, sediment, and water; and accumulating in plants, animals, and humans. These pollutants are known as persistent organic pollutants (POPs) (Breivik et al. 2004).

Thousands of xenobiotic POPs are present in the environment that have a long half-life, consisting of many days in the air and up to a decade in soil and other living organisms. These pollutants consist of a series of chemicals, e.g., 209 different congeners of polychlorinated biphenyls (PCBs) have large differences in chlorination and substitution position. POPs are lipophilic and hydrophobic in nature, and they bind to organic matter strongly. This allows the binding of these chemicals to the fatty tissue present in animals. Such properties allow their persistence in living things and cause bioaccumulation at higher trophic levels (Jones and de Voogt 1999). Persistent organic pollutants have a markedly high tendency to transport the compounds in vaporous form in the presence of ambient conditions. This allows their volatilization from water bodies, the pedosphere, and flora, and their persistence prevents their degradation, making them able to travel great distances before re-deposition. And the cycle continues, allowing the deposition of POPs a distance from where they were emitted (Jones and de Voogt 1999).

Organic pesticides are a class of xenobiotics that commonly have agricultural and industrial uses, and are added to the environment in large quantities each year. From a source point of view, POPs like other organic contaminants can be derived from two broad categories: (a) they can be internationally produced for a single or many purposes, (b) they can be a by-product (un-intentional) of industrial processes or of anthropogenic activities. In addition, POPs can also be produced during natural processes (Breivik et al. 2004).

The POPs produced intentionally can be divided into many subgroups. These chemicals belong to many families of compounds which may be chlorinated, brominated, or aromatic, e.g. PCBs, furans, polybrominated diphenyl ethers (PBDEs), polychlorinated naphthalenes (PCNs), and organochlorines pesticides (OCPs), which include DDTs and other related compounds including toxaphene , chlordane, etc. Very rarely these substances originate from multiple sources like hexachlorobenzene (HCB), which is produced both intentionally and as a by-product of other activities (Bailey 2001).

Polychlorinated biphenyls are one other type of compound produced by both the above-mentioned processes (Brown et al. 1995; Lohmann et al. 2000). Although the comparative significance of the formation of by-products is not clear, according to initial assessment, it has some value with regard to the historic balance of mass of PCBs around the world (Breivik et al. 2002). It is noteworthy that the Stockholm Convention lists HCBs and PCBs as both deliberately and non-deliberately manufactured chemicals, and thus their identification and quantification from both origins and the instatement of discharge inventories from un-intentional production has been requested (Breivik et al. 2004).

9.3 Toxicological Behavior and Quantitative Analysis of Xenobiotics

Xenobiotics like organic pollutants can become bioaccumulated and biomagnified in the ecosystem, so there is concern about their impact on species at the higher trophic levels in the environment, including humans. Recently, the appreciation of such pollution has increased because many compounds have been recognized as causes of endocrine disruption affecting functioning of reproductive and hormonal systems in animals and humans. Such contaminants can survive in lipids for a long time and can cause chronic diseases, including birth defects, a reduction in immunity, disturbance of growth patterns, brain damage, mutations, learning problems, diseases of the respiratory system such as asthma, and behavior anomalies in humans and animals (Harrison et al. 1995). Other effects include carcinogenicity, a greater chance of developing diabetes or endometriosis, and neurobehavioral impairment, along with problems with learning and intellectual deficits. Some researchers also consider them to be a potential risk factor for human breast cancer (Safe 1994; Ross et al. 1995).

The findings of ecological impact studies reveal that persistent organic pollutants are of concern in the areas of dysfunction of the reproductive and immune systems, hormonal problems, cancer, and neurobehavioral disorders (Kelce et al. 1995; Kavlock et al. 1996). In youngsters and newborn babies, a reduction in immunity, infections, neurobehavioral dysfunction, developmental abnormalities, and induction of tumors can be due to contamination by these pollutants. Children are more vulnerable as they pass through their stages of development, as cells during developmental stages are at greater risk of being affected by environmental pollutants and hence are more prone to being exposed to and affected by these organic pollutants. The human brain is of great concern here as exposure to POPs in infancy has resulted in lower scoring for exposed children in assessment of intelligence (Bouwman 2003). Hence, a collective effect of POPs is discussed at the screening stage of risk valuation procedures.

For appraisal of the effectiveness of remedial routes, risk assessment is very much needed and it can help in developing goals for a "clean-up" which are realistic in practice. The standards of life are intended to guarantee the safety of organisms, both terrestrial and aquatic, from the harmful effects of exposure to chemicals which can cause acute and chronic diseases. The criterion is built upon the degree of toxicity and is regulated to protect living organisms from premature death, problems with reproduction, stunted growth, and the buildup of harmful chemicals in the organisms, which eventually may lead to effects at the consumer level. Criteria for life have now been developed by some departments in the form of numeric limits on the acceptable level of chemicals, including heavy metals, organochlorines, and other highly toxic chemicals, that can be present in both marine- and land-dwelling organisms. Some of the departments taking action in this regard are the European Union Directorate, Canadian Council of Ministers of the Environment, United States Environmental Protection Agency, Florida Department of Environment Conservation of America, and the Agency of Toxic Substances and Disease Registry (CCME 2002; ASTDR 2005; Long et al. 1995; Sun et al. 2010).

9.4 Historical Overview of Xenobiotic Usage in the Food Chain

Human beings started using pesticides to control pests as early as 2500 BC. The Samarians used sulfur as an insecticide. In 1000 BC sulfur was used to fumigate homes, and in 900 BC the Chinese controlled garden insects using arsenic (CIPM 2003). The history of pesticide development and usage dates back to Greece and Rome around AD 79; detailed information has been found from the 16th century (Hassall 1982). For centuries, mankind has been fighting pests, i.e., any animal or plant that threatens our food chain and health (Delaplane 1996). Until the Second World War, inorganic plus biological substances like turpentine, petroleum, pyrethrum, acetoarsenite, etc. were used to combat pests (CIPM 2003; Delaplane 1996). Farmers used pesticides widely, to protect valuable crops such as cotton, wheat, and rice between 1860 and 1942. During this period, power-compelled sprayers were invented that resulted in easier and larger scale application of pesticides in agriculture (CIPM 2003). With the advent of new pesticides that stemmed from chemical warfare used in Second World War, pest control became more effectual (CIPM 2003; Delaplane 1996).

A new era of effective and inexpensive pest control started with the advent of the chemicals lindane, benzenehexachloride (BHC), dichloro-diphenyl-trichloroethane (DDT), endrine, aldrine, and dialdrine. In the 1940s, DDT became a standard against insects due to its comprehensive effectiveness against mosquitoes, the Japanese beetle, fire ants, and other insects (Smeltzer 2003). At about that time the production of the phenoxyalkanoic acid group (herbicide) was in progress in Britain. In 1945 British workers discovered the first soil-acting carbamate herbicide, and the chlordane and organochlorine insecticides were introduced by Germany and the USA. During 1950–1955, the USA introduced the herbicidal urea derivatives – captan, glyodin (fungicide), and malathion. In 1955–1960 triazines (herbicidal) were introduced by Switzerland and quaternary ammonium (herbicidal) was developed by Britain. The post-Second World War era saw a boom in chemical pesticide production and thousands of other compounds having pesticidal activities. Discovery of DDT and its relative compounds was said to be a marvel and a failsafe solution for getting rid of pests. In 1962 Rachel Carson's "Silent

Spring" first drew our attention to the retention of persistent organochlorine pesticides in the food chain. As a result of better efficiency of pesticides a rise in crop yield followed shortly after. Furthermore, the efficacy and popularity of these chemicals led to the synthesis of many other pesticides (CIPM 2003).

Since 1970s most of these chemicals have been banned, but are still used in the Third World (Eqani et al. 2011). Pesticides still involve low costs, and in developing countries are used extensively for their effectiveness. In the recent past, DDT and HCH were used broadly in southern Asian countries for agricultural purposes. More than 900 chemical pesticides are used worldwide, legally or illegally, for agricultural purposes (Thurman et al. 2000). The augmented use of innumerable pesticides, mainly POPs, has led to concern regarding the potential for environmental media (i.e., air, water, surface soil, sediment, and biota) contamination and associated risks for animal and human health.

References

Bailey RE (2001) Global hexachlorobenzene emissions. Chemosphere 43:167–182

Bouwman H (2003) POPs in Southern Africa. In: Fiedler H (ed) The hand book of environmental chemistry, persistent organic pollutants, 3rd edn. Springer, Berlin (Chapter 11)

Breivik K, Sweetman A, Pacnya JM, Jones KC (2002) Towards a global historical inventory for selected PCB congeners a mass balance approach 2. Emissions. Sci Total Environ 290:199–224

Breivik K, Alcock R, Li YF, Bailey RE, Fiedler H, Pacyna JM (2004) Primary sources of selected POPs: regional and global scale emission inventories. Environ Pollut 128:3–16

Brown JF, Frame GM, Olson DR, Webb JL (1995) The sources of the coplanar PCBs. Organohalogen Compd 26:427–430

CCME (Canadian Council of Ministers of the Environment) (2002) Canadian sediment quality guidelines for the protection of aquatic life. Canadian environmental quality guidelines. Canadian Council of Ministers of the Environment, Winnipeg

CIPM Center for Integrated Pest Management (2003) Pesticide usage in the United States: trends during the 20th century. North Carolina State University, Raleigh, pp. 1–17

Delaplane KS (1996) Pesticide usage in the United States: history, benefits, risks, and trends. The University of Georgia, Athens, pp. 1–6

Eqani SA, Malik RN et al (2011) The level and distribution of selected organochlorine pesticides in sediments from River Chenab, Pakistan. Environ Geochem Health 33(1):33–47

Eqani SAMAS, Malik RN, Alamdar A, Faheem H (2012) Status of organochlorine contaminants in the different environmental compartments of Pakistan: a review on occurrence and levels. Bull Environ Contam Toxicol 88:303–310

Guo L, Uiu Y, Zhang G, Zheng GJ, Lam PKS, Li X (2008) Levels and bioaccumulation of organochlorine pesticides (OCPs) and polybrominated diphenyl ethers (PBDEs) in fishes from the Pearl River estuary and Daya Bay. South China Environ Pollut 152:604–611

Harrison PTC, Humfrey CDN, Litcheld M, Peakall D, Shuker LK (1995) Environmental oestrogens: consequences to human health and wildlife, IEH-Med. Research Council, Leicester

Hassall KA (1982) The chemistry of pesticides: their metabolism, mode of action and uses in crop protection. The MacMillan, London, p. 372

Jones KC, de Voogt P (1999) Persistent organic pollutants (POPs): state of the science. Environ Pollut 100:209–221

Kalyoncu L, Agca I, Aktumsek A (2009) Some organochlorine pesticide residues in fish species in Konya, Turkey. Chemosphere 74:885–900

Kavlock RJ, Daston GP, DeRosa C, Fenner-Crisp P, Gray LE, Kaattari S, Lucier G, Luster M, Mac MJ, Maczka C, Miller R, Moore J, Rolland R, Scott G, Sheehan DM, Sinks T, Tilson HA (1996) Research needs for the risk assessment of health and environmental e€ects of endocrine disruptors. Environ Health Perspect 104:715–740

Kelce WR, Stone CR, Laws SC, Gray LE, Kemppainen JA, Wilson EM (1995) Persistent DDT metabolite p,p-DDE is a potent androgen receptor antagonist. Nature 375:581–585

Larsen JC, Binderup ML, Dalgaard M, Dragsted LO, Hossaini A, Ladefoged O, Lam HR, Madsen C, Meyer O, Rasmussen ES, Reffstrup TK, Søborg I, Vinggaard AM, Ostergård G (2003) Combined actions and interactions of chemicals in mixtures. In: Larsen JC (ed) The toxicological effects of exposure to mixtures of industrial and environmental chemicals. Danish Veterinary and Food Administration, Fødevare Rapport, p 12

Lohmann R, Northcott GL, Jones KC (2000) Assessing the contribution of diffuse domestic burning as a source of PCDD/Fs, PCBs, and PAHs to the UK atmosphere. Environ Sci Technol 34:2892–2899

Long ER, MacDonald DD, Smith SL, Calder F (1995) Incidence of adverse biological effects within ranges of chemical concentrations in marine and estuary sediments. Environ Manage 19:81–97

Ross PS, De Swart RL, Reijnders PJH, Van Loveren H, Vos JG, Osterhaus ADME (1995) Contaminant related suppression of delayed-type hypersensitivity and antibody responses in harbor seals fed herring from the Baltic Sea. Environ Health Perspect 103:162–167

Safe SH (1994) Polychlorinated biphenyls (PCBs)-environmental impact, biochemical and toxic responses, and implications for risk assessment. Crit Rev Toxicol 24:87–149

Smeltzer, J (2003) Pesticides: a Short history of their use. http://www.protector.pair.com/articles/pesticds/PesticidesBackground.html. Accessed 29 Jan 2009

Sun J, Feng J, Liu Q, Li Q (2010) Distribution and sources of organochlorine pesticides (OCPs) in sediments from upper reach of Huaihe River. East China J Hazard Mater 184:141–146

Thurman EM, Bastian KC, Mollhagen T (2000) Occurrence of cotton herbicides and insecticides in playa lakes of the High Plains of west Texas. Sci Total Environ 248:189–200

Chapter 10
Biphasic Dose–Response Phenomenon Induced by Xenobiotics and Its Application in Soil Risk Assessment

Muhammad Zaffar Hashmi, Adeel Mehmood, Sohaib Khan, Ahmad Hasnain, and Zulkifl Ahmed

10.1 Introduction

The central pillar of toxicology is that of the dose–response relationship. The field of toxicology has been dominated by the use of two dose–response models, the threshold and linear models. These two models essentially determine strategies for animal model selection, study design (including the number and range of doses employed), and the means to estimate population-based risks (Calabrese 2005; Hashmi et al. 2014). This dominance has been manifest since the 1930s in the case of the threshold model and since the early 1970s in the case of the linear model approach when it was applied to assessing risks to carcinogens (Calabrese and Baldwin 2002; Hashmi et al. 2014). Toxicology is the study of chemical toxic effects on the ecosystem or environment (Gallo and Doull 1996), but many chemicals as a factor of dose have opposite effects. For example, antibiotics such as streptomycin, penicillin, and erythromycin at low doses stimulate bacterial

M.Z. Hashmi (✉)
Department of Meteorology, COMSATS Institute of Information Technology, Park road, Islamabad, Pakistan
e-mail: hashmi_qau@yahoo.com

A. Mehmood
Department of Environmental Sciences, Government College Women University, Sialkot, 51310, Pakistan

S. Khan
Department of Anthropology, Bahauddin Zakariya University, Multan, Pakistan

A. Hasnain
Department of Geography, Bahauddin Zakariya University, Multan, Pakistan

Z. Ahmed
Department of Building and Architecture Engineering, Bahauddin Zakariya University, Multan, Pakistan

© Springer International Publishing AG 2017

M.Z. Hashmi et al. (eds.), *Xenobiotics in the Soil Environment*, Soil Biology 49,
DOI 10.1007/978-3-319-47744-2_10

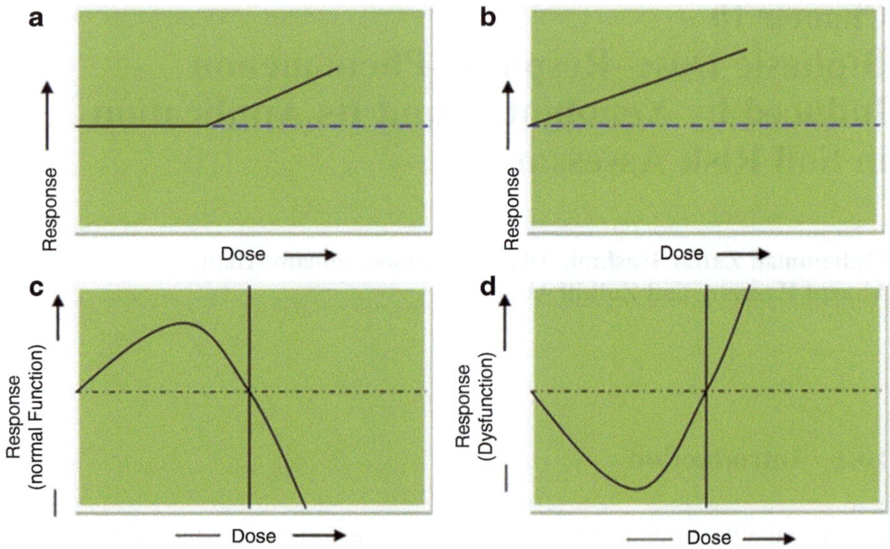

Fig. 10.1 Dose–response relationships described by (**a**) the threshold model, (**b**) the linear non-threshold model, (**c**) the inverted U-shaped hormetic model, and (**d**) the J-shaped hormetic model (Davis and Svendsgaard 1990)

growth and are inhibitory at higher doses. Additionally, it has been documented that mild chemical stress can enhance the growth and respiration of yeast, but intense stress shows inhibitory effects (Cabral et al. 2003). These examples of small quantities having positive effects in contrast to large quantities are commonly termed hormesis (Calabrese and Baldwin 2002; Hashmi et al. 2014).

The quantitative features of the hormetic dose response are a modest stimulation at low doses where the maximum stimulation is typically 30–60 % greater than the controls and a range of stimulation that can be variable but is typically less than 10–20-fold, although ~5–7 % of hormetic dose responses can exceed 100-fold (Calabrese and Baldwin 2002). Many terms have been—and currently are being—used for what appears to be the dose–response relationship, which is called hormesis. A typical dose–response curve shape depends on the end point of interest, which may be either U-shaped or inverted U-shaped (Fig. 10.1). The shape of the dose–response curve is an inverted U-shape if longevity or growth is the end point; if disease is the end point, it would be U-shaped and J-shaped. Some of these terms include biphasic, bitonic, overshoot, preconditioning, non-monotonic, bell-shaped, inverted U-shaped, U-shaped, J-shaped, rebound affect, functional antagonism, and adaptive response (Calabrese 2008). Other terms have been used that indicate this phenomenon has been considered to be the equivalent of a biological law. This is seen with the terms Yerkes–Dodson law (Broadhurst 1957), Hueppe's rule, and the Arndt–Schulz law (Hueppe 1899).

10.2 Biphasic Dose–Response Universality

Hormesis (from Greek *hórmēsis* "eagerness, rapid motion"; from ancient Greek *hormáein* "impel, to set in motion, urge on") is the term for generally stimulatory biological responses to chemicals and other stressors at low exposures. The concept of hormesis in toxicology originated from the research of Schulz (1888) over a century ago who reported that yeast respiration and growth showed stimulatory responses at low doses, while higher levels were inhibitory. In 1896, this finding was confirmed by a microbiologist Hueppe and then, in the field of radiology, also observed a similar phenomenon (Calabrese and Baldwin 2000). The second stage is a preliminary study on hormesis biphasic dose–response relationship in the twentieth century; in "1902, The British physiologist Starling discovered an acid extract of the duodenum in the blood that can stimulate the pancreas secretion; followed in 1904, he created the hormones (hormone) to name the word for such a small dose under the influence of a substance by metabolism," and in 1912, Schulz named this phenomenon the Arndt–Schulz law (Calabrese and Baldwin 2000). This concept of a generalized low-dose stimulation–high-dose inhibition was gradually supported by similar observations with other chemicals and eventually became known as the Arndt–Schulz law.

The third stage is the development of research in the late twentieth century. Back in the 1830s, the hormesis biphasic dose–response relationship of research that had been taken has a large number of experimental data (Calabrese and Baldwin 2000), but because the effect is very slight, sometimes not necessarily observed, together with the experimental reproducibility which was poor, this theory gradually faded from view in 1943. Southam and Ehrlich observed effects on fungi in the study of red cedar extract to low-dose stimulation. Biphasic dose–response curves of high-dose inhibition were named as hormesis, published in 5 Phyt knock anthology 6 magazine, in which the hormesis word first appeared in academic journals (Southam 1943). The fourth stage is the stage of the twenty-first century in-depth study of hormesis biphasic dose–effect relationship. Hormesis phenomenon has become a hot chemical risk assessment, and related fields of study, published in academic journals, increased dramatically the amount of related articles, such as nature, science, and other magazines which have also published many articles on hormesis duplex dose–effect relationship comment (Calabrese and Baldwin 2003). Biphasic dose response is extremely generalizable according to the chemical/ physical agent, end point selection, and measurement. Biphasic dose response is apparent in both toxicology and biology (Calabrese 2008, 2013).

10.3 Biphasic Dose Response and Its Significance in Environmental Science and Risk Assessment Practices

Toxic chemical and biological dose responses show relationships that usually come in two classes. The first is the threshold model, the most extensively used and central model in toxicology, which can influence several aspects of research. The

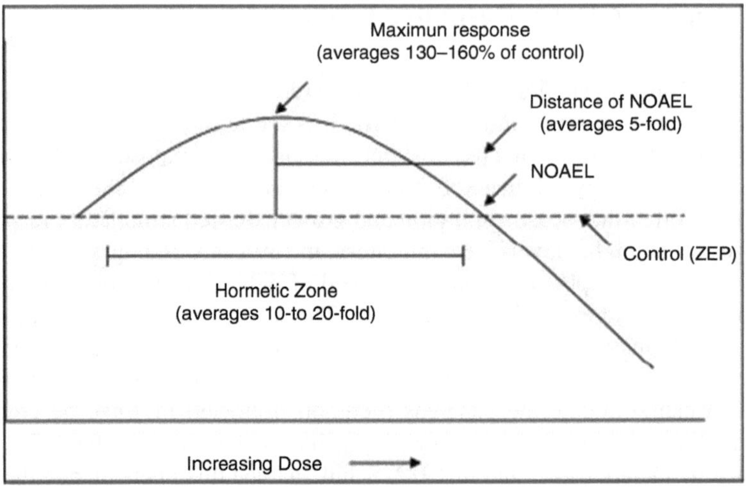

Fig. 10.2 Dose–response curve depicting the quantitative features of hormesis, adapted from Calabrese (2013)

second, the linear non-threshold model (LNT), shows increased biological effect with higher concentrations, typically seen with carcinogenic compounds (Davis and Svendsgaard 1990). However, neither the threshold model nor the linear non-threshold model is completely reliable, because various responses affected by low dose cannot be explained by these models (Fig. 10.2). For a variety of xenobiotics such as polychlorinated biphenyls, heavy metals, organic arsenic compounds, cyanide, polycyclic aromatic hydrocarbons, pesticides, and some antibiotics and different biological systems (animals, plants, and microorganisms), the hormetic model is quite common (Calabrese and Baldwin 2001; Hashmi et al. 2014). The new hormetic model challenges both the threshold model and the LNT model and suggests that lower doses enhance quantitative and qualitative changes in the response assessed. That is, as the dose of a carcinogenic compound decreases, it attains a position where the toxic chemical really might lessen the threat of cancer compared to the control group. Although hormesis is a very commonly observed phenomenon, the presumption of a biphasic response has been crippled by a less understood and accurate underlying vigorous mechanism and the partial correlation between in vitro and in vivo studies (Bae et al. 2008).

10.4 Progress in Biphasic Dose Response Using Growth as End Point

Growth rates are fundamental life-history traits of organisms (West et al. 2001). Variations in growth help differentiate the evolutionary success of individuals and population growth and play a vital role in ecosystem function, e.g., production. This end point not only provides a good picture of the ecological integrity of a system but also provides an early warning of impending ecological change under chemical stress (Harwell 1993; Suter II 2006). Using growth as end points means that one can easily figure out the potential beneficial or harmful effects of chemicals on organisms, their aggregate toxic effects, chemical bioavailability, and also characterize the nature of a beneficial or toxic effect, relatively simply and cheaply (Suter II 2006). Such methods are reproducible, responsive, representative, robust, and relevant and have practical applications in toxicity testing and ecological risk assessment (Spurgeon 2002). Biphasic dose response has been studied for inorganic and organic chemicals using growth as end point in both in vivo and in vitro test model (Hashmi et al. 2014, 2015a, b). Much of the in vivo studies focused on birds (Love et al. 2003; Ostrowski-Meissner 1984), algae (De Nicola et al. 2007), rats (Houshmand et al. 2009), bacteria (Christofi et al. 2002), etc. using PCB and aflatoxins, tannins, oxytocin, and heavy metals, respectively. Biphasic dose response using growth as end point under in vitro conditions is reported by fish cell lines (RTG-2) (Hahn et al. 1996), for Vero cells (Chen et al. 2010), human kidney cells (Ghosh et al. 2010), human liver cells, breast cells, embryonic neural stem cells (Hashmi et al. 2014), etc. exposed to PCBs were studied.

10.5 Cellular and Molecular Mechanisms of Biphasic Responses

In toxicology the biphasic dose response has gained advancement in terms of research; however, mechanistic foundations of the hormetic dose/concentration response are still lacking. General model of the biphasic dose response of cell growth with cell proliferation and apoptosis mechanisms is given in Fig. 10.3. However, it has been reported that biphasic dose response follows the single agonist mechanism, acting via two receptor subtypes that mediated/activated opposing stimulatory and inhibitory pathways (Szabadi 1977). The receptors which indicate high affinity for the agonist but relatively few receptors mediate stimulatory responses, while the inhibitory response typically mediated for those receptors with lower agonist affinity and having greater receptor capacity. The research and idea of Szabadi (1977) was supported and widely extended (Accomazzo et al. 2002; Alfonzo et al. 1998). In complementary fashion, biphasic dose responses followed a cybernetic feedback process which at low doses showed net stimulatory response (Stebbing 1982, 1998, 2003).

10.6 Receptor-Mediated and Cell Signaling-Mediated Biphasic Mechanisms

Receptor-based mechanism for biphasic dose–response relationships provides conceptual formation (Szabadi 1977). After that mechanistic advances in hormetic dose/concentration response began in the 1980s and have continued with increasing activity to the present. Many receptors were figured out to activate biphasic dose/

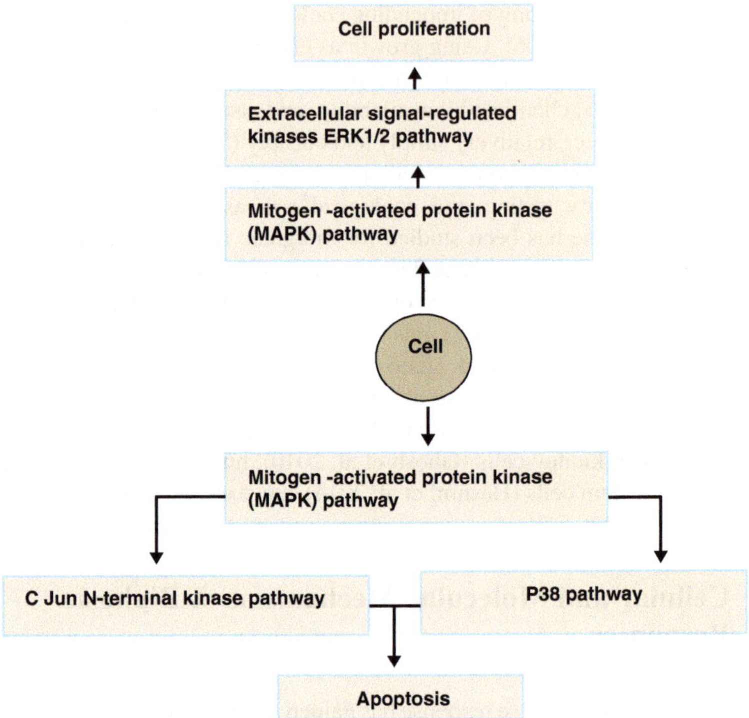

Fig. 10.3 General model of the biphasic dose response of cell growth with cell proliferation and apoptosis mechanisms (Adapted from Hashmi et al. 2014)

concentration responses, in the decade following Szabadi (1977). Whether biphasic dose–response mechanisms are receptor based and/or mediated, antagonists for many receptors were developed in the decades of the 1980s and 1990s. During the decade of the 1990s, receptor antagonists' application, synthesis, and assessment became progressively stronger to evaluate biphasic dose/concentration response mechanism, and similar developments occurred for cell signaling pathway mechanisms (Fig. 10.3).

10.6.1 Antagonist-Mediated Enhancement of Inhibitory Responses

In a number of places, it has been reported that sometimes failure of the antagonist to affect the high-dose/concentration inhibition occurs. Daidzein as an antagonist could enhance the high-dose/concentration inhibitory effect in a consistent fashion but not for genistein using the same cellular model (Dang and Löwik 2004). $CaCa_2$

cell model showed consistent enhancement of the high concentration inhibitory effect with genistein (Chen and Donovan 2004).

10.6.2 Xenobiotic-Mediated Biphasic Dose–Response Mechanisms Using In Vivo Model

In vitro system has been extensively used to study the hormetic mechanism, and wholesome reports used in vivo models (Homayoun et al. 2002; Houshmand et al. 2009; Hashmi et al. 2014). For example, male Sprague Dawley rats as in vivo model showed that a preconditioning treatment of oxytocin protected against IR-induced myocardial damage (Houshmand et al. 2009). Oxytocin antagonist blocked the protective effect of the oxytocin. Similarly, it was reported that pretreatment with ethanolamine decreased infarct size in a hormetic fashion (Kelly et al. 2010). Further, Kawabata et al. (1994) reported that L-arginine induced a biphasic nociception dose response in male mice.

10.6.3 Xenobiotic-Mediated Biphasic Dose–Response Mechanisms Using In Vitro Model

Xenobiotics such as PCBs mediate toxic effects through the aryl hydrocarbon receptor (AhR) signal transduction pathway followed by oxidative events, altered gene expression, and tumor promotion (Safe 1994). Previously it has been reported that PCBs activate AhR and subsequently the cytochrome P450 1A1 (*CYP1A1*) system through the production of reactive oxygen species (ROS) stress (Hennig et al. 2002; Ramadass et al. 2003). Uncoupling of electron transfer and oxygen reduction from monooxygenation by *CYP1A1* can result in the release of ROS and the subsequent formation of oxidative stress (Ramadass et al. 2003; Schecter et al. 2006). On the other hand, the toxic pathways arising from exposure to nonplanar PCBs are believed to be independent of the AhR. Nonplanar PCBs induced the overexpression of the cytochrome P450 2B (*CYP2B*) subfamily. Constitutive androstane receptor (CAR) is involved in nonplanar PCB-mediated induction of *CYP2B* (Honkakoski et al. 1998). At the cellular level, exogenous substances are often involved in oxidative stress, which results from the production of reactive oxygen species (ROS). Due to the severe nature of oxidative damage, ROS are kept as low as possible in organisms by an efficient ROS-scavenging system, which includes enzymatic antioxidants such as superoxide dismutase (SOD) and catalase (CAT). Several studies have shown that the hormetic effect may be linked to levels of oxidative stress (Yang et al. 2007; Zhang et al. 2009). More recently, it was reported that chemical exposure induces oxidative stress (the cellular production of ROS), which mediated activation of AKT, ERK1/2, and JNK signaling pathways, leading to the cell proliferation and apoptosis (Liu et al. 1996).

The well-known mitogen-activated protein kinase (MAPK) pathways transduce signals from environmental stimuli and in turn induce many cellular responses such as stress responses, cell proliferation, cell growth, differentiation, and apoptosis (Fig. 10.4) (Hao et al. 2009; He et al. 2007; Waskiewicz and Cooper 1995). There are three main pillars of MAPK pathways, and every three-tiered cascade provokes a phosphorylating protein pathway (PPP). From a range of extracellular signals to boost the expression of specific genes, the PPPs mediate signal transduction pathways (Samet et al. 1998; Xia et al. 1995). Growth-enhancing factors are the main cause of cell differentiation or proliferation and are regulated by the extracellular signal-regulated kinase 1/2 (ERK1/2) pathway (Marais and Marshall 1995; Minden et al. 1994), while cytokines and ultraviolet irradiation, heat, or synthesis inhibitors (stress-like signals) trigger apoptosis or growth arrest pathways such as the c-Jun N-terminal kinase (JNK) and p38 MAPK pathways (Beyaert et al. 1996; Kyriakis and Avruch 1996).

10.7 Hormesis Application in Xenobiotic-Contaminated Soil Risk Assessment

The key potential application of the hormesis concept is its adoption as the default model used in risk assessment. Recently, Calabrese (2004) proposed that hormesis could be considered as the default model in risk assessment on the basis of objective criteria such as generalizability of the model, its frequency in the toxicological literature, capacity to quantify false positives and negatives, relevance to end points of public health interest, capacity for validation, and ability to harmonize both noncarcinogen and carcinogen risk assessment. The implications of hormesis for ecotoxicology and ecological risk assessment have become even more widespread as suggested by previous literature (Chapman 2001, 2002; Forbes 2000; Gentile 2001; Gentile and van der Schalie 2000; Parsons 2000). Nonetheless, most texts on environmental toxicology and ecological risk assessment have generally not explored the hormesis concept even to a cursory degree.

Toxicology has long been dominated by an emphasis on very high doses and the assessment of toxic responses. In comparison with human health risk assessment, ecological risk assessment is poorly understood. In the environment, it is important to know what happens with a low dose, corresponding to real soil, air, and water contamination. Ecological/ecotoxicological risk assessment is an iterative process, a framework providing a basis for eventual risk management. It typically involves three tiers: *problem formulation* or *hazard assessment* (initial planning and

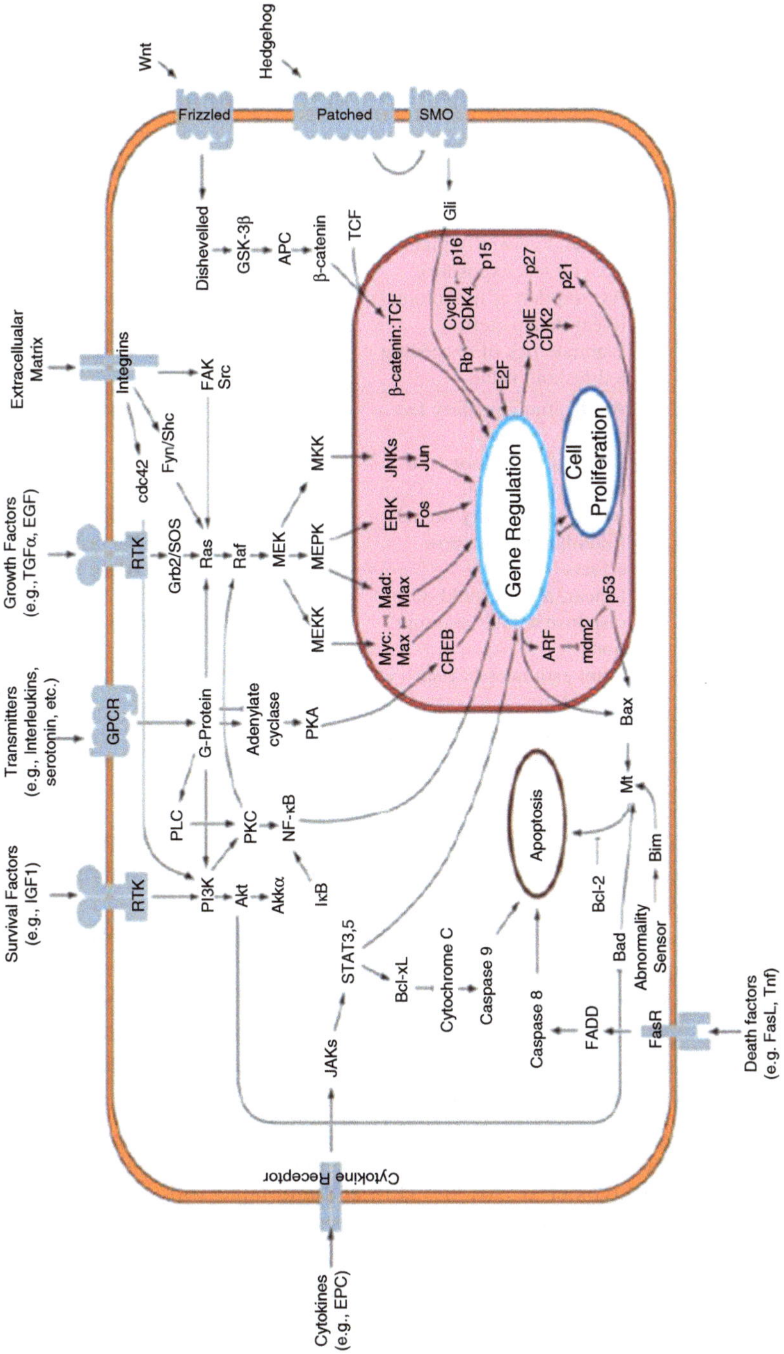

Fig. 10.4 Signal transduction pathways (Source: Wikimedia.org. fil in public domain)

Table 10.1 Implications of hormesis for ERA; similar implications apply to the incorporation of essentiality (e.g., essential elements) into ERA adapted from Chapman (2002)

Step	Present focus	Future focus
Problem formulation/hazard assessment	Few and generally high toxicity test exposures	More exposures; both high and low exposure concentrations
Exposure assessment	Upper exposure bounds	Distribution of actual doses/concentrations from exposure (both upper and lower bounds possible)
Effects assessment	Few and generally high toxicity test exposures Assumption of monotonic/linear data pattern Estimate risk of exceeding a threshold value	More exposures; both high and low exposure concentrations No preconceived assumptions (or confining statistics) Effects estimated by exposure–response modeling
Risk characterization	Application of safety factors Effect of a stressor on a defined end point Upper uncertainty bounds Likelihood of adverse effects Linear, no threshold dose–response default assumptions Point estimates of risk possible	No safety factors (if hermetic thresholds can be determined) Net effect of a stressor on organisms and communities Upper and lower uncertainty bounds Likelihood of adverse and beneficial effects No linear, no threshold dose–response default assumptions Risk always characterized as a range
Risk management	Chemical concentrations at or below background	Chemical concentrations at or below hormetic level

information gathering), *effects and exposure assessment* (data gathering and analysis), and *risk characterization* (assimilation and integration) (Calabrese and Ricci 2013) (Table 10.1).

Estrogenic endocrine-disrupting chemicals (i.e., xenoestrogens) often induce an inverted U-shaped dose response, based on low-dose stimulation. The inverted U-shaped dose response of endocrine-disrupting agents displays the same quantitative features as do those described for hormetic dose responses. Because hormesis is defined as a dose–response phenomenon characterized by low-dose stimulation and a high-dose inhibition, xenoestrogens inducing such biphasic dose responses clearly would be considered examples of hormesis. Literature suggested the use of hormetic model for the ecological risk assessment for certain chemicals such as pesticides in soil (Alves et al. 2013), Cd in contaminated soil (Zhang et al. 2009), and ciprofloxacin, tamoxifen, and cyclophosphamide in hospital wastewater and surface water (Mater et al. 2014). However, there is lack of data for some POPs for ecotoxicological/ecological risk assessment practices, suggesting that more studies should be conducted for better understanding the risk of these chemicals.

10.8 Conclusion

The present chapter provides a brief overview of less considered but more accurate phenomenon hormesis in toxicology with the main emphasis in xenobiotic-contaminated soil. The literature suggested that xenobiotics could also exhibit biphasic dose response; however, more research is needed in the future. The hormetic dose–response model is more common than the threshold model in toxicology. The widespread consideration of hormesis and essentiality into application of ecotoxicology first requires a value judgment of what is the normal response of a species to a toxicant. The hormetic phenomenon has the capacity to affect profound changes in the biomedical sciences as toxicological training and education, hazard assessment as reflected in animal model selection, end points measured, study design, type of statistical analyses used, and risk assessment practices (Calabrese 2005). However, the toxicology and risk assessment communities need to confront the reality that their discipline requires intellectual reform and has to adapt hormetic model as a default model for soil risk assessment practices. The chapter will provide information to university students, researchers, and management authorities for the application of hormetic model in risk assessment practices.

Acknowledgement The research was funded by TWAS-COMSTECH Research Grant Award_15-384 RG/ENG/AS_C and HEC Start Up Research grant 21-700/SRGP/R&D/HEC/2015.

References

Accomazzo MR, Cattaneo S, Nicosia S, Rovati GE (2002) Bell-shaped curves for prostaglandin-induced modulation of adenylate cyclase: two mutually opposing effects. Eur J Pharmacol 454:107–114

Alfonzo MJ, de Becemberg IL, de Villaroel SS, de Herrera VN, Misle AJ, de Alfonzo RG (1998) Two opposite signal transducing mechanisms regulate a G-protein-coupled guanylyl cyclase. Arch Biochem Biophys 350:19–25

Alves PRL, Cardoso EJ, Martines AM, Sousa JP, Pasini A (2013) Earthworm ecotoxicological assessments of pesticides used to treat seeds under tropical conditions. Chemosphere 90:2674–2682

Bae ON, Lim KM, Han JY, Jung BI, Lee JY, Noh JY, Chung SM, Lee MY, Lee JY, Chung JH (2008) U-shaped dose response in vasomotor tone: a mixed result of heterogenic response of multiple cells to xenobiotics. Toxicol Sci 103:181–190

Beyaert R, Cuenda A, Vanden Berghe W, Plaisance S, Lee JC, Haegeman G, Cohen P, Fiers W (1996) The p38/RK mitogen-activated protein kinase pathway regulates interleukin-6 synthesis response to tumor necrosis factor. The EMBO Journal 15:1914

Broadhurst PL (1957) Emotionality and the Yerkes-Dodson law. J Exp Psychol 54:345

Cabral M, Viegas C, Teixeira M, Sá-Correia I (2003) Toxicity of chlorinated phenoxyacetic acid herbicides in the experimental eukaryotic model *Saccharomyces cerevisiae*: role of pH and of growth phase and size of the yeast cell population. Chemosphere 51:47–54

Calabrese EJ (2004) Hormesis: from marginalization to mainstream: a case for hormesis as the default dose-response model in risk assessment. Toxicol Appl Pharmacol 197:125–136

Calabrese EJ (2005) Paradigm lost, paradigm found: the re-emergence of hormesis as a fundamental dose response model in the toxicological sciences. Environ Pollut 138:378–411
Calabrese EJ (2008) Hormesis: why it is important to toxicology and toxicologists. Environ Toxicol Chem 27:1451–1474
Calabrese EJ (2013) Hormesis: toxicological foundations and role in aging research. Exp Gerontol 48:99–102
Calabrese E, Baldwin L (2000) Chemical hormesis: its historical foundations as a biological hypothesis. Hum Exp Toxicol 19:2–31
Calabrese EJ, Baldwin LA (2001) Hormesis: U-shaped dose responses and their centrality in toxicology. Trends Pharmacol Sci 22:285–291
Calabrese EJ, Baldwin LA (2002) Applications of hormesis in toxicology, risk assessment and chemotherapeutics. Trends Pharmacol Sci 23:331–337
Calabrese EJ, Baldwin LA (2003) Hormesis: the dose-response revolution. Annu Rev Pharmacol Toxicol 43:175–197
Calabrese EJ, Ricci PF (2013) Hormesis and risk assessment. Gen Appl Syst Toxicol. doi:10.1002/9780470744307
Chapman P (2001) The implications of hormesis to ecotoxicology and ecological risk assessment. Hum Exp Toxicol 20:499–505
Chapman PM (2002) Ecological risk assessment (ERA) and hormesis. Sci Total Environ 288:131–140
Chen A-C, Donovan SM (2004) Genistein at a concentration present in soy infant formula inhibits Caco-2BBe cell proliferation by causing G2/M cell cycle arrest. J Nutr 134:1303–1308
Chen Y, Shen K, Shen C, Chen L, Chen X (2010) Comparison of structure-dependent hormetic cytotoxicity induced by coplanar and non-coplanar PCB congeners. J Hazard Mater 180:773–776
Christofi N, Hoffmann C, Tosh L (2002) Hormesis responses of free and immobilized light-emitting bacteria. Ecotoxicol Environ Saf 52:227–231
Dang Z, Löwik CW (2004) The balance between concurrent activation of ERs and PPARs determines daidzein-induced osteogenesis and adipogenesis. J Bone Miner Res 19:853–861
Davis JM, Svendsgaard DJ (1990) U-shaped dose-response curves: their occurrence and implications for risk assessment. J Toxicol Environ Health A 30:71–83
De Nicola E, Meriç S, Gallo M, Iaccarino M, Della Rocca C, Lofrano G, Russo T, Pagano G (2007) Vegetable and synthetic tannins induce hormesis/toxicity in sea urchin early development and in algal growth. Environ Pollut 146:46–54
Forbes VE (2000) Is hormesis an evolutionary expectation? Funct Ecol 14:12–24
Gallo MA, Doull J (1996) History and scope of toxicology. In: CD K (ed) Casarett and Doull's toxicology: the basic science of poisons, 5th edn. McGraw-Hill, New York, pp 3–10
Gentile J (2001) The implications of hormesis to ecotoxicology and ecological risk assessment (ERA). Hum Exp Toxicol 20:513–515
Gentile JH, van der Schalie WH (2000) Hormesis and ecological risk assessment: fact or fantasy? Hum Ecol Risk Assess Int J 6(2):227–236
Ghosh G, Subramanian IV, Adhikari N, Zhang X, Joshi HP, Basi D, Chandrashekhar YS, Hall JL, Roy S, Zeng Y, Ramakrishnan S (2010) Hypoxia-induced microRNA-424 expression in human endothelial cells regulates HIF-α isoforms and promotes angiogenesis. J Clin Invest 120:4141
Hahn ME, Woodward BL, Stegeman JJ, Kennedy SW (1996) Rapid assessment of induced cytochrome P4501 a protein and catalytic activity in fish hepatoma cells grown in multiwell plates: response to TCDD, TCDF, and two planar PCBs. Environ Toxicol Chem 15:582–591
Hao C, Hao W, Wei X, Xing L, Jiang J, Shang L (2009) The role of MAPK in the biphasic dose-response phenomenon induced by cadmium and mercury in HEK293 cells. Toxicol In Vitro 23:660–666
Harwell M (1993) Ecological risk assessment. In: 14th annual meeting(SETAC)-ecological risk assessment: lessons learned? Abstract Book
Hashmi MZ, Shen H, Zhu S, Yu C, Shen C (2014) Growth, bioluminescence and shoal behavior hormetic responses to inorganic and/or organic chemicals: a review. Environ Int 64:28–39

Hashmi MZ, Khan KY, Hu J, Su X, Abbas G, Yu C, Shen C (2015a) Hormetic effects of noncoplanar PCB exposed to human lung fibroblast cells (HELF) and possible role of oxidative stress. Environ Toxicol 30:1385–1392

Hashmi MZ, Shen C, Yu C (2015b) Hormetic responses of food-supplied PCB 31 to Zebrafish (*Danio rerio*) growth Dose-Response 13:dose-response. 14–013. Chaofeng

He X-Q, Chen R, Yang P, Li A-P, Zhou J-W, Liu Q-Z (2007) Biphasic effect of arsenite on cell proliferation and apoptosis is associated with the activation of JNK and ERK1/2 in human embryo lung fibroblast cells. Toxicol Appl Pharmacol 220:18–24

Hennig B, Hammock BD, Slim R, Toborek M, Saraswathi V, Robertson LW (2002) PCB-induced oxidative stress in endothelial cells: modulation by nutrients. Int J Hyg Environ Health 205:95–102

Homayoun H, Khavandgar S, Dehpour AR (2002) Anticonvulsant effects of cyclosporin A on pentylenetetrazole-induced seizure and kindling: modulation by nitricoxidergic system. Brain Res 939:1–10

Honkakoski P, Moore R, Washburn KA, Negishi M (1998) Activation by diverse xenochemicals of the 51-base pair phenobarbital-responsive enhancer module in the CYP2B10Gene. Mol Pharmacol 53:597–601

Houshmand F, Faghihi M, Zahediasl S (2009) Biphasic protective effect of oxytocin on cardiac ischemia/reperfusion injury in anaesthetized rats. Peptides 30:2301–2308

Hueppe F (1899) The principles of bacteriology. Open Court, Chicago

Kawabata A, Manabe S, Manabe Y, Takagi H (1994) Effect of topical administration of l-arginine on formalin-induced nociception in the mouse: a dual role of peripherally formed NO in pain modulation. Br J Pharmacol 112:547–550

Kelly RF, Lamont KT, Somers S, Hacking D, Lacerda L, Thomas P, Opie LH, Lecour S (2010) Ethanolamine is a novel STAT-3 dependent cardioprotective agent. Basic Res Cardiol 105:763–770

Kyriakis JM, Avruch J (1996) Protein kinase cascades activated by stress and inflammatory cytokines. Bioessays 18:567–577

Liu Z-g, Hsu H, DV G, Karin M (1996) Dissection of TNF receptor 1 effector functions: JNK activation is not linked to apoptosis while NF-κB activation prevents cell death. Cell 87:565–576

Love OP, Shutt LJ, Silfies JS, Bortolotti GR, Smits JE, Bird DM (2003) Effects of dietary PCB exposure on adrenocortical function in captive American kestrels (*Falco sparverius*). Ecotoxicology 12:199–208

Marais R, Marshall C (1995) Control of the ERK MAP kinase cascade by ras and raf. Cancer Surv 27:101–125

Mater N, Geret F, Castillo L, Faucet-Marquis V, Albasi C, Pfohl-Leszkowicz A (2014) In vitro tests aiding ecological risk assessment of ciprofloxacin, tamoxifen and cyclophosphamide in range of concentrations released in hospital wastewater and surface water. Environ Int 63:191–200

Minden A, Lin A, McMahon M, Lange-Carter C, Dérijard B, Davis RJ, Johnson GL, Karin M (1994) Differential activation of ERK and JNK mitogen-activated protein kinases by Raf-1 and MEKK. Science 266:1719–1723

Ostrowski-Meissner HT (1984) Biochemical and physiological responses of growing chickens and ducklings to dietary aflatoxins. Comp Biochem Physiol C: Comp Pharmacol 79:193–204

Parsons PA (2000) Hormesis: an adaptive expectation with emphasis on ionizing radiation. J Appl Toxicol 20:103–112

Ramadass P, Meerarani P, Toborek M, Robertson LW, Hennig B (2003) Dietary flavonoids modulate PCB-induced oxidative stress, CYP1A1 induction, and AhR-DNA binding activity in vascular endothelial cells. Toxicol Sci 76:212–219

Safe SH (1994) Polychlorinated biphenyls (PCBs): environmental impact, biochemical and toxic responses, and implications for risk assessment. Crit Rev Toxicol 24:87–149

Samet JM, Graves LM, Quay J, Dailey LA, Devlin RB, Ghio AJ, Wu W, Bromberg PA, Reed W (1998) Activation of MAPKs in human bronchial epithelial cells exposed to metals. Am J Physiol-Lung Cell Mol Physiol 275:L551–L558

Schecter A, Birnbaum L, Ryan JJ, Constable JD (2006) Dioxins: an overview. Environ Res 101:419–428

Schulz H (1888) Uber Hefegifte. Pflügers Arch Gesamte Physiol Menschen Tiere 42:517–541

Southam CM (1943) Effects of extract of western red-cedar heartwood on certain wood-decaying fungi in culture. Phytopathology 33:517–524

Spurgeon D (2002) Review of sublethal ecotoxicological tests for measuring harm in terrestrial ecosystems. Environment Agency, Bristol

Stebbing A (1982) Hormesis—the stimulation of growth by low levels of inhibitors. Sci Total Environ 22:213–234

Stebbing A (1998) A theory for growth hormesis. Mutat Res Fundam Mol Mech Mutagen 403:249–258

Stebbing A (2003) Adaptive responses account for the β-curve—hormesis is linked to acquired tolerance nonlinearity in biology, toxicology, medicine. Nonlinearity Biol Toxicol Med 1:493–511. doi:10.1080/15401420390271100

Suter GW II (2006) Ecological risk assessment. CRC Press, Boca Raton

Szabadi E (1977) A model of two functionally antagonistic receptor populations activated by the same agonist. J Theor Biol 69:101–112

Waskiewicz AJ, Cooper JA (1995) Mitogen and stress response pathways: MAP kinase cascades and phosphatase regulation in mammals and yeast. Curr Opin Cell Biol 7:798–805

West GB, Brown JH, Enquist BJ (2001) A general model for ontogenetic growth. Nature 413:628–631

Xia Z, Dickens M, Raingeaud J, Davis RJ, Greenberg ME (1995) Opposing effects of ERK and JNK-p38 MAP kinases on apoptosis. Science 270:1326–1331

Yang P, He XQ, Peng L, Li AP, Wang XR, Zhou JW, Liu QZ (2007) The role of oxidative stress in hormesis induced by sodium arsenite in human embryo lung fibroblast (HELF) cellular proliferation model. J Toxicol Environ Health A 70:976–983

Zhang Y, Shen G, Yu Y, Zhu H (2009) The hormetic effect of cadmium on the activity of antioxidant enzymes in the earthworm *Eisenia fetida*. Environ Pollut 157:3064–3068

Chapter 11
Agrochemicals and Soil Microbes: Interaction for Soil Health

Zaffar Malik, Maqshoof Ahmad, Ghulam Hassan Abassi, Muhammad Dawood, Azhar Hussain, and Moazzam Jamil

11.1 Introduction

Soil is a complex and heterogeneous system consisting of both living and nonliving components. The nonliving components of soil include soil texture (sand, salt, and clay) and organic matter, while soil living components are composed of soil flora (algae, fungi, and bacteria) and fauna (nematodes, arthropods, earthworm, and mollusks). Soils have a large number of bacteria, fungi, and animal species. It is reported that one cm^3 of fertile soil may have up to 10^{10} bacteria belonging to an estimated 10^4 species (Anderson and Weigel 2003). Soil microbes are important biological components of soil and perform functions in soil fertility through nutrient cycling and organic matter degradation (Tejada et al. 2014). Plant species get nutritional benefits from association with soil microbial mutualism and often provide sugars to these microbes, e.g., legumes and ectomycorrhizal tree species respond positively to rhizobia and fungi, respectively, compared to nonlegumes and non-ectomycorrhizal plants (Abbott et al. 2015). Soil microbes have a critical role in the functioning of terrestrial ecosystems and act as drivers of plant diversity (Van Der Heijden et al. 2008). Free-living soil microbes promote plant diversity by increasing the nutrient pool and alter the plant dominance due to symbiotic and pathogenic interactions (Prober et al. 2015). Similarly, plant diversity promotes soil microbes by providing food resources (exudates and litter), physical microhabitats, and environmental conditions (Millard and Singh 2010; Prober et al. 2015). Ecologists are well aware of the significance of microbes in soil functions, but it is due to

Z. Malik (✉) • M. Ahmad • G.H. Abassi • A. Hussain • M. Jamil
Department of Soil Science, University College of Agriculture and Environmental Sciences, The Islamia University of Bahawalpur, Bahawalpur 63100, Pakistan
e-mail: zafar_agrarian@yahoo.com

M. Dawood
Department of Environmental Sciences, Bahauddin Zakriya University, Multan, Pakistan

© Springer International Publishing AG 2017
M.Z. Hashmi et al. (eds.), *Xenobiotics in the Soil Environment*, Soil Biology 49,
DOI 10.1007/978-3-319-47744-2_11

advancements in the technology that enabled scientists to extract and characterize microbial communities to assess their functions in the soil (Bardgett et al. 2005).

The interactions between plant and microbe in the rhizospheric soil determine many processes including plant growth, pollutant degradation, and nutrient mobilization through atmospheric nitrogen fixation, solubilization of nutrients and control movements, and availability of pollutants to plants by the production of chelating agents, biosurfactants, acidification, redox changes, and phosphate solubilization (Glick 2010; Rajkumar et al. 2013). Bacteria and fungi transform and degrade different types of pollutants including pollutants from agriculture origin. The term degradation describes a variety of processes undertaken by the microbe: for example, the pollutant substances may either be completely mineralized, or it may undergo the partial transformation before a nontoxic metabolite excreted in the soil and become immobilized by bonding to matrix component like humic acid (Kiel and Engesser 2015; Murphy 2016).

Agrochemical is a generic term for the chemical products used in agriculture which may include pesticides (herbicides, insecticides, and fungicides), synthetic fertilizers, growth agents, and raw manures. Balanced application of fertilizers can promote soil organic matter and soil microbial diversity and hence improve soil functions and ultimately increases the soil resistance against environmental stresses. The use of these agrochemicals may increase the yield of crops, but their excessive use poses a great threat to environment, especially soil biology. Long-term and excessive uses of synthetic fertilizers (N, P, and K) along with organic fertilizers (manures) were reported to seriously affect soil microbial and soil enzymatic activities (Zhang et al. 2015b). Pesticides when entering into soil can cause environmental hazards and greatly alter the soil biochemical and microbial aspects. Soil microbes on the other hand overcome the toxic effects of these chemicals by performing different functions in the soils and make them less toxic for the environment.

The soil enzymatic activities are used as indicator for soil fertility and quality. These activities are important to predict microbiological and biochemical processes involved in organic matter synthesis and decomposition, nutrient cycling and availability, and biodegradation of toxic organic pollutants especially pesticide utilization in agriculture (Raiesi and Beheshti 2014). Enzymes are also important to monitor sudden changes in the soil environment because of their rapid response to any change within the soil. Studies showed that long-term supply of P and N can affect the soil microbial activities, and absolute and specific acid phosphate activities decreased as mineral P application rates and ratios increase (Zhang et al. 2015b). Apart from the degradation of agrochemicals, soil enzymatic activities also inhibited due to exposure to external pollutants like heavy metals, pesticides, and certain antibiotics used in agriculture. Soil enzymes like urease, alkaline phosphatase, and invertase are considered to be the indicators of soil disruption under stress (Jin et al. 2015). Certain microbial enzymes (urease and protease) are involved in the organic matter decomposition and N dynamics by participating in the hydrolysis of peptide bonds and release of NH^{+4} (Wang et al. 2008).

11.2 Agrochemical and Soil Microbes

Soil microbes perform crucial functions for the restoration of polluted soils and therefore are considered important component of ecosystem restoration (Harris-Hellal et al. 2009; Liu et al. 2016). Soil microorganisms are closely associated with plant roots, contributing to nutrient cycling, and show resistance and tolerance to pests, diseases, drought, and heavy metals and improve the soil structure. In this regard, soil microbial diversity and community composition are important indicators of soil health (Liu et al. 2016). The biodiversity and composition of soil microbes are influenced by many factors like pH, organic matter, and the additives such as pesticides and fertilizers. *Actinobacteria* and *Proteobacteria* were reported to be most abundant phyla in relation to soil containing agrochemicals and had a positive relationship with atrazine in soil (Liu et al. 2016). Similarly, *Arthrobacter* communities found in contaminated soils showed atrazine-degrading capabilities by having trzN, atzB, and atzC genes (Zhang et al. 2015b). Soil management practices such as type of rotation, proportion of organic material, and intensity of tillage can alter the physicochemical characteristics of the soil. These practices can manage the biogeochemical cycles and the dynamics of the microbiota, including colonization of arbuscular mycorrhizal fungi. The conventional tillage practices where crop residues are removed along with the synthetic inputs of herbicides exaggerate the negative effect of herbicides on the microbial population. Synthetic fertilizers and herbicides along with the type of organic material had greatly affected the enzymatic activities and mycorrhizal colonization in the soil (Mariela et al. 2016). Addition of soil organic matter in the form of compost provides several microorganisms and nutrients to indigenous degraders. Pentachlorophenol (PCP) is used in the wood treatment and persists in the environment and classified as priority contaminant. Many fungi such as *Phanerochaete chrysosporium*, *Antracophyllum discolor*, *Trametes versicolor*, *Ganoderma lucidum*, *Armillaria mellea*, and *Gloeophyllum striatum* showed potential to degrade and mineralize PCP (Bosso et al. 2015a, b). Bosso et al. (2015a) also found that compost and fungal strains had a synergistic effect on the reduction of more than 95 % of the extractable PCP. The use of organic fertilizers can enhance the microbial diversity and soil health. Long-term use of organic fertilizer contributed to the improvement of soil C and N contents and most enzyme activities (Zhang et al. 2015a) (Table 11.1).

11.3 Effect of Agrochemicals on Soil Biological Process in Soil

Agrochemicals are used to enhance or protect the agricultural crops, vegetables, and livestock through the world. Agrochemicals include all chemical products which are manufactured or processed for use in the agriculture sector and allied industries. According to The American Heritage Science Dictionary,

Table 11.1 Relationship of agrochemical/manures and soil microbes on the health of soil

Reference and location	Agrochemicals/manures	Microbial species	Effects
Liu et al. (2016) China	Atrazine	*Actinobacteria*, β-*Proteobacteria*, *Firmicutes*, *Acidobacteria*, *Verrucomicrobia*	To convert atrazine-polluted farmland to secondary forest. *Actinobacteria*, *Firmicutes*, and β-*Proteobacteria* species remain unchanged, while community structure of *Acidobacteria* and *Verrucomicrobia* decreased
Bosso et al. (2015b) Switzerland	Pentachlorophenol (PCP)	*Byssochlamys nivea*, *Scopulariopsis brumptii*	The compost and the fungal strains, *B. nivea* and *S. brumptii*, showed good capability to tolerate and degrade PCP
Zhang et al. (2015a) China	Organic fertilizers	Microbial community structure like G^+ and G^- bacteria and fungal communities	Improvement in enzyme activities and diverse community composition
Li et al. (2015) China	Compost manures, antibiotics (tylosin and vancomycin), and Cu	Soil microorganisms	Inhibited enzymatic activities. Selective pressures on both the microbial tolerance to Cu and the co-tolerance to antibiotics including tylosin and vancomycin. Greater risk for ecology due to animal wastes contaminated by the heavy metal Cu
Postma and Schilder (2015) Netherlands	Organic amendments (yeast and chitin)	*Lysobacter* populations	Suppression of *Rhizoctonia* disease by *Lysobacter* populations in soil
Guo et al. (2015) China	Azoxystrobin	Cultivable bacteria, fungi, and actinomycetes	Azoxystrobin showed significant negative effect on cultivable bacteria, fungi, and actinomycetes and enzymes assays (urease, protease, and dehydrogenase)

(continued)

Table 11.1 (continued)

Reference and location	Agrochemicals/manures	Microbial species	Effects
Sabale et al. (2015) India	Kresoxim methyl	Soil microbes	Dissipation kinetics of kresoxim methyl and its residual effect on soil extracellular (acid phosphatase, alkaline phosphatase, and β-glucosidase), and intracellular (dehydrogenase) enzyme activities
Malhotra et al. (2015) India	Fertilizers and pesticides	*Actinobacteria* Ammonifiers Rhizobium 16S rRNA	Adverse effect on soil microbial diversity and function was found at high fertilizer and pesticide usage. Diversity of nitrogen fixers was changed. Reduction in rhizobial *nifH* sequences
Majumder and Das (2016) India	Organophosphate insecticides (monocrotophos, profenophos, quinalphos, and triazophos)	Phosphate solubilizing microorganisms	Phosphate activities (acid phosphatase and alkaline phosphatase) significantly increased in the soil
Taheri et al. (2015) Canada	Chlorothalonil, pyraclostrobin, and boscalid	*Fusarium*, *Olpidium*, *Alternaria*, and *Cryptococcus*	Relative abundance of *Fusarium* increased after the application of fungicides
Zhang et al. (2015b) China	Swine manures and NPK fertilizers	Actinomycetes and G $^{+}$ bacterium enzymatic activities	Enzyme activities were positively correlated with actinomycetes and G+ bacterium. To improve microbial activity, P fertilizer should be applied with inorganic and organic forms

"agrochemicals" are defined as a chemical, such as a hormone, a fungicide, or an insecticide, that improves the production of crops. Pesticides are used to protect the crops from insects and diseases, and fertilizers are used to fulfill the deficiency of nutrients.

Agrochemical (fertilizer, pesticides, and insecticide) application is considerably increased in the developing world. Agrochemicals perform an important role in intensive agriculture. The use of these agrochemicals is a low-cost method of increasing production and gives the farmer a high economic return for his labor and investment. In our agricultural systems, pesticides and insecticides are introduced to protect crops against insects, pests, and weeds to minimize harmful effects and to get maximum yield (Yang et al. 2007). The ultimate objective of using fertilizer and pesticides is to enhance the crop yield for maximum benefit.

Agrochemicals are very important component of an agricultural industry for increasing crop yield from viewpoint of economic; therefore, their continued use is essential (Yang et al. 2007). Insecticides and pesticides are part of insect and pest control strategies which help in improving public health.

No doubt, agrochemicals (fertilizer, pesticide, and insecticides) have tremendous role in the increase of crop production (Blain 1989; Hashmi et al. 2004); on the other hand, these agrochemicals have generated many environmental and health issues. Agrochemicals directly affect the soil microbes which ultimately deteriorate the soil health. Pesticides may drastically affect the proliferation of beneficial soil microorganisms and their associated biotransformation in the soil. The best pesticide is the one, which only toxic to targeted organism and degrade it and not contaminate the soil environment. When nontarget organism is exposed to the applied pesticide, the pesticide may have harmful effects on that nontarget organism. (Odenkirchen and Eisler 1988; Bretaud et al. 2000; Galloway and Handy 2003).

Total amount of applied pesticide which reaches to targeted organism is about 0.1%, while remaining 99.9% contaminates soil environment. (Pimentel 1995; Ardley 1999; Chenseng et al. 2006; Carriger et al. 2006). Hence, these harmful compounds enter the food chain through food crops and soil and cause toxicity to the biodiversity (Araújo et al. 2003; CFTRI 2003). The use of pesticides is increasing day by day in agriculture sector; toxic chemicals in agrochemical directly affect the soil microorganisms (Baxter and Cummings 2008). A pesticide (Simethoate) applied to the agriculture crops also has negative effects on the microbial diversity and the soil enzyme activity in soil (Begum and Rajesh 2015). The indigenous microorganisms are affected by the application of an herbicide Bromoxynil (Baxter and Cummings 2008).

Pesticide application caused reduction in microorganisms present in the soil (Ubuoh et al. 2012). Some pesticides disturbed molecular interactions between plants and nitrogen-fixing rhizobia and consequently reduced the vital process of biological nitrogen fixation. Inactivation of phosphorus-solubilizing and nitrogen-fixing microorganisms is observed in pesticide-contaminated soils (Hussain et al. 2009a, b). A great difference has been observed between microbial population in pesticide-treated and nontreated soil confirming the effects of microbial deduction and extinction due to indiscriminate ways of applying pesticide in the soil (Ubuoh et al. 2012). Pesticides also influence soil biochemical processes driven by microbial and enzymatic reactions (Demanou et al. 2004). Nitrogenous fertilizers decrease the mycorrhizal fungus spore number and root colonization. Application of fertilizer adversely affects soil microorganisms. Negative effect of pesticide on the symbiotic efficiency and nitrogenase activities was observed on *Rhizobium leguminosarum* bv, Trifolii KGL, *Sinorhizobium meliloti* Bp, *Bradyrhizobium sp.*, and *Ornithopus B* bacteria (Demanou et al. 2004). Pesticides have adversely affected the microbial mineralization of organic compounds and related biotransformation like nutrient dynamics and their bioavailability to plants.

In general, fungi, actinomycetes, and the phosphorus-solubilizing microbes were more affected by the agrochemicals (fertilizers, pesticides, insecticide, etc.) with reductions in their abundance. Microorganisms presented varying behavior depending on the agrochemical and the nitrogen fixers were stimulated. Chemicals may affect in different ways the microorganisms that are responsible for the decomposition of organic matter in the rhizospheric soil.

11.4 Effect of Agrochemicals on Soil Enzymes

The living microbial fractions of soil are the governing bodies of a soil ecosystem, and their diversity dictates the potential of a soil to support life. These soil microbes exudate several enzymes performing many fundamental processes in soils. The activities of these enzymes in soil are keenly important for sustainable use of a soil as natural resource. These enzymes are involved in geochemical cycling of all chemical elements, which continue the flow of nutrients in ecosystem not only in soil but, in fact, of the whole planet Earth. However, the emergence of industrial era and the use of chemicals in crop production severely deteriorated the soil health, especially the biological conditions (Trasar-Cepeda et al. 2000; Chu et al. 2003).

In a modern agriculture system, a huge amount of agrochemicals are being incorporated in soil as amendments, fertilizers, and pesticides during the production of a crop. The residual and hazardous effect of these chemicals on living beings is well documented. From a soil microbiologist's point of view, these chemicals are also equally harmful for the microbial populations of the soil resulting in severe destruction of soil enzymatic activity (Dong et al. 2005; Qiu et al. 2006). These chemicals eradicate the beneficial soil microbes which ultimately remove important enzymatic components from a chain of reactions that play vital role in synchronizing important chemical processes in soil (Hernandez-Rodriguez et al. 2006).

There are numerous enzymes that are present in a soil ecosystem. The most important microbial enzymes are dehydrogenase, fluorescein diacetate hydrolase, and enzymes involved in biochemical cycles of nutrients, i.e., enzymes involved in C cycle, P cycle, N cycle, and S cycle (Ronhede et al. 2007). These microbial enzymes are the indicators of soil biological health, fertility, as well as the chemical status of the soil (Trasar-Cepeda et al. 2000; Chu et al. 2003). Consequently, the assessment of a particular enzyme activity plays an important part in understanding the biogeochemical cycles of nutrients. The effect of agrochemicals on soil microbes is assessed by determining several functional attributes that mainly includes the monitoring of enzymes related to the mineralization of nutrients (N, P, S, etc.). These parameters are important as the biogeochemical cycling and transformation of nutrients require microbial enzymes (Antonious 2003; Bending et al. 2004). Agrochemicals incorporated in the soil interrupt the metabolism or enzymatic activities (Hussain et al. 2009a, b). Adverse influences of agrochemicals on soil microbial enzymatic, e.g., hydrolases, oxidoreductases, and dehydrogenase, activities have been reported in the literature (Gianfreda and Bollag 1994; Kalam et al. 2004; Menon et al. 2005; Gil-Sotres et al. 2005; Hussain et al. 2009a, b).

Dehydrogenases are important enzymes synthesized by almost all living cells. These are involved in the respiration reactions in organisms. In soil, its presence regulates the decomposition of organic matter. This microbial enzyme is the indicator of inclusive activity of soil microbial population. The literature showed that agrochemicals significantly reduce the activity of dehydrogenase enzyme (Sebiomo et al. 2012). Several insecticides, fungicides, and herbicides are reported to suppress the activity of this enzyme (Burrows and Edwards 2004; Bending et al. 2007; Rasool and Reshi 2010; Sebiomo et al. 2012). However, the impact of these chemicals on soil microbes is regulated by the physiochemical properties of agrochemicals as well as by the nature of the soil and prevalent environmental circumstances (Monkiedje and Spiteller 2002). In general, the activity of dehydrogenase is severely affected by fungicides that are formulated to act on metabolic machinery of pathogen (Dick et al. 2000; Bello et al. 2008).

The fluorescein diacetate hydrolase (FDH) is also an important functional enzyme in soil ecosystem. It is an important enzymatic parameter to measure the activity of microbial enzymatic activities. These enzymes are related to hydrolyze fluorescein diacetate compounds present in soil and take part in nutrient cycling. Measuring the FDH activity has the potential to roughly denote soil enzymatic activities and collected biological impacts (Janvier et al. 2007; Bishnu et al. 2012). Fluorescein diacetate compounds are hydrolyzed by several FDH enzymes including protease, esterase, amylase, etc. The studies showed that FDH activity is stimulated by the application of imidazolines, organophosphate, and organochlorines (Kalyani et al. 2010; Riah et al. 2014).

The literature showed that microbes involved in the cycling of C, P, N, and S had a variety of response toward pesticide application. The enzymes involved in C cycling, i.e., cellulase and b-glucosidase, are not widely reported to be affected by agrochemicals. Most of the pesticides have neutral effect on their activity (Deng and Tabatabai 1994; Niemi et al. 2009). Omar and Abdel-Sater (2001) reported that even ten times higher doses of herbicides have no pronounced effect on cellulase and b-glucosidase activities. However, some fungicides from amide group are reported to have inhibitory effect on cellulase and b-glucosidase activities (Monkiedje and Spiteller 2002; Niemi and Vepsalainen 2005). Phosphorus cycle enzymes are grouped into five major classes, i.e., the phosphomonoesterases (PME), the phosphodiesterases (PDE), the phosphotriesterases (PTE), the pyrophosphatases (PP), and the phosphoamidases (Riah et al. 2014). Acid phosphatase and alkaline phosphatase function in low and high pH, respectively, are key enzymes involved in P cycling in soil, and are included in PME. These are diverse enzymes adapted to catalyze hydrolysis organic phosphoric acid (Monkiedje and Spiteller 2002; Riah et al. 2014). The activity of phosphatase enzymes are negatively affected by the higher concentrations of pesticides which hamper the velocity of P cycle in the ecosystem (Schneider et al. 2001). Studies showed that this decrease in phosphatases directly related to P stress and plant growth (Dick et al. 2000; Sharma et al. 2010; Monkiedje and Spiteller 2002). Urease is an important enzyme involved in N cycling by catalyzing the hydrolysis of organic urea. This enzyme is integral N transformation in soil, and it is mainly derived from plant and soil microbes (Riah et al. 2014). It is reported that agrochemicals have negative impact on urease activity (Sukul 2006; Caceres et al. 2009; Tejada 2009). On the other hand, the activities of enzymes involved in S cycling are positively affected by most of the agrochemicals, and literature shows that the enzymes show accelerated activities under high concentration of pesticides (Ganeshamurthy and Takkar 1997; Kertesz and Mirleau 2004; Riah et al. 2014).

The incorporation of high concentrations of chemicals in soil severely disturbs the natural equilibrium of soil enzymes that affect the natural cycles otherwise going on a constant speed. The literature showed that these chemicals, in most cases, may speed up or inhibit the activities of soil enzymes that alter the nature of soil processes. However, there is a huge gap present in literature for understanding the chemical, physical, and biological nature of effected processes and enzymatic activities.

11.5 Role of Soil Microbes for Detoxification of Agrochemicals

The use of living organisms for the removal of agrochemical pollutant from the soil is a common process for the remediation of soil because of its low cost and environment-friendly approach (Mackay and Frasar 2000). Biotic degradation is the use of microorganisms, plants, or their enzymes to detoxify contaminants in the soil and water environments (Lu et al. 2006). Most living organisms are capable of directly degrading the agrochemical pollutants, and some of the microbes have the ability of metabolizing even very complex compounds. Biotic degradation is divided in to two different types: microbes and plants.

Microorganisms present in the soil also degrade the agrochemicals. Degradation of a persistent organochlorine pesticide like endosulfan by the soil microbes is an effective approach. Three bacterial strains, *Pseudomonas spinosa*, *P. aeruginosa*, and *Burkholderia cepacia*, were reported as efficient degraders of endosulfan (Hussain et al. 2007). Microbial technology (using plant growth-promoting bacteria) has received increasing attention to detoxify the hazardous waste from the environment (Radhika et al. 2006).

Different bacterial species, including members of the genera *Flavobacterium*, *Alcaligenes*, *Rhodococcus*, and *Pseudomonas*, metabolize toxic compound which is found in pesticide and insecticides (Aislabie and Lloyd-Jones 1995; Richins et al. 1997; Mulchandani et al. 1999). Soil microorganisms have the ability to degrade the toxic compounds of pesticides in the soil (Hussain et al. 2007). The use of bacteria for the degradation and detoxification of toxic chemicals released from agrochemicals (pesticides and fertilizers) is an effective and environment-friendly tool to decontaminate the polluted areas. Indigenous bacteria are capable of metabolizing toxic compounds of pesticides and provide environmentally friendly means of on-site detoxification. Environmental factors such as temperature, pH, water potential, nutrient, amount of pesticides and fertilizer directly influence the rate of biodegradations. Presences of inductive enzyme are also necessary for the degradation of toxic compounds (Singh et al. 2009).

Best suitable pathway of degradation is the most important factor for the remediation of the agrochemical-contaminated soil. Degradation of toxic agrochemical compounds affected by different physical, chemical, and biological factors to a variable extent in the soil environment is a highly complex process (Lu et al. 2006).

Remaining products of pesticides after fully degradation by soil microbial assimilation result to enhance soil microorganisms population and their activities (Das and Mukherjee 2000). Different insecticides with different doses affected the chemical and microbiological properties of rhizospheric soil in rice field. Carbofuran strongly stimulated the mineralization of organic carbon. Microorganisms have similar herbicide-degraded processes and an intrinsic nature for rapid genetic adaptation to chemicals in the environment. Pollutants of diverse chemical nature could be remediated from soil and water with the use of potential microorganisms in the soil.

11.6 Conclusions and Future Perspectives

The interaction of plants and microbes in rhizospheric soil controlled many process like plant growth, nutrient mobilization, nitrogen fixation, pollutant degradation, and control movement and availability of pollutants through chelating agents, biosurfactants, and acidification. Agrochemicals such as fertilizers, herbicides, and pesticides contributed a lot in the production of food and fiber but had profound effects on the soil fertility and health. Extensive use of these chemicals caused a stress over natural resources along with the increase in water and air pollution. Soil microbes are under great threat by the application of diverse nature of agrochemicals which interrupt large biochemical reactions performed by the soil microbes. Many functions like C, P, S, and N cycle, metabolism, and enzymatic activities performed by soil microbes are disrupted by the application of agrochemicals. The use of organic/natural resources should be promoted against agrochemicals which enhance the microbial activities in soil and hence can improve soil health. Moreover, soil microbes have the potential to remediate the agrochemical-contaminated soils thus improving soil fertility. Therefore, researchers and scientists should develop and follow technologies for sustainable use of natural resources.

References

Abbott KC, Karst J, Biederman LA, Borrett SR, Hastings A, Walsh V, Bever JD (2015) Spatial heterogeneity in soil microbes alters outcomes of plant competition. PLoS One 10 (5):e0125788

Aislabie J, Lloyd-Jones G (1995) A review of bacterial-degradation of pesticides. Soil Res 33(6): 925–942

Anderson T, Weigel H (2003) On the current debate about soil biodiversity. Landbauforschung Volkenrode 53(4):223–234

Antonious GF (2003) Impact of soil management and two botanical insecticides on urease and invertase activity. J Environ Sci Health B 38(4):479–488

Araújo AD, Monteiro RTR, Abarkeli RB (2003) Effect of glyphosate on the microbial activity of two Brazilian soils. Chemosphere 52:799–804
Ardley JH (1999) Pesticide considerations: an environmental concern. Agric Sci 12:21–24
Bardgett RD, Bowman WD, Kaufmann R, Schmidt SK (2005) A temporal approach to linking aboveground and belowground ecology. Trends Ecol Evol 20:634–641
Baxter J, Cummings SP (2008) The degradation of the herbicide bromoxynil and its impact on bacterial diversity in a top soil. J Appl Microbiol 104(6):1605–1616
Begum SM, Rajesh G (2015) Impact of microbial diversity and soil enzymatic activity in dimethoate amended soils series of Tamil Nadu
Bello D, Trasar-Cepeda C, Leiros MC, Gil-Sotres F (2008) Evaluation of various tests for the diagnosis of soil contamination by 2,4,5-trichlorophenol (2,4,5-TCP). Environ Pollut 156(3): 611–617
Bending GD, Turner MK, Rayns F, Marx MC, Wood M (2004) Microbial and biochemical soil quality indicators and their potential for differentiating areas under contrasting agricultural management regimes. Soil Biol Biochem 36:1785–1792
Bending GD, Rodriguez-Cruz MS, Lincoln SD (2007) Fungicide impacts on microbial communities in soils with contrasting management histories. Chemosphere 69:82–88
Bishnu A, Chakraborty A, Chakrabarti K, Saha T (2012) Ethion degradation and its correlation with microbial and biochemical parameters of tea soils. Biol Fertil Soils 48:19–29
Blain PG (1989) Aspects of pesticide toxicology. Adverse Drug React Acute Poisoning Rev 9(1): 37–68
Bosso A, Salmaso D, De Faveri E, Guaita M, Franceschi D (2015a) The use of carboxymethylcellulose for the tartaric stabilization of white wines, in comparison with other oenological additives. VITIS-J Grapevine Res 49(2):95
Bosso L, Scelza R, Testa A, Cristinzio G, Rao MA (2015b) Depletion of pentachlorophenol contamination in an agricultural soil treated with byssochlamys nivea, scopulariopsis brumptii and urban waste compost: a laboratory microcosm study. Water Air Soil Pollut 226(6):1–9
Bretaud S, Toutant JP, Saglio P (2000) Effects of carbofuran, diuron and nicosulfuronon acetyl cholinesterase activity in goldfish (*Carassius auratus*). Ecotoxicol Environ Saf 47:17–124
Burrows LA, Edwards CA (2004) The use of integrated soil microcosms to assess the impact of carbendazim on soil ecosystems. Ecotoxicology 13(1):143–161
Caceres TP, He WX, Megharaj M, Naidu R (2009) Effect of insecticide fenamiphos on soil microbial activities in Australian and Ecuadorean soils. J Environ Sci Health B 44:13–17
Carriger JF, Rand GM, Gardinali PR, Perry WB, Tompkins MS, Fernandez AM (2006) Pesticides of potential ecological concern in sediment from south Florida canals: an ecological risk prioritization for aquatic arthropods. Soil Sediment Contam 15(1):21–45
Central Food Technological Research Institute (CFTRI) (2003) Report of analysis of pesticide residue in soft drink samples, Mysore
Chenseng L, Toepel K, Irish R, Richard AF, Dana BB, Bravo R (2006) Organic diets significantly lower children's dietary exposure to organophosphorous pesticides. Environ Health Perspect 114(2):260–263
Chu HY, Zhu JG, Xie ZB, Zhang HY, Cao ZH, Li ZG (2003) Effects of lanthanum on dehydrogenase activity and carbon dioxide evolution in a haplic acrisol. J Soil Res 41:731–739
Das A, Mukherjee D (2000) Soil application of insecticides influences microorganisms and plant nutrients. Appl Soil Ecol 14(1):55–62
Demanou J, Monkiédjé A, Njiné T, Foto SM, Nola M, Zebaze Togouet SH, Kemka N (2004) Changes in soil chemical properties and microbial activities in response to the fungicide Ridomil Gold plus copper. Int J Environ Res Public Health 1(1):26–34
Deng SP, Tabatabai MA (1994) Cellulase activity of soils. Soil Biol Biochem 26:1347–1354
Dick WA, Cheng L, Wang P (2000) Soil acid and alkaline phosphatase activity as pH adjustment indicators. Soil Biol Biochem 32:1915–1919

Dong YJ, Bartlam M, Sun L, Zhou YF, Zhang ZP, Zhang CG, Rao ZH, Zhang XE (2005) Crystal structure of methyl parathion hydrolase from *Pseudomonas* sp. WBC-3. J Mol Biol 353: 655–663

Galloway T, Handy R (2003) Immunotoxicity of organophosphorus pesticides. Ecotoxicology 12: 345–363

Ganeshamurthy AN, Takkar PN (1997) Residual management of sulfur applied to soybean or wheat in a soybean-wheat system on vertisols. J Soil Res 35:199–208

Gianfreda L, Bollag JM (1994) Effect of soils on the behavior of immobilized enzymes. Soil Sci Soc Am J 58:1672–1681

Gil-Sotres F, Trasar-Cepeda C, Leirós MC, Seoane S (2005) Different approaches to evaluating soil quality using biochemical properties. Soil Biol Biochem 37:877

Glick BR (2010) Using soil bacteria to facilitate phytoremediation. Biotechnol Adv 28(3): 367–374

Guo P, Zhu L, Wang J, Wang J, Xie H, Lv D (2015) Enzymatic activities and microbial biomass in black soil as affected by azoxystrobin. Environ Earth Sci 74(2):1353–1361

Harris-Hellal J, Vallaeys T, Garnier-Zarli E, Bousserrhine N (2009) Effects of mercury on soil microbial communities in tropical soils of French Guyana. Appl Soil Ecol 41:59–68

Hashmi I, Khan MA, Kim JG (2004) Malathion degradation by *Pseudomonas* using activated sludge treatment system (biosimulator). Biotechnology 3:82–89

Hernandez-Rodriguez D, Sanchez JE, Nieto MG, Marquez-Rocha FJ (2006) Degradation of endosulfan during substrate preparation and cultivation of *Pleurotus pulmonarius*. World J Microbiol Biotechnol 22:753–760

Hussain S, Arshad M, Saleem M, Khalid A (2007) Biodegradation of α-and β-endosulfan by soil bacteria. Biodegradation 18(6):731–740

Hussain S, Siddique T, Saleem M, Arshad M, Khalid A (2009a) Impact of pesticides on soil microbial diversity, enzymes, and biochemical reactions. Adv Agron 102:159–200

Hussain S, Siddique T, Arshad M, Saleem M (2009b) Bioremediation and phytoremediation of pesticides: recent advances. Crit Rev Environ Sci Technol 39:843–907

Janvier C, Villeneuve F, Alabouvette C, Edel-Hermann V, Mateille T, Steinberg C (2007) Soil health through soil disease suppression: which strategy from descriptors to indicators? Soil Biol Biochem 39:1–23

Jin Z, Li Z, Li Q, Hu Q, Yang R, Tang H, Li M, Huang B, Zhang J, Li G (2015) Canonical correspondence analysis of soil heavy metal pollution, microflora and enzyme activities in the Pb–Zn mine tailing dam collapse area of Sidi village, SW China. Environ Earth Sci 73: 267–274

Kalam A, Tah J, Mukherjee AK (2004) Pesticide effects on microbial population and soil enzyme activities during vermicomposting of agricultural waste. J Environ Biol 25:201–208

Kalyani SS, Sharma J, Dureja P, Singh S, Lata (2010) Influence of endosulfan on microbial biomass and soil enzymatic activities of a tropical alfisol. Bull Environ Contam Toxicol 84: 351–356

Kertesz MA, Mirleau P (2004) The role of soil microbes in plant sulphur nutrition. J Exp Bot 55: 1939–1945

Kiel M, Engesser KH (2015) The biodegradation vs. biotransformation of fluorosubstituted aromatics. Appl Microbiol Biotechnol 99:7433–7464

Li Y, Liu B, Zhang X, Gao M, Wang J (2015) Effects of Cu exposure on enzyme activities and selection for microbial tolerances during swine-manure composting. J Hazard Mater 283: 512–518

Liu X, Hui C, Bi L, Romantschuk M, Kontro M, Strömmer R, Hui N (2016) Bacterial community structure in atrazine treated reforested farmland in Wuying China. Appl Soil Ecol 98:39–46

Lu C, Barr DB, Pearson M, Bartell S, Bravo R (2006) A longitudinal approach to assessing urban and suburban children's exposure to pyrethroid pesticides. Environ Health Perspect 1: 1419–1423

Mackay D, Frasar A (2000) Bioaccumulation of persistent organic chemicals: mechanisms and models. Environ Pollut 110:375–391
Majumder SP, Das AC (2016) Phosphate-solubility and phosphatase activity in Gangetic alluvial soil as influenced by organophosphate insecticide residues. Ecotoxicol Environ Saf 126:56–61
Malhotra J, Aparna K, Dua A, Sangwan N, Trimurtulu N, Rao DL, Lal R (2015) Microbial and genetic ecology of tropical Vertisols under intensive chemical farming. Environ Monit Assess 187(1):1–7
Mariela FP, Pável MEI, Manuel SRL, Jesús FGM, Reyes LO (2016) Dehydrogenase and mycorrhizal colonization: tools for monitoring agrosystem soil quality. Appl Soil Ecol 100:144–153
Menon P, Gopal M, Parsad R (2005) Effects of chlorpyrifos and quinalphos on dehydrogenase activities and reduction of Fe^{3+} in the soils of two semi-arid fields of tropical India. Agric Ecosyst Environ 108:73–83
Millard P, Singh B (2010) Does grassland vegetation drive soil microbial diversity? Nutr Cycl Agroecosyst 88:147–158
Monkiedje A, Spiteller M (2002) Effects of the phenylamide fungicides, mefenoxam and metalaxyl on the microbiological properties of a sandy loam and a sandy clay soil. Biol Fertil Soils 35:393–398
Mulchandani P, Mulchandani A, Kaneva L, Chen W (1999) Biosensor for direct determination of organophosphate nerve agents. 1. Potentiometric enzyme electrode. Biosens Bioelectron 14(1): 77–81
Murphy CD (2016) Microbial degradation of fluorinated drugs: biochemical pathways, impacts on the environment and potential applications. Appl Microbiol Biotechnol 100(6):2617–2627
Niemi RM, Vepsalainen M (2005) Stability of the fluorogenic enzyme substrates and pH optima of enzyme activities in different Finnish soils. J Microbiol Methods 60:195–205
Niemi RM, Heiskanen I, Ahtiainen JH, Rahkonen A, Mäntykoski K, Welling L, Laitinen P, Ruuttunen P (2009) Microbial toxicity and impacts on soil enzyme activities of pesticides used in potato cultivation. Appl Soil Ecol 41:293–304
Odenkirchen EW, Eisler R (1988) Chlorpyrifos hazards to fish, wildlife and invertebrates: a synoptic review. Report no. 85(1.13) U.S. fish and wildlife services. Biological report of chlorpyrifos in pure culture and in soil. Bull Environ Contam Toxicol 62:48–55
Omar SA, Abdel-Sater MA (2001) Microbial populations and enzyme activities in soil treated with pesticides. Water Air Soil Pollut 127:49–63
Pimentel D (1995) Amounts of pesticides reaching target pests: environmental impacts and ethics. J Agric Environ Ethics 8(1):17–29
Postma J, Schilder MT (2015) Enhancement of soil suppressiveness against *Rhizoctoniasolani* in sugar beet by organic amendments. Appl Soil Ecol 94:72–79
Prober SM, Leff JW, Bates ST, Borer ET, Firn J, Harpole WS, Lind EM, Seabloom EW, Adler PB, Bakker JD (2015) Plant diversity predicts beta but not alpha diversity of soil microbes across grasslands worldwide. Ecol Lett 18:85–95
Qiu XH, Bai WQ, Zhong QZ, Li M, He FQ, Li BT (2006) Isolation and characterization of a bacterial strain of the genus *Ochrobactrum* with methyl parathion mineralizing activity. J Appl Microbiol 101:986–994
Radhika V, Subramanian S, Natarajan KA (2006) Bioremediation of zinc using *Desulfotomaculum* nigrificans: bioprecipitation and characterization studies. Water Res 40:3628–3636
Raiesi F, Beheshti A (2014) Soil specific enzyme activity shows more clearly soil responses to paddy rice cultivation than absolute enzyme activity in primary forests of northwest Iran. Appl Soil Ecol 75:63–70
Rajkumar M, Prasad MNV, Swaminathan S, Freitas H (2013) Climate change driven plant–metal–microbe interactions. Environ Int 53:74–86
Rasool N, Reshi ZA (2010) Effect of the fungicide Mancozeb at different application rates on enzyme activities in a silt loam soil of the Kashmir Himalaya, India. Trop Ecol 51:199–205
Riah W, Laval K, Laroche-Ajzenberg E, Mougin C, Latour X, Trinsoutrot-Gattin I (2014) Effects of pesticides on soil enzymes: a review. Environ Chem Lett. doi:10.1007/s10311-014-0458-2

Richins RD, Kaneva I, Mulchandani A, Chen W (1997) Biodegradation of organophosphorus pesticides by surface-expressed organophosphorus hydrolase. Nat Biotechnol 15:984–987

Ronhede S, Sorensen SR, Jensen B, Aamand J (2007) Mineralization of hydroxylatedisoproturon metabolites produced by fungi. Soil Biol Biochem 39:1751–1758

Sabale RP, Shabeer TPA, Utture SC, Banerjee K, Oulkar DP, Adsule PG, Deshmukh MB (2015) Kresoxim methyl dissipation kinetics and its residue effect on soil extra-cellular and intra-cellular enzymatic activity in four different soils of India. J Environ Sci Health B 50(2):90–98

Schneider K, Turrion MB, Grierson PF, Gallardo JF (2001) Phosphatase activity, microbial phosphorus and fine root growth in forest soils in the Sierra de Gata, western central Spain. Biol Fertil Soils 34:151–155

Sebiomo A, Ogundero VW, Bankole SA (2012) Effect of four herbicides on microbial population, soil organic matter and dehydrogenase activity. Afr J Biotechnol 10:770–778

Sharma ID, Chandel RS, Kaushal R, Patyal SK, Sharma PL (2010) Effect of pesticides on biochemical properties of tomato (*Solanum lycopersicon*) cropped soil. Indian J Agric Sci 80:640–644

Singh PB, Sharma S, Saini HS, Chadha BS (2009) Biosurfactant production by *Pseudomonas* sp. and its role in aqueous phase partitioning and biodegradation of chlorpyrifos. Lett Appl Microbiol 49(3):378–383

Sukul P (2006) Enzymatic activities and microbial biomass in soil as influenced by metalaxyl residues. Soil Biol Biochem 28:320–326

Taheri AE, Hamel C, Gan Y (2015) Pyrosequencing reveals the impact of foliar fungicide application to chickpea on root fungal communities of durum wheat in subsequent year. Fungal Ecol 15:73–81

Tejada M (2009) Evolution of soil biological properties after addition of glyphosate, diflufenican and glyphosate + diflufenican herbicides. Chemosphere 76:365–373

Tejada M, Rodríguez-Morgado B, Gómez I, Parrado J (2014) Degradation of chlorpyrifos using different biostimulants/biofertilizers: Effects on soil biochemical properties and microbial community. Appl Soil Ecol 84:158–165

Trasar-Cepeda C, Leiros MC, Seoane S, Gil-Sotres F (2000) Limitations of soil enzymes as indicators of soil pollution. Soil Biol Biochem 32:1867–1875

Ubuoh EA, Akhionbare SMO, Akhionbare WN (2012) Effects of pesticide application on soil microbial spectrum: case study-fecolart demonstration farm, Owerri-West, Imo state, Nigeria. Int J Multidisc Sci Eng 3(2):34

Van Der Heijden MG, Bardgett RD, Van Straalen NM (2008) The unseen majority: soil microbes as drivers of plant diversity and productivity in terrestrial ecosystems. Ecol Lett 11:296–310

Wang SL, Wang CY, Huang TY (2008) Microbial reclamation of squid pen for the production of a novel extracellular serine protease by *Lactobacillus* paracasei subsp paracasei TKU012. Bioresour Technol 99:3411–3417

Yang CL, Sun TH, He WX, Zhou QX, Chen S (2007) Single and joint effects of pesticides and mercury on soil urease. J Environ Sci 19(2):210–216

Zhang Q, Zhou W, Liang G, Sun J, Wang X, He P (2015a) Distribution of soil nutrients, extracellular enzyme activities and microbial communities across particle-size fractions in a long-term fertilizer experiment. Appl Soil Ecol 94:59–71

Zhang X, Dong W, Dai X, Schaeffer S, Yang F, Radosevich M, Xu L, Liu X, Sun X (2015b) Responses of absolute and specific soil enzyme activities to long term additions of organic and mineral fertilizer. Sci Total Environ 536:59–67

Chapter 12
Soil Microbial and Enzymatic Diversity as Affected by the Presence of Xenobiotics

Liliana Gianfreda and Maria A. Rao

12.1 Introduction

A healthy soil is a prerequisite for a healthy life. Health, quality, and sustainability of soils depend on their physical, chemical, and biological diversity. Therefore, soil biodiversity that extremely exceeds aboveground biodiversity is essential for ecosystem stability and services. A precise correlation exists between soil biodiversity and agricultural soil management (Thiele-Bruhn et al. 2012). Since soil is a nonrenewable resource and its degradation is not recoverable within a human life span, any event or disturbance leading to more or less changes in its biodiversity may alter, even irreversibly, the normal equilibrium existing among different soil components. A direct consequence on the safety of the environment and human life will derive.

A large plethora of different microorganisms (more than 1010 bacteria and likely more than thousands of different species g^{-1} soil) and thousands of enzymes, the main effectors of microbial correct functioning, contribute to soil biodiversity. As defined by Nannipieri et al. (2003), biodiversity can be simply defined as the number of species present in the studied system. More accurately, microbial diversity includes "genetic diversity, that is, the amount and distribution of genetic information within microbial species, diversity of bacterial and fungal species in microbial communities, and ecological diversity, that is variation in community structure, complexity of interactions, number of trophic levels, and number of guilds" (Nannipieri et al. 2003). To understand the relationship existing between biodiversity and soil biological functions is of particular interest to maintain this biodiversity and its role for a functional biosphere (Nannipieri et al. 2003). Matulich et al. (2013) outlined that three fundamental questions exist about

L. Gianfreda (✉) • M.A. Rao
Department of Agricultural Sciences, University of Naples Federico II, Naples, Italy
e-mail: liliana.gianfreda@unina.it; mariarao@unina.it

© Springer International Publishing AG 2017

M.Z. Hashmi et al. (eds.), *Xenobiotics in the Soil Environment*, Soil Biology 49,
DOI 10.1007/978-3-319-47744-2_12

microbial diversity: "How do we measure it? How much is there? What is its role in ecosystems?" As reported below reliable methodologies should be employed "to characterize microbial diversity, to survey the magnitude of global microbial diversity, and to review some of the ecological and evolutionary forces important in maintaining microbial diversity" (Matulich et al. 2013).

In the modern era, several industrial and/or agricultural activities have caused the release into the environment of many xenobiotic chemicals (i.e., strange synthetic chemicals foreign to biosphere) of different chemical and structural complexities. Many of them are toxic, carcinogenic, and mutagenic for living organisms. Serious and most often irreversible alterations of the natural environmental balance will derive (Gianfreda and Rao 2008, 2011).

Soil can be considered the natural and preferred sink for contamination by xenobiotics, and the majority of soil processes are reactions mediated by microorganisms. Therefore, several investigations and studies have been dedicated to elucidate the effect of xenobiotics on the life and functioning of microorganisms and their resilience to xenobiotic's presence. When xenobiotics are not recognized by existing degradative enzymes, they accumulate in soil (and in water). One of the most deriving effects is the suppression of soil's microbial life and/or substantial changes of the composition of microbial communities. Important negative effects may also occur on the biochemical activities as evidenced by different levels of production and activity of enzymes involved in the biogeochemical cycles of the main nutrients (Gianfreda and Rao 2008 and references therein).

To understand and study the impact of xenobiotics on the soil's microbial community structure, diversity, and functionality, sensitive and accurate techniques and methodologies are necessary to measure the richness (e.g., the number of different bacteria and fungi) and the evenness (e.g., their relative abundance) of the soil microflora. Evaluation of microbial active fraction is also necessary (Blagodatskaya and Kuzyakov 2013). Similarly, reliable assays should be available to test the activity of enzymes and their performance in soil. Nowadays, several traditional and advanced methodologies are available to measure both soil microflora and their activity (Table 12.1) (Gianfreda and Rao 2008).

Complementary measurements, i.e., polyphasic approaches combining culture-based methods with culture-dependent and culture-independent molecular methods, and comparative data analyses rather than one method are necessary to have a reliable biological soil characterization (Cappa et al. 2014; Cycoń et al. 2013; Kozdrój and van Elsas 2001; Gianfreda and Rao 2008; Zhang et al. 2012).

It should also take into account that soil is a particular dynamic microhabitat, where organic and inorganic components, microorganisms, enzymes, nutrients, and environmental factors interact with each other and change with time and space. Obviously, these interactions can determine spatial heterogeneity of soil microbial communities and enzyme activities and affect their expression and organization levels, in turn depending on different soil properties (Kirk et al. 2004). Consequently, difficulties in the estimation of the entire microbial community and its active part, represented by enzyme activities, may arise even if advanced approaches have been used in their measurement. Therefore, to achieve truthful

Table 12.1 Methods to study microbial diversity (adapted from Kirk et al. 2004; Nannipieri et al. 2003)

Method	Advantages	Disadvantages
Culture-dependent methods (rely on growing prokaryotes)		
Plate count Direct counting by fluorescent microscopy	Very fast; inexpensive From 100 to 1000 more microorganisms detected	Laborious; only a limited number of microorganisms can be cultured (uncultivable microorganisms are not detected); no count of specific species; no discrimination between lived and dead cells; non-adapted for fast-growing species or fungal species, producing huge amounts of spores
Community level physiological profiling (CLPP)	Fast; highly reproducible; quite inexpensive	Fast-growing organisms are favored; only available C-utilizing organisms are detected
Culture-independent but nonmolecular methods		
Fatty acid methyl ester analysis (FAME)	Direct measure of viable biomass in addition to a biochemical profile of the microbial community; extraction from soil without culturing microorganisms; specific for some organisms or species	No identification of specific microorganisms; estimation only of relevant changes in community structure; influence of external factors on results; great amounts of material are needed for fungi
Culture-independent methods (rely on molecular methods to study microorganisms within their environment; little prior knowledge about which microorganisms are of interest is required; useful for emerging pollutants; in general not quantitative)		
DNA microarrays and hybridization G + C (guanine + cysteine), nucleic acid reassociation and hybridization	Extraction of total DNA; not influenced by PCR; DNA or RNA studied	Low sensitivity; need of high copy number of sequences to detection; the efficiency of lysis and extraction strongly influences the method
DGGE and TGGE (denaturing or temperature gradient gel electrophoresis)	Rapid, reproducible and reliable; simultaneous analysis of several samples; identification of bands made by hybridization with specific probes or by extraction and sequencing	Errors by PCA; different fingerprints generated by the same DNA; the lysis and extraction efficiency influences the method; storage (long) of samples before extraction can change community; more than one species can be represented by one band; only dominant species are detected; inhibition due to the presence of contaminants; production of chimeric or heteroduplex DNA molecules
Single-strand conformation polymorphism (SSCP)	Similar to DGGE and TGGE	Errors by PCR; more than one conformation can be formed by some ssDNA

comprehension of such events, methods must be capable not only of accurately measuring the presence and activity of the chosen microorganisms and enzymes but also of detecting their changes, following the application of a disturbing factor. As outlined by Blagodatskaya and Kuzyakov (2013), it is necessary to intensify studies to achieve "the standardization, further elaboration, and broad application of approaches focused on the portion of active microorganisms in soil and their functions."

Soil quality is a basic requirement for sustainable land use. In this respect, soil microbial and enzymatic diversity, both sensitive to human-induced changes, are an important factor contributing to it. Assessing their response to xenobiotic presence may provide a helpful tool to evaluate soil status, its quality, and its productivity. The present chapter will briefly survey the mutual interactions establishing in soil among xenobiotic substances and microbial and enzymatic soil activities.

12.2 Xenobiotics in Soil

Xenobiotic is a comprehensive term (derived from the Greek words ξένος (xenos) = foreigner, stranger and βίος (bios, vios) = life, plus the Greek suffix for adjectives -τικός, -ή, -ό) used to indicate a man-made, foreign chemical substance, found within an organism that is not naturally produced by or expected to be present within that organism. Indeed, it is not recognizable by the enzymatic systems of living organisms, and if introduced into the environment, it can accumulate and reach concentrations such to cause undesirable effects.

The term xenobiotic is very often used as synonymous of pollutant or contaminant, although these two words have different meanings. Pollutant is a chemical or material out of place that has adverse effects on any organism; contaminant is a substance, which usually does not cause hazard of any type, but it is present at concentrations higher than would occur naturally. Typical examples of contaminants are some heavy metals. Many of them are present at very low amount in the environment and living organisms, performing important functions like cofactors of enzymes or other biological molecules or taking part to metabolic pathways. However, when their concentration exceeds some safe thresholds, they become dangerous to the environment and organisms. Extensive and often not reversible injures may arise, as, for instance, by arsenic or aluminum.

The number of xenobiotic substances is very high. Figure 12.1 shows the most common xenobiotics now possibly present into the environment.

Typical organic xenobiotics are pesticides, fuels, solvents, alkanes, polyaromatic, chlorinated and nitro-aromatic compounds, azo compounds, dyes, *s*-triazines, organic sulfonic acids, and nitrogen and phosphorus compounds, while inorganic pollutants are mainly represented by toxic heavy metals. Synthetic fibers and polymers such as polyethylene and polypropylene as well as plasticizers and softeners, contained in textiles, may be also released into the environment and accumulate in it.

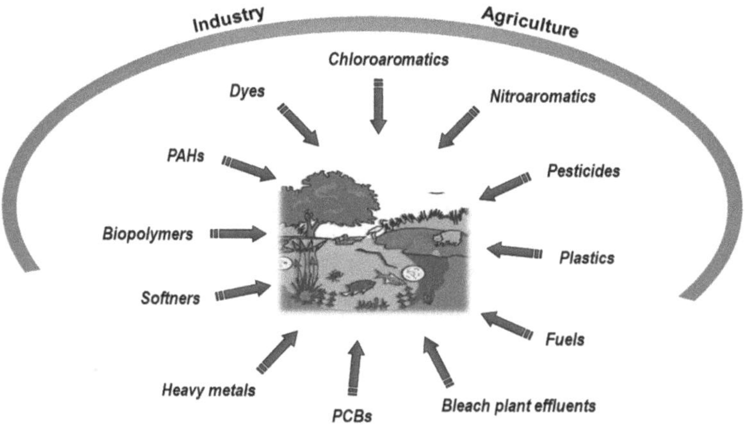

Fig. 12.1 Inorganic and organic compounds polluting the environment (modified from Gianfreda and Rao 2004)

The majority of these compounds are "persistent organic pollutants" (POPs) and among them particular attention has been devoted to the so-called emerging pollutants. They represent a heterogeneous class of chemical compounds produced by anthropic activities, not subjected to specific regulation. For some of them, effects on human health, terrestrial, and aquatic environments in the middle or long term are still unknown (Deblonde et al. 2011; Gavrilescu et al. 2015). They include personal care products, fragrances, plasticizers, steroids and hormones, illicit drugs, gasoline additives, flame retardants, metals, pesticides, phthalates, PAHs, pharmaceuticals, and endocrine-disrupting compounds (EDCs) (Gavrilescu et al. 2015). These latter are of great concern. They include bisphenol A (2,2-bis (4-hydroxyphenyl), propane (BPA), nonylphenol (NP), their alkylphenolic derivatives, triclosan, genistein, and others, broadly used in several industrial and residential applications. Their chemical structure and properties render them mostly active and toxic toward the hormone systems in mammals (Gavrilescu et al. 2015).

Several phenomena will affect the fate of a xenobiotic in soil. Indeed, a xenobiotic entering the soil is subjected to transfer and degradation processes. Transfer processes relocate the substance without altering its structure. They include adsorption, retention by crops, runoff movements, diffusion, and mainly sorption and desorption to soil colloids. On the other hand, degradation processes alter the chemical structure of the chemical and occur through chemical, biological, and photochemical transformations. Several factors including both soil properties (e.g., organic matter, clay content, cation exchange capacity, acidity), and the physical and chemical characteristics of the compound may influence both processes.

In particular, to be transformed by biological degradation, a xenobiotic has to be bioavailable. Usually, a xenobiotic can be not recognizable as substrate by microbes to act upon and degrade it. Its large molecular nature can make it difficult to enter microbial cell. It can be highly stable and insoluble in water. All these factors can render the xenobiotic recalcitrant and mainly not bioavailable.

As outlined by Katayama et al. (2010) "soil bioavailability of chemicals is a complex topic, and is affected by chemical properties, soil properties, species exposed, climate, and interaction processes". Bioavailability can change with time and space. For instance, aging phenomena, e.g., the xenobiotic sorbed to soil colloids, may lead to bound residues, no more available to soil microbiota. Specific assays have been developed to evaluate the real bioavailability of the target xenobiotic (Dean and Scott 2004). Mild extraction methods, largely utilized to predict xenobiotic bioavailability, however, might fail in evaluating the changes of bioavailability over time (Katayama et al. 2010). When a process to eliminate partially or totally xenobiotics from soil should be implemented, it is necessary to consider these factors.

12.3 Response of Soil to Xenobiotics' Presence

When xenobiotics are in soil, interrelated phenomena involving xenobiotics and either soil microorganisms or enzymes may occur (Gianfreda and Rao 2008). Xenobiotics may influence soil microbial activity and diversity as well as soil enzymatic activity and diversity. Simultaneously, xenobiotics can be transformed by soil microbial activity or soil enzymatic activity. The first two phenomena are briefly discussed in the following, whereas xenobiotic transformation by soil biotic components is the object of other chapters of the present book.

As cited above, the effects of xenobiotics on soil microbial and enzymatic diversity are influenced by the interactions at interfaces between organic and inorganic soil colloids and xenobiotics through sorption/desorption mechanisms. These interactions will affect the movement of xenobiotics, their availability for plant or microbial uptake, their transformation by abiotic or biotic agents, and their influence on soil processes. The resulting combination of all these factors will lead to the observed change, if any, of the soil microbial and/or enzymatic diversity.

12.3.1 Influence of Xenobiotics on Microflora Diversity and Activity

Microbial activity and diversity have an important role in soil economy. Indeed, favorable, balanced, and active soil biota could help to provide soil conditions necessary for sustainable crop production with very little negative environmental effects. Any disturbance affecting this equilibrium will result in unsafe effects on microbial activity, carbon turnover, and nutrient-supplying potential of soils. Consequently, important functions driven by soil microflora will be altered. They include (1) preventing aggressive plant pathogens taking hold and improving plant's ability to withstand disease effect, (2) reducing the loss of inorganic fertilizers through erosion and leaching by short-term immobilization, (3) stabilizing soil structure, and (4) reducing the reliance for agrochemicals and reduced persistence of pesticides in soil and thus lessening off-site impacts.

The most evident effect of xenobiotics on soil microflora is probably the death of it or a different distribution of bacterial and/or fungal groups, with consequent alteration of their functionality. Several reports support that the lowest the concentration of the xenobiotic, the lesser the damage of the soil microbial diversity. In other words, when the amount of xenobiotics is very low, no or negligible effects are detected. On the contrary, in heavily contaminated soil, even the absence of microflora or enzymatic activity can occur.

A negative significant correlation between soil hydrocarbon (PAHs and PCBs) concentrations and substrate-induced respiration (SIR), total bacteria and fungi counts, and most of measured enzyme activities was established (Pérez-Leblic et al. 2012). Moreover, the most polluted samples in a municipal solid waste (MSW) landfill placed in Torrejón de Ardoz (Madrid, Spain) showed the lowest microbial diversity, only formed by six phyla being Proteobacteria and Acidobacteria the most representative (Pérez-Leblic et al. 2012). Similarly, genetic diversity of three soil samples, characterized by different properties, hydrocarbon-pollution history, and enrichment cultures were measured by denaturing gradient gel electrophoresis (DGGE) as bands of amplified 16S rDNA sequences from the soils and enrichment community DNAs (Andreoni et al. 2004). When analyzed by Shannon index (H), the highest genetic biodiversity ($H = 2.87$) was found in the soil with a medium-term exposition to PAHs and the poorest biodiversity ($H = 0.85$) in the soil with a long-term exposition to alkanes and PAHs, where the absence or lower levels of enzyme activities were measured (Andreoni et al. 2004). A high Shannon index (2.13) was also found in the soil containing negligible amounts of organic pollutants but a highest Cu content (Andreoni et al. 2004).

Molecular fingerprinting technique and terminal restriction fragment length polymorphism (TRFLP) were utilized to examine the profiles of soil bacterial communities in PAH-contaminated soil taken from a timber treatment facility in southern Ireland (Muckian et al. 2007). Results, interpreted using sophisticated multivariate statistical analysis, suggested a strict correlation between structure and type of PAHs and bacterial community composition. In particular, bacterial community in soil contaminated with two-ringed PAHs was different from that of soil contaminated with multicomponent PAH mixtures (Muckian et al. 2007).

Culture-based methods were combined with molecular methods in a polyphasic approach to increase knowledge on microbial diversity in contaminated soil. Culture-dependent and culture-independent methods were useful to examine microbial diversity in soil samples from a former industrial site in the European Alps (mainly used for aluminum production and heavily contaminated with petroleum hydrocarbons) (Zhang et al. 2012). The majority of cultivable bacterial isolates (34/48) belonged to the Proteobacteria (with a predominance of Alphaproteobacteria and Gammaproteobacteria), while the remaining isolates were affiliated with the Actinobacteria, Cytophaga–Flavobacterium–Bacteroides, and Firmicutes. Most of them were able to utilize hydrocarbons (Zhang et al. 2012).

Experiments made by spiking soil with anthracene were performed to evaluate if and how much the presence of the hydrocarbon could modify bacterial population (Núñez et al. 2012). An initial PCR–DGGE analysis indicated that the bacterial population changed in the contaminated soil over time and in the soil profile after 28 days, while no effects were detected in the uncontaminated soil. Phylogenetic analysis revealed that sequences belonged to six different phyla, i.e., Acidobacteria, Actinobacteria, Chloroflexi, Gemmatimonadetes, Nitrospirae, and Proteobacteria. Contamination of soil changed temporarily by strongly reducing the percentage of Alphaproteobacteria, i.e., Sphingomonadales, and Gammaproteobacteria, i.e., Xanthomonadales, while the percentage of Actinobacteria and Acidobacteria grew up more than doubled. However, the effect of the contaminant was transient, since changes in the bacterial population with depth were only detectable in the contaminated soil after 28 days (Núñez et al. 2012).

PCR–TGGE profiles, combined with band sequencing and phylogenetic analysis of the prominent phylotypes, showed shifts in the *Mycobacterium* spp. community during incubation of soil with pyrene and phenanthrene (Cheung and Kinkle 2005). Reductions in species diversity and enrichments of specific populations were observed in all pyrene- and phenanthrene-treated soils, in contrast to the relatively stable control soil profiles.

The effects on microbial diversity may be amplified by the co-occurrence of different types of xenobiotics like PAHs and heavy metals. Experiments were made by a holist approach (combination of physicochemical, biological, and advanced molecular methods) on soils collected from a former manufactured gas plant site with a very high concentration of both PAHs and toxic heavy metals (lead, cadmium, and zinc) (Thavamani et al. 2012). The presence of the two contaminants not only reduced the diversity of microbial population but also showed a few distinctive species by exerting selective pressure. Furthermore, a severe inhibition of enzyme activities occurred (Thavamani et al. 2012).

Methods based on analyses of signature biomarkers (nucleic acids and fatty acids) permitted to identify reduction of microbial diversity in a soil contaminated by heavy metals and organic pollutants (Kozdrój and van Elsas 2001). Significant shifts in the microbial community structure occurred. Only a few populations (species) dominated and others disappeared. Some of these species were never isolated by conventional methods (e.g., an increase in *Acidobacterium* or a decrease in terrestrial non-thermophilic *Crenarchaeota*) (Kozdrój and van Elsas 2001).

Pesticide application had also deleterious effects on soil microbial diversity. Although these chemicals, designed primarily to control insect, weed, fungal, or nematode pest, have solved several agricultural problems, their indiscriminate use has led to a severe contamination of soil and environment (Gianfreda and Rao 2008, 2011). In some cases, however, soil microorganisms developed adaptation mechanisms to tolerate the presence of such compounds. It depends, however, by the number of exposures to the target pesticide. For instance, Yu et al. (2009) demonstrated that soil microorganisms become adapted to carbendazim, a systemic benzimidazole fungicide, after repeated applications. Indeed, the richness and dominant character of soil microorganisms remained unchanged after repeated application as revealed by Simpson and Shannon indexes of soil microbial community from carbendazim-treated soil, similar to those from the control soil (Yu et al. 2009). After a longer exposure, however, McIntosh indexes indicated that the balance of soil microorganisms was altered, due to the enrichment of the specific carbendazim-adapting strains in soil (Yu et al. 2009).

Similar evolution of bacteria capable of degrading the pesticide among indigenous microflora occurred when the genetic and functional diversity of microbial communities in soil treated with napropamide was evaluated (Cycoń et al. 2013). Napropamide is a very common herbicide in agricultural practice and has toxic effect to soil microorganisms. DGGE and the community level physiological profile (CLPP) methods were used. Moreover, the effect of the herbicide on the community structure of the cultivable soil bacteria was evaluated by r/K-strategy approach. The herbicide affected the structure of microbial community, although no significant changes occurred of the richness (S) and genetic diversity (H) values at a field rate (FR) of 2.25 mg kg^{-1} soil. When a 10 FR dose of herbicide was applied, significant changes in the S and H values of dominant soil bacteria occurred with a shift toward degrading napropamide bacteria. Further analyses (average well-color development and bacterial growth strategy) revealed that napropamide significantly affected the physiological state of cultivable bacteria and caused a reduction in the rate of colony formation as well as a prolonged time of growth rate (Cycoń et al. 2013; Guo et al. 2009).

Enrichment of triclosan-resistant *Pseudomonas* strains and 22-fold decrease of cultivable microbial populations were observed in a soil supplemented with 4 mg kg^{-1} of triclosan (Svenningsen et al. 2011).

Combined toxic effects may arise when pesticide and heavy metals are simultaneously present in soil (Gianfreda and Rao 2008, 2011). Wang et al. (2009) studied the combined effects of cadmium (Cd, 10 mg kg^{-1} soil) and three different amounts of herbicide butachlor (10, 50, and 100 mg kg^{-1} soil) by random amplified polymorphic DNA (RAPD) analysis. The diversity of the microbial community was strongly affected by high concentration butachlor and the combined impacts of Cd and butachlor. Indeed, RAPD analysis showed loss of original bands and appearance of new bands when compared with the control soil (Wang et al. 2009). Significant differences were observed for biochemical activities and microbial community structures in soil microcosms contaminated with crude oil with or without chromium and copper (dos Santos et al. 2012).

12.3.2 Changes of Enzyme Activities

As reported above, any alteration of the soil microflora diversity can result in a change of soil biological functionality. Both the production and the activity of enzymes linked to a particular group of microorganisms and serving for a given biological function may be altered with consequent negative effects on the correct performance of nutrient's biogeochemical cycles (Gianfreda and Bollag 1996; Gianfreda and Rao 2008).

The massive presence of PAHs and PCBs in municipal solid waste (MSW) landfill not only affected microbial diversity (see above) but also reduced the activities of acid and alkaline phosphatases, β-glucosidase, β-N-acetylglucosaminidase, invertase, cellulase, and urease activities, thus profoundly acting on the biological functionality of the area (Pérez-Leblic et al. 2012).

Enzyme activities related to the cycling of the main nutrients C, N, P, and S were investigated in completely different soil types (Gianfreda et al. 2005). Agricultural and uncultivated polluted soils from various parts of Europe were selected. Pollutants included PAHs, alkanes, BTEX, phenols, and heavy metals. As compared to agricultural soils, uncultivated polluted soils in general showed lower values of enzymatic activities. Very low level of urease and dehydrogenase or undetectable levels of both enzymes were measured in the polluted soils. Significant and negative correlation coefficients between phenanthrene and PAHs contents and dehydrogenase and urease activities were also obtained, thus suggesting the two enzymes are the most sensitive to this kind of pollution. Therefore, these enzymes could be proposed as sensitive indicators in soil polluted with these compounds (Gianfreda et al. 2005).

Laboratory experiments were performed to evaluate the effect of phenanthrene or PCP when a soil with no history of contamination was supplemented with the two contaminants (Scelza et al. 2007, 2008). The activities of β-glucosidase, acid phosphatase, arylsulfatase, urease, and dehydrogenase, along with chemical, physical, and other biological properties (e.g., microbial biomass and respiration) were measured over time (Scelza et al. 2007, 2008). As a general response, enzymatic activities were severely depressed, although at different extents, in the presence of pollutants. The disappearance of the pollutant for natural adsorption phenomena on soil components as well as for natural degradation by indigenous microorganisms or by addition of compost and/or competent-degrading bacterial cultures (to simulate a biostimulation and/or bioaugmentation decontamination process) led to a detectable recovery of the soil biological properties (Scelza et al. 2007, 2008).

Complementary measurements of bioassays and microbial activity, made in a soil polluted with hydrocarbons and heavy metals, indicated that urease and dehydrogenase were sensitive to the presence of metals (31 % inhibition of urease and 50 % inhibition of dehydrogenase in the most contaminated soil) (Brohon et al. 2001). Moreover, in the presence of the highest concentration of metals, the microbial activities were low, and the bioassays revealed a high potential toxicity (e.g., IC50 for Microtox obtained with a 15 % dilution of soil, 90 % inhibition of β-galactosidase activity) (Brohon et al. 2001).

Contrasting results on enzyme activities were obtained in the presence of napropamide (Guo et al. 2009). Napropamide applied at 2–80 mg kg^{-1} soil had inhibitory effects on the activity of urease and invertase. By contrast, catalase activity enhanced during the initial 7 days of napropamide application but soon recovered to the basal level. The toxicity of napropamide to the soil microbial populations likely depressed enzyme activities. A PCR–DGGE-based experiment and cluster analysis of 16S rDNA community profiles revealed an apparent difference in bacterial community composition between the napropamide treatments and control, thus suggesting that napropamide-induced toxicity was responsible for the disturbance of the microbial populations in soil (Guo et al. 2009).

Among insecticides, imidacloprid is one of the most commonly used in agricultural practices. Application at FR (field rate) of 1 mg kg^{-1} soil had negative effects on SIR, the number of total bacteria, dehydrogenase, both acid and alkaline phosphatases, and urease at the beginning of the experiment (Cycoń and Piotrowska-Seget 2015). Ten times the FR, instead, caused a detectable decrease of all investigated parameters throughout the experimental period. Imidacloprid negatively affected also nitrifying and N_2-fixing bacteria as well as the physiological state of cultivable bacteria, causing a reduction in the rate of colony formation and prolonging the time for growth (Cycoń and Piotrowska-Seget 2015).

Different results were obtained when the effect of some fungicides (azoxystrobin, tebuconazole, and chlorothalonil) was assessed in soils having identical mineralogical composition but contrasting microbial populations and organic matter contents and arising from different management histories (Bending et al. 2007). Although any of the three fungicides depressed microbial biomass in either soil, significant reduction of dehydrogenase activity to varying extents occurred in the low OM/biomass soil, but not in the high OM/biomass soil (Bending et al. 2007). However, none of the fungicides affected bacterial community structure as assessed by 16S rRNA PCR–denaturing gradient gel electrophoresis (DGGE) (Bending et al. 2007).

By contrast, increases of dehydrogenase, urease, β-glucosidase, and phosphatase activity levels were measured in the presence of prochloraz fungicide, probably because bacterial communities used the fungicide as a source of energy and nutrients. Any measurable change of soil bacterial biodiversity was determined by 16S rDNA–DGGE profiles (Tejada et al. 2011).

Wang et al. (2009) analyzed also by traditional enzyme assays the effect of cadmium (Cd, 10 mg kg^{-1} soil) and three different amounts of herbicide butachlor (10, 50, and 100 mg kg^{-1} soil). The highest butachlor concentration significantly reduced urease and phosphatase activities, and detectable decreases were measured also in the presence of Cd and butachlor added at a ratio of 1:10. Conversely, both activities greatly improved at the ratio of 1:5, thus suggesting that the addition concentration ratios of the two contaminants largely dominated the consequent effect on enzyme activities.

According to early findings by Schaffer (1993), further summarized by Gianfreda and Rao (2008, 2011), Floch et al. (2011) showed that response patterns of various soil enzyme activities differed in their sensitivities to the addition of the pure active ingredients of ten pesticides over time (i.e., stimulation, inhibition, or no effect). Arylamidase, arylsulfatase, cellulase, fluorescein diacetate hydrolase, β-galactosidase, β-glucosidase, phenol oxidase, alkaline and acid phosphatase, phosphodiesterase, and phosphotriesterase were studied. Added pesticides were 2,4D, atrazine, azinphos-methyl, carbaryl, diuron, glyphosate, linuron, mancozeb, parathion methyl, and prometryne. As concluded by the authors, the response of soil enzyme activities to the pesticide action depends not only on the type of enzyme and pesticide but also on the soil under observation, the experimental procedure adopted, and the rate of pesticide application (Gianfreda and Rao 2008, 2011).

12.4 How to Preserve Soil Microbial and Enzymatic Diversity in Contaminated Soils

Microorganisms are versatile organisms. Consequently, they try to adapt to the changed environment. Many of them will develop, through various genetic mechanisms, new catabolic pathways to convey xenobiotic compounds in natural biogeochemical cycles (Ojo 2007). For instance, enzymes catalyzing the degradation of xenobiotics can be often produced by induction process, thus contributing to determine the acclimation time to the xenobiotic substrates (Ojo 2007). A more or less complete biodegradation (even mineralization) of the xenobiotic compound may occur with partial or total restoration of the contaminated site. Indeed, specific enzymes may serve as efficient catalysts for the transformation of xenobiotic substances (Rao et al. 2014). Moreover, the presence of xenobiotics, and much more of a mixed combination of different chemicals, can exert a selective pressure on the existing microflora, shifting it toward those microorganisms much able to transform the contaminants.

Microorganisms present in the contaminated soil could be unable to perform a natural attenuation of contamination. Indeed, several factors may reduce or even prevent the degradation of the xenobiotic compound. They include adverse physicochemical conditions like temperature, pH, oxygen level, redox, etc., more preferred nutrients, low bioavailability of the xenobiotic, presence of predators, or also absence of appropriate genetic information for coding enzymes able to catabolize the compound. In this case, i.e., when the xenobiotic compound cannot be transformed in natural situation, it remains recalcitrant, and human intervention must occur to facilitate and help the soil to remediate itself.

Several chapters in the present book will address different aspects of (bio) remediation of soils contaminated by xenobiotics. Therefore, here just few examples (mainly among those of our research group) of enhancing xenobiotic microbial transformation will be discussed. Commonly, in order to not alter the natural equilibrium of the system, bioremediation of contaminated soils is attempted by stimulation in situ of microbial population. Some of the findings previously discussed have demonstrated that soil microbial population and their biological activity may change in the presence of contaminants, shifting the microbial composition toward those microorganisms much more able to transform the contaminant (Cycoń et al. 2013; Thavamani et al. 2012; Yu et al. 2009; Zhang et al. 2012).

One of the most common methodologies to enhance xenobiotic transformation by the existing soil microflora is to apply organic amendments. Compost has been largely confirmed to be able to improve biological fertility of soil and suppress soilborne pathogens (Scotti et al. 2015). Indeed, one of the most prominent effect of compost is its capability of increasing, even in some case temporarily, the content of soil organic matter and thus of nutrients, with beneficial effects on microbial life. Composting or compost addition is believed to be one of the most cost-effective methods for soil remediation. An exhaustive review on the application of composting/compost for bioremediation of soil contaminated with polycyclic aromatic hydrocarbons, petroleum, pesticides, chlorophenols, and heavy metals has been recently published by Chen et al. (2015). A critical view on the effects of this technology on microbial aspects in contaminated soils is also provided.

Studies performed under laboratory conditions confirmed that the addition of compost to soil artificially contaminated by xenobiotics, like phenanthrene or PCP, or anthracene, had beneficial effect on the removal, although partial, of the xenobiotic and the levels of enzymatic activities, when compared to the contaminated soil non-supplemented with the compost (Ma et al. 2003; Scelza et al. 2007, 2008; Zhan et al. 2010). For instance, in fresh agricultural soil with no history of PAH, contamination spiked with phenanthrene. Scelza et al. (2007) demonstrated that the addition of compost led to a recovery of the activities of β-glucosidase, phosphatase, arylsulfatase, urease, and dehydrogenase as well as of microbial biomass and soil respiration. All these parameters substantially decreased in the presence of the pollutant. Similar results occurred when experiments were carried out in the presence of PCP (Scelza et al. 2008). The development of fungal colonies belonging to the taxonomic group of Ascomycetes and identified as *Byssochlamys fulva* in the PCP-contaminated samples occurred over time. Growth of *B. fulva* in vitro in the presence of PCP showed that the isolate was tolerant to 12.5 and 25 mg l^{-1} PCP and degraded 20 % of its initial concentration in 8 days (Scelza et al. 2008).

Composting of an anthracene-spiked soil mixed with kitchen waste was performed for 42 days, and microbial succession, microbial enzyme activity, microbial diversity, and anthracene removal rate were analyzed (Ma et al. 2003). The results demonstrated an increase of thermophilic microorganisms but no significant increase of anthracene removal. Irregular variation of phenol oxidase and catalase were detected. A drastic increase of microbial diversity with a shift from mesophilic to thermophilic bacteria occurred when temperature elevated from 35 to 56 °C (Ma et al. 2003).

The addition of dissolved organic matter (DOM) can also have effects similar to compost, although its application to contaminated soils showed contrasting responses. Its presence in the PCP-contaminated soil investigated by Scelza et al. (2008) had no effects on the activities of studied enzymes (see above), whereas it improved inhibition of phenanthrene on soil urease and catalase activity during the initial period of applying DOM to a phenanthrene-contaminated soil (Zhan et al. 2010).

12.5 Concluding Remarks and Perspectives

The presence of xenobiotics in soil is a high risk to damage the correct functioning of all biological soil functions. As reported above, microorganisms cannot survive or can have serious changes in their community composition with consequent negative effects on the expression, production, and activity of their enzymatic components. For example, studies on the influence of pesticides on soil microbial activity (Gianfreda and Rao 2008, 2011) have concluded that "pesticides may modify and regulate the synthesis of proteins and enzymes by repression or induction, or they may affect some physiological functions like membrane's functionality. Pesticides may also influence the dynamics of soil populations and consequently soil biodiversity, because they may induce the death of responsive organisms with the consequent formation of organic products usable by the surviving organisms. In all cases, an influence on the activity of enzymes involved in the processes will occur" (Gianfreda and Rao 2008, 2011). Similar effects can derive from the presence of xenobiotics of different nature.

Efforts have been made and are continuously attempted to find ecological and inexpensive methodologies for eliminating xenobiotics from the polluted environments (Gianfreda and Nannipieri 2001; Gianfreda and Bollag 2002; Gianfreda and Rao 2004; Gianfreda et al. 2006). An effective implementation of any bioremediation strategy, however, needs that all factors "governing the growth, metabolism, dynamics and functions of indigenous microbial communities at contaminated sites" (Desai et al. 2010) are well understood. As reviewed by Desai et al. (2010) "innovative breakthroughs in genotypic profiling, ultrafast genome pyrosequencing, metagenomics, metatranscriptomics, metaproteomics and metabolomics along with bioinformatics tools" are now available and "have provided crucial in-sights of microbial communities and their mechanisms in bioremediation of environmental pollutants" (Desai et al. 2010). Application of molecular and "-omics" technologies may help not only in improving "the process of efficacy determination and implementation of microbial bioremediation strategies" but also in evaluating "the innate microbial community structures, dynamics and functions at contaminated sites" (Desai et al. 2010).

Furthermore, advances in soil proteomic studies, still in their infancy, may provide useful information on the presence of proteins in soil. The simultaneous assay of soil enzymatic activities may help in assigning a given enzymatic activity to a certain protein molecule. For instance, comparison of results obtained by classical enzymological studies and by proteomic studies might clarify whether a xenobiotic has influenced the production of the enzymatic protein or has inhibited its activity by inhibition mechanisms.

In conclusion, soil microbial and enzymatic diversity are significant to a safe soil biological life. Their changes in a xenobiotic-polluted soil may serve as useful indicators of pollution. Any advance in the understanding of their behavior and performance in the presence of xenobiotic substances can assist in designing and improving effective and successful bioremediation technologies to restore the polluted site and regain its initial quality characteristics.

References

Andreoni V, Cavalca L, Rao MA, Nocerino G, Bernasconi S, Dell'Amico E, Colombo M, Gianfreda L (2004) Bacterial communities and enzyme activities of PAHs polluted soils. Chemosphere 57:401–412

Bending GD, Rodríguez-Cruz MS, Lincoln SD (2007) Fungicide impacts on microbial communities in soils with contrasting management histories. Chemosphere 69:82–88

Blagodatskaya E, Kuzyakov Y (2013) Active microorganisms in soil: critical review of estimation criteria and approaches. Soil Biol Biochem 67:192–211

Brohon B, Delolme C, Gourdon R (2001) Complementarity of bioassays and microbial activity measurements for the evaluation of hydrocarbon-contaminated soils quality. Soil Biol Biochem 33:883–891

Cappa F, Suciu N, Trevisan M, Ferrari S, Puglisi E, Cocconcelli PS (2014) Bacterial diversity in a contaminated Alpine glacier as determined by culture-based and molecular approaches. Sci Total Environ 497–498:50–59

Chen M, Xu P, Zeng G, Yang C, Huang D, Zhang J (2015) Bioremediation of soils contaminated with polycyclic aromatic hydrocarbons, petroleum, pesticides, chlorophenols and heavy metals by composting: applications, microbes and future research needs. Biotechn Adv 33:745–755

Cheung P-Y, Kinkle BK (2005) Changes in *Mycobacterium* spp. population structure and pyrene mineralization in polycyclic aromatic hydrocarbon-amended soils. Soil Biol Biochem 37:1929–1937

Cycoń M, Piotrowska-Seget Z (2015) Biochemical and microbial soil functioning after application of the insecticide imidacloprid. J Environ Sci 27:147–158

Cycoń M, Markowicz A, Piotrowska-Seget Z (2013) Structural and functional diversity of bacterial community in soil treated with the herbicide napropamide estimated by the DGGE, CLPP and r/K-strategy approaches. Appl Soil Ecol 72:242–250

Dean JR, Scott WC (2004) Recent development in assessing the bioavailability of persistent organic pollutants in the environment. Trends Anal Chem 23:609–618

Deblonde T, Cossu-Leguille C, Hartemann P (2011) Emerging pollutants in wastewater: a review of the literature. Int J Hyg Environ Health 214:442–448

Desai C, Hilor P, Datta M (2010) Advances in molecular and "-omics" technologies to gauge microbial communities and bioremediation at xenobiotic/anthropogen contaminated sites. Bioresour Technol 101:1558–1569

dos Santos EC, Silva IS, Simões THN, Simioni KCM, Oliveira VM, Grossman MJ, Durrant LR (2012) Correlation of soil microbial community responses to contamination with crude oil with and without chromium and copper. Int Biodeter Biodegr 70:104–110

Floch C, Chevremont A-C, Joanico K, Capowiez Y, Criquet S (2011) Indicators of pesticide contamination: soil enzyme compared to functional diversity of bacterial communities via Biolog® Ecoplates. Eur J Soil Biol 47:256–263

Gavrilescu M, Demnerova K, Aamand J, Agathos S, Fava F (2015) Emerging pollutants in the environment: present and future challenges in biomonitoring, ecological risks and bioremediation. New Biotechnol 32:147–156

Gianfreda L, Bollag J-M (1996) Influence of natural and anthropogenic factors on enzyme activity in soil. In: Stotzky G, Bollag J-M (eds) Soil biochemistry, vol 9. Marcel Dekker, New York, pp 123–194

Gianfreda L, Bollag J-M (2002) Isolated enzymes for the transformation and detoxification of organic pollutants. In: Burns RG, Dick R (eds) Enzymes in the environment: activity, ecology and applications. Marcel Dekker, New York, pp 491–538

Gianfreda L, Nannipieri P (2001) Basic principles, agents and feasibility of bioremediation of soil polluted by organic compounds. Minerva Biotechnol 13:5–12

Gianfreda L, Rao MA (2004) Potential of extra cellular enzymes in remediation of polluted soils: a review. Enzyme Microbiol Technol 35:339–354

Gianfreda L, Rao MA (2008) Interaction between xenobiotics and microbial and enzymatic soil activity. Crit Rev Environ Sci Technol 38:269–310

Gianfreda L, Rao MA (2011) The influence of pesticides on soil enzymes. In: Shukla G, Varma A (eds) Soil enzymology. Soil biology, vol 22. Springer, Berlin, pp 293–312

Gianfreda L, Rao MA, Piotrowska A, Palumbo G, Colombo C (2005) Enzyme activities of soils affected by different anthropogenic alterations. Sci Total Environ 341:265–279

Gianfreda L, Mora M, Diez CM (2006) Restoration of polluted soils by means of microbial and enzymatic processes: a review. J Soil Sci Plant Nutr 6:20–40

Guo H, Chen G, Lv Z, Zhao H, Yang H (2009) Alteration of microbial properties and community structure in soils exposed to napropamide. J Environ Sci 21:494–502

Katayama A, Bhula R, Burns GR, Carazo E, Felsot A, Hamilton D, Harris C, Kim YH, Kleter G, Koedel W, Linders J, Peijnenburg JG, Sabljic A, Stephenson RG, Racke DK, Rubin B, Tanaka K, Unsworth J, Wauchope RD (2010) Bioavailability of xenobiotics in the soil environment. Rev Environ Contam Toxicol 203:1–86

Kirk JL, Beaudette LA, Hart M, Moutoglis P, Klironomos JN, Lee H, Trevors JT (2004) Methods of studying soil microbial diversity. J Microbiol Methods 58:169–188

Kozdrój J, van Elsas JD (2001) Structural diversity of microorganisms in chemically perturbed soil assessed by molecular and cytochemical approaches. J Microbiol Methods 43:197–212

Ma Y, Zhang JY, Wong MH (2003) Microbial activity during composting of anthracene-contaminated soil. Chemosphere 52:1505–1513

Matulich KL, Devon JM, Martiny JBH (2013) Microbial biodiversity. In: Levin SA (ed) Encyclopedia of biodiversity, 2nd edn. Academic, Waltham, pp 252–258

Muckian L, Grant R, Doyle E, Clipson N (2007) Bacterial community structure in soils contaminated by polycyclic aromatic hydrocarbons. Chemosphere 68:1535–1541

Nannipieri P, Ascher J, Ceccherini MT, Landi L, Pietramellara G, Renella G (2003) Microbial diversity and soil functions. Eur J Soil Sci 54:665–670

Núñez EV, Valenzuela-Encinas C, Alcántara-Hernández RJ, Navarro-Noya YE, Luna-Guido M, Marsch R, Dendooven L (2012) Modifications of bacterial populations in anthracene contaminated soil. Appl Soil Ecol 61:113–126

Ojo OA (2007) Molecular strategies of microbial adaptation to xenobiotics in natural environment. Biotechnol Molec Biol Rev 2:001–013

Pérez-Leblic MI, Turmero A, Hernández M, Hernández AJ, Pastor J, Ball AS, Rodríguez J, Arias ME (2012) Influence of xenobiotic contaminants on landfill soil microbial activity and diversity. J Environ Manage 95:S285–S290

Rao MA, Scelza R, Acevedo F, Diez MC, Gianfreda L (2014) Enzymes as useful tools for environmental purposes. Chemosphere 107:145–162

Scelza R, Rao MA, Gianfreda L (2007) Effects of compost and phenanthrene-degrading cultures on the properties of an artificially phenanthrene-contaminated soil. Soil Biol Biochem 39: 1303–1317

Scelza R, Rao MA, Gianfreda L (2008) Response of an agricultural soil to pentachlorophenol (PCP) contamination and the addition of compost or dissolved organic matter. Soil Biol Biochem 40:2162–2169

Schaffer A (1993) Pesticide effects on enzyme activities in the soil ecosystem. In: Bollag J-M, Stotzky G (eds) Soil biochemistry, vol 8. Marcel Dekker, New York, pp 273–340

Scotti R, D'Ascoli R, Gonzalez Caceres M, Bonanomi G, Sultana S, Cozzolino L, Scelza R, Zoina A, Rao MA (2015) Combined use of compost and wood scraps to increase carbon stock and improve soil quality in intensive farming systems. Eur J Soil Sci 66:463–475

Svenningsen H, Henriksen T, Priemé A, Johnsen AR (2011) Triclosan affects the microbial community in simulated sewage-drain-field soil and slows down xenobiotic degradation. Environ Pollut 159:1599–1605

Tejada M, Gómez I, García-Martínez AM, Osta P, Parrado J (2011) Effects of Prochloraz fungicide on soil enzymatic activities and bacterial communities. Ecotox Environ Saf 74: 1708–1714

Thavamani P, Malik S, Beer M, Megharaj M, Naidu R (2012) Microbial activity and diversity in long-term mixed contaminated soils with respect to polyaromatic hydrocarbons and heavy metals. J Environ Manage 99:10–17

Thiele-Bruhn S, Bloem J, de Vries FT, Kalbitz K, Wagg C (2012) Linking soil biodiversity and agricultural soil management. Curr Opin Environ Sustain 4:523–528

Wang J-H, Ding H, Lu Y-T, Shen G-Q (2009) Combined effects of cadmium and butachlor on microbial activities and community DNA in a paddy soil. Pedosphere 19:623–630

Yu Y, Chu X, Pang G, Xiang Y, Fang H (2009) Effects of repeated applications of fungicide carbendazim on its persistence and microbial community in soil. J Environ Sci 21:179–185

Zhan X, Wu W, Zhou L, Liang J, Jiang T (2010) Interactive effect of dissolved organic matter and phenanthrene on soil enzymatic activities. J Environ Sci 22:607–614

Zhang D-C, Mörtelmaier C, Margesin R (2012) Characterization of the bacterial archaeal diversity in hydrocarbon-contaminated soil. Sci Total Environ 421–422:184–196

Chapter 13
Current Approaches for the Assessment of In Situ Remediation of Xenobiotics

Safdar Bashir, Zeshan, Waseem Hassan, Nabeel Khan Niazi, Irshad Bibi, and Niaz Ahmad

13.1 Introduction

The quality of life on earth is linked inextricably to the overall quality of the environment. In early times, we believed that we had an unlimited abundance of land and resources; today, however, the resources in the world are running short, in greater or lesser degree, and our carelessness and negligence are creating the worst scenarios. Pollution of the biosphere and the problems associated with contaminated sites now assume increasing prominence in many countries (Hassan et al. 2016, 2014). The problem is worldwide, and the estimated number of contaminated sites is significant (Wuana and Okieimen 2011). It has been estimated that only in

S. Bashir
Institute of Soil and Environmental Sciences, University of Agriculture Faisalabad, Faisalabad 38040, Pakistan

Zeshan (✉)
School of Civil and Environmental Engineering (SCEE), Institute of Environmental Sciences and Engineering (IESE), National University of Sciences and Technology (NUST), Islamabad, Pakistan
e-mail: zeshansheikh@iese.nust.edu.pk; zeshan90@gmail.com

W. Hassan
Department of Soil Science, Muhammad Nawaz Shareef University of Agriculture, Multan, Pakistan

N.K. Niazi • I. Bibi
Institute of Soil and Environmental Sciences, University of Agriculture Faisalabad, Faisalabad 38040, Pakistan

MARUM and Department of Geosciences, University of Bremen, Bremen 28359, Germany

N. Ahmad
Department of Soil Science, Faculty of Agricultural Sciences and Technology, Bahauddin Zakariya University, Multan, Pakistan

© Springer International Publishing AG 2017
M.Z. Hashmi et al. (eds.), *Xenobiotics in the Soil Environment*, Soil Biology 49,
DOI 10.1007/978-3-319-47744-2_13

Western Europe 300,000 sites have been contaminated with heavy metals, metalloids and other xenobiotics (Mahmood 2010; Lone et al. 2008). In the USA, 600,000 heavy metal contaminated sites need reclamation (Mahmood 2010; Lone et al. 2008), whereas 100,000 ha of croplands, 55,000 ha of pasture and 50,000 ha of forests have been lost due to heavy metal pollution (Mahmood 2010; Ragnarsdottir and Hawkins 2005). In China, one-sixth of the total arable land has been contaminated (Tang et al. 2013). Soil and water pollution is also severe in developing countries, e.g. India, Pakistan and Bangladesh, where small industrial units are pouring their untreated effluents in the surface drains, which spread over near agricultural and forest fields (Murtaza et al. 2010). The main sources of contamination are heavy metals (e.g. cadmium, arsenic, zinc, lead, copper, chromium, mercury, etc.), organochlorines (e.g. DDT, HCH, DDE, PCP, PCBs, dieldrin, aldrin, chlordane, dioxins, etc.), solvents (e.g. industrial cleaning agents), corrosives (e.g. acids and alkalis), hydrocarbons (e.g. oil derivatives such as petrol, diesel, tar and creosote), asbestos and cyanides (Wuana and Okieimen 2011). Fate of xenobiotics in the environment has been shown in Fig. 13.1. Major issue today is to remediate numerous sites contaminated by different xenobiotics. Traditional remediation methods are time-consuming, expensive and frequently not efficient to remediate the contaminated site within a rational duration. Thus, natural attenuation (NA) methods, which help to decontaminate without human interferences, have great potential for site remediation. Natural attenuation processes include volatilization, dilution, sorption and microbial degradation (Bombach et al. 2010). Bioremediation is one of the effective and sustainable methods as microbial processes reduce exponential amounts of xenobiotics by transforming

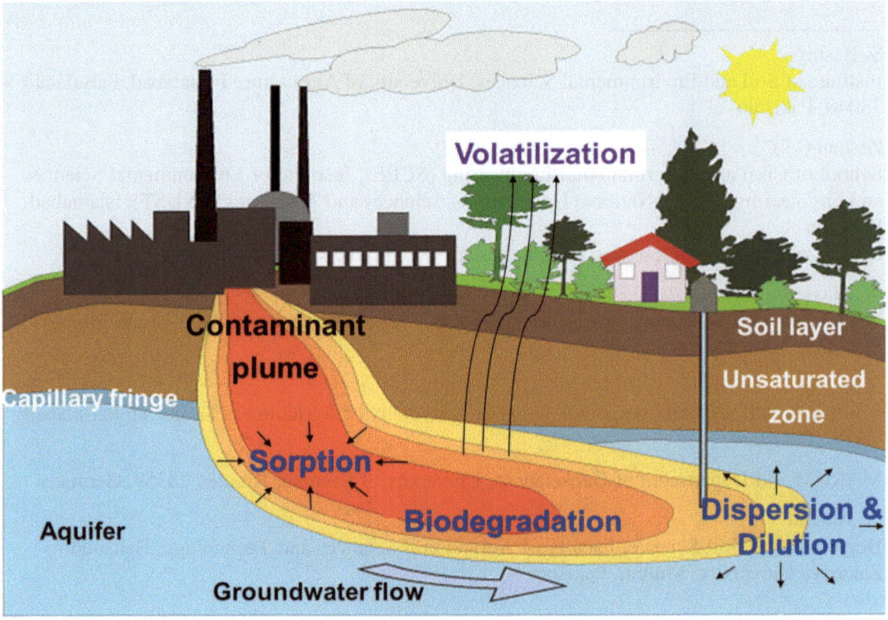

Fig. 13.1 Fate of xenobiotics in the environment

them into less harmful compounds. Therefore, the development of tools for the assessment of in situ bioremediation has become an important field of research which helps site managers to see the progress of remediation.

In the process of accepting natural attenuation as an effective remediation technology by the site managers, the biggest problem is to actually prove that the reduction in the xenobiotic concentration in a contaminated site is mainly due to bioremediation instead of other physical processes such as dispersion which takes the xenobiotic from one place to another, sorption which fixes the xenobiotics and volatilization which transports the xenobiotics into the atmosphere. Selection of natural attenuation as an effective remediation process at contaminated sites needs complete characterization of the bioremediation processes of the specific site. The important questions which need to be addressed before employing a natural attenuation strategy are: (1) Is bioremediation occurring? (2) How effective are the remediation processes? (3) How will the degradation processes continue for a longer period of time, and (4) is the estimated duration for complete remediation of the field site reasonable? To answer these questions, different techniques including geochemical analyses, tracer tests, detection of metabolites, compound-specific isotope analysis (CSIA), in situ microcosms and microbial communities, populations and monitoring of functional genes are used to assess bioremediation (Bombach et al. 2010).

The purpose of this chapter is to give an overview of the recent technologies used for the assessment of in situ remediation and to demonstrate their prospective and limitations. Primary attention is given to the anaerobic degradation of organic groundwater xenobiotics. The first part of the chapter addresses the advantages and limitations of current chemical and microbiological methods. In the second part, effectiveness of integrated technologies to understanding in situ remediation processes is discussed. This chapter will mainly focus on the current approaches for the assessment of in situ remediation of xenobiotics present in groundwater.

13.2 Tools for the Assessment of In Situ Remediation of Xenobiotics

Considering the costs and technical difficulties linked to other conventional remediation techniques, currently, in situ biodegradation is considered as an effective approach for contaminants removal from the soil and water bodies. In situ degradation is the key process for NA strategies for managing contaminated sites (Schirmer et al. 2006). However, complete characterization of site-specific biodegradation processes is essential to validate the effectiveness of in situ biodegradation of organic contaminants at a contaminated site and to investigate the efficiency of in situ remediation processes to replace conventional clean-up technologies (Pope et al. 2004). Recently, several methods or combination of various techniques have received great attention to verify the in situ remediation processes. Currently, there are several techniques which are applied alone or in combination, to monitor in situ remediation of xenobiotics at a contaminated site. The following section will discuss several current methods that can be applied to assess transformation of organic xenobiotics in groundwater, their applicability and limitations associated with their application.

13.3 Evaluation of Xenobiotic Concentration

One of the oldest techniques for verifying in situ xenobiotic remediation involves removing samples of soil and water from the site and bringing them to the laboratory for concentration analysis. In this, temporal comparison of samples is done, and concentration change is evaluated (MacDonald 1994). But the reliability of this method is not authenticated to assess the removal of the xenobiotic as the changes in concertation may be due to other physical processes. In the past decades, there have been great analytical developments which help to assess quantitative and qualitative removal of xenobiotics. List of various analytical tools aiding to assess in situ remediation of xenobiotics is given in Table 13.1.

13.4 Hydrogeochemical Methods to Assess In Situ Transformation of Xenobiotics

Hydrogeochemical characterization involves the quantitative or qualitative analysis of electron donors or accepters during the process of remediation. The preferential electron acceptor is oxygen for the microorganism, but as the oxygen is depleted in the contaminated aquifer, other ions such as manganese (IV) (Mn^{4+}), nitrate (NO^{3-}), ferric iron (Fe^{3+}), sulphate (SO_4^{2-}) and carbon dioxide (CO_2) can be used as electron acceptors. Microbially catalysed redox reactions result in specific redox potential in the contaminated aquifer which ultimately affect the energy gained from each reaction and control the rate of biotransformation under different redox processes (Wiedemeier et al. 1999). So the knowledge of groundwater geochemical parameters, temporal and spatial distribution of xenobiotics and electron acceptors is very important for the characterization of microbial remediation (Chapelle 2000; Christensen et al. 2001). Decrease in xenobiotic concentrations over time might be from abiotic/physical processes such as dispersion, sorption and volatilization and will not always represent natural attenuation. Continuous monitoring of the various redox-sensitive oxidized species (NO^{3-}, Mn^{4+}, Fe^{3+}, SO_4^{2-} and CO_2) reduction as well as the production of their reduced forms (NO^{2-}, NH^{4+}, Mn^{2+}, Fe^{2+}, S^{2-} and CH_4) gives important information about dominant redox reaction responsible for remediation in the groundwater aquifer (Vroblesky and Chapelle 1994; Ludvigsen et al. 1998; Richmond et al. 2001). But this is not the case for all, for example, ferrous iron or sulphide as both redox species react to form precipitates of iron sulphide which result in an underestimation of ferric iron and sulphate reduction (Christensen et al. 2000). McMahon and Chapelle (2008) documented well the assessment of

Table 13.1 List of most common analytical methods for assessing in situ transformation of xenobiotics

Analytical technique	Class of xenobiotics	Advantages	Disadvantages
Gas chromatography (GC)	Volatile and semi-volatile	High-resolution power compared to other methods High sensitivity when used with thermal detectors, good accuracy and precision Separation and analysis of sample very quickly Samples required in low quantities	Only volatile samples or the sample which can be made volatile are separated by this method During injection of the gaseous sample, proper attention is required
Thin layer chromatography (TLC)	Non-volatile compounds	Less equipment required Simple and sensitive method	The plate length is limited, and hence separation takes place only up to certain length Separation takes place in an open system or in open condition, and hence there are chances that sample may be affected by the humidity and temperature
Solid-phase micro extraction (SPME)	Both volatile and non-volatile	Does not require organic solvent Highly sensitive No cross contamination	Limited to aqueous samples Cannot be applied for highly concentrated samples
Spectroscopy	Both volatile and non-volatile	Powerful tool for qualitative and quantitative analysis	Sample preparation is time-consuming often not efficient at low concentrations
Nuclear magnetic resonance (NMR)	Broad range of application from natural products to pharmaceuticals, organic, inorganic	Works for organic and inorganic, qualitative and quantitative, versatile	Very expensive, time-consuming, spectra take long time to interpret

redox processes associated with in situ transformation. Quantifying removal of electron acceptors can be used to assess the bioremediation potential of xenobiotics using mass balances of electron donor and electron acceptor balances (Wiedemeier et al. 1999). True quantification of the removal of a single organic xenobiotic is limited by the co-contamination of other organic xenobiotics as the reduction of electron acceptors represents the oxidation of all organic materials in the aquifer instead of a single xenobiotic. But sometimes other by-products can be used to

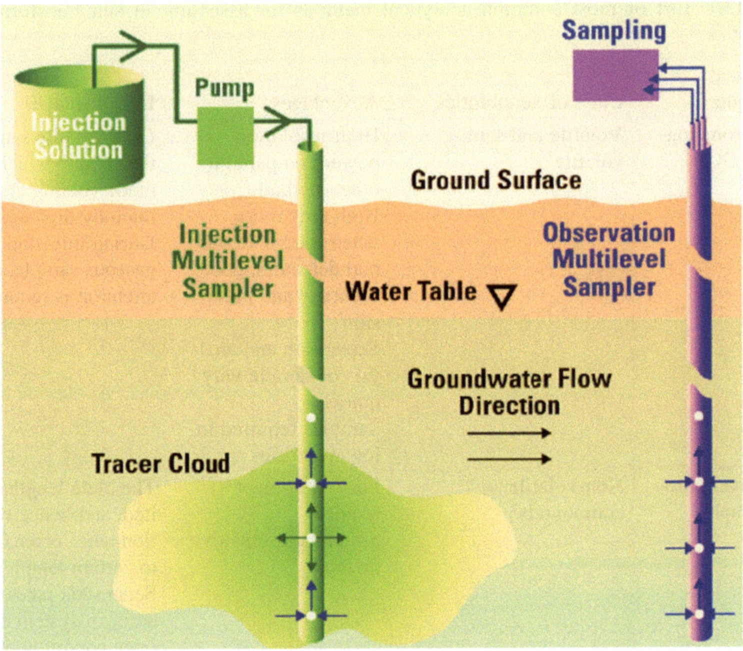

Fig. 13.2 A conceptual diagram of the setup of the subsurface tracer test (Figure modified from USGS)

validate the bioremediation of xenobiotic, for example, in the case of reductive dechlorination processes where the chlorinated solvents (tetrachloroethene (PCE)) act as electron acceptors, the increase in chloride concentration and daughter product presence (trichloroethene (TCE), dichloroethene (DCE), vinyl chloride (VC) and ethene) in the contaminated aquifer can be used as indicators for microbial activity. Thus, geochemical data are vital for understanding natural attenuation processes in the subsurface (Fig. 13.2), but these data have no potential to provide a direct indication of specific compound degradation. However, continuous temporal monitoring of the concentrations electron acceptor delivers valuable information on their current and future bioavailability, which is essential for assessing the sustainability of natural attenuation processes as well as the scope of applying a successful enhanced natural attenuation (ENA) plan (Bombach et al. 2010).

13.5 Application of Tracer Technology to Assess Natural Attenuation

In recent years, tracer technology has been applied extensively to distinguish between in situ microbial remediation and physical processes responsible for the depletion of xenobiotics. Usually, two types of tracer experiments are being performed. The first includes test solutions containing a conservative tracer such as chloride or bromide, and the pollutant of interest as a reactive tracer is injected (Ptak et al. 2004). The second type of tracer test includes application of stable isotopes to trace the bioremediation, for example, stable isotope probing (SIP) test. Two methods are employed to perform tracer experiments in the saturated zone of an aquifer. In first, tracers are injected in a groundwater well, and its distribution is observed along a flow path (Thierrin et al. 1995; Schirmer and Barker 1998; Rügge et al. 1999; Eganhouse et al. 2001; Fischer et al. 2006; Gödeke et al. 2006). In second, tracers due to the naturally low abundance of the heavier stable isotope, e.g. carbon, have a ^{12}C:^{13}C ratio of 98.9:1.1; stable isotopes can be applied in a tracer approach. Thus, using a ^{13}C-labelled substance, product formation can be followed, intermediates can be identified and, additionally, due to the incorporation of ^{13}C into biomass, the microorganisms and enzymes responsible for the biotransformation can be identified using total lipid fatty acids (TLFA), DNA-, RNA- or protein-stable isotope probing (SIP) approaches (Nijenhuis 2016; Reusser et al. 2002; Hageman et al. 2004). In assessing in situ pollutant bioremediation and also electron acceptor processes, tracer approaches are valuable (Schroth et al. 2001; Kleikemper et al. 2005). Extensive monitoring and construction of groundwater monitoring wells is the major restriction for the application of tracer technology. Also, this is an expensive and labour-requiring approach. Other limitation for tracer experiments is that the tracers should not be present or should be present in very low concentrations within the monitored aquifer to allow the injected tracers differentiation from already existing tracer compounds (Bombach et al. 2010). To overcome this issue, stable isotope tracer can be more effective in which isotopically labelled analogues of the pollutant of interest are applied as the reactive tracer. But again the cost of the technology is limitated here as these isotopically labelled analogues are expensive and often not available. Yet, application of isotopically labelled substances not only gives direct quantitative proof of in situ xenobiotic remediation but also gives a clear clue of parent compounds and their by-products. Currently, Adrian and Marco (2016) reviewed challenges and examples of stable isotope applications to characterize geochemical process to understand in situ remediation.

13.6 Compound-Specific Stable Isotope Analysis

Carbon stable isotope analysis (CSIA) has been shown to be an effective tool to evaluate in situ bioremediation processes (Bashir et al. 2015). Stable isotope ratio measurements of specific compounds can be used to distinguish bioremediation from non-destructive processes affecting concentration only and have been reviewed extensively (Thullner et al. 2012). Isotope effects can also be used to explain reaction mechanisms and, in combination with spatial/temporal data, can be applied in situ to estimate reaction rates.

Moreover, in recent years, attempts have been made to use stable isotope fractionation data to discriminate different reaction pathways of xenobiotic degradation in groundwater (Bashir et al. 2013; Zhang et al. 2014). Carbon and hydrogen are the backbone of organic xenobiotics, and both elements have two stable isotopes. The effectiveness of stable isotopes for evaluating bioremediation is based on the fact that chemical bonds formed by a heavy isotope of an element naturally are stronger than those formed by a light isotope of the same element in the same compound. Thus, when a specific bond is broken, molecules containing heavy isotopes of elements involved in the bond (or neighbouring to a lesser degree) generally react more slowly than molecules containing light isotopes of those elements (Fig. 13.3). This process is called as isotope fractionation, and, when it is related to an irreversible reaction (Fig. 13.4), as a kinetic isotope effect is within contaminated aquifers, shifts in the isotope ratios of xenobiotics have been shown to result only from bioremediation processes and somewhat from abiotic nondegrading mechanisms like mixing (Dempster et al. 1997; Slater et al. 1999), vaporization (Harrington et al. 1999; Poulson and Drever 1999; Wang and Huang 2001, 2003), diffusion (Abe and Hunkeler 2006) and sorption (Slater et al. 2000; Schüth et al. 2003). Consequently, variations in isotope signatures are a reliable clue of in situ xenobiotic remediation. For most abundant groundwater xenobiotics such as CHCs (Wanner et al. 2016; Nijenhuis 2016; Bashir et al. 2015, 2013; Zhang et al. 2014, Griebler et al. 2004a; Lee et al. 2007; Cichocka et al. 2008; Badea et al. 2009; Fletcher et al. 2009), BTEX (Meckenstock et al. 2004; Morasch et al. 2004;

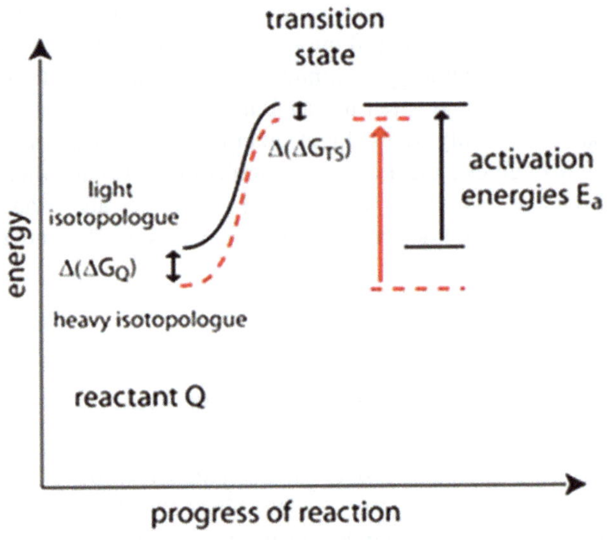

Fig. 13.3 Principle of isotopic fractionation in bioremediation reactions. Lighter isotopologues require less activation energy (Ea) to achieve the transition state and will therefore react faster than heavy isotopologues. Δ(ΔGQ) are the energetic differences between the isotopologues of the reactant at the beginning of the reaction, and Δ(ΔGTS) are energetic differences between the isotopologues in the transition state (modified from Elsner 2010)

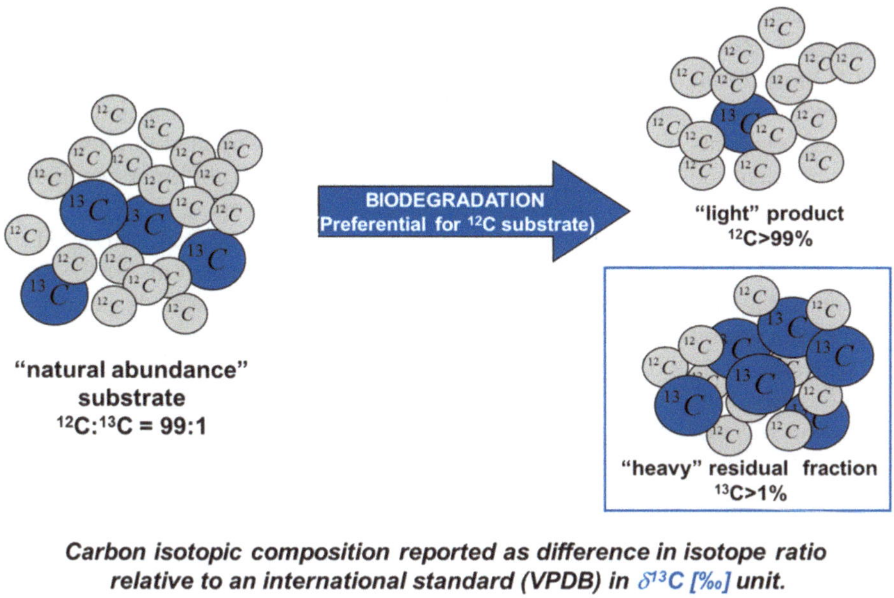

Fig. 13.4 Principle of compound-specific stable isotope analysis process

Fischer et al. 2008, 2009; Mancini et al. 2008; Vogt et al. 2008; Herrmann et al. 2009) and MTBE (Gafni et al. 2016; Kuder et al. 2005; Somsamak et al. 2005, 2006; Rosell et al. 2007), isotope fractionation has been observed. To prove in situ bioremediation using CSIA is known to be more effective for small molecules than for larger molecules, as only the first irreversible step of the reaction pathway responsible for bond breakage has a substantial effect on isotopes. Non-reactive components of a molecule do not change the isotopic signatures of a molecule. But sometimes carbon isotope fractionation during bioremediation, e.g. of higher PAHs, is too low to be analysed by CSIA (Bombach et al. 2010; Trust et al. 1995; Mazeas et al. 2002). In such a situation, two-dimensional isotope analysis will be more helpful to characterize bioremediation reactions, for example, in case of PAHs, measuring hydrogen isotope fractionation not only provides qualitative evidence for microbial degradation, but it has potential to quantify in situ bioremediation processes. The main requirement for applying this technique is the knowledge of suitable enrichment factors obtained by laboratory experiments according to the Rayleigh equation which mathematically describes the isotope fractionation processes (Rayleigh 1896; Mariotti et al. 1981). Pollutant bioremediation can be quantified in contaminated aquifers using the Rayleigh equation-based approach if enrichment factors are known (Meckenstock et al. 2004; Elsner et al. 2005). The reliability of this approach has been shown by biodegradation experiments (Bashir et al. 2013; Zhang et al. 2014; Peter et al. 2004; Fischer et al. 2006) and modelling (Manna and Dybala 2013; Abe and Hunkeler 2006; Van Breukelen 2007; Thullner et al. 2009). Successful application of this approach has been efficiently helped to quantify natural (e.g. Bashir et al. 2015; Sherwood et al.

2001; Richnow et al. 2003; Vieth et al. 2005, 2003; Griebler et al. 2004b; Hunkeler et al. 2005; Zwank et al. 2005; Fischer et al. 2007) and enhanced (Chartrand et al. 2005; Morrill et al. 2005; Hirschorn et al. 2007; McKelvie et al. 2007) groundwater xenobiotic bioremediation. In laboratory experiments, isotope fractionation for different elements (C, H, Cl and N) has been successfully applied to prove CSIA as an efficient tool to characterize degradation pathways of different xenobiotics such as BTEX (Fischer et al. 2008; Mancini et al. 2008; Vogt et al. 2008), CHCs (Elsayed et al. 2014; Bashir et al. 2013; Zhang et al. 2014; Abe et al. 2009), MTBE (Kuder et al. 2005; McKelvie et al. 2009) and atrazine (Meyer et al. 2009). Dual-carbon and hydrogen isotope signatures or triple element isotope analysis has been applied for characterizing bioremediation in contaminated aquifers (Palau et al. 2016; Kuder et al. 2005; Zwank et al. 2005). Two-dimensional isotope analyses can also describe redox condition responsible for remediation, for example, sulphate reduction was identified as the primary respiratory process driving the bioremediation of benzene in an anoxic aquifer using this two-dimensional isotope fractionation approach (Fischer et al. 2007, 2009). Recently, application of CSIA to assess bioremediation and characterization of pharmaceuticals has been successfully applied in various studies (Souchier et al. 2016). Barriers to apply CSIA method is that it cannot be applied if only small proportion of xenobiotic is removed, which mostly occurs within the source zone or adjacent area of contaminated sites and also if the transformation is accompanied by negligible isotope fractionation (Fischer et al. 2008; Vogt et al. 2008). Mixing of groundwater from aquifer zones that exhibit a different degree of bioremediation is another process that hinders the CSIA applicability. This mixing effect results in decrease in the observed isotope fractionation of a xenobiotic and thus an underestimation of bioremediation by the Rayleigh equation-based approach (Kopinke et al. 2005; Thullner et al. 2009).

Additionally, a suitable selection of isotope enrichment factors is required for precise quantification of xenobiotic transformation using the Rayleigh equation-based approach. Isotope enrichment factors are estimated within laboratory experiments, for example, summary of enrichment factors calculated in different laboratory studies for hexachlorocyclohexane isomers is shown in Table 13.2. Consequently, there might be uncertainty as isotope fractionation in the field may be dependent on specific microbial degradation pathways and on environmental factors that cannot be fully regenerated in laboratory systems. To achieve a conservative assessment of bioremediation, the enrichment factor representing the highest isotope fractionation for bioremediation of a pollutant should be chosen for the Rayleigh equation-based approach (Meckenstock et al. 2004; Elsner et al. 2005). Interpretation of isotope data requires special care when the filed sites have long history of contamination, as isotope signatures of different xenobiotic sources may vary depending on their production processes and the variability in the isotope composition of raw materials used during production. In addition, the isotope signature of the xenobiotic source can be influenced due to bioremediation (Bashir et al. 2015). Although these make a bit difficult to interpret in situ remediation using isotope technique yet, CSIA has been proven to be an efficient technique to assess qualitatively and quantitatively the in situ transformation of xenobiotics in the environment.

Table 13.2 Summary of enrichment factors ε (‰) calculated for hexachlorocyclohexane isomers

Laboratory study	Xenobiotic	Initial transformation step	ε (‰)	References
Aerobic				
Sphingobium indicum (B90A)	γ-HCH	Dehydrochlorination	−1.5 ± 0.1	Bashir et al. (2013)
	α-HCH		−1.6 ± 0.3	
Sphingobium japonicum UT26	γ-HCH		−1.7 ± 0.2	
	α-HCH		−1 ± 0.2	
Anaerobic				
(*C. pasteurianum*)	α-HCH	Dechlorination	−3.7 ± 0.8	Badea et al. (2011)
Desulfovibrio gigas	γ-HCH	Dechlorination	−3.9 ± 0.6 ‰	Badea et al. (2009)
Desulfococcus multivorans			−3.4 ± 0.5 ‰	
Abiotic				
UV/H_2O_2 (≥280 nm)	α-HCH	Hydrogen abstraction	−1.9 ± 0.2	Zhang et al. (2014)
Alkaline hydrolysis (pH 9.78)	α-HCH	Dehydrochlorination	−7.6 ± 0.4	Zhang et al. (2014)
UV (≥200 nm)	α-HCH	Dechlorination	−2.8 ± 0.2	
Electrochemical reduction	α-HCH		−3.8 ± 0.4	
Fe^0 reduction	α-HCH	Dichloroelimination	−4.9 ± 0.1	

13.7 Enantiomer Fraction

The enantiomeric fraction (EF) is used to explain the relationship between enantiomers during biodegradation. The EF (+) is defined as A+/(A++ A−), where A+ and A− correspond to the peak area or concentrations of (+) and (−) enantiomers (Harner et al. 2000). Racemic compounds have an EF (+) equal to 0.5. An EF (+) > 0.5 shows the preferential degradation of (−) enantiomer, and an EF (+) < 0.5 indicates the preferential degradation of (+) enantiomer. EF (−) is defined as A−/(A++ A−) (Bashir et al. 2013). But this technique is limited to only chiral contaminants, but 25 % of worldwide applied organic chemicals, e.g. pharmaceuticals or pesticides, are chiral and were applied as mixtures of isomers and/or enantiomers. When enantiomer fraction is combined with enantiomer stable isotope analysis, then this provides two-dimensional verification of the xenobiotics remediation. The potential of this technique has been studied in various investigations, e.g. Bashir et al. (2013) and Badea et al. (2011). Also, the successful application of enantiomeric fractionation has been successfully applied to assess transformation of pharmaceuticals (Souchier et al. 2016).

13.8 Metabolite Analysis

Another technique to assess in situ bioremediation is the detection of metabolites produced during biotransformation of xenobiotics. To conform in situ remediation these intermediates/degradation products have to possess certain properties which mainly include biochemical relationship to the parent substrate, which must be the results of metabolic process rather than the result of co-metabolism or its natural presence in the environment (Beller 2000). Preferably, the metabolite might be associated with the biotransformation of only one compound, but there is also a range of intermediates which indicate the transformation of a certain group of compounds. For example, anaerobic transformation of HCH results in monochlorobenzene and benzene as accumulating metabolites via reductive beta-elimination (Quintero et al. 2005), while aerobic degradation proceeds via dehydrochlorination. The aerobic transformation of γ-HCH is characterized by two initial dehydrochlorination reactions producing the putative product 1,3,4,6-tetrachloro-1,4-cyclohexadiene (1,3,4,6-TCDN), via the intermediate γ-pentachlorocyclohexene (γ-PCCH) (Imai et al. 1991; Nagasawa et al. 1993a, b), whereby on the other side, end products 1,2,4-trichlorobenzene (1,2,4-TCB) and 2,5-dichloropenol (2,5-DCP) may be formed depending on the different reaction routes (Geueke et al. 2012). α-HCH biotransformation via dehydrochlorination proceeds similarly, resulting in β-PCCH, as the respective enantiomer from (+) and (−)-α-HCH, as the first intermediate which are then converted to 1,2,4-TCB (Suar et al. 2005). Metabolite analysis can be an efficient tool to assess in situ remediation, but still several issues are associated with its validation. To interpret metabolites as an indicator of in situ remediation, knowledge of the transformation pathway of the xenobiotic is required, which is not the case for all common xenobiotics (Chakraborty and Coates 2004). In the case of substrates that are mineralized or act as carbon source, it is hard to detect intermediates because of their very low concentrations (Caldwell and Suflita 2000; Griebler et al. 2004b). So to apply this technique, the development in analytical techniques has to some extent overcome its limitations. For example, direct sample injection and solid-phase extraction methods coupled with liquid chromatography–mass spectrometry (LC-MS/MS) can be applied instead of the traditional method of liquid–liquid extraction followed by derivatization of the extracts and analysis by gas chromatography–mass spectrometry (GC-MS) (Beller 2002; Alumbaugh et al. 2004). Mass balance of substrate and intermediates can help to quantify the transformation, but co-contamination of other compounds with similar transformation pathways makes this technique ambiguous to apply in the field. But still this technique can be an indicator of in situ remediation qualitatively.

13.9 Microbiological Methods

13.9.1 Laboratory Methods

Site material can be used to validate the in situ remediation and characterization of xenobiotics transformation pathways (Strevett et al. 2002; Madsen 2005). Laboratory microcosms are produced with the contaminated site. Material, either groundwater or aquifer sludge (Kasai et al. 2006; Kleikemper et al. 2002), is used as inoculum to set up microcosms and artificially spiked with xenobiotic present in situ. Xenobiotic concentration reduction gives an evidence for its removal and degradation rates, and by-products are determined (Aronson and Howard 1997; Suarez and Rifai 1999). Laboratory microcosm studies help to distinguish between biotic and abiotic chemical transformation of a specific xenobiotic. It also helps to characterize single biogeochemical process, monitoring toxic effects of xenobiotics and assessing the impact of different electron acceptors or donors augmentation on transformation rates which ultimately help in developing enhanced natural attenuation strategies. The enriched microcosms are further processed to isolate active players (microorganisms) by suitable cultivation techniques and thus allow a linkage between the microbial identification and the detected biogeochemical process (Madsen 2005). The main limitation of laboratory microcosm method is that the collected samples used for the laboratory microcosms may not represent the actual site conditions, and there are certain chances of change in microbial community structure, and there are possibilities of change in their activity during microcosm preparation and incubation. The other disadvantage of this technique is that complex interactions happening in the environment are simplified and that the availability of substrates, nutrients and electron acceptors may differ from the laboratory and field sites due to the batch system that does not enable a continuous substrate flux (Bombach et al. 2010; Roeling and van Verseveld 2002). Additionally, microcosms may not be able to reproduce the exact biochemical conditions as well as the heterogeneity of an aquifer, where both probably govern in situ degradation kinetics. Another issue with laboratory experiments using site samples is that less than 1 % of bacteria are cultivable (Amann et al. 1995). Therefore, degradation results obtained from laboratory may represent the potential of the microbial community to degrade and give qualitative clues of transformation, but quantitative transformation is not sure with this technology.

13.9.2 In Situ Microcosms Technique

Often it is not possible to reproduce the in situ conditions in laboratory, and ambiguities arise in the validation of in situ remediation under laboratory conditions. Thus, the interpretation of laboratory results to in situ conditions is often criticized. In situ microcosms provide a more confined study site for measurements of microbial reactions yet closer to natural conditions than laboratory microcosms. Two basic

Fig. 13.5 BACTRAP with activated carbon bead figure (Courtesy from André Künzelmann/UFZ)

types of in situ aquifer microcosm have been described in recent years, and both originated from in situ instruments initially designed for geochemical measurements. Gillham et al. (1990) constructed an instrument that isolates a portion of an aquifer for in situ biochemical rate measurements. Later on, Shati et al. (1996) modified a multilayer sampler for studying the activity of inoculated bacteria in a contaminated aquifer. Keeping in mind recent advances in environmental microbiology methodologies such as immunofluorescence direct counts, oligonucleotide and PCR probes, fatty acid methyl ester analysis for the detection and characterization of bacterial communities, measurement of mRNA and expression of proteins, it is evident that much new information can now be gained from in situ work. The concept of in situ microcosms takes into consideration that free-living microorganisms are certainly present in contaminated aquifers, but favours species preferring a sessile lifestyle. In situ microcosms provide a surface, which can be rapidly colonized by floating bacteria or fragments released from indigenous biofilms of the aquifer matrix during the incubation period of the in situ microcosmos. If the microcosms are loaded with a compound that certain bacteria can use as a carbon source, degraders will take advantage of these substrates and grow on the provided surfaces. Thus, a biofilm enriched with microbes from the indigenous microbial community able to use the carbon source under the given biogeochemical condition of the groundwater will develop within the microcosms (Kästner and Richnow 2010).

Another development of in situ microcosm is Biotrap in which in situ microcosms are loaded with ^{13}C-labelled substrates (BACTRAP®) (Fig. 13.5), which are well suited for providing xenobiotic-specific evidence of bioremediation within contaminated sites. The composition of Biotrap includes a perforated Teflon tube or stainless steel cage filled with activated carbon pellets, for instance, Bio-Sep® beads (Bombach et al. 2010; Peacock et al. 2004). ^{13}C-labelled compound is loaded on in situ microcosms and subsequently exposed in a contaminated aquifer for several weeks. Native microorganisms colonize the microcosms and ingest the amended ^{13}C-labelled compound which will be transformed into the biomass and can be traced within biomarker molecules such as phospholipid fatty acids, amino acids or nucleic acids using CSIA (Fig. 13.6). The concept of this type of in situ microcosm

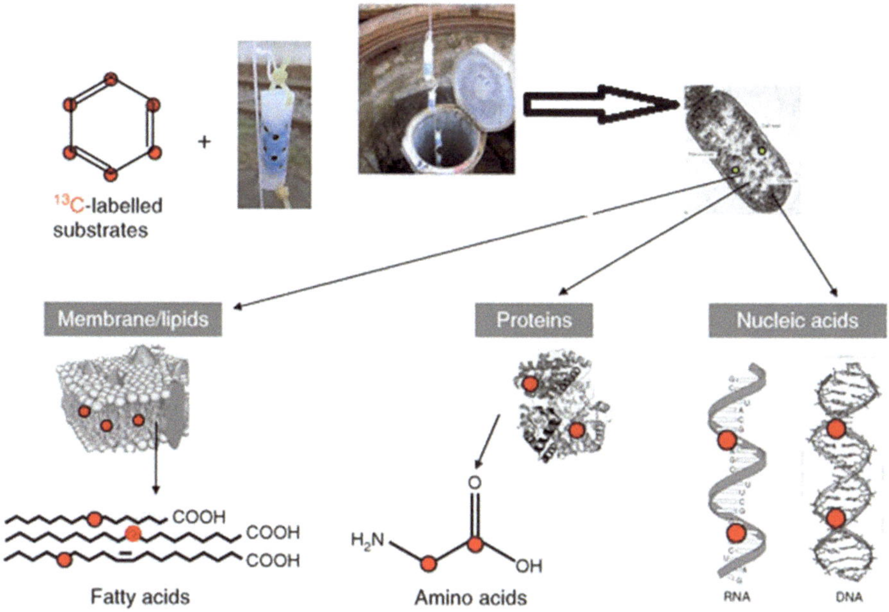

Fig. 13.6 Concept of isotopically labelled BACTRAP® in subsurface monitoring wells (Figure modified from Kästner and Richnow 2010)

has been successfully applied to prove anaerobic bioremediation of BTEX compounds (Geyer et al. 2005; Kästner et al. 2006; Stelzer et al. 2006), monochlorobenzene (MCB; Nijenhuis et al. 2007; Stelzer et al. 2009) and MTBE (Busch-Harris et al. 2008). Furthermore, this technique gives dual proof of in situ transformation as metabolites formed during the bioremediation of the compound may be enriched due to adsorption to the activated carbon pellets which provide an additional information about xenobiotic bioremediation pathways.

13.10 Molecular Methods

With the development in technology, techniques to assess in situ transformation of xenobiotics have been developed. There has been great scientific development in the field of molecular biology, and various techniques have been developed to assess the fate and in situ transformation of xenobiotics; summary of various techniques is given in Table 13.3. Advantages and disadvantages are also described in the Table 13.3.

Table 13.3 Summary of molecular techniques with their advantages and disadvantages for in situ assessment

Technique	Indicator	Advantages	Disadvantages
Plating/CFU counting	Estimation of bacterial cells survival during in situ bioremediation	A quick and easy method	Subjective to errors based on the inherent biases of culture-dependent methods
BIOLOG	Based on assimilation/hydrolysis of different carbon substrate for analysis of microbial activity	Allows estimation of similarity between microbial communities from various environments and habitats	Successful use of this method is sensitive to inoculum sizes
Active cell staining	Direct comparative enumeration of active microbial population during in situ bioremediation	A culture-independent method	The regular application of this method is limited due to the inability to distinguish the desired catabolic activity in the environmental background
Most probable number-PCR	Targets amount of DNA in environmental sample and results from this method can be correlated to the amount of bacterial cells present in the environment	A culture-independent method Provides selective advantage based on the sequence specificity of PCR primers	It may not differentiate amongst live or dead bacterial cells
DNA hybridization	Estimates the abundance of a target gene fragments characteristic to degradative microorganism	A culture-independent method	It may not differentiate amongst live or dead bacterial cells
Colony hybridization	The method allows a differentiation between live and dead cells but it is	A cultivation-based method that is one of the most frequently use method for monitoring the bacterial cell survival during in situ bioremediation studies	Subjective to biases associated with culture-dependent method
Soil enzyme analysis	Soil enzyme analysis for constitutively expressed bacterial enzymes such as dehydrogenase, lipase, etc.	Have been used in some of the in situ bioremediation studies as indicators of impacts of technological intervention on the indigenous microbial community	Time-consuming increased labour requirements
Immunochemical enumeration	The high-affinity binding of bacterial cell surface antigen with antibody	Very strong method monitoring of survival and activity of bacterial cells used for in situ	Subjective to biases associated with culture-dependent method

(continued)

Table 13.3 (continued)

Technique	Indicator	Advantages	Disadvantages
		bioremediation Provides high degree of sensitivity to the method	
FISH	Spatial and temporal monitoring of microbial cell based on the visualization of fluorescence	The method is based on specificity of DNA–DNA hybridization and ease of visual observation of emitted fluorescence	Not yet very common amongst the in situ bioremediation studies due to high cost and labour required
Genome tagging	This method is based on integration of 'non-natural' DNA sequence (s) in the genome of the microorganism before in situ application, followed by PCR-based monitoring of integrated DNA sequence as an indirect measure of bacterial cell survival	Deeper understanding of active microbial communities and reduce bias of functional diversity	PCR-based method, biases inherent to PCR amplification
Bacterial sensors	This method uses fusion constructs of a reporter gene to promoter element induced by the target compound	Offer the possibility to characterize the biodegradability of specific contaminants present in a complex mixture without pretreatment of the environmental sample	Short lifetime, Performance dependent on environmental procedures
Microarray	Rapid method for automated determination of transcriptional activity allows justifications for the pollution-removal kinetics as well along with monitoring the bacterial cell survival	A high-throughput method; it is being increasingly used for in situ bioremediation studies because of its speed, specificity and reproducibility	Expensive and less flexible technique
Metabolic gene probing	Detect gene with function of interest, mRNA detection can reveal information about expression	Directly indicate microbial metabolic pathways	Limited to known genes activity cannot be inferred from the presence of genes alone
2D gel electrophoresis	It allows the analysis of comparative bacterial cell behaviour during bioremediation at total proteome level	Hundreds to thousands of polypeptides can be analysed in a single run	Technically difficult method to standardize Large amount of sample handling and limited reproducibility

(continued)

Table 13.3 (continued)

Technique	Indicator	Advantages	Disadvantages
Amplified rDNA restriction (ARDRA)	Characterization of diversity and richness of the microbial community under analyses	Method allows the downstream confirmation by DNA sequencing	PCR-based method ARDRA has also been reported to be subjective biases inherent to PCR amplification
Terminal restriction fragment length polymorphism (T-RFLP)	Analysis of microbial community structure and its dynamics during in situ bioremediation studies	The major advantage of T-RFLP is the use of an automated sequencer which gives highly reproducible results for repeated samples	The interpretation of T-RFLP data can be difficult and requires complementary analysis with multivariate statistical analysis to draw meaningful information
Denaturing/thermal gradient gel electrophoresis (D/TGGE)	PCR-based fingerprinting method that makes use of slight differences in denaturation profile of DNA fragments occurring as a consequence of base pair difference in the DNA sequence	It also offers the advantage of downstream DNA sequencing for confirmation of preliminary observation	PCR-based method so subjective biases inherent to PCR amplification
Single-strand confirmation polymorphism (SSCP)	Like D/TGGE, this fingerprinting method also makes use of differential electrophoretic mobility of DNA strands with difference at nucleotide composition	One of the commonly used method for assessment of microbial community structure and dynamics during in situ bioremediation	PCR-based method so electrophoretic behaviour of single-stranded molecule is unpredictable

13.11 Integration of Various Approaches to Assess In Situ Remediation of Xenobiotics

In assessing bioremediation processes in a contaminated aquifer, with respect to the selection of an appropriate remediation strategy, a single method can hardly understand the complex interactions between observed biogeochemistry and microbial remediation activities. Therefore, the integration of different technologies is recommended to overcome the limitations associated with a particular method. Integrated approaches aim to get an overall understanding of subsurface microbial processes and create a more solid basis for the evaluation of in situ bioremediation processes in contaminated aquifers. In many scenarios, integration of microbiological and geochemical methods has been used to narrate the microbiology associated with the geochemistry (Weiss and Cozzarelli 2008). In general, there is no universal comprehensive method for the in situ assessment of bioremediation. The selection of appropriate techniques always depended mainly on the specific questions needed to be addressed as well as the compound of interest. Griebler et al. (2004b) revealed in situ bioremediation of BTEX and PAHs in an aquifer by combined application of CSIA and metabolite analysis. While the detection of metabolites provided evidence of degradation of PAHs, CSIA data were not providing any significant information about bioremediation. For the BTEX compounds, however, CSIA was proven to be suitable as limited amount of metabolites were detected during biodegradation of BTEX compounds. The usefulness of combining CSIA with metabolite analysis has also been demonstrated in another bioremediation study, in which the metabolite analysis was found to be conclusive at higher toluene concentrations when analyses were not constrained by detection limits, whereas CSIA verified the bioremediation at lower toluene concentrations (McKelvie et al. 2005). Strong evidence of toluene and o-xylene (TX) in situ bioremediation was provided by the combination of analysis of metabolites analysis, CSIA and quantitative PCR analysis of bssA in another investigation (Beller et al. 2008). Another example of integrative, i.e. combining geochemical analysis, CSIA, in situ microcosm analysis and laboratory microcosms, study which assessed bioremediation of chlorobenzenes in an anoxic aquifer was done by Stelzer et al. (2009). Stable isotope probing of microbial biomass within in situ microcosms provided evidence for the assimilation of MCB-derived carbon into microbial biomass. Additionally, the relationship between hydrogeochemistry and degradation activity was investigated emphasizing flexible hydrogeochemical conditions related with explicit degradation activity within the xenobiotic plume (Bombach et al. 2010).

13.12 Conclusion

Sustainable removal of xenobiotics from the environment has created strong recent interest in applying natural attenuation strategies for the remediation of contaminated sites. This has ultimately led to the rapid development of various methods for the assessment and monitoring of in situ xenobiotic bioremediation. The assessment of in situ biotransformation and the description of the related reactions are presently still a big task in situ. Development in equipment, approaches and methods has improved our understanding of in situ reactions. However, application of these methods is strictly limited for their application in field; therefore, integrated application of several methods and approaches can give authenticated clue for in situ remediation of various xenobiotics. In particular, the development of compound-specific stable isotope analytical approaches, which include tracer and fractionation concepts, allows the evaluation of in situ microbial element cycles, as well as the characterization of pathways and mechanisms involved. These concepts are still in progress phase and are challenging to use for field investigations, and, in many cases, reference laboratory studies are required for a reliable and conclusive analysis. As discussed in this chapter, there is not a single universally reliable method to assess in situ bioremediation processes. Mostly the decision for the selection of the method is site specific, and this must be decided by the site manager which technique is best suitable for which kind of scenarios and also specific to type of xenobiotics.

References

Abe Y, Aravena R, Zopfi J, Shouakar-Stash O, Cox E, Roberts JD, Hunkeler D (2009) Carbon and chlorine isotope fractionation during aerobic oxidation and reductive dechlorination of vinyl chloride and cis-1,2-dichloroethene. Environ Sci Technol 43:101–107

Abe Y, Hunkeler D (2006) Does the Rayleigh equation apply to evaluate field isotope data in contaminant hydrogeology? Environ Sci Technol 40:1588–1596

Alumbaugh RE, Gieg LM, Field JA (2004) Determination of alkylbenzene metabolites in groundwater by solid-phase extraction and liquid chromatography-tandem mass spectrometry. J Chromatogr 1042:89–97

Adrian L, Marco UE (2016) Isotopes in geobiochemistry: tracing metabolic pathways in microorganisms of environmental relevance with stable isotopes. Curr Opin Biotechnol 41:19–25

Amann RI, Ludwig W, Schleifer KH (1995) Phylogenetic identification and in situ detection of individual microbial cells without cultivation. Microb Rev 59:143–169

Aronson D, Howard PH (1997) Anaerobic bioremediation of organic chemicals in groundwater: a summary of field and laboratory studies. Environmental Science Center, North Syracuse

Badea SL, Vogt C, Weber S, Danet AF, Richnow HH (2009) Stable isotope fractionation of gamma-hexachlorocyclohexane (lindane) during reductive dechlorination by two strains of sulfate-reducing bacteria. Environ Sci Technol 43:3155–3161

Badea SL, Vogt C, Gehre M, Fischer A, Danet AF, Richnow HH (2011) Development of an enantiomer-specific stable carbon isotope analysis (ESIA) method for assessing the fate of α-hexachlorocyclohexane in the environment. Rapid Comm Mass Spect 25(10):1363–1372

Bashir S, Hitzfeld KL, Gehre M, Richnow HH, Fischer A (2015) Evaluating degradation of hexachlorcyclohexane (HCH) isomers within a contaminated aquifer using compound-specific stable carbon isotope analysis (CSIA). Water Res 71:187–196

Bashir S, Fischer A, Nijenhuis I, Richnow HH (2013) Enantioselective carbon stable isotope fractionation of hexachlorocyclohexane during aerobic bioremediation by Sphingobium spp. Environ Sci Technol 47(20):11432–11439

Beller HR (2000) Metabolic indicators for detecting in situ anaerobic alkylbenzene degradation. Bioremediation 11:125–139

Beller HR (2002) Analysis of benzylsuccinates in groundwater by liquid chromatography/tandem mass spectrometry and its use for monitoring in situ BTEX bioremediation. Environ Sci Technol 36:2724–2728

Beller HR, Kane SR, Legler TC, McKelvie JR, Sherwood LB, Pearson F, Balser L, Mackay DM (2008) Comparative assessments of benzene, toluene, and xylene natural attenuation by quantitative polymerase chain reaction analysis of a catabolic gene, signature metabolites, and compound-specific isotope analysis. Environ Sci Technol 42(16):6065–6072

Bombach P, Richnow HH, Kästner M, Fischer A (2010) Current approaches for the assessment of in situ biodegradation. Appl Microbiol Biotechnol 86(3):839–852

Busch-Harris J, Sublette K, Roberts KP, Landrum C, Peacock AD, Davis G, Ogles D, Holmes WE, Harris D, Ota C, Yang XM, Kolhatkar A (2008) Bio-traps coupled with molecular biological methods and stable isotope probing demonstrate the in situ bioremediation potential of MTBE and TBA in gasoline-contaminated aquifers. Ground Water Monit Remediat 28:47–62

Caldwell ME, Suflita JM (2000) Detection of phenol and benzoate as intermediates of anaerobic benzene bioremediation under differ-ent terminal electron-accepting conditions. Environ Sci Technol 34:1216–1220
Chakraborty R, Coates JD (2004) Anaerobic degradation of mono-aromatic hydrocarbons. Appl Microbiol Biotechnol 64:437–446
Chapelle FH (2000) The significance of microbial processes in hydrogeology and geochemistry. Hydrol J 8:41–46
Chartrand MMG, Morrill PL, Lacrampe CG, Sherwood LB (2005) Stable isotope evidence for bioremediation of chlorinated ethenes at a fractured bedrock site. Environ Sci Technol 39:4848–4856
Christensen TH, Bjerg PL, Banwart SA, Jakobsen R, Heron G, Albrechtsen HJ (2000) Characterization of redox conditions in groundwater contaminant plumes. J Contam Hydrol 45:165–241
Christensen TH, Kjeldsen P, Bjerg PL, Jensen DL, Christensen JB, Baun A, Albrechtsen HJ, Heron C (2001) Biogeochemistry of landfill leachate plumes. Appl Geochem 16:659–718
Cichocka D, Imfeld G, Richnow HH, Nijenhuis I (2008) Variability in microbial carbon isotope fractionation of tetra-and trichloroethene upon reductive dechlorination. Chemosphere 71:639–648
Dempster HS, Sherwood LB, Feenstra S (1997) Tracing organic contaminants in groundwater: a new methodology using compound-specific isotopic analysis. Environ Sci Technol 31:3193–3197
Eganhouse RP, Cozzarelli IM, Scholl MA, Matthews LL (2001) Natural attenuation of volatile organic compounds (VOCs) in the leachate plume of a municipal landfill: using alkylbenzenes as process probes. Ground Water 39:192–202
Elsner M, Zwank L, Hunkeler D, Schwarzenbach RP (2005) A new concept linking observable stable isotope fractionation to transformation pathways of organic pollutants. Environ Sci Technol 39:6896–6916
Elsner M (2010) Stable isotope fractionation to investigate natural transformation mechanisms of organic contamiants: principles, prospects and limitations. J Environ Monit 12(11):2005–2031
Elsayed OF, Maillard E, Vuilleumier S, Nijenhuis I, Richnow HH, Imfeld G (2014) Using compound-specific isotope analysis to assess the degradation of chloroacetanilide herbicides in lab-scale wetlands. Chemosphere 99:89–95
Fischer A, Bauer J, Meckenstock R, Stichler W, Griebler C, Maloszewski P, Kästner M, Richnow HH (2006) A multitracer test proving the reliability of stable isotope fractionation analysis for assessing anaerobic degradation in a BTEX contaminated aquifer. Environ Sci Technol 40:4245–4252
Fischer A, Gehre M, Breitfeld J, Richnow HH, Vogt C (2009) Carbon and hydrogen isotope fractionation of benzene during biodegra-dation under sulfate-reducing conditions: a laboratory to field site approach. Rapid Commun Mass Spectrom 23:2439–2447
Fischer A, Herklotz I, Herrmann S, Thullner M, Weelink SAB, Stams AJM, Schlömann M, Richnow HH, Vogt C (2008) Combined carbon and hydrogen isotope fractionation investigations for elucidating benzene bioremediation pathways. Environ Sci Technol 42:4356–4363
Fischer A, Theuerkorn K, Stelzer N, Gehre M, Thullner M, Richnow HH (2007) Applicability of stable isotope fractionation analysis for the characterization of benzene bioremediation in a BTEX-contaminated aquifer. Environ Sci Technol 41:3689–3696
Fletcher KE, Löffler FE, Richnow HH, Nijenhuis I (2009) Stable carbon isotope fractionation of 1,2-dichloropropane during dichloroelimination by Dehalococcoides populations. Environ Sci Technol 43:6915–6919
Gafni A, Rosenzweig R, Gelman F, Ronen Z (2016) Anaerobic biodegradation of MTBE in a field site above the Israeli Coastal Aquifer: evidence from $\delta^{13}C$ compound specific isotope analysis. J Chem Technol Biotechnol 91(6):1638–1645
Geyer R, Peacock AD, Miltner A, Richnow HH, White DC, Sublette KL, Kästner M (2005) In situ assessment of bioremediation potential using biotraps amended with 13C-labeled benzene or toluene. Environ Sci Technol 39:4983–4989

Gillham RW, Starr RC, Miller DJ (1990) A device for in situ determination of geochemical transport parameters 2: biochemical reactions. Ground Water 28:858–862

Gödeke S, Richnow H-H, Weiss H, Fischer A, Vogt C, Borsdorf H, Schirmer M (2006) Multi tracer test for the implementation of enhanced in-situ bioremediation at a BTEX-contaminated mega-site. J Contam Hydrol 87:211–236

Geueke B, Garg N, Ghosh S, Fleischmann T, Holliger C, Lal R, Kohler HPE (2012) Metabolomics of hexachlorocyclohexane (HCH) transformation: ratio of LinA to LinB determines metabolic fate of HCH isomers. Environ Micro 15(4):1040–1049

Griebler C, Adrian L, Meckenstock RU, Richnow HH (2004a) Stable carbon isotope fractionation during aerobic and anaerobic transformation of trichlorobenzene. FEMS Microbiol Ecol 48:313–321

Griebler C, Safinowski M, Vieth A, Richnow HH, Meckenstock RU (2004b) Combined application of stable isotope analysis and specific metabolites determination for assessing in situ degradation of aromatic hydrocarbons in a tar oil-contaminated aquifer. Environ Sci Technol 38:617–631

Hageman KJ, Field JA, Istok JD, Semprini L (2004) Quantifying the effects of fumarate on in situ reductive dechlorination rates. J Contam Hydrol 75:281–296

Harner T, Wiberg K, Norstrom R (2000) Enantiomer fractions are preferred to enantiomer ratios for describing chiral signatures in environmental analysis. Environ Sci Technol 34(1):218–220

Harrington RR, Poulson SR, Drever JI, Colberg PJS, Kelly EF (1999) Carbon isotope systematics of monoaromatic hydrocarbons: vaporization and adsorption experiments. Org Geochem 30:765–775

Hassan W, Bano R, Bashir F, David J (2014) Comparative effectiveness of ACC-deaminase and/or nitrogen fixing rhizobacteria in promotion of maize (*Zea mays* L.) growth under lead pollution. Environ Sci Pollut Res 21:10983–10996

Hassan W, Bashir S, Ali F, Ijaz M, Hussain M, David J (2016) Role of ACC-deaminase and/or nitrogen fixing rhizobacteria in growth promotion of wheat (*Triticum aestivum* L.) under cadmium pollution. Environ Earth Sci 75(3):1–14

Herrmann S, Vogt C, Fischer A, Kuppardt A, Richnow HH (2009) Characterization of anaerobic xylene bioremediation by two-dimensional isotope fractionation analysis. Environ Microbiol Rep 1:535–544

Hirschorn SK, Grostern A, Lacrampe-Couloume G, Elizabeth AEB, MacKinnon L, Repta C, Major DW, Sherwood LB (2007) Quantification of biotransformation of chlorinated hydrocarbons in a biostimulation study: added value via stable carbon isotope analysis. J Contam Hydrol 94:249–260

Hunkeler D, Aravena R, Berry-Spark K, Cox E (2005) Assessment of degradation pathways in an aquifer with mixed chlorinated hydrocarbon contamination using stable isotope analysis. Environ Sci Technol 39:5975–5981

Imai R, Nagata Y, Fukuda M, Takagi M, Yano K (1991) Molecular cloning of a pseudomonas paucimobilis gene encoding a 17-kilodalton polypeptide that eliminates HCl molecules from γ-hexachlorocyclohexane. J Bacteriol 173(21):6811–6819

Kasai Y, Takahata Y, Manefield M, Watanabe K (2006) RNA-based stable isotope probing and isolation of anaerobic benzene-degrading bacteria from gasoline-contaminated groundwater. Appl Environ Microbiol 72:3586–3592

Kästner M, Richnow HH (2010) In situ microcosm studies to characterize microbial processes in the field. Handb Hydrocarb Lipid Microbiol:3503–3511

Kästner M, Fischer A, Nijenhuis I, Geyer R, Stelzer N, Bombach P, Tebbe CC, Richnow HH (2006) Assessment of microbial in situ activity in contaminated aquifers. Eng Life Sci 6:234–251

Kleikemper J, Pelz O, Schroth MH, Zeyer J (2002) Sulfate-reducing bacterial community response to carbon source amendments in contaminated aquifer microcosms. FEMS Microbiol Ecol 42:109–118

Kleikemper J, Pombo SA, Schroth MH, Sigler WV, Pesaro M, Zeyer J (2005) Activity and diversity of methanogens in a petroleum hydrocarbon-contaminated aquifer. Appl Environ Microbiol 71:149–158
Kopinke FD, Georgi A, Voskamp M, Richnow HH (2005) Carbon isotope fractionation of organiccontaminatss due to retardation on humic substances: implications for natural attenuation studies in aquifers. Environ Sci Technol 39:6052–6062
Kuder T, Wilson JT, Kaiser P, Kolhatkar R, Philp P, Allen J (2005) Enrichment of stable carbon and hydrogen isotopes during anaerobic bioremediation of MTBE: microcosm and field evidence. Environ Sci Technol 39:213–220
Lee PKH, Conrad ME, Alvarez-Cohen L (2007) Stable carbon isotope fractionation of chloroethenes by dehalorespiring isolates. Environ Sci Technol 41:4277–4285
Lone MI, He ZL, Stoffella PJ, Yang XE (2008) Phytoremediation of heavy metal polluted soils and water: progresses and perspectives. J Zhejiang Uni Sci B 9(3):210–220
Ludvigsen L, Albrechtsen HJ, Heron G, Bjerg PL, Christensen TH (1998) Anaerobic microbial redox processes in a landfill leachate contaminated aquifer (Grindsted, Denmark). J Contam Hydrol 33:273–291
Madsen EL (2005) Identifying microorganisms responsible for ecologically significant biogeochemical processes. Nat Rev Microbiol 3:439–446
Mahmood T (2010) Phytoextraction of heavy metals: the process and scope for remediation of contaminated soils. Soil Environ 29(2):91–109
Mancini SA, Devine CE, Elsner M, Nandi ME, Ulrich AC, Edwards EA, Sherwood LB (2008) Isotopic evidence suggests different initial reaction mechanisms for anaerobic benzene bioremediation. Environ Sci Technol 42:8290–8296
Manna RN, Dybala DA (2013) Insights into the elimination mechanisms employed for the degradation of different hexachlorocyclohexane isomers using kinetic isotope effects and docking studies. J Phys Org Chem 26(10):797–804
Mariotti A, Germon JC, Hubert P, Kaiser P, Letolle R, Tardieux A, Tardieux P (1981) Experimental determination of nitrogen kinetic isotope fractionation: some principles; illustration for the denitri-fication and nitrification processes. Plant Soil 62:413–430
Mazeas L, Budzinski H, Raymond N (2002) Absence of stable carbon isotope fractionation of saturated and polycyclic aromatic hydrocarbons during aerobic bacterial bioremediation. Org Geochem 33:1259–1272
MacDonald J (1994) In situ bioremediation: when does it work? Ind Waste 2(6):32–38
McKelvie JR, Hyman MR, Elsner M, Smith C, Aslett DM, Lacrampe CG, Sherwood LB (2009) Isotopic fractionation of methyl tert-butyl ether suggests different initial reaction mechanisms during aerobic bioremediation. Environ Sci Technol 43:2793–2799
McKelvie JR, Lindstrom JE, Beller HR, Richmond SA, Sherwood LB (2005) Analysis of anaerobic BTX bioremediation in a subarctic aquifer using isotopes and benzylsuccinates. J Contam Hydrol 81:167–186
McKelvie JR, Mackay DM, de Sieyes NR, Lacrampe CG, Sherwood LB (2007) Quantifying MTBE bioremediation in the Vandenberg air force base ethanol release study using stable carbon isotopes. J Contam Hydrol 94:157–165
McMahon PB, Chapelle FH (2008) Redox processes and water quality of selected principal aquifer systems. Ground Water 46:259–271
Meckenstock RU, Morasch B, Griebler C, Richnow HH (2004) Stable isotope fractionation analysis as a tool to monitor bioremediation in contaminated aquifers. J Contam Hydrol 75:215–255
Meyer AH, Penning H, Elsner M (2009) C and N isotope fractionation suggests similar mechanisms of microbial atrazine transformation despite involvement of different enzymes (AtzA and TrzN). Environ Sci Technol 43:8079–8085
Morasch B, Richnow HH, Vieth A, Schink B, Meckenstock RU (2004) Stable isotope fractionation caused by glycyl radical enzymes during bacterial degradation of aromatic compounds. Appl Environ Microbiol 70:2935–2940

Morrill PL, Lacrampe CG, Slater GF, Sleep BE, Edwards EA, McMaster ML, Major DW, Lollar BS (2005) Quantifying chlorinated ethene degradation during reductive dechlorination at Kelly AFB using stable carbon isotopes. J Contam Hydrol 76(3):279–293

Murtaza G, Ghafoor A, Qadir M, Owens G, Aziz MA, Zia MH (2010) Disposal and use of sewage on agricultural lands in Pakistan: a review. Pedosphere 20:23–34

Nagasawa S, Kikuchi R, Matsuo M (1993a) Indirect identification of an unstable intermediate in γ-HCH degradation by Pseudomonas paucimobilis UT26. Chemosphere 26(12):2279–2288

Nagasawa S, Kikuchi R, Nagata Y, Takagi M, Matsuo M (1993b) Aerobic mineralization of γ-HCH by Pseudomonas paucimobilis Ut26. Chemosphere 26(9):1719–1728

Nijenhuis I (2016) Characterisation of microbial transformation of halogenated organic contaminants using compound-specific stable isotope analysis (Doctoral dissertation, Halle (Saale), Martin-Luther-Universität Halle-Wittenberg, Habil.-Schr, 2016)

Nijenhuis I, Stelzer N, Kästner M, Richnow HH (2007) Sensitive detection of anaerobic monochlorobenzene degradation using stable isotope tracers. Environ Sci Technol 41:3836–3842

Palau J, Jamin P, Badin A, Vanhecke N, Haerens B, Brouyère S, Hunkeler D (2016) Use of dual carbon–chlorine isotope analysis to assess the degradation pathways of 1, 1, 1-trichloroethane in groundwater. Water Res 92:235–243

Peacock AD, Chang YJ, Istok JD, Krumholz L, Geyer R, Kinsall B, Watson D, Sublette KL, White DC (2004) Utilization of microbial biofilms as monitors of bioremediation. Microb Ecol 47:284–292

Peter A, Steinbach A, Liedl R, Ptak T, Michaelis W, Teutsch G (2004) Assessing microbial degradation of o-xylene at field-scale from the reduction in mass flow rate combined with compound-specific isotope analyses. J Contam Hydrol 71:127–154

Pope DF, Acree SD, Levine H, Mangion S, Van EJ, Hurt K, Wilson B, Burden DS (2004) Performance monitoring of MNA remedies for VOCs in ground water. US Environmental Protection Agency, National Risk Management Research Laboratory, Cincinnati

Poulson SR, Drever JI (1999) Stable isotope (C, Cl, and H) fractionation during vaporization of trichloroethylene. Environ Sci Technol 33:3689–3694

Ptak T, Piepenbrink M, Martac E (2004) Tracer tests for the investigation of heterogeneous porous media and stochastic modelling of flow and transport-a review of some recent developments. J Hydrol 294:122–163

Quintero JC, Moreira MT, Feijoo G, Lema JM (2005) Anaerobic degradation of hexachlorocyclohexane isomers in liquid and soil slurry systems. Chemosphere 61(4):528–536

Ragnarsdottir KV, Hawkins D (2005) Trace metals in soils and their relationship with scrapie occurrance. Geochim Cosmo Acta 69:196

Rayleigh JWS (1896) Theoretical considerations respecting the separation of gases by diffusion and similar processes. Philos Mag 42:493–498

Reusser DE, Istok JD, Beller HR, Field JA (2002) In situ transformation of deuterated toluene and xylene to benzylsuccinic acid analogues in BTEX-contaminated aquifers. Environ Sci Technol 36:4127–4134

Richmond SA, Lindstrom JE, Braddock JF (2001) Assessment of natural attenuation of chlorinated aliphatics and BTEX in subarctic groundwater. Environ Sci Technol 35:4038–4045

Richnow HH, Annweiler E, Michaelis W, Meckenstock RU (2003) Microbial in situ degradation of aromatic hydrocarbons in a contaminated aquifer monitored by carbon isotope fractionation. J Contam Hydrol 65:101–120

Roeling WFM, Van VHW (2002) Natural attenuation: what does the subsurface have in store? Bioremediation 13:53–64

Rosell M, Barcelo D, Rohwerder T, Breuer U, Gehre M, Richnow HH (2007) Variations in 13C/12C and D/H enrichment factors of aerobic bacterial fuel oxygenate degradation. Environ Sci Technol 41:2036–2043

Rügge K, Bjerg PL, Pedersen JK, Mosbaek H, Christensen TH (1999) An anaerobic field injection experiment in a landfill leachate plume, Grindsted, Denmark 1. Experimental setup, tracer movement, and fate of aromatic and chlorinated compounds. Water Resour Res 35:1231–1246

Schirmer M, Dahmke A, Dietrich P, Dietze M, Gödeke S, Richnow HH, Schirmer K, Weiß H, Teutsch G (2006) Natural attenuation research at the contaminated megasite Zeitz. J Hydro 328 (3):393–407

Schirmer M, Barker JF (1998) A study of long-term MTBE attenuation in the Borden aquifer, Ontario, Canada. Ground Water Monit Rem 18:113–122

Schroth MH, Kleikemper J, Bolliger C, Bernasconi SM, Zeyer J (2001) In situ assessment of microbial sulfate reduction in a petroleum-contaminated aquifer using push-pull tests and stable sulfur isotope analyses. J Contam Hydrol 51:179–195

Schüth C, Taubald H, Bolano N, Maciejczyk K (2003) Carbon and hydrogen isotope effects during sorption of organic contaminants on carbonaceous materials. J Contam Hydrol 64:269–281

Shati MR, Ronen D, Mandelbaum RT (1996) Method for in situ study of bacterial activity in aquifers. Environ Sci Technol 30:2646–2653

Sherwood LB, Slater GF, Sleep B, Witt M, Klecka GM, Harkness M, Spivack J (2001) Stable carbon isotope evidence for intrinsic bioremediation of tetrachloroethene and trichloroethene at area 6, Dover Air Force Base. Environ Sci Technol 35:261–269

Slater GF, Ahad JME, Sherwood Lollar B, Allen-King R, Sleep B (2000) Carbon isotope effects resulting from equilibrium sorption of dissolved VOCs. Anal Chem 72:5669–5672

Slater GF, Dempster HS, Sherwood Lollar B, Ahad J (1999) Headspace analysis: a new application for isotopic characterization of dissolved organic contaminants. Environ Sci Technol 33:190–194

Somsamak P, Richnow HH, Häggblom MM (2005) Carbon isotopic fractionation during anaerobic biotransformation of methyl tert-butyl ether and tert-amyl methyl ether. Environ Sci Technol 39:103–109

Somsamak P, Richnow HH, Häggblom MM (2006) Carbon isotope fractionation during anaerobic degradation of methyl tert-butyl ether under sulfate-reducing and methanogenic conditions. Appl Environ Microb 72:1157–1163

Souchier M, Benali RD, Casellas C, Ingrand V, Chiron S (2016) Enantiomeric fractionation as a tool for quantitative assessment of biodegradation: the case of metoprolol. Water Res 95:19–26

Stelzer N, Büning C, Pfeifer F, Dohrmann AB, Tebbe CC, Nijenhuis I, Kästner M, Richnow HH (2006) In situ microcosms to evaluate natural attenuation potentials in contaminated aquifers. Org Geochem 37:1394–1410

Stelzer N, Imfeld G, Thullner M, Lehmann J, Poser A, Richnow H-H, Nijenhuis I (2009) Integrative approach to delineate natural attenuation of chlorinated benzenes in anoxic aquifers. Environ Pollut 157:1800–1806

Strevett K, Davidova I, Suflita JM (2002) A comprehensive review of the screening methodology for anaerobic biodegradability of surfactants. Rev Environ Sci Biotechnol 1:143–167

Suar M, Hauser A, Poiger T, Buser HR, Müller MD, Dogra C, Raina V, Holliger C, van der Meer JR, Lal R (2005) Enantioselective transformation of α-hexachlorocyclohexane by the dehydrochlorinases LinA1 and LinA2 from the soil bacterium Sphingomonas paucimobilis B90A. App Environ Micro 71(12):8514–8518

Suarez MP, Rifai HS (1999) Bioremediation rates for fuel hydro-carbons and chlorinated solvents in groundwater. Biorem J 3:337–362

Tang W, Zhao Y, Wang C, Shan B, Cui J (2013) Heavy metal contamination of overlying waters and bed sediments of Haihe Basin in China. Ecotox Environ Saf 98:317–323

Thierrin J, Davis GB, Barber C (1995) A ground-water tracer test with deuterated compounds for monitoring in situ-bioremediation and retardation of aromatic hydrocarbons. Ground Water 33:469–475

Thullner M, Richnow HH, Fischer A (2009) Characterization and quantification of in situ bioremediation of groundwater contaminants using stable isotope fractionation analysis:

advantages and limitations. In: Gallo D, Mancini R (eds) Environmental and regional air pollution. Nova Science, Hauppage, pp 41–81

Thullner M, Centler F, Richnow HH, Fischer A (2012) Quantification of organic pollutant degradation in contaminated aquifers using compound specific stable isotope analysis–review of recent developments. Org Geochem 42(12):1440–1460

Trust BA, Mueller JG, Coffin RB, Cifuentes LA (1995) The bioremediation of fluoranthene as monitored using stable carbon isotopes. In: Hinchee RE, Douglas GS, Ong SK (eds) Monitoring and verification of bioremediation. Batelle Press, Columbus, pp 233–239

Van Breukelen BM (2007) Quantifying the degradation and dilution contribution to natural attenuation of contaminants by means of an open system Rayleigh equation. Environ Sci Technol 41:4980–4985

Vieth A, Kästner M, Schirmer M, Weiss H, Gödeke S, Meckenstock RU, Richnow HH (2005) Monitoring in situ bioremediation of benzene and toluene by stable carbon isotope fractionation. Environ Toxicol Chem 24:51–60

Vieth A, Müller J, Strauch G, Kästner M, Gehre M, Meckenstock RU, Richnow HH (2003) In-situ bioremediation of tetrachloroethene and trichloroethene in contaminated aquifers monitored by stable isotope fractionation. Isot Environ Health Stud 39:113–124

Vogt C, Cyrus E, Herklotz I, Schlosser D, Bahr A, Herrmann S, Richnow HH, Fischer A (2008) Evaluation of toluene degradation pathways by two-dimensional stable isotope fractionation. Environ Sci Technol 42:7793–7800

Vroblesky DA, Chapelle FH (1994) Temporal and spatial changes of terminal electron-accepting processes in a petroleum hydrocarbon-contaminated aquifer and the significance for contaminant bioremediation. Water Resour Res 30:1561–1570

Wang Y, Huang YS (2001) Hydrogen isotope fractionation of low molecular weight n-alkanes during progressive vaporization. Org Geochem 32:991–998

Wang Y, Huang YS (2003) Hydrogen isotopic fractionation of petroleum hydrocarbons during vaporization: implications for assessing artificial and natural remediation of petroleum contamination. Appl Geochem 18:1641–1651

Wanner P, Parker B, Chapman SW, Aravena R, Hunkeler D (2016) Quantification of degradation of chlorinated hydrocarbons in saturated low permeability sediments using compound-specific isotope analysis (CSIA). Environ Sci Technol. doi:10.1021/acs.est.5b06330

Weiss JV, Cozzarelli IM (2008) Bioremediation in contaminated aquifers: incorporating microbial/molecular methods. Ground Water 46:305–322

Wiedemeier TD, Rifai HS, Newell CJ, Wilson JT (1999) Natural attenuation of fuels and chlorinated solvents in the subsurface. Wiley, New York

Wuana RA, Okieimen FE (2011) Heavy metals in contaminated soils: a review of sources, chemistry, risks and best available strategies for remediation. ISRN Ecol 2011:20

Zhang N, Bashir S, Qin J, Schindelka J, Fischer A, Nijenhuis I, Herrmann H, Wick LY, Richnow HH (2014) Compound specific stable isotope analysis (CSIA) to characterize transformation mechanisms of α-hexachlorocyclohexane. J Hazard Mater 280:750–757

Zwank L, Berg M, Elsner M, Schmidt TC, Schwarzenbach RP, Haderlein SB (2005) New evaluation scheme for two-dimensional isotope analysis to decipher bioremediation processes: application to groundwater contamination by MTBE. Environ Sci Technol 39:1018–1029

Chapter 14
Transgenic Approaches for Building Plant Armor and Weaponry to Combat Xenobiotic Pollutants: Current Trends and Future Prospects

Umesh B. Jagtap and Vishwas A. Bapat

14.1 Introduction

Xenobiotics are chemical compounds that are foreign to the biological systems. A large amount of xenobiotic compounds were released into the environment from extensive anthropogenic activities such as agriculture, industries (textile, pharmaceutical, petroleum, etc.), military operations, wastewater treatments, constructions, the transport sector, and electronic waste deposition which can cause soil, water, and air pollution, affecting humans, animals, and ecosystems across the globe (Manzetti et al. 2014). Human population increases rapidly and it will reach 9.5 billion in 2050. The demand for land is likely to rise in parallel for the purpose of agriculture, housing, energy, leisure, and other activities. However, most of the land that becomes contaminated with harmful organic and inorganic xenobiotic compounds require a remediation process to bring the contaminated land back into the use for agriculture, housing, and recreation (Batty and Dolan 2013).

The various conventional (e.g., soil excavation, land filling, soil washing) and physicochemical (e.g., thermal treatment, chemical extraction, encapsulation) methods were employed to remediate the contaminated soil. These methods are expensive as these require movement of contaminated soil material off-site, are inefficient especially for large-scale cleanup, need to design different remediation processes for mixed contaminant sites, and are non-eco-friendly as they destroy the natural habitat and leave unsightly scars on the landscape. Besides these, some

U.B. Jagtap (✉)
Department of Botany, Government Vidarbha Institute of Science and Humanities, Amravati 444604, Maharashtra, India
e-mail: umeshjagtap93@gmail.com

V.A. Bapat
Department of Biotechnology, Shivaji University Vidyanagar, Kolhapur 416004, Maharashtra, India

© Springer International Publishing AG 2017

M.Z. Hashmi et al. (eds.), *Xenobiotics in the Soil Environment*, Soil Biology 49,
DOI 10.1007/978-3-319-47744-2_14

remediation treatments involve application of other harmful chemicals that negatively affects the soil fertility by altering physical and chemical nature of the soil. Therefore, alternative biological methods such as bioremediation (use of living organisms), phytoremediation (use of plants), and zoo remediation (use of animals) (Gifford et al. 2007) are developed for environmental remediation. Among all the available technologies, phytoremediation has emerged as an inexpensive, eco-friendly, and publicly acceptable technology for remediation of contaminated sites as it offers a number of benefits such as help in CO_2 sequestration, soil stabilization, watershed management, biodiversity improvement, providing diverse sources of energy, and aesthetics (Dickinson et al. 2009).

14.2 Phytoremediation Strategies

Phytoremediation is defined as the use of plants and their associated microorganisms for removal/detoxification of environmental pollutants to facilitate soil or water reclamation. Plants can deal with xenobiotic compounds present in the contaminated sites by different ways. Plants take up organic contaminants (e.g., gasoline, trinitrotoluene, tetrachloroethylene, polycyclic aromatic hydrocarbons, herbicides/pesticides, etc.) through roots inside their tissues, where it can be degraded with intracellular enzymes known as phytodegradation. Whereas, rhizospheric degradation of contaminants by extracellular enzymes secreted by the plant roots or microbes in rhizosphere is called as rhizodegradation. Unlike organic contaminants, inorganics (heavy metals/metalloids, radionuclides, plant fertilizers) do not undergo degradation and persist in the soil for a long period and can only be moved. The use of plants prevents/reduces the migration and bioavailability of pollutants in the environment through immobilization/stabilization of soil by plant roots called as phytostabilization. However, in phytoextraction (phytoaccumulation), plants absorb inorganic pollutants from contaminated sites and accumulate them in harvestable plant parts (shoot, leaves, etc.), and such plants are termed as hyperaccumulators. Some plants were involved in absorption or adsorption of inorganics by plant roots in hydroponic systems known as rhizofiltration and commonly used for treatment of industrial discharge, agricultural runoff, metals, and radioactive contamination. Phytovolatilization involves the use of plants to take up the volatile organic compounds and certain metal(loid)s such as mercury (Hg) and selenium (Se) from the soil, transforming them into volatile form, and release them through leaf stomata into the atmosphere (Arthur et al. 2005; Doty 2008; Pilon-Smits 2005).

14.3 Wild Versus Transgenic Plants

The number of xenobiotic compounds present in the soil has led to the soil pollution affecting the plant growth, metabolism, and yield. As a sessile and immovable organism, plants have developed the ability to cope up with xenobiotic compounds and colonize the contaminated soils. Some wild plant species are able to deal with xenobiotics by obtaining resistance, acquiring tendency to accumulate, and increasing its ability to degrade xenobiotic compounds (Viehweger 2014). However, the small habitat range or size of plants expressing remediation potential and insufficient abilities of wild plants to tolerate and accumulate contaminants limit widespread utilization of wild plants for phytoremediation (Arthur et al. 2005). In addition to this, wild plants should have to possess some special traits/characters such as ability to grow in different climatic conditions, extensive root system, and fast growth rate, accumulate high amounts of heavy metals in their easily harvestable parts, tolerate soil pollution, and also produce a great quantity of biomass in contamination condition in order to expand its exploitation in phytoremediation of xenobiotic compounds (Pilon-Smits 2005), but all these properties are not present in single plant species. When plants do not show enough or not at all show resistance/tolerance/degradation ability toward xenobiotic compounds, the genetic engineering of plants offers a viable option to enhance phytoremediation ability of plants.

On the other hand, recent advances in plant physiology, biochemistry, and molecular biology studies provide enough insights to decipher the mechanisms of plant tolerance to xenobiotic stresses. Furthermore, advent of genetic engineering and molecular biology techniques offers various methods to transfer one or several structural genes from different organisms into a desired plant to improve and modulate its phytoremediation potential toward various xenobiotic compounds.

Transgenic plants provide several advantages over wild-type plants with respect to elevated phytoremediation capacities, modulating the accumulation inside the plant tissue in order to avoid accumulation of toxic compounds in edible plant parts or accelerate accumulation of essential compounds in edible plant parts which in turn could prevent and reduce chemical contamination and convert contaminated sites into a safe agricultural or recreational land (Macek et al. 2008).

14.4 Approaches for Xenobiotic Remediation

The genetic engineering approaches for phytoremediation of xenobiotics mainly consist of genetic engineering of plants and plant-associated microorganisms to improve their *in planta* and *ex planta* metabolism that may lead to enhanced removal of xenobiotics.

14.4.1 Transgenic Plants

Recent progress in molecular biology, various "omics" platforms, and sequencing technology offer a molecular toolbox not only for the identification, isolation, and characterization of structural genes involved in xenobiotic remediation but also allows its interkingdom transfer. The myriad genes, conferring resistance to xenobiotic compounds, were identified from different organisms and then transferred to the intended crop plants to develop transgenic plants with enhanced phytoremediation capacity (Table 14.1).

Table 14.1 Some examples of transgenic plants developed for phytoremediation of xenobiotic contaminants

Type of xenobiotics	Gene	Product	Origin	Target plant	Effect	Reference
Herbicides/pesticides	CYP1A1	Cytochrome P450 monooxygenase	Human	*Oryza sativa*	Tolerance and metabolism of different herbicides	Kawahigashi et al. (2007, 2008)
	CYP1A1	Cytochrome P450 monooxygenase	Human	*Oryza sativa*	Remediation of atrazine and simazine	Kawahigashi et al. (2005a)
	CYP1A1, CYP2B6 and CYP2C19	Cytochrome P450 monooxygenase	Human	*Oryza sativa*	Remediation of atrazine and metolachlor	Kawahigashi et al. (2006)
	CYP1A1, CYP2B6 and CYP2C19	Cytochrome P450 monooxygenase	Human	*Solanum tuberosum*, *Oryza sativa*	Resistance to sulfonylurea and other herbicides	Inui and Ohkawa (2005)
	CYP105A1	Cytochrome P450 monooxygenase	*Streptomyces griseolus*	*Nicotiana tabacum*	Resistance to sulfonylurea	O'keefe et al. (1994)
	CYP2C9	Cytochrome P450 monooxygenase	Human	*Oryza sativa*	Tolerance to sulfonylurea	Hirose et al. (2005)
	CYP2B6	Cytochrome P450 monooxygenase	Human	*Oryza sativa*	Remediation of metolachlor	Kawahigashi et al. (2005b)
	CYP2B6	Cytochrome P450 monooxygenase	Human	*Oryza sativa*	Metabolism of ethofumesate and benfuresate	Kawahigashi et al. (2005c)
	CYP2B22, CYP2C49	Cytochrome P450 monooxygenase	*Sus scrofa*	*Oryza sativa*	Tolerance to several herbicides	Kawahigashi et al. (2005c)
	CYP71A10	Cytochrome P450 monooxygenase	*Glycine max*	*Nicotiana tabacum*, *Arabidopsis thaliana*	Tolerance to phenyl urea herbicide	Siminszky et al. (1999)

	CYP76B1	Cytochrome P450 monooxygenase	*Helianthus tuberosus*	*Nicotiana tabacum*	Tolerance to herbicide	Didierjean et al. (2002)
	GstI-6His	Glutathione S-transferases (GST I)	Maize	*Nicotiana tabacum*	Higher tolerance to alachlor	Karavangeli et al. (2005)
	atzA	Atrazine chlorohydrolase	Bacteria	Alfalfa, *Nicotiana tabacum*	Enhanced metabolic activity against atrazine	Wang et al. (2005)
	p-AtzA	Atrazine chlorohydrolase	Modified bacterial gene	Grasses and alfalfa	Bioremediation of atrazine in the environment	Vail et al. (2015)
	Protox	Protoporphyrinogen IX oxidase	*Bacillus subtilis*	*Oryza sativa*	Tolerance to diphenyl ether herbicide oxyfluorfen	Jung et al. (2008)
	ophc2	Organophosphorus hydrolase (OPH)	*Pseudomonas pseudoalcaligenes*	*Nicotiana tabacum*	Enhanced degradation of organophosphorus (methyl parathion)	Wang et al. (2008)
	γ-ECS, GS	γ-Glutamylcysteine synthetase; glutathione synthetase	*Brassica juncea*	*Brassica juncea*	Enhanced tolerance to atrazine, 1-chloro-2,4-dinitrobenzene, phenanthrene, metolachlor	Flocco et al. (2004)
	γ-ECS	γ-Glutamylcysteine synthetase	Poplar	*Populus trichocarpa*	Increased tolerance to chloroacetanilide herbicides	Gullner et al. (2001)
	CYP450E1	Cytochrome P450 monooxygenase	Human	*Nicotiana tabacum*	Enhanced degradation of anthracene and chlorpyrifos	Dixit et al. (2008)
	GST	Glutathione-S-transferase	*Trichoderma virens*			
Explosives	onr	Pentaerythritoltetranitrate reductase (PETN)	*Enterobacter cloacae*	*Nicotiana tabacum*	Enhanced denitration of glyceryl trinitrate (GTN) and TNT	French et al. (1998)
	NfsI	Nitroreductase	*Enterobacter cloacae*	*Nicotiana tabacum*	The transgenic plants removed high amount of TNT from the test solution and reduction of TNT to 4-hydroxylamino-2,6-dinitrotoluene	Hannink et al. (2001, 2007)

(continued)

Table 14.1 (continued)

Type of xenobiotics	Gene	Product	Origin	Target plant	Effect	Reference
	NfsA	Nitroreductase	*E. coli*	*Arabidopsis thaliana*	Higher nitroreductase activity and higher uptake	Kurumata et al. (2005).
	XplA and XplB	Cytochrome P450 monooxygenase	*Rhodococcus rhodochrous*	*Arabidopsis thaliana*	Increased degradation of RDX	Jackson et al. (2007)
	pnrA	Nitroreductase	*Pseudomonas putida*	Hybrid aspen (*Populus tremula* × *Populus tremuloides*)	Tolerate and take up greater amounts of TNT from contaminated water and soil	Van Dillewijn et al. (2008)
Heavy metals	AtMHX1	Vacuolar transporter	*Arabidopsis thaliana*	Tobacco	Reduced tolerance to Mg and Zn	Shaul et al. (1999)
	AtNramp3	Fe transporter		*Arabidopsis*	Increased accumulation of Fe, on Cd^{2+} treatment, and Cd hypersensitivity	Thomine et al. (2000)
	PgIREG1	Vacuolar transporter	*Psychotria gabriellae*	*Arabidopsis*	Confers Ni tolerance when expressed in transgenic plants	Merlot et al. (2014)
	MerE	Mercury transporter	*E. coli*	*A. thaliana*	Increased accumulation of methylmercury and mercuric ions into plants	Sone et al. (2013)
	TaPCS1	Phytochelatin synthase	Wheat	Tobacco	Tolerance to Pb and Cd	Gisbert et al. (2003)
	CdPCS1	Phytochelatin synthase	*Ceratophyllum demersum*	*Arabidopsis*	Enhanced accumulation of heavy metal(loid)s	Shukla et al. (2013)
	CUP1	Metallothionein	Yeast	Tobacco	2–3 × higher Cu content than the control but no Cd tolerance	Thomas et al. (2003)
	ScMTII	Metallothionein	*Saccharomyces cerevisiae*	*Nicotiana tabacum*	Accumulated more Cd	Daghan et al. (2013)

	merA and merB	Mercuric reductase and organomercurial reductase	*E. coli* Tn21	*Nicotiana tabacum* chloroplasts	Doubled biomass yield with seedlings grown on medium with 400μM phenyl–Hg^+	Ruiz et al. (2003)
	CYP2E1 and GST	Human CYP2E1 and glutathione S-transferase (GST)	Human	Alfalfa	Resistance toward the mixtures of Cd and trichloroethylene	Zhang and Liu (2011)
	StGCS-GS	γ-Glutamylcysteine synthetase-glutathione synthetase	*Streptococcus thermophilus*	Sugar beet (*Beta vulgaris* L.)	Enhanced tolerance and accumulated more Cd, Zn, and Cu ions	Liu et al. 2015
Polychlorinated biphenyls (PCBs)	bphc	2,3, Dihydroxybiphenyl-1,2-dioxygenase	PCB-degrading bacteria	*Nicotiana tabacum*	Enhanced degradation of PCBs	Chrastilova et al. (2007)
	bphC	Biphenyl catabolic enzymes	*Pandoraea pnomenusa*	*Nicotiana tabacum*	Enhanced degradation of PCBs	Francova et al. (2003), Novakova et al. (2010)
	bphC	2,3-Dihydroxybiphenyl-1,2-dioxygenase	Bacteria	*Nicotiana tabacum*	Efficient transformation 2,3-dihydroxybiphenyl	Viktorová et al. (2014)
	BphC.B	2,3-Dihydroxybiphenyl-1,2-dioxygenase	Soil metagenomic library	Alfalfa	Enhanced tolerance and remediation to mixed contaminates of PCBs and 2,4-dichlorophenol (2,4-DCP)	Wang et al. (2015)
Solvents	CYP4502E1	Cytochrome P450 monooxygenase	Rabbit	Hybrid poplar (*Populus tremula* × *Populus alba*)	Increased removal of TCE, vinyl chloride, carbon tetrachloride, benzene, and chloroform from hydroponic solution and air	Doty et al. (2007)
Petroleum pollutants	CYP4502E1	Cytochrome P450 monooxygenase	Human	*Nicotiana tabacum*	Oxidation of TCE and ethylene dibromide	Doty et al. (2000)

14.4.1.1 Which Crops Should Be Targeted?

The selection of an appropriate crop species for the phytoremediation is a prerequisite step in designing successful phytoremediation strategy. Several factors, viz., type (organic or inorganic), concentration (high or low) and characteristic of contaminants, type of land, biotic and abiotic factors affecting plant growth, and weather conditions at particular site, are taken into consideration while selecting plant species for phytoremediation (Surriya et al. 2014). In addition to that, certain characteristics of plants like type of root system, depth of root, plant growth rate, transpiration rate, seed and plant source, and allelopathic effects of plant are also taken into account. This is achieved with the coordinated efforts from agronomists, plant physiologists, soil scientists, environmental engineers, and governmental regulators having expertise in a particular area.

Beside these, the plant system exploited for phytoremediation should possess certain characteristics such as tolerance to contaminants, ease of genetic transformation, stability of gene expression, and amenability to breeding procedures, should not be a food or fodder crop, and has an advantage if it has an economic value (aromatic plants producing essential oils, energy crops) (Chang et al. 2014; Gupta et al. 2013; Kotrba et al. 2009).

14.4.1.2 Choosing the Target Gene for Transgenic Expression/Xenobiotic Remediation

Modern developments in genetics, cell imaging, omics technologies, next-generation sequencing technology bioinformatics, as well as cell and molecular biology have helped us not only to fast screening and identification of novel genes/proteins but also to understand its role in xenobiotic tolerance and degradation in great molecular details. The number of genes belonging to the membrane transporters, phytochelatins, metallothioneins, metal chelators, and enzymes responsible for uptake, translocation, and sequestration of inorganic contaminants was isolated from bacteria, mammals, and plants (Table 14.1). The successful transformation of these genes to the ideal plants suitable for phytoremediation shows enhanced phytoremediation. The plants are an autotrophic system and lack the catabolic genes responsible for the degradation of organic compounds. Since the catabolic genes isolated from bacteria and fungus have been transferred to the plants, plants have been able to express the catabolic genes/pathways resulting into an improved phytoremediation process for organic xenobiotics (Van Aken 2008) (Table 14.1).

14.4.1.3 Fine-Tuning of Transgene in Transgenic Plants for Better Response

Simply, transformation of catabolic genes/pathways to the plants doesn't promise the better phytoremediation capability of transgenic plants due to low levels of transgene expression in heterologous systems. Therefore, fine-tuning of transgene expression in transgenic plants is essential to make phytoremediation more efficient, and this will depend on the ability to apply tighter control over the transgene expression with respect to both cell/tissue/organ location and time. The engineering of plant regulatory systems and optimization of transgene expression by using additional transcriptional and translational enhancers and constitutive and tissue-specific promoters are crucial to design adaptable plants that will enable efficient phytoremediation.

Constitutive Expression When manipulating biochemical pathways for optimal phytoremediation capacity, the expression of respective genes in all plant tissues is important. The constitutive promoters direct transgene expression in virtually all tissues. Phytoremediation-related plant transformation studies have largely utilized the cauliflower mosaic virus 35S (CaMV 35S) (Thomine et al. 2000) and actin promoters (Dhankher et al. 2002) which have been employed to drive constitutive high level of expression of integrated genes in most of the plant tissues.

When to Express? Sometimes high-level constitutive expression of transgene was not desirable and affects the plant survival and growth. Therefore, dynamic regulation of transgene expression under external stimuli and at specific time normally results in more efficient phytoremediation. This was achieved by using different light-/stress-inducible and suitable promoter sequences. The light-inducible soybean RuBisCO small subunit 1 (SRS1) promoter was used to carry out light driven and leaves specific expression of *E. coli* arsenate reductase, arsC, gene in *Arabidopsis thaliana* that results in arsenate reduction in leaves (Dhankher et al. 2002). Similarly, copC gene from *Pseudomonas* sp. Az13, encoding a periplasmic Cu-binding protein that was engineered under the control of the light-inducible and tissue-specific cab1 (chlorophyll a/b binding protein 1) promoter, allowed higher Cu accumulation in *A. thaliana* shoots (Rodríguez-Llorente et al. 2012). Furthermore, the spatiotemporal regulation of transgene in transgenic plants was used to enhance the phytoremediation capability of plants.

Where to Express? The optimal engineering of spatial regulation of transgene is crucial to the effective design of plants for phytoremediation of xenobiotic contaminants. For effective removal of xenobiotic contaminants, it is necessary to express the respective gene in specific tissues (e.g., roots, leaves, tubers, fruits and seeds, etc.), and this was achieved by using tissue-specific promoters. The constitutive expression of metal hyper accumulator genes leads to the accumulation of metal in all tissues of the plant. Heavy metal accumulations in palatable plant parts (fruits/seeds/tuber) are not desirable because these are eaten by birds and many other organisms, resulting into biomagnification. However the accumulation of heavy metals in easily harvestable unpalatable plant parts like leaves has an advantage in phytomining and phytoextraction (Pandey et al. 2015).

The constitutive expression of phytochelatin synthase (PCS) gene, OsPCS1, in rice leads to accumulation of Cd throughout the plant including seeds, making it unsuitable for human consumption. Whereas, constitutive silencing of OsPCS1 gene makes rice plant sensitive to the Cd exposure and affects its growth on Cd-contaminated land. However, the knockdown of OsPCS1 gene by RNAi under the control of the seed-specific promoter ZMM1 (from maize) in rice showed reduced Cd accumulation in rice seeds and enhanced tolerance to Cd as compared to the wild plants (Li et al. 2007). In another studies, the leaf-specific expression of arsC gene and copC in *A. thaliana* plant showed hyperaccumulation of arsenic (As) and Cu, respectively (Dhankher et al. 2002; Rodríguez-Llorente et al. 2012). Hence, tissue-specific transgene expression leads to the increased phytoremediation ability of plants and would probably reduce the undesirable side effects.

Besides this, an aspect that should not be underestimated is the organelle targeting of transgene expression. Transgene integration in the various targeted organelle compartments like chloroplast, mitochondria, and vacuoles facilitates sequestration/detoxification of xenobiotic contaminants in the organelle. This prevents adverse interaction of xenobiotic compounds with cytoplasmic environment (Jagtap and Bapat 2015).

For instance, the organomercurial compounds are highly toxic and accumulate in membrane-bound organelles like plastids where they disrupt essential oxidative and photosynthetic pathways. As chloroplast is a main target of Hg poisoning, it is necessary to protect the chloroplast from Hg toxicity. Therefore, the native bacterial merA and merB genes were integrated into the tobacco chloroplast genome, and resulting transgenic plants were resistant to very high concentrations (up to 400 mM) of phenylmercuric acetate (Ruiz et al. 2003). In addition to this, chloroplast transformation provides several advantages over nuclear transformation such as restriction of transgene movement in environment by the maternal inheritance/cytoplasmic male sterility. Therefore, organelle targeting/compartmentalization should be considered seriously for the building of efficient plants for phytoremediation.

14.4.2 Genetic Engineering of Plant-Associated Microbes: Additional Troops

Plants are surrounded by their own personal cloud of microorganisms both below and above the ground. The plant-associated bacteria including endophytic, phyllospheric, and rhizospheric bacteria are involved in the promotion of plant growth, development, and plant protection. Recently, the potential of plants and their associated microorganisms in degradation and removal of xenobiotic pollutants has been recognized. However, not every plant-associated bacterium possesses the ability to degrade every toxic compound, and not every bacterium that has a degrading capacity toward a contaminant limits their widespread application in phytoremediation (Newman and Reynolds 2005).

Therefore, plant-associated bacteria engineered with the appropriate characteristics to achieve microbe-powered phytoremediation could help address all of these challenges in ways that are both environmentally and economically sound (Menn et al. 2000).

14.4.2.1 Engineered Rhizospheric Bacterium

The microorganisms living on or in the soil around the plant roots (rhizosphere) are called as rhizospheric microorganisms have been received considerable attention and studied extensively for their role in degradation of xenobiotic pollutants. Recently, rhizospheric microorganisms have been also engineered for improvement of phytoremediation capacities (Table 14.2).

14.4.2.2 Engineering Endophytes

Endophytes are microorganisms that live inter- and/or intracellularly within plant tissues without causing symptoms of disease (Porras-Alfaro and Bayman 2011). They are presumably ubiquitous in the plant kingdom and important components of plant microbiomes. The dozens of symbiotic microorganisms inhabited within roots and stems of plants, often providing essential services (nitrogen fixation, essential nutrient supply, production of plant growth regulators, defense mechanisms) to their host plants (Taghavi et al. 2011). However, this endophyte community varies according to the plant and environmental conditions in which particular plant species grow. For instance, plants growing in soil contaminated with petroleum had endophyte communities with higher frequencies of hydrocarbon-degrading bacteria than plants on uncontaminated soil (Siciliano et al. 2001). However, some plants had ability to recruit, or selectively augment, the necessary bacteria to remove pollutants, while other plants in the same area were unable to do so (Doty 2008). However, in some plants endophytic bacteria are unable to grow well enough and lost its pollutant degradation ability.

Since, genetic engineering of endophytes offers a viable option by making microbes capable of growing in required plant species with increased pollutant degradation capacity, thus boosting up overall phytoremediation process. The number of endophytic bacteria was engineered and used in association with plants to improve remediation of xenobiotic pollutants (Table 14.3).

Table 14.2 Some examples of plant and engineered plant rhizospheric association for xenobiotic remediation

Class of pollutant	Gene/ plasmid	Product	Source	Targeted rhizospheric bacterium	Association with plant	Effect	References
Heavy metal	MT1	Metallothionein I	Mouse	*Ralstonia eutropha* CH34	Tobacco (*Nicotiana benthamiana*)	Cd tolerance	Valls et al. (2000)
	MTL4	Tetrameric human metallothionein	Synthetic	*Mesorhizobium huakuii*	*Astragalus sinicus*	Cd accumulation	Sriprang et al. (2002)
	MTL4 and AtPCS	Metallothionein and phytochelatin synthase	Synthetic and *Arabidopsis thaliana*	*Mesorhizobium huakuii* subsp. rengei B3	*Astragalus sinicus*	Enhanced accumulation of Cd	Ike et al. (2007)
	MBP-EC20	Metal-binding peptide	–	*Pseudomonas putida* 06909	Sunflower	Cd tolerance	Wu et al. (2006)
	pTOM-TCE plasmid	Toluene and tri-chloroethylene (TCE) degradation	*Burkholderia vietnamiensis* BU61	*Burkholderia* sp. HU001, *Pseudomonas* sp. HU002	Willow	Production of siderophores, organic acids, and indoleacetic acid and showed increased Cd resistance	Weyens et al. (2013)
	arsRDABC	Arsenic resistant operon	Plasmid R773	*Pseudomonas fluorescens* F113rifPCB	Alfalfa (*Medicago sativa*)	As resistance	Ryan et al. (2007)
Polychlorinated biphenyls	Ohb	*Ortho*-halobenzoate 1,2-dioxygenase	Plasmid pE43 carrying *Pseudomonas aeruginosa* strain 142 ohb gene	*Sinorhizobium meliloti*	Alfalfa (*Medicago sativa*)	Degradation of 2′,3,4-trichloro-biphenyl	Chen et al. (2005)
Solvents	$tomA^+$	Toluene *o*-monooxygenase	*Burkholderia* cepacia PR123	*Pseudomonas fluorescens*	Wheat	Remediation of trichloroethylene	Yee et al. (1998)

Table 14.3 Some examples of plant and engineered plant entophyte association for xenobiotic remediation

Class of pollutant	Plant	Endophytic bacterium	Effect	References
Heavy metal	*Lupinus luteus* and *Lolium perenne*	*Burkholderia cepacia* and *Herbaspirillum seropedicae*	Phytoremediation of heavy metals	Lodewyckx et al. (2001)
Solvents	Yellow lupin	*Burkholderia cepacia* VM1330	Toluene degradation	Barac et al. (2004)
	Poplar	*B. cepacia* VM1468	Toluene degradation	Taghavi et al. (2005)
	Poplar	*Pseudomonas putida* W619-TCE	TCE degradation	Weyens et al. (2010)
Herbicide	Pea (*Pisum sativum*)	*Pseudomonas putida* VM1450	2,4 Dichlorophenoxyacetic acid degradation	Germaine et al. (2006)
Polyaromatic hydrocarbons	*Pisum sativum* var. early onward	*Pseudomonas putida* strain VM1441 (pNAH7)	Naphthalene phytoprotection and phytoremediation	Germaine et al. (2009)

14.5 Translation From Lab to Land: Unmet Challenges

Genetic engineering of plants and plant-associated microorganisms offers a promising means for remediation of xenobiotic pollutants that arise from extensive anthropogenic activities. To date, most of the phytoremediation studies are carried out in a number of model plants grown under controlled laboratory conditions. Translation of transgenic plants from lab to land remains a formidable challenge, even though some of the transgenic plants show remarkable phytoremediation abilities in the laboratory/greenhouse trials as compared to the non-transgenic plants. Each transgenic plant/product undergoes environmental risk assessment strategies, multiple regulatory loops prior to commercialization to assess potential harmful effects on the environment (Prado et al. 2014; Paoletti et al. 2008). In spite of these, the public is still unconvinced about transgenic plants and their safety. The possible risk associated with transgenic plants for phytoremediation of xenobiotic compounds is biomagnification, creation of super weeds due to uncontrolled spread of transgene to their wild relatives. Meanwhile, the several transgene bioconfinement strategies have been developed to restrict transgene movement in the environment (Daniell 2002; Hills et al. 2007).

Despite great efforts, these remediation strategies remain difficult and sometimes become impossible to translate from lab to land. In light of this fact, a comprehensive research agenda is needed to move lab discoveries into environmental practice in a way that maximizes remediation of contaminated sites and minimizes environmental pollution.

14.6 Advanced Techniques

To date, the myriad transgenic plants were generated by using well-established *Agrobacterium*-mediated and biolistic transformation methods. These methods are crude as these integrate transgene at random position in the genome; hence these plants have encountered a strong public opposition. Therefore, there is need to develop new techniques aimed at precise and site-specific modification in plant genomes which are acceptable to public and government regulators so that translation of transgenic plants from lab to land becomes possible (Kathiria and Eudes 2014).

Recent advances in genome editing tools including zinc finger nucleases (ZFNs) of transcription activator-like effector nucleases (TALENs), meganucleases, and clustered regularly interspaced short palindromic repeats-associated (CRISPR/Cas) genes provide emerging ways to targeted genome modifications in wider range of organisms (Liu et al. 2013).

On the other hand, modern scientific and technological advances in bioengineering and synthetic biology are making this possible and offer potential for great improvements in this area.

14.7 Future Perspectives: Synthetic Biology for Designing Ultimate Plant Genome

Modern developments in omics (genomics, transcriptomics, proteomics, metabolomics, secretomics, interactomics) approaches and high-throughput and next-generation sequencing technologies in plant combined with bioinformatics method allow us to understand the plant system biology which provides the basis for the plant synthetic biology (Bell et al. 2014). However, synthetic biology is nothing but putting engineering into bioengineering to create new biological systems (user-designed plants/plant cells/microbes) that do not occur naturally. For instance, synthetic yeast has been developed and employed for production of high-value secondary metabolites for medicine and industry (Klein et al. 2014). Similarly, synthetic biology projects focusing on development of user-designed plants including modifying cereals for nitrogen fixing, transferring C_4 photosynthesis pathway to rice, and introducing synthetic signal transduction systems that respond to external cues are underway (Baltes and Voytas 2015).

However, synthetic biology has been unexploited for the designing plants for phytoremediation. A further understanding of plant system biology and of the availability of standardized genetics will increase our knowledge on how to design synthetic plant systems for remediation of xenobiotic contaminants. To achieve this aim, the integration of knowledge and coordinated efforts from plant physiologists, agronomists, soil scientists, molecular biologists, microbiologists, systems biologists, bioinformatics, chemists, environmental engineers, and government regulators is required. This opens up unprecedented opportunities for the rational design and development of plants for remediation of xenobiotic. Meanwhile, the plant synthetic biology research raises questions about its future use in the remediation of xenobiotic contaminants and their perception by society. It is our hope that by exploiting nature's molecular design for creating genetically varied plant systems, synthetic biology can be used as a powerful platform for building new ultimate plants. Hitherto, it is long way to go to apply designer plant concepts in a concerted fashion on a large scale for phytoremediation of xenobiotic.

Acknowledgment UBJ thanks the Department of Biotechnology, government of India, for providing assistant professorship under the DBT-IPLS program. VAB thanks the Indian National Science Academy, New Delhi, India, for providing INSA Senior Scientist Fellowship.

References

Arthur EL, Rice PJ, Rice PJ, Anderson TA, Baladi SM, Henderson KL, Coats JR (2005) Phytoremediation-an overview. Crit Rev Plant Sci 24:109–122

Baltes NJ, Voytas DF (2015) Enabling plant synthetic biology through genome engineering. Trends Biotechnol 33:120–131

Barac T, Taghavi S, Borremans B, Provoost A, Oeyen L, Colpaert JV, Vangronsveld J, van der Lelie D (2004) Engineered endophytic bacteria improve phytoremediation of water-soluble, volatile, organic pollutants. Nat Biotechnol 22:583–588

Batty LC, Dolan C (2013) The potential use of phytoremediation for sites with mixed organic and inorganic contamination. Crit Rev Environ Sci Technol 43:217–259

Bell TH, Joly S, Pitre FE, Yergeau E (2014) Increasing phytoremediation efficiency and reliability using novel omics approaches. Trends Biotechnol 32:271–280

Chang FC, Ko CH, Tsai MJ, Wang YN, Chung CY (2014) Phytoremediation of heavy metal contaminated soil by *Jatropha curcas*. Ecotoxicology 23:1969–1978

Chen Y, Adam A, Toure O, Dutta SK (2005) Molecular evidence of genetic modification of *Sinorhizobium meliloti*: enhanced PCB bioremediation. J Ind Microbiol Biotechnol 32:561–566

Chrastilova Z, Mackova M, Novakova M, Macek T, Szekeres M (2007) Transgenic plants for effective phytoremediation of persistent toxic organic pollutants present in the environment. J Biotechnol 131:S38

Daghan H, Arslan M, Uygur V, Koleli N (2013) Transformation of tobacco with ScMTII gene-enhanced cadmium and zinc accumulation. Clean Soil Air Water 41:503–509

Daniell H (2002) Molecular strategies for gene containment in transgenic crops. Nat Biotechnol 20:581–586

Dhankher OP, Li Y, Rosen BP, Shi J, Salt D, Senecoff JF, Sashti NA, Meagher RB (2002) Engineering tolerance and hyperaccumulation of arsenic in plants by combining arsenate reductase and γ-glutamylcysteine synthetase expression. Nat Biotechnol 20:1140–1145

Dickinson NM, Baker AJ, Doronila A, Laidlaw S, Reeves RD (2009) Phytoremediation of inorganics: realism and synergies. Int J Phytoremediat 11:97–114

Didierjean L, Gondet L, Perkins R, Lau SMC, Schaller H, O'Keefe DP, Werck-Reichhart D (2002) Engineering herbicide metabolism in tobacco and *Arabidopsis* with CYP76B1, a cytochrome P450 enzyme from Jerusalem artichoke. Plant Physiol 130:179–189

Dixit P, Singh S, Mukherjee PK, Eapen S (2008) Development of transgenic plants with cytochrome P4502E1 gene and glutathione-S-transferase gene for degradation of organic pollutants. J Biotechnol 136:S692–S693

Doty SL (2008) Enhancing phytoremediation through the use of transgenics and endophytes. New Phytol 179:318–333

Doty SL, Shang TQ, Wilson AM, Tangen J, Westergreen AD, Newman LA, Strand SE, Gordon MP (2000) Enhanced metabolism of halogenated hydrocarbons in transgenic plants containing mammalian cytochrome P450 2E1. PNAS 97:6287–6291

Doty SL, James CA, Moore AL, Vajzovic A, Singleton GL, Ma C, Khan Z, Xin G, Kang JW, Park JY, Meilan R, Strauss SH, Wilkerson J, Farin F, Strand SE (2007) Enhanced phytoremediation of volatile environmental pollutants with transgenic trees. PNAS 104:16816–16821

Flocco CG, Lindblom SD, Elizabeth AH, Smits P (2004) Overexpression of enzymes involved in glutathione synthesis enhances tolerance to organic pollutants in *Brassica juncea.* Int J Phytoremediat 6:289–304

Francova K, Sura M, Macek T, Szekeres M, Bancos S, Demnerova K, Sylvestre M, Mackova M (2003) Preparation of plants containing bacterial enzyme for degradation of polychlorinated biphenyls. Fresen Environ Bull 12:309–313

French CE, Nicklin S, Bruce NC (1998) Aerobic degradation of 2, 4, 6-trinitrotoluene by *Enterobacter cloacae* PB2 and by pentaerythritol tetranitrate reductase. Appl Environ Microbiol 64:2864–2868

Germaine KJ, Liu X, Cabellos GG, Hogan JP, Ryan D, Dowling DN (2006) Bacterial endophyte-enhanced phytoremediation of the organochlorine herbicide 2, 4-dichlorophenoxyacetic acid. FEMS Microbiol Ecol 57:302–310

Germaine KJ, Keogh E, Ryan D, Dowling DN (2009) Bacterial endophyte-mediated naphthalene phytoprotection and phytoremediation. FEMS Microbiol Lett 296:226–234

Gifford S, Dunstan RH, O'Connor W, Koller CE, MacFarlane GR (2007) Aquatic zooremediation: deploying animals to remediate contaminated aquatic environments. Trends Biotechnol 25:60–65

Gisbert C, Ros R, De Haro A, Walker DJ, Bernal MP, Serrano R, Navarro-Aviñó J (2003) A plant genetically modified that accumulates Pb is especially promising for phytoremediation. Biochem Biophys Res Commun 303:440–445

Gullner G, Kömives T, Rennenberg H (2001) Enhanced tolerance of transgenic poplar plants overexpressing γ-glutamylcysteine synthetase towards chloroacetanilide herbicides. J Exp Bot 52:971–979

Gupta AK, Verma SK, Khan K, Verma RK (2013) Phytoremediation using aromatic plants: a sustainable approach for remediation of heavy metals polluted sites. Environ Sci Technol 47:10115–10116

Hannink N, Rosser SJ, French CE, Basran A, Murray JA, Nicklin S, Bruce NC (2001) Phytodetoxification of TNT by transgenic plants expressing a bacterial nitroreductase. Nat Biotechnol 19:1168–1172

Hannink NK, Subramanian M, Rosser SJ, Basran A, Murray JA, Shanks JV, Bruce NC (2007) Enhanced transformation of TNT by tobacco plants expressing a bacterial nitroreductase. Int J Phytoremediat 9:385–401

Hills MJ, Hall L, Arnison PG, Good AG (2007) Genetic use restriction technologies (GURTs): strategies to impede transgene movement. Trends Plant Sci 12:177–183

Hirose S, Kawahigashi H, Ozawa K, Shiota N, Inui H, Ohkawa H, Ohkawa Y (2005) Transgenic rice containing human CYP2B6 detoxifies various classes of herbicides. J Agric Food Chem 53:3461–3467

Ike A, Sriprang R, Ono H, Murooka Y, Yamashita M (2007) Bioremediation of cadmium contaminated soil using symbiosis between leguminous plant and recombinant rhizobia with the MTL4 and the PCS genes. Chemosphere 66:1670–1676

Inui H, Ohkawa H (2005) Herbicide resistance in transgenic plants with mammalian P450 monooxygenase genes. Pest Manage Sci 61:286–291

Jackson RG, Rylott EL, Fournier D, Hawari J, Bruce NC (2007) Exploring the biochemical properties and remediation applications of the unusual explosive-degrading P450 system XplA/B. PNAS 104:16822–16827

Jagtap UB, Bapat VA (2015) Genetic engineering of plants for heavy metal removal from soil. In: Sherameti I, Varma A (eds) Heavy metal contamination of soils. Springer, Zurich, pp 433–470

Jung S, Lee HJ, Lee Y, Kang K, Kim YS, Grimm B, Back K (2008) Toxic tetrapyrrole accumulation in protoporphyrinogen IX oxidase-overexpressing transgenic rice plants. Plant Mol Biol 67:535–546

Karavangeli M, Labrou NE, Clonis YD, Tsaftaris A (2005) Development of transgenic tobacco plants overexpressing maize glutathione S-transferase I for chloroacetanilide herbicides phytoremediation. Biomol Eng 22:121–128

Kathiria P, Eudes F (2014) Nucleases for genome editing in crops. Biocatal Agric Biotechnol 3:14–19

Kawahigashi H, Hirose S, Ohkawa H, Ohkawa Y (2005a) Phytoremediation of metolachlor by transgenic rice plants expressing human CYP2B6. J Agric Food Chem 53:9155–9160

Kawahigashi H, Hirose S, Ohkawa H, Ohkawa Y (2005b) Transgenic rice plants expressing human CYP1A1 remediate the triazine herbicides atrazine and simazine. J Agric Food Chem 53:8557–8564

Kawahigashi H, Hirose S, Ozawa K, Ido Y, Kojima M, Ohkawa H, Ohkawa Y (2005c) Analysis of substrate specificity of pig CYP2B22 and CYP2C49 towards herbicides by transgenic rice plants. Transgenic Res 14:907–917

Kawahigashi H, Hirose S, Ohkawa H, Ohkawa Y (2006) Phytoremediation of the herbicides atrazine and metolachlor by transgenic rice plants expressing human CYP1A1, CYP2B6, and CYP2C19. J Agric Food Chem 54:2985–2991

Kawahigashi H, Hirose S, Ohkawa H, Ohkawa Y (2007) Herbicide resistance of transgenic rice plants expressing human CYP1A1. Biotechnol Adv 25:75–84

Kawahigashi H, Hirose S, Ohkawa H, Ohkawa Y (2008) Transgenic rice plants expressing human P450 genes involved in xenobiotic metabolism for phytoremediation. J Mol Microbiol Biotechnol 15:212–219

Klein J, Heal JR, Hamilton WD, Boussemghoune T, Tange TO, Delegrange F, Jaeschke G, Hatsch A, Heim J (2014) Yeast synthetic biology platform generates novel chemical structures as scaffolds for drug discovery. ACS Synth Biol 3:314–323

Kotrba P, Najmanova J, Macek T, Ruml T, Mackova M (2009) Genetically modified plants in phytoremediation of heavy metal and metalloid soil and sediment pollution. Biotechnol Adv 27:799–810

Kurumata M, Takahashi M, Sakamoto A, Ramos JL, Nepovim A, Vanek T, Hirata T, Morikawa H (2005) Tolerance to, and uptake and degradation of 2, 4, 6-trinitrotoluene (TNT) are enhanced by the expression of a bacterial nitroreductase gene in *Arabidopsis thaliana*. Z Naturforsch C 60:272–278

Li JC, Guo JB, Xu WZ, Ma M (2007) RNA interference-mediated silencing of phytochelatin synthase gene reduce cadmium accumulation in rice seeds. J Integr Plant Biol 49:1032–1037

Liu W, Yuan JS, Stewart CN Jr (2013) Advanced genetic tools for plant biotechnology. Nat Rev Genet 14:781–793

Liu D, An Z, Mao Z, Ma L, Lu Z (2015) Enhanced heavy metal tolerance and accumulation by transgenic sugar beets expressing *Streptococcus thermophilus* StGCS-GS in the Presence of Cd, Zn and Cu alone or in combination. PLoS One 10:e0128824

Lodewyckx C, Taghavi S, Mergeay M, Vangronsveld J, Clijsters H, Lelie DVD (2001) The effect of recombinant heavy metal-resistant endophytic bacteria on heavy metal uptake by their host plant. Int J Phytoremediat 3:173–187

Macek T, Kotrba P, Svatos A, Novakova M, Demnerova K, Mackova M (2008) Novel roles for genetically modified plants in environmental protection. Trends Biotechnol 26:146–152

Manzetti S, van der Spoel ER, van der Spoel D (2014) Chemical properties, environmental fate, and degradation of seven classes of pollutants. Chem Res Toxicol 27:713–737

Menn FM, Easter JP, Sayler GS (2000) Genetically engineered microorganisms and bioremediation. In: Klein J (ed) Biotechnology. Wiley, New York, pp 441–463

Merlot S, Hannibal L, Martins S, Martinelli L, Amir H, Lebrun M, Thomine S (2014) The metal transporter PgIREG1 from the hyperaccumulator *Psychotria gabriellae* is a candidate gene for nickel tolerance and accumulation. J Exp Bot 65:1551–1564

Newman LA, Reynolds CM (2005) Bacteria and phytoremediation: new uses for endophytic bacteria in plants. Trends Biotechnol 23:6–8

Novakova M, Mackova M, Antosova Z, Viktorova J, Szekeres M, Demnerova K, Macek T (2010) Cloning the bacterial bphC gene into *Nicotiana tabacum* to improve the efficiency of phytoremediation of polychlorinated biphenyls. Bioeng Bugs 1:419–423

O'Keefe DP, Tepperman JM, Dean C, Leto KJ, Erbes DL, Odell JT (1994) Plant expression of a bacterial cytochrome P450 that catalyzes activation of a sulfonylurea pro-herbicide. Plant Physiol 105:473–482

Pandey VC, Pandey DN, Singh N (2015) Sustainable phytoremediation based on naturally colonizing and economically valuable plants. J Cleaner Prod 86:37–39

Paoletti C, Flamm E, Yan W, Meek S, Renckens S, Fellous M, Kuiper H (2008) GMO risk assessment around the world: some examples. Trends Food Sci Technol 19:S70–S78

Pilon-Smits E (2005) Phytoremediation. Annu Rev Plant Biol 56:15–39

Porras-Alfaro A, Bayman P (2011) Hidden fungi, emergent properties: endophytes and microbiomes. Annu Rev Phytopathol 49:291–315

Prado JR, Segers G, Voelker T, Carson D, Dobert R, Phillips J, Cook K, Cornejo C, Monken J, Grapes L, Reynolds T, Martino-Catt S (2014) Genetically engineered crops: from idea to product. Annu Rev Plant Biol 65:769–790

Rodríguez-Llorente ID, Lafuente A, Doukkali B, Caviedes MA, Pajuelo E (2012) Engineering copper hyperaccumulation in plants by expressing a prokaryotic copC gene. Environ Sci Technol 46:12088–12097

Ruiz ON, Hussein HS, Terry N, Daniell H (2003) Phytoremediation of organomercurial compounds via chloroplast genetic engineering. Plant Physiol 132:1344–1352

Ryan RP, Ryan D, Dowling DN (2007) Plant protection by the recombinant, root-colonizing *Pseudomonas fluorescens* F113rifPCB strain expressing arsenic resistance: improving rhizoremediation. Lett Appl Microbiol 45:668–674

Shaul O, Hilgemann DW, de-Almeida-Engler J, Van Montagu M, Inzé D, Galili G (1999) Cloning and characterization of a novel Mg^{2+}/H^+ exchanger. The EMBO J 18:3973–3980

Shukla D, Kesari R, Tiwari M, Dwivedi S, Tripathi RD, Nath P, Trivedi PK (2013) Expression of *Ceratophyllum demersum* phytochelatin synthase, CdPCS1, in *Escherichia coli* and *Arabidopsis* enhances heavy metal (loid) s accumulation. Protoplasma 250:1263–1272

Siciliano SD, Fortin N, Mihoc A, Wisse G, Labelle S, Beaumier D, Ouellette D, Roy R, Whyte LG, Banks MK, Schwab P, Lee K, Greer CW (2001) Selection of specific endophytic bacterial genotypes by plants in response to soil contamination. Appl Environ Microbiol 67:2469–2475

Siminszky B, Corbin FT, Ward ER, Fleischmann TJ, Dewey RE (1999) Expression of a soybean cytochrome P450 monooxygenase cDNA in yeast and tobacco enhances the metabolism of phenylurea herbicides. PNAS 96:1750–1755

Sone Y, Nakamura R, Pan-Hou H, Sato MH, Itoh T, Kiyono M (2013) Increase methylmercury accumulation in *Arabidopsis thaliana* expressing bacterial broad-spectrum mercury transporter MerE. AMB Express 3:52

Sriprang R, Hayashi M, Yamashita M, Ono H, Saeki K, Murooka Y (2002) A novel bioremediation system for heavy metals using the symbiosis between leguminous plant and genetically engineered rhizobia. J Biotechnol 99:279–293

Surriya O, Saleem SS, Waqar K, Kazi AG (2014) Phytoremediation of soils: prospects and challenges. In: Hakeem K, Sabir M, Ozturk M, Mermut A (eds) Soil remediation and plants: prospects and challenges. Academic, Heidelberg, pp 1–36

Taghavi S, Barac T, Greenberg B, Borremans B, Vangronsveld J, van der Lelie D (2005) Horizontal gene transfer to endogenous endophytic bacteria from poplar improves phytoremediation of toluene. Appl Environ Microbiol 71:8500–8505

Taghavi S, Weyens N, Vangronsveld J, van der Lelie D (2011) Improved phytoremediation of organic contaminants through engineering of bacterial endophytes of trees. In: Endophytes of forest trees. Springer, Dordrecht, pp 205–216
Thomas JC, Davies EC, Malick FK, Endreszl C, Williams CR, Abbas M, Petrella S, Swisher K, Perron M, Edwards R, Osenkowski P, Urbanczyk N, Wiesend WN, Murray KS (2003) Yeast metallothionein in transgenic tobacco promotes copper uptake from contaminated soils. Biotechnol Progr 19:273–280
Thomine S, Wang R, Ward JM, Crawford NM, Schroeder JI (2000) Cadmium and iron transport by members of a plant metal transporter family in Arabidopsis with homology to Nramp genes. PNAS 97:4991–4996
Vail AW, Wang P, Uefuji H, Samac DA, Vance CP, Wackett LP, Sadowsky MJ (2015) Biodegradation of atrazine by three transgenic grasses and alfalfa expressing a modified bacterial atrazine chlorohydrolase gene. Transgenic Res 24:475–488
Valls M, Atrian S, de Lorenzo V, Fernández LA (2000) Engineering a mouse metallothionein on the cell surface of *Ralstonia eutropha* CH34 for immobilization of heavy metals in soil. Nat Biotechnol 18:661–665
Van Aken B (2008) Transgenic plants for phytoremediation: helping nature to clean up environmental pollution. Trends Biotechnol 26:225–227
Van Dillewijn P, Couselo JL, Corredoira E, Delgado A, Wittich RM, Ballester A, Ramos JL (2008) Bioremediation of 2, 4, 6-trinitrotoluene by bacterial nitroreductase expressing transgenic aspen. Environ Sci Technol 42:7405–7410
Viehweger K (2014) How plants cope with heavy metals. Bot Stud 55:1–12
Viktorová J, Novakova M, Trbolova L, Vrchotova B, Lovecka P, Mackova M, Macek T (2014) Characterization of transgenic tobacco plants containing bacterial bphc Gene and study of their phytoremediation ability. Int J Phytoremediat 16:937–946
Wang L, Samac DA, Shapir N, Wackett LP, Vance CP, Olszewski NE, Sadowsky MJ (2005) Biodegradation of atrazine in transgenic plants expressing a modified bacterial atrazine chlorohydrolase (atzA) gene. Plant Biotechnol J 3:475–486
Wang X, Wu N, Guo J, Chu X, Tian J, Yao B, Fan Y (2008) Phytodegradation of organophosphorus compounds by transgenic plants expressing a bacterial organophosphorus hydrolase. Biochem Biophys Res Commun 365:453–458
Wang Y, Ren H, Pan H, Liu J, Zhang L (2015) Enhanced toleranc and remediation to mixed contaminates of PCBs and 2, 4-DCP by transgenic alfalfa plants expressing the 2, 3-dihydroxybiphenyl-1, 2-dioxygenase. J Hazard Mater 286:269–275
Weyens N, Croes S, Dupae J, Newman L, van der Lelie D, Carleer R, Vangronsveld J (2010) Endophytic bacteria improve phytoremediation of Ni and TCE co-contamination. Environ Pollut 158:2422–2427
Weyens N, Schellingen K, Beckers B, Janssen J, Ceulemans R, van der Lelie D, Taghavi S, Carleer R, Vangronsveld J (2013) Potential of willow and its genetically engineered associated bacteria to remediate mixed Cd and toluene contamination. J Soil Sediment 13:176–188
Wu CH, Wood TK, Mulchandani A, Chen W (2006) Engineering plant-microbe symbiosis for rhizoremediation of heavy metals. Appl Environ Microbiol 72:1129–1134
Yee DC, Maynard JA, Wood TK (1998) Rhizoremediation of trichloroethylene by a recombinant, root-colonizing pseudomonas fluorescensstrain expressing toluene ortho-monooxygenase constitutively. Appl Environ Microbiol 64:112–118
Zhang Y, Liu J (2011) Transgenic alfalfa plants co-expressing glutathione S-transferase (GST) and human CYP2E1 show enhanced resistance to mixed contaminates of heavy metals and organic pollutants. J Hazard Mater 189:357–362

Chapter 15
Bioavailability/Phytostabilization of Xenobiotics in Soil

Biljana Balabanova

15.1 Introduction

Increased population, industrialization and urbanization are responsible for environmental contamination. Environmental decontamination is an enigma. All substances originated into the environment either by biogenic or anthropogenic sources. Anthropogenic compounds describe synthetic compounds and compound classes as well as elements and naturally occurring chemical entities which are mobilized by man's activities. A substance foreign to biological system is known as xenobiotic compound (Varsha et al. 2011). Most of the xenobiotic compounds are degraded by microorganism. However, few of them may persist longer in the environment and not easily degraded, known as recalcitrant compound (Thakur 2008). A xenobiotic is a chemical which is found in an organism but which is not normally produced or expected to be present in it. It can also cover substances which are present in much higher concentrations than are usual. The term xenobiotic is also used to refer to organs transplanted from one species to another.

The word xenobiotic is a combination of two different roots, "xeno" and "biotic." *Xeno* is from the Greek and means strange, unnatural, or different. *Biotic* is a word that implies life. Xenobiotic, therefore, refers to an organic compound that mimics natural biochemicals that are essential for life but which have characteristics about them that are strange and unnatural (Thakur 2008, 2011). Xenobiotics are often toxic to life. Also, they may not be recognized by biochemical processes in plants and microorganisms and are thus resistant to degradation in the environment. Xenobiotics include many compounds that are involved in both industrial and agricultural activities.

B. Balabanova (✉)
Faculty of Agriculture, University "Goce Delčev", Krste Misirkov bb, 2000 Štip, Republic of Macedonia
e-mail: biljana.balabanova@ugd.edu.mk

© Springer International Publishing AG 2017

M.Z. Hashmi et al. (eds.), *Xenobiotics in the Soil Environment*, Soil Biology 49,
DOI 10.1007/978-3-319-47744-2_15

The fate of industrial solvents and other industrial chemicals in the soil environment is an important domain of soil biochemistry (Alexander 2000; Bohn et al. 2001). Research is desperately needed to determine the biochemical reactions in soil that transform these xenobiotic compounds to their mineral components. The fates of xenobiotics in soil include complete mineralization or stabilization of the parent compound, or some metabolite of the compound, in soil. If the xenobiotics are toxic and the rate of degradation is very slow, adverse effects on human and on ecological health are possible. Therefore it is desirable to maintain xenobiotic concentrations in soil at as low a level as possible. Because of these concerns, considerable research effort has been spent in trying to understand the metabolic degradation pathways of xenobiotics in soil (Bollag et al. 1992).

15.2 Xenobiotics in the Environment

Xenobiotic substances are becoming an increasingly large problem in sewage treatment systems, since they are relatively new substances and are very difficult to remove from the environment. These substances are released into the environment in amounts that are unnatural due to human activity. The substances are composed of only about a hundred fundamental kinds of matter called element. Elements may be of environmental concern. The heavy metals are well recognized toxic substances in the environment. Elemental forms of essential elements may be very toxic or cause environmental damages (Thakur 2011). Compounds may be inorganic or organic compounds. Inorganic and organic compounds may be divided depending upon presence of elements or group. Anthropogenic inorganic and organic pollutants are dispersed throughout the atmosphere, hydrosphere, and lithosphere, and they have tendency to transform into another compound which may be toxic, less toxic, and not toxic to flora and fauna (Thakur 2008, 2011).

15.2.1 Sources of Xenobiotic

According to Kaur and Parihar (2014), xenobiotic sources include agricultural practices, cigarette smoking, electronic waste, energy generation resulting from burning of fuels, and also leaks of transformer oils from electrical installations, industry (textile, agrochemical, paints, etc.), mining of precious minerals, natural emissions, oil and gas production and processing, pharmaceuticals and hospital effluents, radioactive materials, transportation, and others.

Thakur (2008) simply cites the sources of xenobiotics, divided into five categories:

- Chemical and pharmaceutical industries that produce a wide array of xenobiotics and synthetic polymers

- Pulp and paper bleaching (which are the main sources of natural and man-made chlorinated organic compounds in the environment)
- Mining (releases heavy metals into biogeochemical cycles)
- Fossil fuels (which may be accidentally released in large amounts into the ecosystem)
- Intensive agriculture (releases massive amounts of fertilizers, pesticides, and herbicides)

Classification of the main xenobiotics sources of contamination, given by Varsha et al. (2011), gives an overview of basic anthropogenic introduction in the environment.

15.2.1.1 Direct Sources

The prime direct source of xenobiotics is wastewater and solid residual releases from the industries like chemical and pharma, plastics, paper and pulp mills, textile mills, and agriculture (enhancement products like pesticides, herbicides, etc.). Some of the common residual compounds in the wastewater and other effluents are phenol, hydrocarbons, different dyes, paint effluents, pesticides and insecticides, etc.

Phenol and Phenol Compounds The natural water sources from the effluents of various chemical and pharma industries like coal refineries, phenol manufacturing, pharmaceuticals, dying, petrochemical, pulp mill etc., include wide variety of organic chemicals like phenol and various substituted phenol.

Plastics Plastics are durable and degrade very slowly due to the molecular bonds and interactions. Plastics are made of polystyrene and polyvinyl chloride, polyethylene, and its derivatives. Nowadays plastics (from crude oil) are used as fuels in industries since it breaks down in to liquid hydrocarbons. Types of bioplastics include starch-based plastics, cellulose-based plastics, polylactic acid plastics, and bio-derived polyethylene.

Hydrocarbons Petroleum effluents mainly contain polycyclic aromatic hydrocarbons, saturated hydrocarbons, and nitrogen-sulfur-oxygen compounds.

Paints Volatile organic compounds and additives like emulsifiers and texturizers in paint are considered harmful which can be degraded by different means like chemicals (water as solvent), hygroscopic stresses, and microbial sources.

Dyes Dye agglomeration is the major cause for the persistence of xenobiotics, and their presence in aquatic bodies will affect photosynthetic activity in aquatic life due to reduced light penetration even at low concentrations. The number of industrial processes, such as textile industries, paper printing, and photography, uses synthetic dyes extensively, which usually have complex aromatic molecular structures. Azo dyes, anthraquinone, and phthalocyanine dyes are commonly used dyes in these industries. The degradation of these dyes produces aromatic amines, which may be carcinogenic and mutagenic.

Pesticides and Insecticides A large number of pesticides and insecticides like organophosphorous compounds, benzimidazoles, methyl parathion, and morpholine are widely used and have contributed to the pollution load due to its slow degradation.

Paper and Pulp Effluents Effluent release from paper mill industries also contributes environmental pollution and cannot be neglected. Many of the chlorinated organic compounds randomly synthesized during pulp bleaching are reason for this. The situation can often be made worse if pulp mill effluents are released to oxygen-limited or oxygen-depleted (anaerobic) waters. The increased public awareness and more restrictive laws against polluting processes have forced paper industries to minimize release of adsorbable organic halides and to search technologies for cleaner productions.

15.2.1.2 Indirect Sources

Indirect sources of xenobiotics include nonsteroidal anti-inflammatory drugs, pharmaceutical compounds, pesticide residues, etc. Pharmaceutically active compounds, being an indirect source of xenobiotics, are discharged directly by manufacturers of the pharmaceuticals or effluents from hospitals which have performed their biologically intended effect and are passed onto the environment in either their complete or a fragmented state. These mainly include hormones, anesthetics, and antibiotics which bioaccumulate in an organism and passed on the other through the common food chain (Varsha et al. 2011). Biomaterials developed from the synthetic polymers have the biocompatibility, but their degradation into toxic substances in the body is a cause for concern. Even though they are the indirect sources, they cause adverse effect on the ecological cycle. Bioaccumulation of pesticides and biomagnification processes lead to toxic behavioral effect on animals and mankind.

15.3 Fate of Xenobiotics in Environment

Xenobiotic compounds are often toxic to life and are also often hard for microorganisms to metabolize (because they contain molecular arrangements that are not normally encountered in nature). Thus, many of these compounds can accumulate in the environment and continue to be a hazard for many years. Microbes (bacteria and fungi) found in natural waters and soils have a very broad ability to utilize (catabolize) virtually all naturally occurring compounds as their sources of carbon and energy, thus recycling the fixed organic carbon back into harmless biomass and carbon dioxide. This capability of microbes has evolved over 3 billion years of the planets history and is responsible for the balance between photosynthesis (by plants and algae), fixing carbon dioxide into biomass, and respiration (by animals and bacteria) and converting organic compounds back to carbon dioxide by oxidation through natural detoxification processes. When compounds are persistent in the environment, their biodegradation often proceeds through multiple steps utilizing different enzyme systems or different microbial populations (Macek et al. 2000).

Several methods, such as physicochemical and biological methods, have been employed in the xenobiotics degradation. The physicochemical methods are costly and often produce undesirable products which are toxic, requiring further treatment steps (Varsha et al. 2011). There are three requirements for biodegradation such as capable organisms, synthesis of requisite enzymes, and suitable environmental conditions.

A huge number of bacterial and fungal genera possess the capability to degrade organic pollutants. *Biodegradation* is defined as the biologically catalyzed reduction in complexity of chemical compounds (Kanaly and Harayama 2000). Biodegradation is the partial simplification or complete destruction of the molecular structure of environmental pollutants by complex, genetically regulated physiological reactions catalyzed largely by microorganisms (Alexander 1999; Madsen 1991, 1998) and plants (Bhadra et al. 1999). It is based on two processes: growth and cometabolism. In the case of growth, organic pollutants are used as sole source of carbon and energy. This process results in a complete degradation (mineralization) of organic pollutants. Cometabolism is defined as the metabolism of an organic compound in the presence of a growth substrate, which is used as the primary carbon and energy source. Many of these reactions are predictable based on established laws of thermodynamics applied under environmental geochemical conditions (temperature, pressure, and oxidation-reduction potential) that prevail in terrestrial and aquatic habitats (Madsen 1998). Biodegradation of xenobiotic compounds primarily dependent upon many of the enzymes used in the pathways for degrading unusual substrates catalyzes novel reactions. The initial intracellular attack of organic pollutants is an oxidative process; the activation and incorporation of oxygen is the enzymatic key reaction catalyzed by oxygenases and peroxidases. After those peripheral degradation pathways convert organic pollutants step by step into intermediates of the central intermediary metabolism.

Microbial biodegradation is routinely measured by applying chemical and physiological assays to laboratory incubations of flasks containing pure cultures of microorganisms, mixed cultures, or environmental samples (Madsen 1998). Plant-mediated biodegradation and/or contaminant accumulation also is routinely documented in appropriately scaled laboratory test systems (Bhadra et al. 1999; Van der Lelie et al. 2001; Tang et al. 2015; Su et al. 2015).

15.4 Bioavailability/Bioaccessibility of Xenobiotics

The term *bioavailability* means many things in many contexts. This section compiles definitions of bioavailability from several recent authoritative sources. The quotes that appear below are intentionally extensive to reveal both the many facets of the term *bioavailability* and the biases of the sources.

There are over 20 different definitions found in literature including those of absolute bioavailability, relative bioavailability, bioaccessibility, bioavailable fraction, etc. This study will review some of them. Scow and Johnson (1996), from a review in the soil sciences literature, reiterate that several processes act on organic and metallic pollutants in soils to reduce their toxicological and microbiological availability. An analogy between the availability of native soil organic material and pollutant (xenobiotic) chemicals is drawn. Bioavailability has been defined as the proportion of the nutrient that is digested, absorbed, and metabolized through normal pathways (Alexander 1999). Another definition from a recent textbook on biodegradation and bioremediation makes the abrupt transition from human nutrition to pollutant chemicals in soils (Alexander 2000). Chemically extractable pollutant compounds are contrasted with biologically available pollutant compounds, as assessed with biodegradation assays. The concept of biological availability was upgraded from a recent international conference on the environmental behavior of organic xenobiotic compounds and evokes the historical, toxicological, and physical-chemical basis for reduced bioavailability (Semple et al. 2003). Bioavailability can be defined as the amount of contaminant present that can be readily taken up by microorganisms and degraded (Semple et al. 2004).

15.4.1 The Role of Organic Matter in the Fate of Xenobiotics in Soil

The soil organic matter consists of a mixture of plant and animal residues at various stages of decomposition and of substances synthesized microbially and/or chemically from the breakdown products (Schnitzer 1991; Diehl et al. 2014). Soil organic matter plays a significant role in the functioning of soil in the environment. Quantity and quality of soil organic matter regulate the processes of sorption, sequestration, and biodegradation of xenobiotics in soil.

Soil organic matter can be subdivided into non-humic and humic substances. First include substances with still recognizable chemical characteristics. These are, for example, carbohydrates, amino acids, proteins, peptides, fatty acids, resins, and other low-molecular weight organic substances, which generally can be easily degraded in soil (Shchegolikhina and Marschner 2013). The main constituents of soil organic matter are humic substances, the products produced through humification of plant residues and other organic materials. Humic substances have large chemical heterogeneity and geographical variability, and they range in molecular weight from a few hundred to several thousand daltons. Humic substances have largely remained uncharacterized at the molecular level and are still operationally defined in terms of the methods used to extract or isolate them (Shchegolikhina et al. 2012). Humic substances are not single molecules, but can rather be interpreted by the concept of loosely bound humic supramolecular associations. In this concept, humic substances are viewed as relatively small and heterogeneous molecules of various origins that self-organize in supramolecular conformations (Piccolo 2001). Hydrophilic and hydrophobic domains of humic molecules can interact with each other and form large molecular size associations, which are stabilized only by weak forces such as dispersive hydrophobic interactions (Shchegolikhina et al. 2012). These forces determine the conformational structure of humic substances and the complexities of the multiple noncovalent interactions. Reduction of molecular mobility and chain segment flexibility may be caused by the increase of humic substances' molecular weight, the degree of unsaturation (due to multiple bonds, aromatic rings), and the degree of chain branching. Introduction of covalent cross-links between chains also leads to the reduction of molecular mobility in humic substances' supramolecular structure and to its transformation to glassy state (Yuan and Xing 2001). Soil organic matter may consist of a mixture of organic matter particles with different structures. Also individual particles of organic matter may contain micro domains, which vary widely in rubbery–glassy character (Shchegolikhina et al. 2012; Diehl et al. 2014).

The rubbery state can reattain equilibrium quickly following an incremental change in temperature or other environmental condition. Unlike the glassy state is a nonequilibrium metastable state, which cannot quickly respond to environmental changes (Zhang et al. 2009; Shchegolikhina et al. 2012). In a rubbery polymer, the diffusion of sorbate (the organic compound) involves a cooperative interchange of the penetrant and polymer matrix. The diffusion through the glassy state is more complex and may be considerably slower. Sorption of organic compounds on glassy parts of soil organic matter is coupled to structural deformation of organic matter. Thus way, the process of sorption itself can alter the soil organic matter structure over a time of soil–xenobiotic interaction (Zhang et al. 2009; Pignatello 2012; Shchegolikhina et al. 2012).

15.4.2 Cationic Effects Occurred in Soil Solution

Cations in soil can be divided into three categories: solid phase and exchangeable and soluble ions (Diehl et al. 2014). Cations in solid phase are presented by soil primary minerals, which are released by weathering and then reprecipitating. The exchangeable cations can be released to the solution, but they tend to associate with surfaces of the solid phase of soil. The soluble cations are poor competitors of exchangeable cations for surface charge (Shchegolikhina and Marschner 2013). When ions have the same valency, the cation with smallest hydrated size is preferably adsorbed by soil organo-mineral matrix. Polyvalent cations are strong complexing agents, which can form up to four linkages with functional groups of one macromolecule and few relatively small molecules of soil organic matter (Marschner and Shchegolikhina 2010).

Numerous studies for dissolved organic matter and humic acids in aqueous soil solution showed that addition of polyvalent cations to the solution increased condensation, matrix rigidity, thermal stability, and stability against biological degradation, as well as hydrophobicity and water repellency of these components of soil organic matter (Buurman et al. 2002; Lu and Pignatello 2004; Polubesova et al. 2007; Scheel et al. 2007, 2008; Marschner and Shchegolikhina 2010).

15.4.3 Basic Ionic Interaction

The most important mode of soil–xenobiotic interaction is adsorption, which can vary from complete reversibility to total irreversibility, and depends on soil features and the compound properties (Bollag et al. 1992). The sorption phenomenon includes physical and chemical interactions between soil and pollutants. The introduction to some possible binding forces, which govern the sorption and entrapment of xenobiotics on soil organic matter, is presented below.

Ionic bonds occur between organic matter and xenobiotics, which exist in cationic form or can act as proton acceptors. Ionic bonding involves ionized, or easily ionizable, carboxylic and phenolic hydroxyl groups of humic substances and thus can be pH dependent. Ionic bonding of xenobiotics on the structure of organic matter is highly stable (Dec and Bollag 1997; Marschner and Shchegolikhina 2010).

Covalent bonding between pollutants in metabolites and organic matter is a very strong forces leading to stable, irreversible retention of xenobiotic in the humic substances matrix. Processes of covalent bonding of xenobiotic onto organic matter are often mediated by chemical, photochemical, or enzymatic catalysts and may occur in soil due to oxidative coupling mechanism. Covalently bound pollutants become integral components of the organic matter and cannot be extracted without changing its properties (Dec and Bollag 1997; Shchegolikhina et al. 2012).

Hydrogen bonding is suggested to play an important role in the adsorption of nonionic polar pesticides on humic substances. Hydrogen bonding can be formed between oxygen- and hydroxyl-containing functional groups of humic substances. Xenobiotic molecules compete with already existing ligands, e.g., water molecules, for these binding sites (Dec and Bollag 1997; Shchegolikhina and Marschner 2013).

Van der Waals forces are relatively weak short-range intermolecular forces, which have different origins. Often, in humic substance chemistry, the term "van der Waals forces" is a synonym for dipole–dipole forces, which occur between nonionic or nonpolar xenobiotics and soil organic matter (Shchegolikhina and Marschner 2013).

Ligand exchange is a process that occurs in soil organic matter complexes with polyvalent metal ions, which are generally also associated with water molecules.

These relatively weak ligands can be replaced by appropriate functional groups of organic pollutants (Shchegolikhina and Marschner 2013).

Charge-transfer complexes can be formed via electron donor–acceptor mechanisms, when molecules with a high electron density react with molecules with electron deficiency. The amount of complexes formed in the organic matter structure is determined, for example, by the aromatic structure of the xenobiotic and quinone structures of the humic substances (Shchegolikhina et al. 2012).

Sequestration is a physicochemical process of slow integration and diffusion of nonpolar and hydrophobic xenobiotics into the structure of soil organic matter. Sequestration is closely related with sorption phenomena, which are generally ascribed to binding mechanisms, occurring between xenobiotics and organic matter surfaces instantaneously upon the first contact (Dec and Bollag 1997; Shchegolikhina et al. 2012).

Xenobiotic fractions in soil can be evaluated using a great number of biological and chemical methods. Various methods allow to distinguish labile, sequestered, and bound forms of the organic contaminants in soil (Northcott and Jones 2000; Shchegolikhina et al. 2012).

15.5 Mechanism of Phytoremediation of Contaminated Soils

Phytoremediation comprises technologies that use higher plants to clean up and revegetate contaminated sites (Robinson et al. 2003; Bolan et al. 2011). Many techniques and applications are included in the term phytoremediation. They differ in the process by which plants can remove, immobilize, or degrade contaminants. For example, the process in which plants are used to remove organic or inorganic contaminants from soil and store them in harvestable tissue is called phytoextraction, rhizoextraction, or phytofiltration (Sinha et al. 2007). The technique in which plants are used to remove contaminants through volatilization is called phytovolatilization (Bolan et al. 2011). Phytostabilization is process due to the contaminants in the soil being immobilized, thereby minimizing their transport in water or dust. This technology may enhance the degradation of organic contaminants such as pesticides and hydrocarbons via microbial activity associated with the plant roots that accelerates the transformation of these contaminants into nontoxic forms (Bolan et al. 2011; Tang et al. 2015). These mechanisms are interrelated and dependent upon plant physiological processes driven by solar energy, rhizospheric processes, and other available precursors. Therefore, in bioremediation application, multiple mechanisms are involved depending on the designed application.

15.5.1 Phytosequestration

Phytochemicals can be exuded into the rhizosphere, leading to the precipitation or immobilization of target contaminants in the root zone. This mechanism of phytosequestration may reduce the fraction of the contaminant that is bioavailable.

- Transport protein inhibition on the root membrane: Transport proteins associated with the exterior root membrane can irreversibly bind and stabilize contaminants on the root surfaces, preventing contaminants from entering the plant.
- Vacuolar storage in the root cells: Transport proteins are also present that facilitate transfer of contaminants between cells. However, plant cells contain a compartment (the vacuole) that acts, in part, as a storage and waste receptacle for the plant. Contaminants can be sequestered into the vacuoles of root cells, preventing further translocation to the xylem.

15.5.2 Phytodegradation

Specifically, phytodegradation, also called *phytotransformation*, refers to the uptake of contaminants with the subsequent breakdown, mineralization, or metabolization by the plant itself through various internal enzymatic reactions and metabolic processes. Depending on factors such as the concentration and composition, plant species, and soil conditions, contaminants may be able to pass through the rhizosphere only partially or negligibly impeded by phytosequestration and/or rhizodegradation. In this case, the contaminant may then be subject to biological processes occurring within the plant itself, assuming it is dissolved in the transpiration stream and can be phytoextracted. Plants catalyze several internal reactions by producing enzymes with various activities and functions. Specifically, oxygenases have been identified in plants that are able to address hydrocarbons such as aliphatic and aromatic compounds (Nam and Alexander 2001).

15.5.3 Phytovolatilization

Phytovolatilization is the volatilization of contaminants from the plant either from the leaf stomata or from plant stems. In some cases, a breakdown product derived from the rhizodegradation and/or phytodegradation of the parent contaminant along the transpiration pathway may be the phytovolatilized constituent. This effect was studied for the uptake and phytovolatilization of trichloroethene or its breakdown products in poplars. Similarly, certain inorganic constituents such as mercury may be volatilized as well. Specifically, tobacco plants have been modified to be able to take up the highly toxic methyl-mercury, alter the chemical speciation, and phytovolatilize relatively safe levels of the less toxic elemental mercury into the atmosphere. Once volatilized, many chemicals that are recalcitrant in the subsurface environment react rapidly in the atmosphere with hydroxyl radicals, an oxidant formed during the photochemical cycle. Phytovolatilization occurs as growing trees and other plants take up water and the contaminants. Some of these contaminants can pass through the plants to the leaves and volatilize into the atmosphere at comparatively low concentrations (Macek et al. 2000).

15.5.4 *Phytostabilization*

Phytostabilization refers to the holding of contaminated soils in place by vegetation and to immobilizing toxic contaminants in soils. Establishment of rooted vegetation prevents windblown dust, an important pathway for human exposure at hazardous waste sites. Hydraulic control is possible, in some cases, due to the large volume of water that is transpired through plants which prevents migration of leachate toward groundwater or receiving waters. Certain plant species immobilize contaminants in the soil through absorption by and adsorption on to roots or precipitation within the root zone-rhizosphere (Pradhan et al. 1998; Sinha et al. 2007). Plants capable of tolerating high level of contaminants and having efficient growth rate with dense root systems and canopies are preferred (Bolan et al. 2011). Trees which transpire large amounts of water for hydraulic control and grasses with fibrous roots to bind and hold soil are best suited for phytostabilization (Macek et al. 2000). Generally, plants suitable for *soil conservation* are useful in phytostabilization of soil contaminants. Phytostabilization reduces the mobility, and therefore the risk, of inorganic contaminants without necessarily removing them from the site. This technology does not generate contaminated secondary waste that needs further treatment. It also enhances soil fertility, thereby achieving ecosystem restoration (Macek et al. 2000; Bolan et al. 2011). However, since the contaminants are left in place, the site requires regular monitoring to ensure that the optimal stabilizing conditions are maintained. If soil amendments are used to enhance immobilization, they may need to be periodically reapplied to maintain their effectiveness (Keller et al. 2005; Bolan et al. 2011). Phytostabilization is particularly suitable for stabilizing radionuclide-contaminated sites, where one of the best alternatives is to hold contaminants in place to prevent "secondary contamination" and exposure to humans and animals. Plants roots also help to minimize water percolation through soil, thus reducing xenobiotics leaching significantly (Keller et al. 2005).

15.5.4.1 Soil Stabilization

Soil can mobilize (vertically and laterally) when exposed to uncontrolled water flows. Soil can also mobilize by blowing wind. Both of these modes of soil migration are known as *erosion* or *leaching*. If the soil is impacted, the migration of the contaminants through these modes is generally considered nonpoint source pollution (Alexander 2000). Phytostabilization covers provide a natural barrier and resistance to erosion and leaching and can be further used to minimize nonpoint source pollution if the soil or sediment is impacted. The main mechanism contributing to stabilizing erosion is the infusion of plant roots into the soil. Typically, plants with fibrous root systems are used, such as many grasses, herbaceous species, and wetland species (Bolan et al. 2011).

Phytostabilization covers are simply soils that are planted with vegetation selected specifically to control bulk soil migration and/or prevent contaminant migration through phytosequestration. In addition to phytosequestering contaminants in the rhizosphere, other plants, such as halophytes and hyperaccumulators, can be selected based on their ability to phytoextract and accumulate contaminants into the aboveground tissues (Casida and Lykken 1969). Obviously, additional risks are involved with moving contaminants into the plant; however, this aspect of a phytostabilization cover application for soil may still be acceptable, depending on the overall human health and ecological risks associated with the site. This is a decision factor to consider when selecting this phytotechnology application as the site remedy. If a harvesting and removal plan is implemented for the application to mitigate the additional risks, then the application is classified as a phytoremediation groundcover (Kaur and Parihar 2014).

15.5.4.2 Infiltration Control

Another method to stabilize contaminants in the subsurface is to prevent water from interacting with the waste, possibly leading to its migration. This is a common approach for landfill covers but can also be applied to minimize surface water recharge of groundwater plumes. Phytostabilization covers for infiltration control, also known as evapotranspiration, water balance, or vegetative covers, use the ability of plants to intercept rain to prevent infiltration and take up and remove significant volumes of water after it has entered the subsurface to minimize the percolation into the contained waste (Macek et al. 2000). The main phytotechnology mechanism for these applications is phytohydraulics. Phytostabilization covers for infiltration control are composed of soil and plants that maximize evaporation from the soil and plant evapotranspiration processes from the system. To allow these time-dependent (and climate-dependent) processes to occur and successfully remove water from the system, the soil component of the cover is specifically designed and installed such that the available water storage capacity in the soil is maximized (Semple et al. 2003). The vegetation component of the cover usually entails specially formulated seed mixes or mixed communities of plants/trees that can access the stored water as well as create the intercepting canopy. Furthermore, the entire cover is often contoured to promote runoff as another significant loss mechanism for the overall water balance (Macek et al. 2000).

When minimizing infiltration, one of the potential outcomes is to create an anaerobic zone underneath the phytostabilization cover. In some cases, the subsurface conditions will be driven into methanogenic (methane-producing) conditions.

These covers may not be appropriate for sites that can lead to the production of chronic, large, or uncontrolled amounts of this landfill gas. While the methane itself may or may not be toxic to the plants, the presence of the gas in the vadose zone may restrict the oxygen transport needed for cell respiration in the root system. Furthermore, these covers have not been shown to be able to prevent the diffusion of landfill gases to the surface. Therefore, these gases are adventitious for the root system, because coupled with the plant growth microbes affects on the plant growth process (Sinha et al. 2007).

15.5.4.3 Successful Cases

Metabolism of herbicides and pesticides was extensively studied many years ago (Casida and Lykken 1969; Dodge 1989; Komosa and Sandermann 1992; Bollag et al. 1992). During recent years, metabolism of nonagricultural xenobiotics such as trichloroethylene (TCE), TNT, glyceryl trinitrate (GTN), polyaromatic hydrocarbons (PAHs), PCBs, and other chlorinated compounds has been studied (Tang et al. 2015; Su et al. 2015). It was shown that most of these compounds are metabolized but only a few chemicals appear to be fully mineralized. Some plant metabolites of pollutants may be more toxic than the original compounds, making plants less attractive compared with bacteria, which totally degrade organic pollutants.

Singleton et al. (2003) studied the bacterial flora associated with the intestine and vermicasts of the earthworms and found species like *Pseudomonas*, *Paenibacillus*, *Azoarcus*, *Burkholderia*, *Spiroplasma*, *Alcaligenes*, and *Acidobacterium*. Some of them, such as *Pseudomonas*, *Alcaligenes*, and *Acidobacterium*, are known to degrade hydrocarbons. *Rhodococcus* can use anthracene, phenanthrene, pyrene, and fluoranthene as sole source of carbon and energy. Some fungi such as *Penicillium*, *Mucor*, and *Aspergillus* have also been found in the intestine of earthworms and they degrade PAHs (Muratova et al. 2003). Biodegradation of PAHs occurs when microorganisms break the aromatic rings of benzene and produce aliphatic compounds that readily enter the tricarboxylic acid cycle (metabolic activity) operating in living cells. *Cunninghamela elegans* and *Candida tropicalis* have been reported to degrade PAHs (Kanaly and Harayama 2000). Degradation products of PAHs are, however, not necessarily less toxic than the parent compounds. *Alcaligenes* can even degrade PCBs and *Mucor* dieldrin (Johnsen et al. 2005; Contreras-Ramos et al. 2006).

Robinson et al. (2007) describe the phytostabilization of this pile using poplar trees. Poplar clones were selected that tolerate high boron concentrations. The pile was heavily fertilized with nitrogen, and a collection pond to store any leachate was installed at the foot of the pile. After 3 years of growth, the trees have reduced leaching events from the pile during the Southern hemisphere winter. The poplar trees accumulated high concentrations (>1000 mg/kg) of boron in the leaves.

15.6 SWOT Analysis: Phytostabilization in Use

The main objectives for successful phytostabilization are (a) to change the contaminant speciation in the soil aiming to reduce the easily soluble and exchangeable fraction of these component, (b) to stabilize the vegetation cover and limit xenobiotic uptake by crops, (c) to reduce the direct exposure of soil heterotrophic living organisms, and (d) to enhance biodiversity. To realize this in situ remediation makes use of contaminant immobilizing soil additives that enhance processes such as precipitation, sorption, ion exchange, and redox reaction. The formation of insoluble contaminant species results in a reduced xenobiotic mobility, and bioavailability reduces leaching through the soil profile and biological interactions with living organisms. Soil amendments also can restore appropriate soil conditions for plant growth by balancing pH, adding organic matter, restoring soil microbial activity, increasing moisture retention, and reducing compaction (Table 15.1).

Table 15.1 SWOT analysis of the positive/negative and internal/external factors characterizing phytostabilization (Alexander 1999; Bohn et al. 2001; Thakur 2008, 2011; Varsha et al. 2011; Bolan et al. 2011)

S – strengths	*W – weaknesses*
• Removal of contaminants through plant uptake and volatilization and this technology can be enhanced by using soil amendments that are effective in the immobilization • Containing the mobility of contaminants through their immobilization within the root zone of plants • Preventing off-site contamination through contaminants migration via wind and water erosion and leaching and soil dispersion • It is far less disruptive to the environment • There is no need for disposal sites • It has a high probability of public acceptance	• The contaminant must be within, or drawn toward, the root zones of plants • Contaminated sites are not conducive for plant growth • Contaminated site must be large enough to make farming techniques appropriate • It must not present an eminent danger to human health or further environmental harm • It may take longer than other technologies • Rainfall and temperature affect phytostabilization through their effects on plant growth, contaminant reactions, and soil erosion
O – opportunities	*T – threats*
• Phytostabilization can be enhanced by increasing plant growth and altering bioavailability • Monitor natural attenuation of contaminated sites which is employed within the context of a carefully controlled site-specific cleanup strategy • Reduces the mobility and, therefore, the risk of contaminants without necessarily removing them from their source location • Phytostabilization does not generate contaminated secondary waste that needs • Phytostabilization provides ecosystem development to achieve biodiversity corridors	• Environmental managers often use changes in surface chemical properties of soils • Application of fertilizer can lead to environmental degradation

15.7 Conclusions

Human activities have brought about widespread pollution of the natural environment. A number of organic pollutants, such as polycyclic aromatic hydrocarbons, polychlorinated aromatic compounds, and nitrogen containing aromatic compounds, are resistant to degradation and represent an ongoing toxicological threat to both wildlife and human beings. Bioremediation is an attractive alternative to traditional physicochemical techniques for the remediation of the contaminated site, as it can be more cost effective, and it can selectively degrade the pollutants without damaging the site or its indigenous flora and fauna. Phytostabilization is primarily aimed at containing the mobility of contaminants through their immobilization within the root zone of plants and "holding" soil, thereby preventing off-site contamination through their migration via wind and water erosion and leaching and soil dispersion. This technique is readily suited to monitor natural attenuation of contaminated sites which is employed within the context of a carefully controlled site-specific cleanup strategy.

References

Alexander M (1999) Biodegradation and bioremediation, 2nd edn. Academic, New York

Alexander M (2000) Aging, bioavailability, and overestimation of risk from environmental pollutants. Critical review. Environ Sci Technol 34(20):4259–4265

Bhadra R, Spanggord RJ, Wayment DG, Hughes JB, Shanks JV (1999) Characterization of oxidation products of TNT metabolism in aquatic phytoremediation systems of *Myriophyllum aquaticum*. Environ Sci Technol 33:3354–3361

Bohn HL, McNeal BL, O'Connor GA (2001) Soil chemistry. Wiley, New York

Bolan NS, Park JH, Robinson B, Naidu R, Huh KY (2011) Phytostabilization: A green approach to contaminant containment. Adv Agron 112:145

Bollag JM, Myers CJ, Minard RD (1992) Biological and chemical interactions of pesticides with soil organic-matter. Sci Total Environ 123:205–217

Buurman P, van Lagen B, Piccolo A (2002) Increase in stability against thermal oxidation of soil humic substances as a result of self association. Org Geochem 33:367–381

Casida JE, Lykken L (1969) Metabolism of organic pesticide chemicals in higher plants. Annu Rev Plant Physiol 20:607–636

Contreras-Ramos SM, Alvarez-Bernal D, Dendooven L (2006) *Eisenia fetida* increased removal of polycyclic aromatic hydrocarbons from soil. Environ Pollut 141(3):396–401

Dec J, Bollag JM (1997) Determination of covalent and noncovalent binding interactions between xenobiotic chemicals and soil. Soil Sci 162:858–874

Diehl D, Schneckenburger T, Krüger J, Goebel MO et al (2014) Effect of multivalent cations, temperature and aging on soil organic matter interfacial properties. Environ Chem 11(6): 709–718

Dodge AD (1989) Herbicides and plant metabolism, vol 38. Cambridge University Press, London

Johnsen AR, Wick LY, Harms H (2005) Principles of microbial PAH-degradation in soil. Environ Pollut 133(1):71–84

Kanaly RA, Harayama S (2000) Biodegradation of high-molecular-weight polycyclic aromatic hydrocarbons by bacteria. J Bacteriol 182(8):2059–2067

Kaur P, Parihar L (2014) Bioremediation: step towards improving human welfare. Annu Res Rev Biol 4(20):3150
Keller C, Marchetti M, Rossi L, Lugon-Moulin N (2005) Reduction of cadmium availability to tobacco (*Nicotiana tabacum*) plants using soil amendments in low cadmiumcontaminated agricultural soils: a pot experiment. Plant Soil 276:69–84
Komosa D, Sandermann H (1992) Plant metabolism of herbicides with C-P bonds: phosphinotricin. Pest Biochem Physiol 43:95–102
Lu YF, Pignatello JJ (2004) Sorption of apolar aromatic compounds to soil humic acid particles affected by aluminum (III) ion cross-linking. J Environ Qual 33:1314–1321
Macek T, Mackova M, Káš J (2000) Exploitation of plants for the removal of organics in environmental remediation. Biotech Adv 18(1):23–34
Madsen EL (1991) Determining *in situ* biodegradation: facts and challenges. Environ Sci Technol 25:1662–1673
Madsen EL (1998) Epistemology of environmental microbiology. Environ Sci Technol 32: 429–539
Marschner B, Shchegolikhina A (2010) Effect of different cation saturations on the sorption and mineralization of the hydrophobic organic compounds nonylphenol and phenanthrene in soils. In: Proceedings of the 19th World Congress of Soil Science: soil solutions for a changing world, Brisbane
Muratova A, Hübner T, Tischer S, Turkovskaya O, Möder M, Kuschk P (2003) Plant–rhizosphere-microflora association during phytoremediation of PAH-contaminated soil. Int J Phytoremediat 5(2):137–151
Nam K, Alexander M (2001) Relationship between biodegradation rate and percentage of a compound that becomes sequestered in soil. Soil Biol Biochem 33:787–792
Northcott GL, Jones KC (2000) Experimental approaches and analytical techniques for determining organic compound bound residues in soil and sediment. Environ Pollut 108:19–43
Piccolo A (2001) The supramolecular structure of humic substances. Soil Sci 166:810–832
Pignatello J (2012) Dynamic interactions of natural organic matter and organic compounds. J Soils Sediments 12:1241–1256
Polubesova T, Sherman-Nakache M, Chefetz B (2007) Binding of pyrene to hydrophobic fractions of dissolved organic matter: effect of polyvalent metal complexation. Environ Sci Technol 41: 5389–5394
Pradhan SP, Conrad JR, Paterek JR, Srivastava VJ (1998) Potential of phytoremediation for treatment of PAHs in soil at MGP Sites. J Soil Contam 7:467–480
Robinson BH, Green SR, Mills TM, Clothier BE, van der Velde M, Laplane R, Fung L, Deurer M, Hurst S, Thayalakumaran T, van den Dijssel C (2003) Phytoremediation: using plants as biopumps to improve degraded environments. Aust J Soil Res 41:599–611
Robinson BH, Green SR, Chancerel B, Mills TM, Clothier BE (2007) Poplar for the phytomanagement of boron contaminated sites. Environ Pollut 150:225–233
Scheel T, Dörfler C, Kalbitz K (2007) Precipitation of dissolved organic matter by aluminum stabilizes carbon in acidic forest soils. Soil Sci Soc Am J 71:64–74
Scheel T, Haumaier L, Ellerbrock RH, Rühlmann J, Kalbitz K (2008) Properties of organic matter precipitated from acidic forest soil solutions. Org Geochem 39:1439–1453
Schnitzer M (1991) Soil organic matter-the next 75 years. Soil Sci 151(1):41–58
Scow KM, Johnson CR (1996) Effect of sorption on biodegradation of soil pollutants. Adv Agro 58:1–56
Semple KT, Morriss AWJ, Paton GI (2003) Bioavailability of hydrophobic organic contaminants in soils: fundamental concepts and techniques for analysis. Eur J Soil Sci 54:809–818
Semple KT, Doick KJ, Jones KC, Burauel P, Craven A, Harms H (2004) Defining bioavailability and bioaccessibility of contaminated soil and sediment is complicated. Environ Sci Technol 38:228–231

Shchegolikhina A, Marschner B (2013) Effects of sterile storage, cation saturation and substrate additions on the degradability and extractability of nonylphenol and phenanthrene in soil. Chemosphere 93(9):2195–2202

Shchegolikhina A, Schulz S, Marschner B (2012) Interacting effects of cation saturation and drying, freezing, or aging on the extractability of nonylphenol and phenanthrene from a sandy soil. J Soils Sediments 12(8):1280–1291

Singleton DR, Hendrix PF, Coleman DC, Whitman WB (2003) Identification of uncultured bacteria tightly associated with the intestine of the earthworm *Lumbricus rubellus* (Lumbricidae; Oligochaeta). Soil Biol Biochem 35(12):1547–1555

Sinha RK, Herat S, Tandon PK (2007) Phytoremediation: role of plants in contaminated site management. In: Singh SN, Tripathi RD (eds) Environmental bioremediation technologies. Springer, New York, pp 315–330

Su XM, Liu YD, Hashmi MZ, Ding LX, Shen CF (2015) Culture-dependent and culture-independent characterization of potentially functional biphenyl-degrading bacterial community in response to extracellular organic matter from *Micrococcus luteus*. Microb Biotechnol 8(3): 569–578

Tang X, Hashmi MZ, Zeng B, Yang J, Shen C (2015) Application of iron-activated persulfate oxidation for the degradation of PCBs in soil. Chem Eng J 279:673–680

Thakur IS (2008) Xenobiotics: pollutants and their degradation-methane, benzene, pesticides, bioabsorption of metals. School of Environmental Sciences, Jawaharal Nehru University, New Delhi

Thakur IS (2011) Environmental biotechnology: basic concept and applications. IK International, New Delhi

Van der Lelie D, Schwitzbuebel JP, Glass DJ, Vangronsveld J, Baker A (2001) Assessing phytoremediation's progress in the United States and Europe. Environ Sci Technol 35:447–452

Varsha YM, Naga Deepthi CH, Chenna S (2011) An emphasis on xenobiotic degradation in environmental cleanup. J Bioremed Biodegrad S 11:1–10

Yuan GS, Xing BS (2001) Effects of metal cations on sorption and desorption of organic compounds in humic acids. Soil Sci 166:107–115

Zhang J, Li Z, Ge G, Sun W, Liang Y, Wu L (2009) Impacts of soil organic matter, pH and exogenous copper on sorption behavior of norfloxacin in three soils. J Environ Sci 21:632–640

Chapter 16
Biodegradation of Xenobiotics in Soil by Fungi

Fariha Khan, Abida Farooqi, and Mazhar Iqbal Zafar

16.1 Introduction

Xenobiotic compounds are among the most problematic pollutants and becoming an increasingly large problem globally. Contamination of soil and terrestrial ecosystems is always of concern to environmental scientists; however, these systems also possess unique capabilities allowing them to eliminate or remediate certain levels of pollutants. The manufacturing, processing, and handling of xenobiotic chemicals on the large scale have led to serious surface and subsurface soil contamination. Soil is exposed to continuous supply of a wide range of hazardous and toxic pollutants because of anthropological activities. As a result, agriculture productions, ecosystem, and quality of soils are getting deteriorated severely. Soil health can be defined as "the capacity of soil to function as a vital living system to sustain biological productivity, promote environmental quality and maintain plant and animal health" (Doran and Zeiss 2000). The cleanup of the contaminated soils is a priority environmental task due to the risks and contaminants posing to the soil fertility and groundwater. Since soil and groundwater are considered sinks for complex contamination, various chemical and biological soil properties are deeply transformed, ultimately which affects biodiversity and soil function. Xenobiotic substances present one of the most pressing problems for bio-treatment of contaminated soils or sediments. These compounds constitute a broad class of chemicals appearing as persistent contaminants in soils and sediments, including petroleum and fuel

F. Khan
Department of Biosciences, COMSATS Institute of Information Technology, Islamabad 45550, Pakistan

A. Farooqi (✉) • M.I. Zafar
Department of Environmental Sciences, Faculty of Biological Sciences, Quaid-i-Azam University, Islamabad 45320, Pakistan
e-mail: afarooqi@qau.edu.pk

© Springer International Publishing AG 2017

M.Z. Hashmi et al. (eds.), *Xenobiotics in the Soil Environment*, Soil Biology 49,
DOI 10.1007/978-3-319-47744-2_16

residues, and creosotes. Such compounds along with typical properties such as low solubility and low volatility exhibit very slow release rates from soil or sediment (Pignatello and Xing 1996). The major soil pollutants are heavy metals and mineral oils, while organic contaminants such as polycyclic aromatic hydrocarbons (PAHs) are also listed in soil contaminants. Organic compounds lethal for atmosphere and human health escape environmental degradation disturbing the food chain and are regarded as persistent organic pollutants (POPs) by the Stockholm Convention. POPs include industrial chemicals and by-products formed in industrial processes, such as polychlorinated biphenyls (PCBs); polychlorinated dibenzo-p-dioxins (PCDD); polychlorinated dibenzofurans (PCDF); carcinogenic PAHs and pesticides, e.g., hexachlorobenzene (HCB); dichlorodiphenyltrichloroethane (DDT); aldrin; chlordane; dieldrin; heptachlor; and a few organometallic chemicals such as tributyltin (TBT) (Convention 2001a). Many toxic industrial chemicals, including polychlorinated biphenyls (PCBs), pentachlorophenol (PCP), chloroethenes, and dichloromethane, threaten the environment and human health and further are classified as hazardous by the US Environmental Protection Agency. Being a wide variety of industrial and agricultural activities, a large number of chemicals are released into the environment, and the resultant accumulations of the various organic chemicals in the environment, particularly in soil, are of significant concern because of their toxicity, including their carcinogenicity, and also because of their potential to bioaccumulate in soil.

16.2 Biodegradation

Biodegradation encompasses the breakdown of organic compounds either through biotransformation into less complex metabolites or through mineralization into inorganic mineral. It has been reported in literature that fungi has been studied well for their ability to degrade a range of environmental pollutants including recalcitrant polycyclic aromatic hydrocarbons, halogenated hydrocarbons and nitro-aromatic compounds. Following the principle of reduction process in the environment, biodegradation is considered to be the most important parameter that influences the toxicity, persistence, and ultimate fate of pollutants in soils.

Biodegradation is a low-cost, effective technology which is less time- and energy-consuming applicable to an array of organic and inorganic contaminants, referred to as the green technology solution which is noninvasive for the ecosystem (Perelo 2010) and permanently destroys the contaminants rather transforming into another medium. Worker's safety is ensured by minimizing exposure to xenobiotics (Okoh and Trejo-Hernandez 2006). Bioremediation has been described as the most promising, economical method for organic pollutants. Biotransformation of the hazardous pollutants to less toxic substances and finally to water, carbon dioxide, and complete mineralization with less or no use of chemicals and energy by utilizing microbes with metabolic capabilities which degrade the xenobiotics represents an economical and powerful alternative to clean up soil and water

(Marzorati et al. 2010). Microorganisms represent a valuable solution to the remediation of polluted environments due to the diverse ecological behaviors and enormous metabolic potential. Utilization of microbes as cell factories under human management to enzymatically detoxify or remove a compound has been redefined as microbial resource management (MRM) (Verstraete et al. 2007). Since the molecular and biotechnological techniques grow fast, identifying the useful microbial strains with catabolic genes is becoming easy, while advance systematic strategies have been lined up for their utilization in situ.

16.3 Fungi in Biodegradation

Fungi represent a non-photosynthetic, diverse kingdom of organisms which includes aerobic organisms such as molds, mildews, rusts, mushrooms, and fermenting organisms, e.g., yeasts, accounting to estimated global species of 1.5–5.1 million mostly existing in tropical rain forests, according to a recent study (Tedersoo et al. 2014). Fungi is classified into seven phyla: *Microsporidia*, *Blastocladiomycota*, *Neocallimastigomycota*, *Chytridiomycota*, *Glomeromycota*, *Ascomycota*, and *Basidiomycota* (Hibbett et al. 2007). Fungi exert a massive influence on humankind and ecosystem, due to their functions in mineralization and humification of soil organic matter, maintenance of biogeochemical cycles, decomposition, and toxin productions, as pathogens and symbionts. In principle, they can degrade any naturally existing biopolymer and some of the synthetic polymers as well. The use of fungi offers promising approach in producing alternative fuels and biodegradation of contaminated soil and water. Extensive research is going on to understand the fungal diversity and to target beneficial fungi for the useful applications. Thousands of fungal genomes are an initiative of the Joint Genome Institute (JGI) of the Department of Energy, USA, in collaboration with a team of researchers to sequence 1000 fungal genomes across the fungi kingdom (http://jgi.doe.gov/our-science/science-programs/fungal-genomics/1000-fungal-genomes).

The bioremediation strategy employing fungi for remediation of xenobiotics is referred to as "mycoremediation." Biodegradation of polymers largely depends on the extracellular enzymes the fungi secrete, which belongs to two classes of enzymes known as oxidoreductases and hydrolases (Mäkelä et al. 2013). Oxidoreductase carries out oxidation or reduction reactions, while hydrolases catalyze hydrolysis reactions, i.e., breaking the chemical bonds accompanied by the addition of water molecule. The fungal lignin-modifying enzymes (LMEs) are classified as oxidative enzymes. Since LMEs are nonspecific, it can degrade a wide variety of organic contaminants and mixtures of chemicals. LMEs include laccase, manganese peroxidase (MnP), lignin peroxidase (LiP), versatile peroxidase (VP), coprinopsis cinerea peroxidase (CiP), and dye-decolorizing peroxidase (DyP) (Harms et al. 2011; Tuomela and Hatakka 2011; Winquist 2014). Exploiting the fungal enzymes might be considered as a promising tool for soil bioremediation even for large-scale contaminations. Various methods including the use of bio-slurry reactor, biopile arrangements, and land farming are the principal methods usually used for bioremediation of contaminated soil and are reliably useful when aided with fungal enzymes (Mougin et al. 2002).

The concept of using fungi for bioremediation of contaminated soil came up in the1980s when, for the first time, the white-rot fungus was observed to break down many organic contaminants (Canet et al. 2001).

White-rot ligninolytic fungal strains are reported to be the major decomposers of biopolymers and have been used for 20 years to degrade pollutants in soils (Barr and Aust 1994). They have extensive applications in the remediation of a wide array of xenobiotics including pesticides, explosives, petrol, diesel and PAHs, dyes, and industrial wastes, while the studies have recognized the key roles of the enzymes, e.g., laccase, Mn-peroxidase, or (sometimes) lignin peroxidases in these remediation applications (Baldrian 2006; Fragoeiro and Magan 2005; Pointing 2001; Riva 2006). The bioremediation capabilities and environmental factors including water, pH, and temperature, affecting the degradation of pesticides by using white-rot fungi, *Trametes versicolor* and *Phanerochaete chrysosporium*, have been reviewed (Magan et al. 2010).

16.4 Degradation of Important Soil Xenobiotics Mediated by Fungal Enzymes

The xenobiotic compounds such as PAHs, nitro-aromatic compounds, and endocrine-disrupting chemicals (EDCs) have huge importance in soil contamination as major pollutants (Singh et al. 2009). The importance of fungal enzymes in biodegradation of these lethal xenobiotic families is highlighted below.

16.4.1 Polycyclic Aromatic Hydrocarbons

Naphthalene, phenanthrene, and benzo(a)pyrene and compounds such as benzene, toluene, and xylene are the PAHs registered as major pollutants in priority list by the USEPA. Due to the characteristics having low solubility in water and high stability, they persistently stay in the environment (Husain 2010).

Fungi do not have the ability to use PAHs as carbon source; rather, they transform PAH co-metabolically. White-rot ligninolytic fungi (able to degrade lignin), *T. versicolor* and *P. ostreatus*, were effective in PAH degradation via laccase-mediated transformation as reported in many studies (Anastasi et al. 2010; Mollea et al. 2005; Rama et al. 2001). Furthermore, PAH degradation and ligninolytic enzymatic activity reflected a positive association (Novotný et al. 1999). However, encouraging results were obtained when free laccases were directly applied to PAH-contaminated soil degrading a combination of 15 PAHs, wherein anthracene degradation was reported as 60 % as the highest effective degradation of any PAH studied (Wu et al. 2008).

An extensive study on ligninolytic fungal strains demonstrated maximum degradation of naphthalene (69 %) by fungal strain carrying enzyme Mn-peroxidase, while strain-producing lignin peroxidase and laccase were also able to degrade naphthalene. In addition, fungus having enzymes such as Mn-peroxidase and laccase degraded phenanthrene (Clemente et al. 2001).

Non-ligninolytic fungi *Cunninghamella elegans*, *Cladosporium sphaerospermum*, and *Penicillium janthinellum* efficiently degraded PAHs (Potin et al. 2004). Especially *Cunninghamella elegans* degraded many PAHs (e.g., naphthalene, acenaphthene, anthracene, phenanthrene, benzo[a]pyrene, fluoranthene, and pyrene). Many non-ligninolytic fungi degrade PAHs via cytochrome P450 monooxygenase and epoxide hydrolase-catalyzed reactions to form transdihydrodiols (Reineke 2001).

16.4.2 Nitro-aromatic Compounds

Nitro-aromatics, a large group of xenobiotics, are the compounds produced as a result of incomplete combustion of fossil fuel and nitration reactions. They are widely used for the synthesis of explosives, pesticides, dyes, or pharmaceuticals with an annual production of 10^8 t (Ye et al. 2004). Since mammalian enzymatic system degrades nitro-aromatics to carcinogenic intermediates, therefore, these compounds are regarded as recalcitrant and have been categorized with hazardous rating 3, as the worst level of toxicity.

Fungi can also tolerate aromatic amines (AA). For instance, fungus *Podospora anserina* producing two *N*-acetyltransferase (NAT) enzymes (PaNAT1 and PaNAT2) was able to acetylate a wide range of AA as a tolerance mechanism in contaminated soil (Martins et al. 2009). Snellinx et al. (2002) reported that remediation of nitro-aromatics-contaminated soil can be carried out with biological treatment systems, utilizing fungus *Phanerochaete chrysosporium* or *Pseudomonas* sp. ST53. Especially the most studied *Phanerochaete chrysosporium* has been reported as a highly beneficial organism in such studies (Hodgson et al. 2000; Jackson et al. 1999). In addition to that, white-rot fungi have the ability of TNT degradation and mineralization to CO_2 due to oxidative enzymes (Pointing 2001). In another study, 100 % 4-amino-2,6-dinitrotoluene (4ADNT) and 80 % 2,4,6-trinitrotoluene (TNT) was transformed in the presence of humic monomer and catechol in oxidative coupling reactions mediated by phenol oxidases, i.e., laccase from *T. villosa* and horseradish peroxidase. However, TNT transformation increased at a pH above 6.8 only because of the polymerization of catechol even with no enzymes added (Wang et al. 2001). Yeasts and molds have been used in bioremediation of metal and chromium-contaminated soil (Kotaś and Stasicka 2000).

16.4.3 Endocrine-Disrupting Phenolic Chemicals

Endocrine-disrupting chemicals (EDCs) are among the pollutants of anthropogenic origin, since they interfere with the endocrine system causing numerous pathological conditions, especially during reproduction and development. Stockholm Convention had listed many potential EDCs including some pesticides, phytoestrogens, phthalate compounds, and other phenolic compounds (Convention 2001b). Alkylphenols (e.g., nonylphenol and octylphenol), and biphenyls like bisphenol A, are used as model compounds to study endocrine disruptors. Fungal laccases have been reported in the transformation of these compounds in many studies (Saito et al. 2004, Tsutsumi et al. 2001). It was proposed that polymerization of phenolic EDCs involved enzymatic conversion to phenoxy radicals. *T. versicolor* catalyzed partially the transformation of nonylphenol into carbon dioxide (Dubroca et al. 2005). In another report, free laccase of *T. versicolor* immobilized on nylon membrane transformed bisphenol A effectively (Diano et al. 2009).

16.5 Conclusions

Soil is constantly unprotected to xenobiotics which are seriously destroying the ecosystems and agriculture. It was three decades ago, when the fungi were used for the first time in bioremediation; now the beneficial use of fungal enzymes require the engineering and production of the novel enzymes at large scale. This is high time to explore new methodologies to effectively utilize enzymes for large-scale contaminations. Discovering the new beneficial fungal strains, sequencing and isolation of new useful enzymes is also highly desirable to further strengthen the biodegradation of contaminated soil.

References

Anastasi A, Spina F, Prigione V et al (2010) Scale-up of a bioprocess for textile wastewater treatment using *Bjerkandera adusta*. Bioresour Technol 101:3067–3075. doi:10.1016/j.biortech.2009.12.067

Baldrian P (2006) Fungal laccases—occurrence and properties. FEMS Microbiol Rev 30:215–242. doi:10.1111/j.1574-4976.2005.00010.x

Barr DP, Aust SD (1994) Pollutant degradation by white rot fungi. Rev Environ Contam Toxicol 138:49–72

Canet R, Birnstingl JG, Malcolm DG et al (2001) Biodegradation of polycyclic aromatic hydrocarbons (PAHs) by native microflora and combinations of white-rot fungi in a coal-tar contaminated soil. Bioresour Technol 76:113–117. doi:10.1016/S0960-8524(00)00093-6

Clemente AR, Anazawa TA, Durrant LR (2001) Biodegradation of polycyclic aromatic hydrocarbons by soil fungi. Brazilian J Microbiol 32:255–261. doi:10.1007/s00128-002-0011-z

Convention SP (2001a) Final act of the conference of plenipotentiaries on the stockholm convention on persistent organic pollutants

Convention SP (2001b) Final act of the conference of plenipotentiaries on the stockholm convention on persistent organic pollutants. Stock POPS Conv 44

Diano A, Peeters J, Dynesen J, Nielsen J (2009) Physiology of *Aspergillus niger* in oxygen-limited continuous cultures: Influence of aeration, carbon source concentration and dilution rate. Biotechnol Bioeng 103:956–965. doi:10.1002/bit.22329

Doran JW, Zeiss MR (2000) Soil health and sustainability: managing the biotic component of soil quality. Appl Soil Ecol 15:3–11. doi:10.1016/S0929-1393(00)00067-6

Dubroca J, Brault A, Kollmann A et al (2005) Environmental chemistry: green chemistry and pollutants in ecosystems. In: Lichtfouse E, Schwarzbauer J, Robert D (eds) Environmental chemistry. Springer, Berlin, pp 305–315

Fragoeiro S, Magan N (2005) Enzymatic activity, osmotic stress and degradation of pesticide mixtures in soil extract liquid broth inoculated with *Phanerochaete chrysosporium* and *Trametes versicolor*. Environ Microbiol 7:348–355. doi:10.1111/j.1462-2920.2004.00699.x

Harms H, Schlosser D, Wick LY (2011) Untapped potential: exploiting fungi in bioremediation of hazardous chemicals. Nat Rev Microbiol 9:177–192. doi:10.1038/nrmicro2519

Hibbett DS, Binder M, Bischoff JF et al (2007) A higher-level phylogenetic classification of the Fungi. Mycol Res 111:509–547. doi:10.1016/j.mycres.2007.03.004

Hodgson J, Rho D, Guiot SR, Ampleman G, Thiboutot S, Hawari J (2000) Tween 80 enhanced TNT mineralization by *Phanerochaete chrysosporium*. Can J Microbiol 46(2):110–118

Husain Q (2010) Peroxidase mediated decolorization and remediation of wastewater containing industrial dyes: a review. Rev Environ Sci Biotechnol 9:117–140

Jackson CJ, Barton RC, Evans EGV (1999) Species identification and strain differentiation of dermatophyte fungi by analysis of ribosomal-DNA intergenic spacer regions. J Clin Microbiol 37:931–936

Kotaś J, Stasicka Z (2000) Chromium occurrence in the environment and methods of its speciation. Environ Pollut 107:263–283

Magan N, Fragoeiro S, Bastos C (2010) Environmental factors and bioremediation of xenobiotics using white rot fungi. Mycobiology 38:238–248. doi:10.4489/MYCO.2010.38.4.238

Mäkelä MR, Lundell T, Hatakka A, Hildén K (2013) Effect of copper, nutrient nitrogen, and wood-supplement on the production of lignin-modifying enzymes by the white-rot fungus *Phlebia radiata*. Fungal Biol 117:62–70. doi:10.1016/j.funbio.2012.11.006

Martins M, Rodriges-Lima F, Dairou J et al (2009) An acetyltransferase conferring tolerance to toxic aromatic amine chemicals: molecular and functional studies. J Biol Chem 284:18726–18733. doi:10.1074/jbc.M109.015230

Marzorati M, Verhelst A, Luta G et al (2010) In vitro modulation of the human gastrointestinal microbial community by plant-derived polysaccharide-rich dietary supplements. Int J Food Microbiol 139:168–176. doi:10.1016/j.ijfoodmicro.2010.02.030

Mollea C, Bosco F, Ruggeri B (2005) Fungal biodegradation of naphthalene: microcosms studies. Chemosphere 60:636–643. doi:10.1016/j.chemosphere.2005.01.034

Mougin C, Kollmann A, Jolivalt C (2002) Enhanced production of laccase in the fungus *Trametes versicolor* by the addition of xenobiotics. Biotechnol Lett 24:139–142. doi:10.1023/A:1013802713266

Novotný Č, Erbanová P, Šašek V, Kubátová A, Cajthaml T, Lang E, Krahl J, Zadražil F (1999) Extracellular oxidative enzyme production and PAH removal in soil by exploratory mycelium of white rot fungi. Biodegradation 10:159–168. doi:10.1023/A:1008324111558

Okoh AI, Trejo-Hernandez MR (2006) Remediation of petroleum hydrocarbon polluted systems: exploiting the bioremediation strategies. J Biotechnol 5:2520–2525

Perelo LW (2010) Review: in situ and bioremediation of organic pollutants in aquatic sediments. J Hazard Mater 177:81–89

Pignatello JJ, Xing B (1996) Mechanisms of slow sorption of organic chemicals to natural particles. Environ Sci Technol 30:1–11. doi:10.1021/es940683g

Pointing SB (2001) Feasibility of bioremediation by white-rot fungi. Appl Microbiol Biotechnol 57:20–33

Potin O, Veignie E, Rafin C (2004) Biodegradation of polycyclic aromatic hydrocarbons (PAHs) by *Cladosporium sphaerospermum* isolated from an aged PAH contaminated soil. FEMS Microbiol Ecol 51:71–78. doi:10.1016/j.femsec.2004.07.013

Rama R, Sigoillot JC, Chaplain V et al (2001) Inoculation of filamentous fungi in manufactured gas plant site soils and PAH transformation. Polycycl Aromat Compd 18:397–414

Reineke W (2001) Aerobic and anaerobic biodegradation potentials. Environ Chem 2:1–161

Riva S (2006) Laccases: blue enzymes for green chemistry. Trends Biotechnol 24:219–226

Saito K, Suyama Y, Sato S, Sugawara K (2004) Defoliation effects on the community structure of arbuscular mycorrhizal fungi based on 18S rDNA sequences. Mycorrhiza 14:363–373. doi:10.1007/s00572-003-0286-x

Singh A, Kuhad RC, Ward OP (2009) Advances in applied bioremediation. Springer, Berlin, pp. 1–19

Snellinx Z, Nepovím A, Taghavi S et al (2002) Biological remediation of explosives and related nitroaromatic compounds. Environ Sci Pollut Res Int 9:48–61. doi:10.1007/BF02987316

Tedersoo L, Bahram M, Põlme S et al (2014) Global diversity and geography of soil fungi. Science 346:1256688. doi:10.1126/science.1256688

Tsutsumi Y, Haneda T, Nishida T (2001) Removal of estrogenic activities of bisphenol A and nonylphenol by oxidative enzymes from lignin-degrading basidiomycetes. Chemosphere 42:271–276. doi:10.1016/S0045-6535(00)00081-3

Tuomela M, Hatakka A (2011) Oxidative fungal enzymes for bioremediation. Elsevier Scientific, London, pp. 183–196

Verstraete W, Wittebolle L, Heylen K et al (2007) Microbial resource management: the road to go for environmental biotechnology. Eng Life Sci 7:117–126

Wang C, Wu J, Mei X (2001) Enhanced taxol production and release in *Taxus chinensis* cell suspension cultures with selected organic solvents and sucrose feeding. Biotechnol Prog 17:89–94. doi:10.1021/bp0001359

Winquist E (2014) The potential of ligninolytic fungi in bioremediation of contaminated soils. Aalto University, Helsinki, p. 124

Wu QS, Xia RX, Zou YN (2008) Improved soil structure and citrus growth after inoculation with three arbuscular mycorrhizal fungi under drought stress. Eur J Soil Biol 44:122–128. doi:10.1016/j.ejsobi.2007.10.001

Ye M, Qu G, Guo H, Guo D (2004) Specific 12β-hydroxylation of cinobufagin by filamentous fungi. Appl Environ Microbiol 70:3521–3527. doi:10.1128/AEM.70.6.3521-3527.2004

Chapter 17
The Use of Enzymes in Bioremediation of Soil Xenobiotics

Anna Piotrowska-Długosz

17.1 Introduction

Environmental pollution with hazardous chemicals that contain compounds resistant to degradation is now one of the most important ecological problems of the world. Three main sources of pollution can be identified: industrial activities, munitions wastes, and agricultural practices (Gainfreda and Rao 2004). The rapid development of the chemical industry has produced various chemical compounds that include pesticides, fuels, plastics, explosives, dyes and solvents. They are a mixture of different types of hydrocarbons that can be harmful for both the environment and human health. Such contamination can be long term and can have a significant impact on decomposition processes and thus nutrient cycling in soil. Some of these compounds have a high persistence and toxicity as well as the ability to accumulate and as a consequence they are resistant to biodegradation (Marchut-Mikołajczak et al. 2013). For example, the widespread incorporation of herbicides into the soil every year constitutes a major concern since they can potentially pose a threat to our health as well as to the quality of the soil, surface water, and groundwater resources (Han et al. 2004).

Cleaning up the environment, including soils, by removing hazardous pollutants is a crucial problem that needs multifaceted approaches in order to reach acceptable solutions. Same strategies have been practiced to remediate and restore polluted environments: physical and chemical methods as well as bioremediation techniques that use biological agents (Eapen et al. 2007). The main bioremediation agents are microorganisms, plants, and associations of plant microorganisms (Gianfreda and

A. Piotrowska-Długosz (✉)
Department of Soil Science and Soil Protection, Division of Biochemistry, Faculty of Agriculture and Biotechnology, UTP University of Science and Technology, 6 Bernardyńska St, 85-029 Bydgoszcz, Poland
e-mail: ap03@wp.pl

© Springer International Publishing AG 2017
M.Z. Hashmi et al. (eds.), *Xenobiotics in the Soil Environment*, Soil Biology 49,
DOI 10.1007/978-3-319-47744-2_17

Rao 2004). The biodegradation method for the removal of xenobiotics is a better alternative than physical and chemical methods, since it is more economical and faster. That is why, over the last sixty years bioremediation has been promoted as a low-cost, effective strategy for reducing the impact of xenobiotics on soil environment (Sinha et al. 2009). Microbial bioremediation may take the form of biostimulation or bioaugmentation. Biostimulation involves modification of the environment in order to stimulate the existing microorganisms that are capable of bioremediation. This can be done through the addition of various forms of rate-limiting nutrients and electron acceptors, such as phosphorus, nitrogen, oxygen, or carbon (e.g., in the form of molasses), while bioaugmentation is the introduction of nonnative microorganisms into the polluted area in order to increase the remediation effect of the environment, e.g., the injection of contaminant-degrading bacteria into a contaminant zone (Tyagi et al. 2011). Finally, since all bioremediation processes are carried out by specific enzymes (microbiological or plant origin), one promising option is the use of cell-free enzymes rather than whole organisms. The use of cell-free enzymes in bioremediation has many advantages as well as significant constraints compared to microorganisms and plants. According to Scott et al. (2011), the activities of enzymes are independent of growth and therefore do not require the addition of growth-enhancing nutrients.

17.2 Soil Enzymes: The Concept of Their Use in Bioremediation

Enzymes are proteins that catalyze biochemical reactions without undergoing any permanent alteration. Enzymes in soil are similar to enzymes in other systems; however, the unique aspect of enzyme activity in soil is that the source may be associated with different locations in soil. The enzymatic activity in soil is mainly of a microbial or plant origin. The same enzyme may be intracellularly located in proliferating or nonproliferating cells, within dead cells or in cell debris; it may also be present in the extracellular environment in the aqueous phase, in which it is temporarily associated with its substrate or is complexed with clay minerals and organic colloids (Gianfreda and Bollag 1996). Soil enzymes are important soil components that are closely associated with the physicochemical and biological characteristics of soil. However, human activities, agricultural practices, and environmental pollution severely influence their existence and activities in soil (Karaca et al. 2011). Depending on their origin, soil enzymes are powerful tools that can be used to assess short-term or long-term changes in soil. The majority of the anthropogenic or environmental factors that affect soil quality can be attributed to the changes in the soil's physical and chemical properties, but at the same time, they cause changes in the soil enzyme pool, which is used as an indicator of soil biological activity. Soil enzymes play key biochemical roles in the overall process of organic matter decomposition in the soil system (Burns et al. 2013). They are important in catalyzing several important reactions that are necessary for the life processes of the microorganisms in soils and the stabilization of the soil structure, organic matter formation, nutrient cycling, and finally the decomposition of wastes of different origins that are introduced into the soil (Dick et al. 1994). Some extensive reviews concerning soil enzymes, their origin, structure and classification, as well as their behavior and function have previously been published (Gianfreda and Bollag 1996; Gianfreda and Ruggiero 2006; Burns et al. 2013).

The concept of using enzyme systems to enhance soil bioremediation has been adopted globally. The effectiveness of enzymes in bioremediation depends on their efficiency and specificity. They are catalysts with either a narrow (chemo-, region-, and stereo-selectivity) or broad specificity, and, therefore, they can be applied to a wide range of different compounds. They may produce extensive transformations of the structural and toxicological properties of contaminants and even their complete conversion into innocuous inorganic end products. They can perform processes for which no efficient chemical transformations have been devised (Rao et al. 2010). Moreover, enzymes can be used under extreme conditions that can limit microbial activity. All of these characteristics have led to enzymes being considered to be eco-friendly catalysts for bioremediation, and thus, enzymatic techniques are considered to be environmentally friendly processes (Gianfreda and Rao 2008). Several extracellular enzymes are efficient catalysts of organic pollutant transformation. The main enzymatic classes are oxidoreductases and hydrolases from both plant and microbiological sources.

17.3 Plant-Derived Enzymes in the Bioremediation of Soil Xenobiotics

Significant progress has been made in recent years in developing native or genetically modified plants for the remediation of environmental contaminants. The enzymes that are released by plant roots into their surrounding environment are important bioremediation agents. These enzymes are usually wall-associated enzymes and partially transform substances into products that are more easily uptaken by the roots of plants or the microorganisms in the rhizosphere (Gianfreda and Rao 2004). Research that was focused on the soil enzymes of a plant origin began about 20 years ago. Soil laccases, dehalogenases, nitroreductases, nitrilases, and peroxidases derived from plant sources were the enzymes that were studied most often (Boyajian and Carreira 1997; Chroma et al. 2002; Harvey et al. 2002). Boyajian and Carreira (1997) showed the ability of nitroreductase to degrade various additional nitroaromatic compounds. Similarly, other studies have shown the ability of nitrilase to degrade 4-chlorobenzonitrile and of halogenases to metabolize hexachloroethane and TCE (Wenzel et al. 1999). In the study of Chroma et al. (2002), several in vitro cultures of different plants were tested for their ability to transform polychlorinated biphenyls (PCBs) and polycyclic aromatic hydrocarbons (PAHs). Liu et al. (2015) indicated that species' extra- and intracellular peroxidase activity after the cultivation of plant cells with and without xenobiotics was correlated with an ability to metabolize them. Cultures that had a good transformation ability exhibited higher levels of peroxidase in the presence of PCBs/PAHs than the controls that had been incubated without contaminants. Cultures with markedly lower peroxidase activity also exhibited lower PCB conversion in the presence of PCBs.

Recently, the study of Liu et al. (2015) on *Echinacea purpurea*, *Festuca arundinacea* Schred, *Fire Phoenix* (a combined *F. arundinacea*), and *Medicago sativa* L. (expect for *Callistephus chinensis*) indicated that they had the potential to remediate PAH-contaminated soils during a 150-day-long cultivation. The study found that the enzymatic reaction of polyphenol oxidase (except for *Fire Phoenix*), dehydrogenase (except for *Fire Phoenix*), and urease (except for *Medicago sativa* L.) was the highest at the cultivation periods of 60 days and 120 days than at 150 days, which was parallel with the highest PAH degradation rate during that period. Meanwhile, the activity of alkaline phosphatase in the plants that were tested was inhibited during the cultivation process (60 and 120 days), thus indicating that the activity of this enzyme has no role in promoting the hydrolysis of PAHs during the phytoremediation of PAH-contaminated soils. The authors concluded that the phytoremediation of organic pollutants in soil is closely related to the type of plant and is promoted by the comprehensive activity of many enzymes that occur in the soil of the plant rhizosphere.

Although plants have the inherent ability to detoxify xenobiotics, they generally lack the catabolic pathway that is necessary for the complete degradation of these compounds compared to microorganisms. There are also concerns about the potential for the introduction of contaminants into the food chain. Another problem is how to dispose of plants that accumulate xenobiotics (Abhilash et al. 2009). For these reasons, research has been done to engineer plants that have genes that can confer additional and enhanced degradation abilities on them (Rao et al. 2010). The efficiency of phytoremediation can be increased directly by overexpressing the genes involved in the metabolism, uptake, or transport of specific pollutants in plants. Some transgenic plants that have been enriched with genes from humans, plants, and animals have been produced and have shown enhanced abilities of metabolizing many xenobiotics (Abhilash et al. 2009). The first attempt to develop transgenic plants for phytoremediation was done in the case of explosives and the halogenated organic compounds in tobacco plants (French et al. 1999; Doty et al. 2000). The efficiency of transgenic plants to degrade chlorinated solvents, explosives, phenolics, etc., was found in the literature (e.g., Eapen et al. 2007; Doty 2008; Macek et al. 2008).

Recently, a screen-house ecological study was carried out to evaluate the remediation potential of three species of the Fabaceae Family (*Peltophorum pterocarpum, Leucaena leucocephala*, and *Crotalaria retusa*) for cleaning up soils contaminated with crude oil petroleum hydrocarbons (Edwin-Wosu and Nkang 2015). An enzymatic analysis and hydrocarbon index assessment of the species and the vegetated and non-vegetated soils showed that *Peltophorum pterocarpum* had a better performance than *C. retusa* and *L. leucocephala* in the uptake of hydrocarbons. The results showed a significant reduction in the enzyme expression and soil hydrocarbon index.

One of the promising developments in transgenic technology is the insertion of multiple genes (cytochrome P450s and GSH, GT, etc.) for the complete degradation of the xenobiotics within a plant system. Cytochrome P450 enzymes comprise a family of proteins crucial for the oxidation, peroxidation, and reduction of a wide range of endobiotics and a diverse group of xenobiotics including most of therapeutic drugs and environmental pollutants (Nelson et al. 1996; Kumar et al. 2012). Many experiments have been done on the overexpression of human and mammalian CYP450 isoenzymes (CYP1, CYP2, CYP3) in higher plants such as *Nicotiana tabaccum, Solanum tuberosum, Oryza sativa*, or *Arabidopsis thaliana* (Doty 2008). The aim of these experiments was to produce either herbicide-resistant plants that have a tolerance to atrazine and simazine or plants that are able to enhance the metabolization of xenobiotics and their subsequent removal from contaminated soils. Many of these plants were overexpressed with a single gene (e.g., CYP1A1 or CYP2E1) or more P450 genes (e.g., CYP1A1, CYP2B6, and CYP2C19). In addition to P450 oxidation, glutathione conjugation is an important mechanism for the transformation of xenobiotics. Glutatione S-transferases (GSTs) (EC. 2.5.1.18) are a family of enzymes involved in the cellular detoxification and excretion of many of the physiological and endogenous substances (Wilce and Parker 1994) found in plants, animals, and microorganisms. GSTs catalyze the nucleophilic addition of the thiol of reduced glutathione to the electrophonic centers in organic compounds. The glutathione conjugates formed in this way are more hydrophilic and therefore more susceptible to exclusions. That is why a GST-catalyzed transformation is one of the early steps in the mercapturic acid pathway in which hydrophobic xenobiotics are detoxified and eliminated from an organism (Habig et al. 1974).

Another promising approach to enhancing the phytoremediation technology is the insertion of xenobiotics-degrading genes in the root system of selected plant species in order to increase rhizospheric secretion and the subsequent degradation of pollutants (Gerhardt et al. 2009; Kawahigashi 2009; Spaczyński et al. 2012). The plants do not need to take up the pollutants in order to detoxify them. Instead these plants secrete enzymes into the rhizosphere zone where pollutants can be degraded (Kawahigashi 2009). Transgenic plants that secrete detoxifying enzymes can be useful in the transformation of a wide range of hydrophobic compounds. Promising results were obtained when the resistance to 2,4,6-trichlorophenol was assessed in *Arabidopsis* seedlings that had a built-in gene that encoded root-specific laccase, which was originally produced by *Gossypium* (Wang et al. 2004). Similarly, Sonoki et al. (2005) studied the ability of the rhizodegradation of biphenol A and PCBs in tobacco seedlings inoculated with a gene-coding laccase obtained from the fungus *Coriouls vericolar*. Transgenic *Arabidopsis thaliana* plants with an extradiol dioxygenase gene significantly increased their capacity for the rhizodegradation of 2.3-dihydroxybiphenol (Uchida et al. 2005). Similarly, transgenic tobacco plants that expressed haloalkane dehydrogenase (DHA) accelerated the detoxification of 1-chlorobutane in the rhizosphere zone (Uchida et al. 2005). Some examples of

Table 17.1 Transgenic plants and their enzymes for enhanced remediation of xenobiotics (Adopted from Abhilash et al. 2009; Gerhardt et al. 2009)

Target plant	Gene (s)	Enzymes	Source	Transgene effects
Oryza sativa	CYP1A1	Cytochrome P450 monooxygenase	Human	Enhanced metabolism of chlorotoluron, nurflurazon
Nicotiana tabacum	CYP2E1	Cytochrome P450 monooxygenase	Human	Enhanced metabolism of trichloroethylene; increased uptake and debromination of ethylene dibromide
Nicotiana tabacum	onr	Pentaerythritol tetranitrate reductase (PETN)	*Enterobacter cloaceae*	Enhanced denitration of glycerol trinitrate
Oryza sativa	CYP2B6	Cytochrome P450 Monoxygenase	Human	Metabolism of ethofumesate and benfuresate
Nicotiana tabacum	GstI-6His	Glutathione S-transferases (GST I)	Maize	Higher tolerance to alachlor
Lycopersicon esculentum	tpxI	Peroxidases (Px)	Roots of *L. esculentum*	The overexpression of tpxI gene in transgenic tomato hairy roots resulted in the enhanced removal of phenol
Brassica juncea	γ-ECS, GS	γ-Glutamycysteine synthetase; Glutathione synthetase	*Brassica juncea*	Overexpression of ECS and GS resulted in enhanced tolerance to atrazine, 1-chloro-2,4-dinitrobenzene, phenanthrene
A. thaliana	743B4, 73C1	Glycosyltransferases (UGTs)	*A. thaliana*	Overexpression of UGT genes resulted in the enhanced detoxification of TNT and enhanced root growth
Oryza sativa	Protox	Protoporphyrinogen IX oxidase	*Bacillus subtilis*	Tolerance to diphenyl ether herbicide oxyflufen
Solanum tuberosum, *O. sativa*	CYP1A1, CYP2B6, CYP2C19	Cytochrome P450 Monoxygenase	Human	Resistance to sulfonylurea and other herbicides
N. tabaccum	bphc	2,3, dihydroxybiphenyl-1, 2-dioxygenase	PCB-degrading bacteria	Enhanced degradation of PCBs
Hybrid aspen (P. tremula x P. tremuloides)	pnrA	Nitroreductase	*Pseudomonas putida*	The transgenic aspen (hybrid) was shown to tolerate and take up greater amounts of TNT from contaminated water and soil

transgenic plants and their enzymes for enhanced phytoremediation of xenobiotics are presented in Table 17.1.

17.4 Microbial Enzymes in the Bioremediation of Soil Xenobiotics

17.4.1 Oxidoreductases (EC 1)

Many oxidoreductases have been described that could potentially help to remove hazardous compounds from the environment (Gianfreda and Rao 2004; Torres et al. 2003). Microorganisms obtain energy via an energy-yielding biochemical reaction mediated by these enzymes to cleave chemical bonds and to assist in the transfer of electrons from a reduced organic substrate (donor) to another chemical compound (acceptor). During such oxidation–reduction reactions, the contaminants are finally oxidized into harmless compounds (Karigar and Rao 2011). Two types of copper-containing enzymes, laccase and tyrosinase, and two kinds of heme enzymes, namely peroxidases and cellobiose dehydrogenase (CDH), are introduced below. In particular, fungal lignin-modifying enzymes (LMEs), laccases and peroxidases, have been extensively investigated for potential biotechnological applications and have been well represented in some reviews (e.g., Strong and Claus 2011; Viswanath et al. 2014). Lignin, a highly complex, stable, and irregular polymer, requires enzymes that have the ability to nonspecifically oxidize substrates that have a high redox potential. The most important oxidoreductases that are involved in xenobiotic transformation are characterized below.

Laccases (Benzenediol:Oxygen Oxidoreductase; EC 1.10.3.2) constitute multicopper proteins that belong to the family of blue-oxidase enzymes, which can oxidize a variety of aromatic compounds along with the concomitant reduction of oxygen to water (Torres et al. 2003). These enzymes have been studied since the nineteenth century due to their ability to oxidize phenolic compounds, and their applications in several industrial sectors have been studied intensively (e.g., Madhavi and Lele 2009; Giardina et al. 2010). Some interesting properties of laccases have fostered intense research in recent years. In addition to their broad substrate specificity, which allows them to transform a wide range of substrates, most laccases are very stable, especially at pH values near neutrality. Their organic substrate oxidation site exhibits a high redox potential and finally they use dioxygen, which is a harmless and abundant compound as a co-substrate instead of oxygen peroxide which is used by other oxidases (like peroxidases) (Mougin et al. 2009). Laccases are therefore involved in the transformation of a wide range of phenolic compounds, including natural substrates such as lignin and humic substances. They can also transform xenobiotics such as trichlorophenols, pesticides, polynitrated aromatic compounds (Ramos et al. 2005), azo dyes, and PAHs, which are a major source of contamination in soil; therefore, their degradation is of

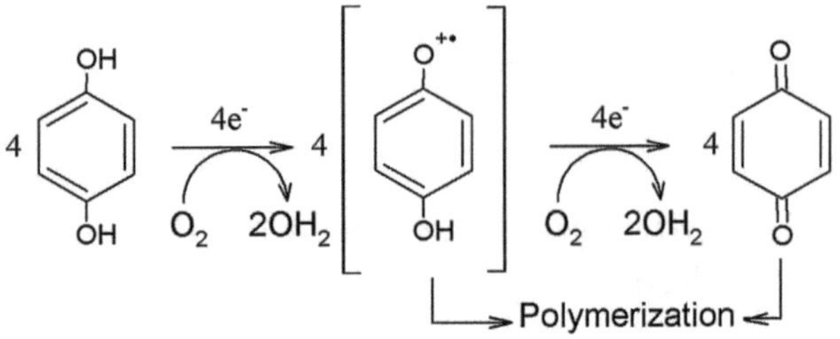

Fig. 17.1 General reaction mechanisms for phenol oxidation by laccase (Reproduced from Karigar and Rao 2011)

great importance for the environment. The general reaction mechanism for phenol oxidation by laccase is shown in Fig. 17.1. The detailed characteristics of action mechanism of laccases and their potential use for the detoxification of organic pollutants have been extensively reviewed (e.g., Durán et al. 2002; Torres et al. 2003; Mougin et al. 2003; Couto and Herrera 2006; Fernández-Fernández et al. 2013; Viswanath et al. 2014).

Tyrosinase (Catechol:Oxygen Oxidoreductase, Monophenol:Oxygen Oxidoreductase; EC 1.10.3.1, EC 1.14.18.1) is an enzyme that is known as a monophenol oxidase or catecholase and is ubiquitously distributed in organisms (Claus and Decker 2006). Tyrosinases catalyze two types of reactions that occur sequentially. The first reaction is the *o*-hydroxylation of phenols with molecular oxygen in order to produce catechols (cresolase activity); the second reaction is the subsequent oxidation of catechols with oxygen in order to form *o*-quinones (catecholase activity); *o*-quinones are unstable compounds and spontaneously polymerize in a nonenzymatic reaction to produce insoluble melanin-like products (Demarche et al. 2012).

Peroxidases (Donor:Hydrogen Peroxide Oxidoreductases; EC 1.1.1.X) use hydrogen peroxide or other organic hydroperoxides as a cosubstrate to oxidize a variety of organic and inorganic substrates. Chloroperoxidase (CPO), lignin peroxidase (LIP), manganese peroxidase (MnP), horseradish peroxidase (HRP), and ascorbate peroxidase (APX) are some examples of peroxidases (Torres et al. 2003). Peroxidases share common structural and catalytic features—they are glycosylated proteins that have an iron protoporphyrin IX (heme) prosthetic group located at the active site (Mougin et al. 2009). In the study of Novotný et al. (1997), the abilities of *Phanerochaete chrysosporium*, *Trametes versicolor*, *Coriolopsis polyzona*, and *Pleurotus ostreatus* that had been grown in a nitrogen-limited mineral medium to degrade PCBs were compared. Mn-dependent peroxidase activity was high and relatively stable during the experiment. Mn-independent and lignin peroxidases characterized efficient PCB degraders. The MnP enzymes are able to oxidize and depolymerize their natural substrate, i.e., lignin, as well as recalcitrant xenobiotics such as nitroaminotoluenes and textile dyes (Knutson et al. 2005; Wesenberg et al. 2003). The remediation of phenol using commercial horseradish peroxidase (Wanger and Nicell 2002) or soybean peroxidase (Ryan et al. 2006) has been reported, but several drawbacks limit its widespread application, including the intolerance to high concentrations of the primary substrate H_2O_2, low enzymatic reusability, and financial costs. The choice of peroxidase for wastewater treatment also depends on the characteristics of the effluent, operational requirements, and costs (Bodalo et al. 2005). The use of peroxidases for soil cleaning has also been studied, specifically for soils contaminated with aromatic hydrocarbons over a long period and then detoxified by autochthonous fungi-producing peroxidases (D'Annibale et al. 2006).

Cellobiose dehydrogenase (CDH) (EC.1.1.99.18) is an extracellular hemoflavoenzyme produced by several wood-degrading fungi. Its probable physiological function is the depolymerization of cellulose and lignin and prevention of cellulose repolymerization (Cameron and Aust 2001). Cellobiose dehydrogenase oxidizes various substrates such as cellobiose, cellodextrins, mannodextrins, and lactose to their corresponding lactones with the concomitant reduction of the flavin (FAD to $FADH_2$). CDH has a potential role in the bioremediation of recalcitrant pollutants, since it can directly reduce munitions such as 2,4,6-trinitrotoluene and can also indirectly degrade many more chemicals, including polyacrylate polymers (Cameron et al. 2000).

17.4.2 Hydrolytic (EC 3) and Other Enzymes

The other important microbial enzymes involved in the transformation of xenobiotics in soil are hydrolases. Hydrolases are classified as EC 3 and further divided into 13 subcategories according to the type of bonds hydrolyzed (Nomenclature Committee of the International Union of Biochemistry and Molecular Biology). They catalyze the hydrolysis of various compounds according to the following reaction:

$$A - B + H_2O \rightarrow A - OH + B - H$$

Both fungi and bacteria produce a group of enzymes that include lipases, proteases, carbohydratases (e.g., cellulases, amylases, and xylanases), esterases, phytases, and phosphatases, which are physiologically important to living microorganisms. Some of these such as proteases and carbohydratases catalyze the hydrolysis of large molecules. Others such as phosphatases contribute to the transfer of organic P compounds to their inorganic forms, which are available to plants and microorganisms (Gianfreda and Rao 2004). Because of their low substrate specificity, hydrolases could play an important role in the bioremediation processes of some xenobiotics. Hydrolytic enzymes disrupt the chemical bonds in harmful molecules, which results in a reduction in their toxicity. Some hydrolases and enzymes from other classes important in the bioremediation of soil xenobiotics are presented below.

Lipases (Triacylglycerol Acylhydrolases, E.C. 3.1.1.3) are ubiquitous enzymes that have a considerable physiological and industrial significance as they catalyze the hydrolysis of triacylglycerol into glycerol and free fatty acids (Suneetha and Khan 2011). Their physiological role involves the breakdown and mobilization of lipids inside cells as well as the transfer of lipids between organisms (Hasan et al. 2006). Lipase activity is related to significant reduction of total hydrocarbon in contaminated soils and therefore was found to be the most useful indicator for testing their degradation in soil (Riffaldi et al. 2006). Microbial lipases have biotechnological applications in organic synthesis, the food and flavor industries, the manufacture of detergents, the removal of oil and grease in wastewaters, and biodiesel refining (Demarche et al. 2012). A technologically interesting dimension of lipases is their tolerance against organic solvent (Doukyu and Ogino 2010).

Proteases (EC 3.4) are another group of enzymes that have a significant economic implication. Proteases (synonyms: peptidases or proteinases) are enzymes that conduct the proteolysis or cleavage of peptide bonds. Numerous proteases are commercially produced and are used in (laundry) detergent formulations, protein synthesis, brewing, the food industry, leather, and dairy processing (Anwar and Saleemuddin 1998). Biodegradation/biotransformation applications include the treatment of various protein-rich effluents from food and beverage industries. Keratinases, collagenases, and pepsins are promising tools for enhancing the low biodegradability of structural animal proteins such as keratin and collagen. Keratynic wastes, which originate from animal breeding, processing, and handling, as well as used paper products that are discarded by the human population, may contribute to the enrichment of the environment with solid waste. Keratin is compact and is strongly stabilized by hydrogen bonds, hydrophobic interactions, and disulfide bonds, which render it resistant to proteolytic degradation (Brandelli 2008). *Pseudomonas aeruginosa* strain, which was isolated from poultry waste, was tested for its ability to hydrolyze feathers due to keratinase activity (Venkata Naga Raju and Divakar 2013). Keratinases (EC 3.4.4.25) are able to hydrolyze insoluble keratins more efficiently than other proteases, and their actions are very specific, i.e., they only act on keratin substrates. Keratinase-producing bacteria (such as *Bacillus subtilis*, *B. cereus*, *B. amyloloquefaciens*, and *B. megaterium*) and fungi (such as *A. niger, Penicillum*, *Cephalosporium*, *Neurospora*, and *Rhizopus*) are major keratinase-producing microorganisms (Venkata Naga Raju and Divakar 2013).

Phosphotriesterases: OPH, Opda (EC 3.1.8) Organophosphorus compounds (OPC) are derivatives of orthophosphoric acid, and alkylphosphonic acid have found widespread use in agriculture as pesticides (Scott et al. 2008). The yearly large-scale use of OPC as well as the low decomposition rates of OPC in the environment results in the accumulation of these compounds in soils from which they are subsequently washed out to enter the groundwater and rivers (Regnery and Trier 2010). Soil contamination and the accumulation of OPC exhibit the toxic effects on living organisms by causing various degenerative disorders of the nervous system in addition to having an influence on the reproductive system

(Colosio et al. 2009; Jokanovic and Prostran 2009). Additionally, OPCs are strong mutagens that cause multiple chromosomal aberrations and carcinogenesis (Hong et al. 2003). Due to this, efficient and environmentally friendly techniques for eliminating OPC from soils are of major practical interest. The decomposition of OPC in soils in situ using biocatalysts, which can include enzymes capable of destructing toxic OPC, could be a promising method.

The bacterial phosphotriesterases are a subgroup of the amidohydrolase metalloenzyme family. The phosphotriesterases primarily catalyze the hydrolysis of OP (organophosphate) triesters (Fig. 17.2). There are two closely related bacterial phosphotriesterases—OpdA from *Agrobacterium radiobacter* (Harcourt et al. 2002) and OPH (organophosphorus hydrolase) from *Pseudomonas diminuta* (Serdar et al. 1985) and *Flavobacterium* (Mulbry et al. 1986). Field trials of OpdA as a bioremediation agent have been conducted, and it is already in use as a commercial product to detoxify OP residues in different contaminated wastes and is sold under the brand name LandGuard™ from Orica Watercare (Australia) (Scott et al. 2011). A few years ago, a genetically modified form of the OPH enzyme containing a hexahistidine sequence (His6-OPH) at the N-terminus of the protein molecule (Efremenko et al. 2007) substantially increased its catalytic potential with respect to the number of substrates as compared to the genetically unmodified OPH enzyme, which makes His6-OPH attractive for the detoxification of OPC in soils. Recently, an immobilized biocatalyst developed based on the hexahistidine-tagged organophosphorus hydrolase (His6-OPH) was tried for the bioremediation of various soil samples heavily contaminated by pesticides (Sirotkina et al. 2012). Various cellulose-containing carriers, which are agricultural wastes, were applied to immobilize His6-OPH, thus enabling the stabilization of the enzyme and its retention in the organophosphorus pesticide destruction zone.

Nitrile-Degrading Enzymes: Nitrilases (EC 3.5.5.1), Nitrile Hydratases (EC 4.2.1.84), Amidase (EC 3.5.1.4) A nitrile is any organic compound that has a –C≡N (nitrile) functional group in which the carbon atom and nitrogen atom are triple bonded together. They are found naturally in plants, insects, and microorganisms, but they are also synthesized and used extensively in the chemical industries to produce a variety of polymers and other chemicals. Other nitrile compounds are used as feedstock, solvents, extractants, pesticides (e.g., dichlobenil, bromoxynil, and buctril), or drug intermediates. Moreover, they are intermediates in the organic synthesis of many of different compounds (e.g.,

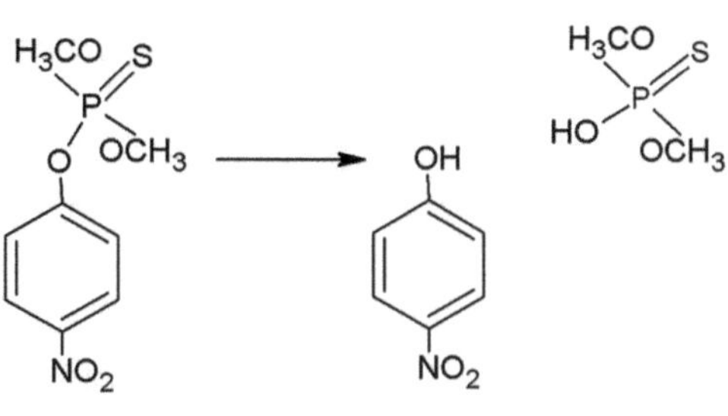

Fig. 17.2 Hydrolysis of the insecticidal phosphotriester parathion by the bacterial phosphotriesterase OpdA (Reproduced from Scott et al. 2008)

amines, amides, carboxylic acids, esters, aldehydes, and ketones) (Banerjee et al. 2002). Most nitriles are highly toxic, mutagenic, and carcinogenic in nature. When they are present at high concentrations in the environment, they may cause several diseases in humans (Brady et al. 2004). That is why, efforts have been made to control their release into the environment and efficient technologies have been proposed to degrade them as well. Thus, the bioremediation of nitriles using microorganisms and their enzymes seems to be a promising method. The microbial hydrolysis of nitriles proceeds through two major enzymatic pathways (Fig. 17.3). In the first, nitrile hydratase catalyzes nitriles to their amides, which are then hydrolyzed to their respective carboxylic acids and ammonia by amidase (Asano et al. 1982). In the second, nitriles undergo direct hydrolysis to their carboxylic acids and ammonia by nitrilase (Kobayashi et al. 1990). Benzonitrile and other aromatic nitriles have been shown to be converted directly into the carboxylic acids and ammonia that are catalyzed by nitrilases. Aliphatic nitriles are broken down into their respective acids and ammonia via the formation of amides. Martinková et al. (2009) characterized the nitrilases by examining the findings on enzyme screening, production, purification, and immobilization as well as their application as a biocatalyst. The authors compared the potential of fungal and bacterial nitrilases and found that the nitrilases of filamentous fungi have high relative activities toward (hetero) aromatic nitriles and that they accept a wide range of aliphatic and alicycling nitriles. Recently, Mukram et al. (2015) have isolated a bacterium from a nitrile-contaminated agricultural soil that was characterized and identified as *Rhodococcus* sp. MTB5, which is capable of utilizing 30 mM of benzonitrile within 42 h and can withstand a concentration of up to 60 mM of hydrolyzing benzonitrile by the nitrile hydratase/amidase pathway. Further, the strain has the ability to utilize mixtures of both aliphatic and aromatic nitriles in different combinations. The authors concluded that this strain may be promising for the remediation of sites that have been contaminated with both aliphatic and aromatic nitriles.

Cyanide-Degrading Enzymes Although certain bacteria, fungi, algae, and plants produce cyanides, a large number of microorganisms are capable of cyanide biodegradation into less toxic compounds. There are many enzymes that are

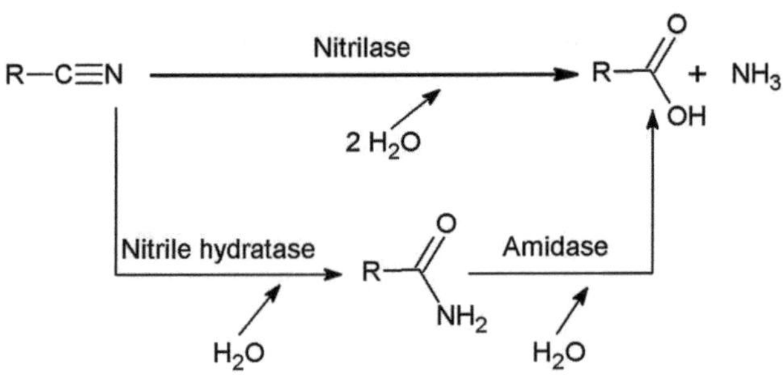

Fig. 17.3 Enzymatic hydrolysis of nitriles (Reproduced from Brady et al. 2004)

produced by microorganisms that utilize cyanides as a substrate to make alanine, glutamic acid, α-amino-butyric acid, β-cyanoalanine, etc. Cyanide compounds are produced as waste products of a number of industrial processes, such as the production of iron and steel, chemicals, drugs, pesticides and plastics, electroplating, photo developing, and others (Gupta et al. 2010). The cyanides that are produced regularly by industries are fast-acting poisons and are hazardous for humans and the ecosystem. Cyanide can be present in the environment as simple cyanides (e.g., HCN, CN^-, NaCN), metal cyanide complexes, cyanates, and nitriles (Ebbs 2004). Biological transformation involves cyanide degradation and assimilation by the microorganisms and plants in the form of amino acids, thiocyanate, cyanoanaline, and vitamins (Trapp et al. 2003). Cyanide is converted into carbon and nitrogen sources by the various enzymes present in microorganisms. Five general pathways for the biodegradation of cyanide can be specified: hydrolytic, oxidative, reductive, substitution/transfer, and syntheses (Table 17.2).

The first three pathways are degradation pathways in which enzymes catalyze the conversion of cyanides into simple organic or inorganic molecules that are further converted into ammonia, methane, CO_2, formic acid, and carboxylic acid. The final two pathways are for the assimilation of cyanide into the microbe as a nitrogen and carbon source.

One of the mechanisms evolved for cyanide detoxification is rhodaneses (thiosulfate: cyanide sulfurtransferases, EC 2.8.1.1), which converts toxic cyanide into less toxic thiocyanate (Banerjee et al. 2002). This enzyme belongs to the class of transferases, specifically to the sulfurtransferases, which transfer sulfur-containing groups (Gupta et al. 2010). Cipollone et al. (2006) examined the role of recombinant *Pseudomonas aeruginosa* rhodanese (r-RhdA) in *E. coli* in cyanide detoxification. The accessibility of thiosulfate to r-RhdA limited the sulfur transfer reaction in the cellular system, although the permeabilization of the bacterial membrane increased the cyanide conversion into thiocyanate. The overall results indicated that the engineered *E. coli* was able to perform cyanide detoxification even under laboratory conditions and suggested that microbial rhodaneses may contribute to cyanide transformation in natural environments (Cipollone et al. 2006). Although the enzyme rhodanese is found ubiquitously in nature and a number of bacterial and mammalian sources exist, there are very few reports on the characterization of rhodanese in fungi. The kinetics of rhodanese in certain *Fusarium* strains, which had been reported to degrade cyanides, was also analyzed and compared to the enzyme from the *Trichoderma*. The rhodanese enzyme in all of the *Trichoderma* strains demonstrated a broad optimum pH ranging from 8.5 to 11.5 and a wide optimum temperature of 35–55 °C. The *Km* of CN degradation and V_{max} values ranged from 7 to 16 mM and from 0.069 to 0.093 mole ml^{-1} min^{-1} mg $protein^{-1}$, respectively, between the strains of *Trichoderma* (Gupta et al. 2010).

Table 17.2 Enzymatic pathways for cyanide degradation

Pathway	Enzymes	Reactions	Examples of microorganisms	References
Hydrolytic	Cyanide hydratase	$HCN + H_2O \leftrightarrow HCONH_2$	*Sternphyliumloti Fusarium solani* IHEM 8026	Huertas et al. (2006), Luque-Almagro et al. (2005a)
	Nitrile hydratase	$R - C \equiv N + H_2O \rightarrow R - CONH_2$	*Bacillus* sp. *Brevibacterium imperalis* CBS489-74 *Pseudonocardia thermophila* JCM3095	Nawaz et al. (1989), Yamaki et al. (1997)
	Thiocyanate hydratase	$R - CN + H_2O \leftrightarrow R - COOH + NH_3$	*Thiobacillus thioparus* THI115	Gupta et al. (2010)
	Nitrilase	$R - CN + H_2O \leftrightarrow R - COOH$	*Nocardia* sp. *Arthrobacter* sp. J1	Chen et al. (2008), Kobayashi et al. (1990)
	Cyanidase	$HCN + 2H_2O \rightarrow HCOOH$	*Bacillus pumilus* C1 *P. stutzeri* AK61	Meyers et al. (1993), Ebbs (2004)
Oxidative	Cyanide monooxygenase and cyanase	$HCN + O_2 + H^+ + NADPH \rightarrow NADP^+ + HOCN + H_2O$	*Pseudomonas pseudoalcaligenes, Pseudomonas putida*	Little and Anderson (1987), Luque-Almagro et al. (2005b)
	Cyanide dioxygenase	$HCN + O_2 + 2H^+ + NADPH \rightarrow NADP + CO_2 + NH_3$	*Trametes versicolor* 200801, *Phanerochaete chrysosporium* ME 496, *Pseudomonas fluorescens* NCIMB 11764	Kunz et al. (1994), Cabuk et al. (2006)
Reductive	Nitrogenase	Step 1. $HCN + 2H^+ + 2e^- \rightarrow CH_2 = NH + H_2O \rightarrow CH_2 = O$ Step 2. $CH_2 = NH + 2H^+ + 2e^- \rightarrow CH_3 - NH + 2H^+ + 2e^- \rightarrow CH_4 + NH_3$	*Streptomyces thermoautotrophicus, Herbaspirillum seropedicae, Azospirillum* spp., *Rhodospirillum rubrum*	Ebbs (2004), Sicking et al. (2005), Gupta et al. (2010)

Substitution/ transfer	Rhodanese	$CN^- + S_2O_3^{2-} \rightarrow SCN^- + SO_3^{2-}$	*Thermobacillus denitrificans*, *E. coli*	Williams et al. 2003
	Mercaptopyruvate sulfurtransferase	Step 1. $HSCH_2COCOO^- + E \leftrightarrow CH_2COCOO^- + ES$ Step 2. $ES + CN^- \leftrightarrow E + SCN^-$	*Leishmania major* (parasite trypanosomatid)	Ezzi et al. (2003), Cipollone et al. (2004), Gupta et al. (2010)
Syntheses	β-Cyanoalanine synthase	1. $HCN + HS - CH_2CH(NH_2)COOH \rightarrow H_2S + NC - CH_2CH(NH_2)COOH$ 2. $HCN + O$ – acetyloserine $\rightarrow NC^- CH_2CH(NH_2)COOH + CH_3COOH$	*E. coli*, *C. violaceum*, *Bacillus megaterium*	Dunnil and Fowden 1965 Akopyan et al. (1975), Ebbs (2004)
	λ-Cyano-α‚- -aminobutyric acid synthase	$–S - (CH_2)_2CH(NH_2)COOH + CN^- \rightarrow HS - (CH_2)_2CH(NH_2)COOH + NCH^-(CH_2)_2CH(NH_2) \rightarrow NC - (CH_2)_2CH(NH_2) + SCN^-$	*Bacillus stearothermophilus*, *C. violaceum* (strain D341)	Brysk and Ressler 1970), Omura et al. (2003)

E enzyme, *ES* enzyme–sulfur intermediate

17.5 Advantages and Limitations of the Use of Enzymes for Soil Bioremediation

The rationale for developing enzymatic cleanup methods is that enzymes are the ultimate cause of pollutant degradation during the bioremediation procedures based on microbial activity. When applied extracellularly, enzymes may eliminate or at least considerably reduce the need to support specific microorganisms in the area that is being treated. Enzymatic treatment may have a number of advantages over microbial treatment (Dick and Tabatabai 1999; Gianfreda and Bollag 2002; Scott et al. 2011):

1. Enzymes possess unique and specialized substrate-transforming capabilities.
2. Enzymes do not require an acclimatization phase as is required by microorganisms.
3. Enzymes can be effective at both low and high concentrations of pollutants, but a high concentration of pollutant may be toxic to microorganisms, thus reducing the degradation efficiency. In turn, at a low pollutant concentration, microbial cultures are generally less effective because other soluble carbon sources are utilized.
4. Enzymes can be used under a wide range of environmental conditions (e.g., pH, moisture, temperature) and may be resistant to many inhibitors that can affect microbial metabolism.
5. Remediation with enzymes can be selected so as not to generate toxic products, as is often the case with chemical and some microbiological processes.
6. The requirement to enhance bioavailability through the introduction of organic cosolvents or surfactants is much more feasible for enzymes than for whole microorganisms.
7. The fate of free enzymatic systems that are introduced into soil is easy to control since the enzymes are digested in situ by the indigenous microorganisms after the treatment. To be efficient in the degradation of the xenobiotics in soil, microorganisms must stay alive, which may cause an imbalance in the ecological equilibrium of the ecosystem, thus preventing any sustainable further use of the soil, e.g., for agricultural purpose.
8. Enzymes act relatively quickly (requiring just minutes or hours for adequate remediation) and have predictable behaviors, so it is possible to develop cost-effective dosing regimens for specific applications.

Even though using enzymes in bioremediation seems to be an attractive method for the detoxification of pollutants, some limitations of enzymes as biocatalysts have been shown (Gianfreda and Bollag 2002; Torres et al. 2003; Gianfreda and Rao 2004):

1. Sometimes the toxicity of the products is higher than that of the parent compounds, so it is important to first investigate whether the products that are

obtained are toxic under real conditions in which many different compounds are present.
2. Enzyme extraction and purification procedures are time-consuming and expensive. It is probable that the development of better enzyme production procedures and the use of molecular biology tools will reduce the enzyme production costs and also improve their activity and stability. Production of animal and plant enzymes could be performed by means of genetically modified microorganisms and plants.
3. The next drawback of enzymatic treatment is the need for cofactors, without which many enzymes do not show any activity.
4. The most desired enzymatic action is the complete mineralization of the toxic compounds. The majority of enzymes, however, are not able to mineralize when applied individually and usually several different enzymes must be applied simultaneously for complete degradation or transformation processes, which causes higher costs of enzymatic bioremediation.
5. Enzymes, especially those used as cell-free enzymes, may exist for only a very short time in an inhospitable soil environment. Enzymes may be inactivated through both nonbiological and biological denaturation processes, such as adsorption on soil colloids, extreme acidity or alkalinity, as well as biodegradation by proteases.
6. The enzymes that are extracted from cells may remain unstable even after they are immobilized on matrices.

17.6 Conclusions and Future Research

Biological agents such as microorganisms and plants have received intense attention because of their potential to remove pollutants from the environment. Finally, however, it is the enzyme systems that transform pollutants during bioremediation procedures based on microbial activity or plant treatment. Therefore, investigating the possibility of the direct application of enzymes on contaminated environmental sites, with particular emphasis on soils, should be of major interest. Their properties render them attractive for the treatment processes of pollutants, and their use might be advantageous over conventional techniques. Although the application of enzymes to degrade pollutants and removing them from the polluted environment is not a new technique, their large-scale application for the remediation of polluted soils is still limited for the reasons given above (Sect. 17.5). A better understanding of the relationship between the activity, kinetic properties, and stability of enzymes in soil is required. As was mentioned above, the extension of the catalytic life of an enzyme may be obtained through its immobilization on a solid support. The immobilization of enzymes offers several improvements for enzyme applications because the storage and operational stabilities are significantly enhanced. Additionally, the reusability of immobilized enzymes represents a greater advantage compared with free enzymes (Fernández-Fernández et al. 2013). Despite the obvious advantages, immobilized enzymes are not widely applied in the remediation of polluted soils.

Enzyme evolution has advanced rapidly over the last three decades. Genetic engineering can be successfully used to improve the ability of enzymes in bioremediation processes. Several enabling technologies that allow the generation of random and rational gene libraries of varying diversity have been developed in this field. Several of these strategies have now been used to improve the performance of potential bioremediation enzymes (Iyer and Iken 2015).

In further studies, special attention should be paid to treatments that require multistep transformations that involve a number of different enzymes acting sequentially. It is necessary to apply multienzyme simultaneously in order to degrade pollutants in the soils. The addition of the extracellular enzymatic complexes produced by the microorganisms that degrade diesel provides better results than the process of bioremediation without these enzymes (Gómez Jiménez et al. 2011). The selection of the most suitable enzymes depends on the requirements for their specificity, their need for cofactors, and their capacity to retain a substantial amount of their activity for an extended period of time. One important challenge for the future of enzymatic bioremediation will therefore be the development of cost-effective cofactor regeneration systems. Enzymes that require inexpensive cofactors (such as O_2) or that do not require any cofactors will be preferred.

The significance of enzymes for soil bioremediation can increase through the application of microorganisms and their enzymes from extreme environments. Thermophiles and psychrophiles, for example, have to adapt themselves to live under extreme environmental conditions (Haki and Rakshit 2003). As a consequence, they produce specialized enzymatic proteins that have an elevated catalytic activity, which are particularly stable with different denaturing agents and other particular features that may render them appropriate for the degradation of polluting compounds. An increasing knowledge of the properties of cold-adapted enzymes is now being developed as well. The use of these enzymes has found many industrial applications to date and special attention should now be paid to their use in biotechnology including soil bioremediation.

References

Abhilash PC, Jamil S, Singh N (2009) Transgenic plants for enhanced biodegradation and phytoremediation of organic xenobiotics. Biotechnol Adv 27:474–488

Akopyan TN, Braunstein AE, Goryachenkova EV (1975) Beta-cyanoalanine synthase: purification and characterization. Proc Natl Acad Sci USA 72:1617–1621

Anwar A, Saleemuddin M (1998) Alkaline proteases: a review. Bioresour Technol 64:175–183

Asano Y, Fujishiro K, Tany Y, Yamada H (1982) Aliphatic nitrile hydratase from *Arthrobacter* sp. J-1: purification and characterization. Agric Biol Chem 46:1165–1174

Banerjee A, Sharma R, Banerjee UC (2002) The nitrile-degrading enzymes: current status and future prospects. Appl Microbiol Biotechnol 60:33–44

Bodalo A, Gomez JL, Gomez E, Bastida J, Maximo MF (2005) Comparison of commercial peroxidases for removing phenol from water solutions. Chemosphere 63:626–632
Boyajian GE, Carreira LH (1997) Phytoremediation: a clean transition from laboratory to marketplace? Nat Biotechnol 15:127–128
Brady D, Beeton A, Zeevaart J, Kgaje C, van Rantwijk F, Sheldon RA (2004) Characterisation of nitrilase and nitrile hydratase biocatalytic systems. Appl Microbiol Biotechnol 64:76–85
Brandelli A (2008) Bacterial keratinases: useful enzymes for bioprocessing agroindustrial wastes and beyond. Food Bioprocess Technol 1:105–116
Brysk MM, Ressler C (1970) γ-Cyano-α-L-aminobutyric acid, a new product of cyanide fixation in *Chromobacterium violaceum*. J Biol Chem 245:1156–1160
Burns RG, DeForest JL, Marxsen J, Sinsabaugh RL, Stromberger ME, Wallenstein D, Weintraub MN, Zoppini A (2013) Soil enzymes in a changing environment: current knowledge and future directions. Soil Biol Biochem 58:216–234
Cabuk A, Taspinar A, Kolankaya UN (2006) Biodegradation of cyanide by a white rot fungus, *Trametes versicolor*. Biotechnol Lett 28:1313–1317
Cameron MD, Aust SD (2001) Cellobiose dehydrogenase - An extracellular fungal flavocytochrome. Enzyme Microb Technol 28:129–138
Cameron MD, Timofeevski S, Aust SD (2000) Enzymology of *Phanerochaete chrysosporium* with respect to the degradation of recalcitrant compounds and xenobiotics. Appl Microbiol Biotechnol 54:751–758
Chen CY, Kao CM, Chen SC (2008) Application of *Klebsiella oxytoca* immobilized cells on the treatment of cyanide wastewater. Chemosphere 71:133–139
Chroma L, Mackova M, Kucerova P, in der Wiesche C, Burkhard J, Macek T (2002) Enzymes in plant metabolism of PCBs and PAHs. Acta Biotechnol 22(1-2):35–41
Cipollone R, Bigotti MG, Frangipani E, Ascenzi P, Visca P (2004) Characterization of a rhodanese from the cyanogenic bacterium *Pseudomonas aeruginosa*. Biochem Biophys Res Commun 325:85–90
Cipollone R, Ascenzi P, Frangipani E, Visca P (2006) Cyanide detoxification by recombinant bacterial rhodanese. Chemosphere 63:942–949
Claus H, Decker H (2006) Bacterial tyrosinases. Syst Appl Microbiol 29:3–14
Colosio C, Tiramani M, Brambilla G, Colombi A, Moretto A (2009) Neurobehavioural effects of pesticides with special focus on organophosphorus compounds: which is the real size of the problem? Neurotoxicology 30(6):1155–1161
Couto SR, Herrera JLT (2006) Industrial and biotechnological applications of laccases: a review. Biotechnol Adv 24:500–513
D'Annibale A, Rosetto F, Leonardi V, Federici F, Petruccioli M (2006) Role of autochthonous filamentous fungi in bioremediation of a soil historically contaminated with aromatic hydrocarbons. Appl Environ Microbiol 72:28–36
Demarche P, Junghanns C, Nair RR, Agathos SN (2012) Harnessing the power of enzymes for environmental stewardship. Biotechnol Adv 30:933–953
Dick WA, Tabatabai MA (1999) Use of immobilized enzymes for bioremediation. In: Segoe S (ed) Bioremediation of contaminated soils. Agronomy monograph 37. Soil Science Society of America, Madison, pp 315–338
Dick RP, Sandor JA, Eash NS (1994) Soil enzyme activities after 1500 years of terrace agriculture in the Colca Valley, Perú. Agric Ecosyst Environ 50:123–131
Doty SL (2008) Enhancing phytoremediation through the use of transgenic plants and entophytes. New Phytol 179:318–333
Doty SL, Shang QT, Wilson AM, Moore AL, Newman LA, Strand SE, Gordon MP (2000) Enhanced metabolism of halogenated hydrocarbons in transgenic plants contain mammalian P 450 2E1. Proc Natl Acad Sci USA 97:6287–6291
Doukyu N, Ogino H (2010) Organic solvent-tolerant enzymes. Biochem Eng J 48:270–282
Dunnil PM, Fowden L (1965) Enzymatic formation of β-cyanoalanine from cyanide by *E. coli* extract. Nature 208:1206–1207

Durán N, Rosa MA, D'Annibale A, Gianfreda L (2002) Application of laccases and tyrosinases (phenoloxidases) immobilized on different supports: a review. Enzyme Microb Technol 31:907–931
Eapen S, Singh S, D'Souza SF (2007) Advances in development of transgenic plants for remediation of xenobiotic pollutants. Biotechnol Adv 25:442–451
Ebbs S (2004) Biological degradation of cyanide compounds. Curr Opin Biotechnol 15:231–236
Edwin-Wosu NL, Nkang AE (2015) Evaluation of enzyme expression in a macrophytic treated crude oil soil habitat: implication for enhanced phytoremediation potential using transgenic botanicals. Insight Ecol 4:13–23
Efremenko E, Lyagin I, Votchitseva Y, Sirotkina M, Varfolomeyev S (2007) Polyhistidine-containing organophosphorus hydrolase with outstanding properties. Biocatal Biotransform 25:103–108
Ezzi IM, Pascual AJ, Gould JB, Lynch MJ (2003) Characterisation of the rhodanese enzyme in *Trichoderma* spp. Enzyme Microbial Technol 32:629–634
Fernández- Fernández M, Sanromán MA, Moldes D (2013) Recent development and applications of immobilized laccase. Biotechnol Adv 31:1808–1825
French CJ, Rosser SJ, Davies GJ, Nicklin S, Bruce NC (1999) Biodegradation of explosives by transgenic plants expressing pentaerythritol tetranitrate reductase. Nat Biotechnol 7:491–494
Gerhardt KE, Huang XD, Glick BR, Greenberg BM (2009) Phytoremediation and rhizoremediation of organic soil contaminants: potential and challenges. Plant Sci 176:20–30
Gianfreda L, Bollag JM (1996) Influence of natural and anthropogenic factors on enzyme activity in soil. In: Stotzky G, Bollag JM (eds) Soil biochemistry. Marcel Dekker, New York, pp 123–193
Gianfreda L, Bollag JM (2002) Isolated enzymes for the bioremediation and detoxyfication of organic pollutants. In: Burns RG, Dick RP (eds) Enzymes in the environment. Marcel Dekker, New York, pp 495–538
Gianfreda L, Rao MA (2004) Potential of extra cellular enzymes in remediation of polluted soils: a review. Enzyme Microb Technol 35:339–354
Gianfreda L, Ruggiero P (2006) Enzyme activities in soil. In: Nannipieri P, Smalla K (eds) Nucleic acids and proteins in soil. Springer, Heidelberg, pp 257–311
Gianfreda L, Rao MA (2008) Bonifica di suoli contaminati e depurazione dell'acqua. In: Gennari M, Trevisan M (eds) Agrofarmaci. Conoscenze per un Uso Sostenibile. Oasi Alberto Perdisa, Bologna, pp 521–564
Giardina P, Faraco V, Pezzella C, Piscitelli A, Vanhulle S, Sannia G (2010) Laccases: a never-ending story. Cell Mol Life Sci 67:369–385
Gómez Jiménez RT, Moliterni E, Rodríguez L, Fernández FJ, Villaseñor J (2011) Feasibility of mixed enzymatic complexes to enhanced soil bioremediation processes. Proc Environ Sci 9:54–59
Gupta N, Balomajumder C, Agrwal VK (2010) Enzymatic mechanism and biochemistry for cyanide degradation: a review. J Hazard Mater 176:1–13
Habig WH, Pabst MJ, Jacoby WB (1974) Glutathione S-transferases. The first step in mercapturic acid formation. J Biol Chem 249:7130–7139
Haki GD, Rakshit SK (2003) Developments in industrially important thermostabile enzymes: a review. Bioresour Technol 89:17–34
Han MJ, Choi HT, Song HG (2004) Degradation of phenanthrene by *Trametes versicolor* and its laccase. J Microbiol 42:94–98
Harcourt RL, Horne I, Sutherland TD, Hammock BD, Russell RJ, Oakeshott JG (2002) Development of a simple and sensitive fluorimetric method for isolation of coumaphos-hydrolysing bacteria. Lett Appl Microbiol 34:263–268
Harvey PJ, Xiang M, Palmer JM (2002) Extra cellular enzymes in the rhizosphere. In: Proceeding of the Inter-Cost Workshop on soil-microbe-root interactions: maximising phytoremediation/bioremediation, Grainau, pp 23–25

Hasan F, Shah AA, Hameed A (2006) Industrial applications of microbial lipases. Enzyme Microb Technol 39:235–251

Hong MS, Hong SJ, Barhoumi R, Burghardt RC, Donnelly KC, Wild JR, Venkatraj V, Tiffany-Castiglioni E (2003) Neurotoxicity induced in differentiated SK-N-SH-SY5Y human neuroblastoma cells by organophosphorus compounds. Toxicol Appl Pharmacol 186:110–118

Huertas MJ, Luque-Almagro VM, Martinez-Luque M, Blasco R, Moreno-Vivia C, Castillo F, Rolda MD (2006) Cyanide metabolism of *Pseudomonas pseudoalcaligenes* CECT5344: role of siderophores. Biochem Soc Trans 34:152–155

Iyer R, Iken B (2015) Protein enginnering of representive hydrolytic enzymes for remediation of organophosphates. Biochem Eng J 94:134–144

Jokanovic M, Prostran M (2009) Pyridinium oximes as cholinesterase reactivators. Structure – activity relationship and efficacy in the treatment of poisoning with organophosphorus compounds. Curr Med Chem 16(17):2177–2188

Karaca A, Cetin SC, Turgay OC, Kizilkaya R (2011) Soil enzymes as indicators of soil quality. In: Shukla G, Varma A (eds) Soil enzymology. Springer, Heidelberg, pp 119–148

Karigar CS, Rao SS (2011) Role of microbial enzymes in the bioremediation of pollutants: a review. Enzyme Res ID 805187:11

Kawahigashi H (2009) Transgenic plants for phytoremediation of herbicides. Curr Opin Biotechnol 20(2):225–230

Knutson K, Kirzan S, Ragauskas A (2005) Enzymatic biobleaching of two recalcitrant paper dyes with horseradish and soybean peroxidase. Biotechnol Lett 27:753–758

Kobayashi M, Yanaka N, Nagasawa T, Yamada H (1990) Purification and characterization of a novel nitrilase of *Rhodococcus rhodochrous* K22 that acts on aliphatic nitriles. J Bacteriol 172:4807–4815

Kumar S, Jin M, Weemhoff JL (2012) Cytochrome P450-mediated phytoremediation using transgenic plants: a need for engineered cytochrome P450 enzymes. J Pet Environ Biotechnol 3(5):1–5

Kunz DA, Wang CS, Che JL (1994) Alternative routes of enzymic cyanide metabolism in *Pseudomonas fluorescens* NCIMB 11764. Microbiology 140:1705–1712

Little RM, Anderson PM (1987) Structural properties of cyanase. J Biol Chem 262:10120–10126

Liu R, Dai Y, Sun L (2015) Effect of rhizosphere enzymes on phytoremediation in PAH-contaminated soil using five plant species. PLoS One 10(3):e0120369. doi:10.1371/journal.pone.0120369

Luque-Almagro VM, Huertas MJ, Martínez-Luque M, Moreno-Vivián C, Dolores Roldan M, Jesus Garcıa-Gil L, Castillo F, Blasco R (2005a) Bacterial degradation of cyanide and its metal complexes under alkaline conditions. Appl Environ Microbiol 71:940–947

Luque-Almagro VM, Blasco R, Huertas MJ, Martínez-Luque M, Moreno-Vivián C, Castillo F, Rolda MD (2005b) Alkaline cyanide biodegradation by *Pseudomonas pseudoalcaligenes* CECT5344. Biochem Soc Trans 33:168–169

Macek T, Kotrba P, Svatos A, Novakova M, Demnerova K, Mackova M (2008) Novel roles for genetically modified plants in environmental protection. Trends Biotechnol 26:146–152

Madhavi V, Lele SS (2009) Laccase: properties and applications. BioResources 4:1694–1717

Martinková L, Vejvoda V, Kaplan O, Kubáč D, Malandra A, Cantarella M, Bezouška K (2009) Fungal nitrilases as biocatalysts: recent developments. Biotechnol Adv 27:661–670

Meyers PR, Rawlings DE, Woods DR, Lindsey GG (1993) Isolation and characterization of a cyanide dihydratase for *Bacillus pumilus* C1. J Bacteriol 175:6105–6112

Mougin C, Jolivalt C, Briozo P, Madzak C (2003) Fungal laccases: from structure activity studies to environmental applications. Environ Chem Lett 1(2):145–148

Mougin C, Boukcim H, Jolivalt C (2009) Soil bioremediation strategies based on the use of fungal enzymes. In: Singh A, Kuhad RC, Ward OP (eds) Advances in applied bioremediation, Soil biology, vol 17. Springer, Heidelberg, pp 123–149

Mukram I, As N, Kirankumar B, Monisha TR, Reddy PV, Karegoudar TB (2015) Isolation and identification of a nitrile hydrolyzing bacterium and simultaneous utilization of aromatic and aliphatic nitriles. Int Biodeterior Biodegrad 100:165–171

Mulbry WW, Karns JS, Kearney PC, Nelson JO, McDaniel CS, Wild JR (1986) Identification of a plasmid-borne parathion hydrolase gene from *Flavobacterium* sp. by southernhybridization with *opd* from *Pseudomonas diminuta*. Appl Environ Microbiol 51:926–930

Marchut-Mikołajczak O, Kwapisz E, Antczak T (2013) Enzymatic bioremediation of xenobiotics. Eng Prot Environ 16(1):39–55

Nawaz SM, Chapatwala DK, Wolfram HJ (1989) Degradation of acetonitrile by *Pseudomonas putida*. Appl Environ Microbiol 55:2267–2274

Nelson DR, Koymans L, Kamataki T, Stegeman JJ, Feyereisen R, Waxman DJ, Waterman MR, Gotog O, Coon MJ, Estabrook RW, Gunsalus IC, Nebert DW (1996) P450 superfamily: update on new sequences, gene mapping, accession numbers and nomenclature. Pharmacogenetics 6:1–42

Novotný C, Vyas BRM, Erbanová P, Kubatová A, Šašek V (1997) Removal of PCBs by various white rot fungi in liquid cultures. Folia Microbiol 42:136–140

Omura H, Ikemot M, Kobayash M, Shimizq S, Yoshida T, Nagasawa T (2003) Purification, characterization and gene cloning of thermostable O-acetyl-L-homoserine sulfhydrylase forming γ-cyano-α-aminobutyric acid. J Biosci Bioeng 96:53–58

Ramos JL, Gonzalez-Perez MM, Caballero A, van Dillewijn P (2005) Bioremediation of polynitrated aromatic compounds: plants and microbes put up a fight. Curr Opin Biotechnol 16:275–281

Rao MA, Scelza R, Scotti R, Gianfreda L (2010) Role of enzymes in the remediation of polluted environments. J Soil Sci Plant Nutr 10(3):333–353

Regnery aus Trier J (2010) Organophosphates in precipitation, lake water, and groundwater from urban and remote areas. Dissertation zur Erlangung des Doktorgrades der Naturwissenschaften. Johann Wolfgang Goethe-Universität in Frankfurt am Main, p 133

Riffaldi R, Levi-Minzi R, Cardelli R, Palumbo S, Saviozzi A (2006) Soil biological activities in monitoring the bioremediation of diesel oil-contaminated soil. Water Air Soil Pollut 170 (1–4):3–15

Ryan BJ, Carolan N, Faga CO (2006) Horseradish and soybean peroxidases: comparable tools for alternative niches? Trends Biotechnol 24:359–363

Scott C, Pandey G, Hartley CJ, Jackson CJ, Cheesman MJ, Taylor MC, Pandey R, Khurana JL, Teese M, Coppin CW, Weir KM, Jain RK, Lal R, Russel RJ, Oakeshott JG (2008) The enzymatic basis for pesticide bioremediation. Indian J Microbiol 48:65–79

Scott C, Begley C, Taylor MJ, Pandey G, Momiroski V, French N, Brearley C, Kotsonis SE, Selleck MJ, Carino FA, Bajet CM, Clarke C, Oakeshott JG, Russel RJ (2011) Free-enzyme bioremediation of pesticides. In: Goh KS, Bret B, Potter T, Gan J (eds) Pesticide mitigation strategies for surface water quality. ACS Symposium Series 1075. American Chemical Society, Washington, DC, p 155-174

Serdar CM, Gibson DT, Munnecke DM, Lancaster JH (1985) Enzymatic hydrolysis of organophosphates: cloning and expression of a parathion hydrolase gene from *Pseudomonas diminuta*. Biotechnology 3:367–371

Sicking C, Brusch M, Lindackers A, Riedel KU, Schubert B, Isakovic N, Krall C, Klipp W, Drepper T, Schneider K, Masepohl B (2005) Identification of two new genes involved in diazotrophic growth via the alternative Fe-only nitrogenase in the phototrophic purple bacterium *Rhodobacter capsulatus*. J Bacteriol 187:92–98

Sinha RK, Valani D, Sinha S, Singh S, Heart S (2009) Bioremediation of contaminated sites: a low-cost nature's biotechnology for environmental clean up by versatile microbes, plants and earthworms. In: Faerber T, Herzog J (eds) Solid waste management and environmental remediation. Nova Science, New York, pp 1–72

Sirotkina M, Lyagin I, Efremenko E (2012) Hydrolysis of organophosphorus pesticides in soil: new opportunities with ecocompatible immobilized His6-OPH. Int Biodeterior Biodegrad 68:18–23

Sonoki T, Kajita S, Ikeda S, Uesugi M, Tatsumi K, Katayama Y et al (2005) Transgenic tobacco expressing fungal laccase promotes the detoxification of environmental pollutants. Appl Microbiol Biotechnol 67:138–142

Spaczyński M, Seta-Koselska A, Patrzylas P, Betlej A, Skórzyńska-Polit E (2012) Phytodegradation and biodegradation in rhizosphere as efficient methods of reclamation of soil contaminated by organic chemicals (a review). Acta Agrophys 19(1):155–169

Strong PJ, Claus H (2011) Laccase: a review of its past and its future in bioremediation. Crit Rev Environ Sci Technol 41:373–434

Suneetha V, Khan ZA (2011) Actinomycetes: sources for soil enzymes. In: Shukla G, Varma A (eds) Soil enzymology. Springer, Heidelberg, pp 259–269

Torres E, Bustos-Jaimes I, Le Borgne S (2003) Potential use of oxidative enzymes for the bioremediation of organic pollutants. Appl Catal B Environ 46:1–15

Trapp S, Larsen M, Pirandello A, Danquah-Boakye J (2003) Feasibility of cyanide elimination using plants. Eur J Min Proc Environ Prot 3:128–137

Tyagi M, da Foseca MM, Carvalho CC (2011) Bioaugmentation and biostimulation strategies to improve the effectiveness of bioremediation processes. Biodegradation 22(2):231–241

Uchida E, Ouchi T, Suzuki Y, Yoshida T, Habe H, Yamaguchi I, Omori T, Nojiri H (2005) Secretion of bacterial xenobiotic-degrading enzymes from transgenic plants by an apoplastic expressional system: an applicability for phytoremediation. Environ Sci Technol 39:7671–7677

Venkata Naga Raju E, Divakar G (2013) Production of keratinase by using *Pseudomonas aeruginosa* isolated from poultry waste. IJPCBS 3(1):9–86

Viswanath B, Rajesh B, Janardhan A, Kumar AP, Narasimha G (2014) Fungal laccases and their applications in bioremediation. Enzyme Res 163242:21

Wang GD, Li QJ, Luo B, Chen XY (2004) Ex planta phytoremediation of trichlorophenol and phenolic allelochemicals via engineered secretory laccase. Nat Biotechnol 22:893–897

Wanger M, Nicell JA (2002) Detoxification of phenolic solutions with horseradish peroxidase and hydrogen peroxide. Water Res 36:4041–4052

Wenzel WW, Adriano DC, Salt D, Smith R (1999) Phytoremediation: a plant-microbe-based remediation system. In: Segoe S (ed) Bioremediation of contaminated soils, agronomy monograph, vol 37. Soil Science Society of America, Madison, pp 315–338

Wesenberg D, Kyriakides I, Agathos SN (2003) White-rot fungi and their enzymes for the treatment of industrial dye effluents. Biotechnol Adv 22:161–187

Wilce MCJ, Parker MW (1994) Structure and function of glutathione S-transferases. Biochem Biophys Acta 1205:1–18

Williams RAM, Kelly SM, Mottram JC, Coombs GH (2003) 3-mercaptopyruvate sulfur transferase of *Leishmania* contains an unusual C-terminal extension and is involved in thioredoxin and antioxidant metabolism. J Biol Chem 278:1480–1486

Yamaki T, Olikawa T, Ito K, Nakamura T (1997) Cloning and sequencing of a nitrile hydratase gene from *Pseudonocardia thermophila*, JCM3095. J Ferment Bioeng 83:474–477

Chapter 18
Soil Xenobiotics and Their Phyto-chemical Remediation

Haythum M. Salem, Ali Abdel-Salam, Mohamed A. Abdel-Salam, and Mahmoud F. Seleiman

18.1 Xenobiotics in Soil

Xenobiotics are foreign chemicals found in organisms which are not naturally produced by or expected to be present within those organisms. Xenobiotics also refer to substances found in exceptionally very high contents (materials such as antibiotics) in organisms.

Soil can be a habitat for many xenobiotic pollutants under certain circumstances including the use of some biocides and pharmaceuticals and dumping of harmful materials, aromatic chemicals, and left-outs of some industries (Schröder 2007; Chapman 2007; Schröder and Collins 2011). Other pollutants which may be found in soils include benzene, toluene, ethylbenzene and xylene, aniline, nitrobenzene, trinitrotoluene, and chlorinated solvents (Haberl et al. 2003; Barac et al. 2004; Schwitzguébel et al. 2011; Schröder 2007; Weyens et al. 2009a, b). Polycyclic aromatic hydrocarbons, chlorobenzenes (CBs), polychlorinated biphenyls (PCBs), and polychlorinated dibenzo-p-dioxins/dibenzofurans (PCDD/Fs) are also other pollutants which may be found in soils (Wild et al. 1992; Duarte-Davidson and Jones 1996).

Natural compounds can be xenobiotics if they enter the soil from excretions, such as human excretes or chemical defense materials produced by some organisms

H.M. Salem • A. Abdel-Salam • M.A. Abdel-Salam
Department of Soils and Water, Faculty of Agriculture, Benha University, P.O. Box 13637, Benha, Egypt

M.F. Seleiman (✉)
Department of Crop Sciences, Faculty of Agriculture, Menoufia University, P.O. Box 32511, Shibin El-Kom, Egypt

Department of Agricultural Sciences, University of Helsinki, P.O. Box 27, 00014 Helsinki, Finland
e-mail: mahmoud.seleiman@agr.menofia.edu.eg; mahmoud.seleiman@gmail.com

© Springer International Publishing AG 2017

M.Z. Hashmi et al. (eds.), *Xenobiotics in the Soil Environment*, Soil Biology 49,
DOI 10.1007/978-3-319-47744-2_18

as protection against predators (Mansuy 2013). Some of the xenobiotic pollutants are hard to degrade as follows:

Halocarbons, which contain halogens, are found in pesticides and in some solvents. They are volatile and accumulate in soils. Polychlorinated biphenyls (PCBs) are halogens with benzene ring; and they were found in insulators. Synthetic polymers, which are mainly used in plastic industry to form materials such as polyester (PE) and polyvinyl chloride (PVC), are insoluble in water and of high molecular weight. Alkyl-benzyl sulfonates are sulfonates found in detergents. Oil mixtures are oils insoluble in water; some of their components are toxic.

Compounds found in the soil/plant system of the soil rhizosphere include polycyclic aromatic hydrocarbons (PAHs), along with the ones mentioned above (CBs, PCBs, and PCDD/Fs), all of which were in high contents in some plant roots (Duarte-Davidson and Jones 1996; Wild et al. 1992). Pollutants such as benzene, toluene, ethylbenzene and xylene (BTEX), aniline, nitrobenzene, trinitrotoluene and leftovers/wastes of military actions, and chlorinated solvents are found in some soils (Barak et al. 1983; Fiorenza et al. 2000; Haberl et al. 2003; Barac et al. 2004; Schwitzguebel and Vanek 2003; Weyens et al. 2009a, b; Schwitzguébel et al. 2011). Synthetic organochlorides such as plastics and pesticides or natural organic chemicals as poly-aromatic hydrocarbons and some fractions of crude oil and coal are found in polluted soils. The organic substances, whether non-harmful or harmful, can exist into the soil rhizosphere through adsorption on the colloidal soil phase as well as on the fine and very fine root hairs of growing plants. The transport of materials within water in plants can be done by mass flow leading to transpiration. Water is transported penetrating a number of plant tissues: the epidermis, cortex, endodermis, and pericycle upward from roots via the xylem by mass flow resulting from a pressure gradient (McFarlane 1995; Trapp and McFarlane 1995; Hellstrom 2004).

18.2 Phyto-remediation of Xenobiotics from Soil

Phyto-remediation is defined as the use of plant to remediate soils from pollutants. The remediation techniques using plants can remove some pollutants, but some of the pollutants will be remained in the soil (Schröder and Collins 2002). Detoxifying enzymes of plants can localize pollutants into the cytosol of plant cells.

Phyto-remediation for removal of pollutants such as xenobiotics can be an alternate effective method, since most of the plant species are vigorous in growth, a renewable resource, and can also be used for in situ remediation (Cunningham and Ow 1996; Suresh and Ravishankar 2004; Parameswaran et al. 2007). In addition, there are some plants which can survive higher concentrations of contaminants than numerous microorganisms used for bioremediation. Phyto-remediation can increase the organic carbon content in the soil; consequently, this can stimulate microorganism activities and supplement the degradation of the contaminants in the rhizosphere. Plant species used in the phyto-remediation can result in soil

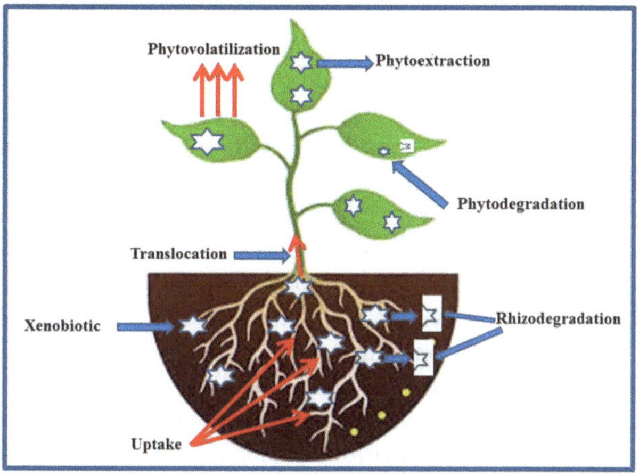

Fig. 18.1 The different mechanisms that are used by plants for remediation of soil. In summary, the xenobiotics can be stabilized or degraded in the rhizosphere, adsorbed on or uptaked by the plant roots and translocated to the aboveground parts, and volatilized or degraded within the plant tissues

stabilization, and their biomass or yield can be used for bioenergy (Seleiman et al. 2013) or fiber production.

Usually, plants use different strategies for remediation of soil such as phytoextraction, phytodegradation, phytovolatilization, and rhizodegradation (Fig. 18.1) (Schnoor 1997). In phytoextraction, the plants uptake the pollutants from the soil into their parts, but these contaminants cannot be metabolized. On the other hand, plants are able to metabolize the uptake contaminants in the phytodegradation. In phytovolatilization, the contaminants release from the plant parts to the atmosphere after their uptake from the soil. This method has been detected for organic and inorganic contaminants (i.e., heavy metals) (Compton et al. 1998). However, rhizodegradation means the transformation of pollutants through resident microbes in the plant rhizosphere.

18.2.1 Uptake of Xenobiotics by Plant

Plants have the ability to take up and absorb many organic and inorganic xenobiotics from soils. This is mainly through the root system. It could also occur through foliage from air. Such phyto-remediation of organic xenobiotics depends on specific ability of plant organs, particularly roots, to uptake organic xenobiotics from its environment. Plant roots and microorganisms associated with them (Alkorta and Garbisu 2001) such as mycorrhiza can work actively in remediating the rhizosphere from organic pollutants. The amount and type of organic substances uptake by plants depend on many factors including the genus and species of the plant as well as many environmental factors.

Plants vary considerably in their ability to uptake xenobiotics from soils (Bell 1992). Among important properties affecting uptake of xenobiotics are the manner and pattern of root-system growth; depth and volume explored by the root;

transpiration rate of plant; growth period of the plant; climatological nature of the environment; type of uptake mechanisms; place of storage in plant such as leaves, tubers, fruits, and stems; and other characteristics (Buckley 1982). Organic xenobiotic contaminants can bind to soil particles and form persistent organic pollutants (POPs) which resist extraction, microbial degradation, or volatilization. Mattina et al. (2002) postulated hypotheses on contamination of soil with POPs which include the following: (1) volatilization POPs (soil-to-air translocation), (2) fallout of POPs (usually in low contents) on soil surface from the atmosphere (air-to-soil translocation), and (3) addition of agrochemicals (anthropic translocation). They concluded that cycling POPs through the biosphere must be done before the full impact of POPs on human health can be a serious risk. Although the rules about POPs are restricted in many developed countries, it is not the case in many underdeveloped countries where organochlorine insecticides, such as DDT and chlordane, remain a primary and permanent health hazard. Global translocation of POPs will probably remain unavoidable in the foreseeable future.

18.2.2 Plants and Their Storing Capacity for Removing Xenobiotics

Plants vary in their capacity to the uptake of xenobiotics from the soil according to a number of factors including plant species type, environment where plants grow, plant roots and its leaves, and magnitudes of plant growth.

18.2.3 Type of Plant

Regarding variation in removing xenobiotics, some plants have big storing organs to store the xenobiotics. Cucurbit plants which include courgette (*Cucurbita pepo*) and pumpkin (*Cucurbita maxima*) have such organs. They accumulate and translocate high amounts of organic pollutants such as polychlorinated dibenzo-p-dioxins (PCDD) and dibenzofurans (PCDF) from soils contaminated with xenobiotics (Hulster et al. 1994; Simonich and Hites 1995). It is assumed (Mattina et al. 2002; White 2002) that exudates of roots of courgette and pumpkin mobilize the xenobiotics rendering them easily taken up by plant roots. Other plants have high lipid contents which enable them to accumulate xenobiotics in their storage organs. Carrots are plants with high lipid contents, which make them absorb up high amounts of xenobiotics from soils (Bell 1992; Topp et al. 1986). Carrots can uptake high level of oil-derived nonionized substances (Smelt and Leistra 1974; O'Connor et al. 1990). Lipophilics in plants would be a part of plant leaf lipids (Buckley 1982), and variability between plants regarding their ability to absorb xenobiotics can be monitored via plant leaf lipids (Simonich and Hites 1995). On the other hand, such variability could be assessed through the lipid content ratio of leaf/air using special models (Bohme et al. 1999). Leaf lipid content is regarded a highly reliable parameter in evaluating the efficiency of plant to uptake xenobiotics (Collins et al. 2006b).

18.2.4 Environment of Plant and Endurance of Contamination

The environment of plants here means the habitat of the plants. Once they are introduced to contaminated soils, they should adapt to their new environment and consequently act through different processes to their new habitat. Their uptake of substances and materials from their habitat includes the needed nutrients for growth and also the harmful substances including xenobiotics. Plants could survive in polluted environments by modifying their chemo-physical processes to endure and/or retard the movement of pollutants. Willow plants (*Salix* spp.) are deciduous trees and shrubs which survive in contaminated wet lands. One tree or shrub can take up contaminated waters from such soils in high amounts which could be 200 liters per day (Susarla et al. 2002). Plants take up substances from the soil solution along with the needed water into their roots, and from the roots, the substances move within the plant. Collins et al. (2012) identified the key processes involved in plant uptake of organic chemicals and assessed important factors in this respect, including uptake of water, plant lipids, growth dilution, and metabolism.

18.2.5 Plant Roots and Their Role

Physiology and composition of roots can affect plant uptake of phytotoxic xenobiotics such as carbamates. There are theories on the relationship between root lipids and their role in uptake of xenobiotics (Briggs et al. 1982; Hsu et al. 1990; Trapp and Pussemeir 1991). Besides the content of lipids in plant, the kind of lipids also plays an important role in the uptake of xenobiotics (Bromilow and Chamberlain 1995; Collins et al. 2006a). *Lipophilicity* is a kind of passive absorption of lipids by plant roots. High lipophilicity causes a high sorption of lipids and low sorption of non-lipids by plants and also low adsorption by soil organic colloids (Karickhoff 1981; Doucette 2003). There is a competition in sorption between plant lipids and soil organic carbon (Schröder and Collins 2011). Amines, carboxylic acids, phenols (such as chlorophenols and nitrophenols), and some pesticides may be found in plants (USEPA 1996). Organic substances could be taken up by plants in marked levels through plant leaves (Collins and Finnegan 2010). Particular importance in the uptake of xenobiotics by plants depends on the nature of transport of substances from roots to aboveground parts of plants (Wild et al. 1992).

18.2.6 Uptake of Substances Through Plant Leaves

Plant leaves can get xenobiotic bodies and particles adhered to them through carriers such as air and rain. This is most evident where plant foliage is as near as possible to the soil or where agents such as rain or air carry the xenobiotics. Hulster and Marschner (1993) reported a marked adherence of PCDD/Fs (polychlorinated dibenzo-p-dioxins/dibenzofurans) on grass fodder in the rural areas. In an area of rural and urban locations, Kao and Venkataraman (1995) estimated fallouts of PCDD/Fs substances on soils and plants. They found that the rural zones acquired up to 40 % of the PCDD/Fs in comparison to the urban parts which acquire up to 90 % of PCDD/Fs. Uptake of substance through plant is affected by various mechanisms including adsorption, fallout, interception, and diffusion, before their removal or degradation (Chamberlain 1991). Chemical and physical properties of plant as well as the organic adhering particles have important effect on the efficiency of foliage uptake of organic substances. Rain drops falling on soil surface under tomato plants caused fine soil particles to be retained by the aboveground plant parts including the wide leaves of plants (Dreicer et al. 1984). A parameter assessing partition of organic substances was presented by McLachlan (1999) designating it as the "octanol-air partition coefficient" (K_{OA}) to serve as an indicator of deposition contamination pathway for semi-volatile organic chemicals. Classification of different substances based on this coefficient was given by Cousins and Mackay (2001). The differences between experimental findings and mathematical models results are assessed by Rikken et al. (2001). Other theories and modeling efforts in this regard have been discussed by other researchers (Trapp and Pussemeir 1991; Trapp and McFarlane 1995; Trapp and Matthies 1995).

18.2.7 Plant Species and Their Capacity to Accumulate Xenobiotics

Plants vary in their ability and capacity to uptake and accumulate organic substances including the xenobiotics, depending on many aspects and factors (Buckley 1982; Bell 1992). Besides the built-in plant properties which enhance uptake by plant, the following are of great importance in this respect: length of growth period, root exudates, shape and growth characteristics, foliage shape and size, and fruit and tuber shape, size, and location (Buckley 1982). Examples of the plants which have high capacity to uptake and absorb soluble and suspended organic materials from the rhizosphere are species of the *Cucurbitaceae* family such as winter squash, pumpkin, and marrow (*Cucurbita maxima*) and summer squash (*Cucurbita pepo*), with their big fruits, expanding roots with effective exudates in mobilizing substances (Hulster et al. 1994; Simonich and Hites 1995; Mattina et al. 2002, 2003; White 2002). Experiments were conducted by Parrish et al. (2006) to investigate the uptake of polycyclic aromatic hydrocarbons (PAHs) added to soil in the level of

40 mg kg^{-1}. Plants were grown for four 4-week cycle periods. During the first period, zucchini (*Cucurbita pepo*) plants contained considerably more PAHs than in the cucumber (*Cucumis sativus*) plants or the squash (*Cucurbita pepo*) ones. The zucchini plants accumulated up to about six times higher PAH than other plant species. During the second to the fourth period, the accumulation of PAH by zucchini was decreased by about 90 %, whereas accumulation by the cucumber and squash plants remained relatively constant. Also removal of PAHs by zucchini was about twice higher than in others species.

High contents of lipids in plants enhance their uptake of substances from soil. Carrots with their high lipids collected high amounts of organic substances (Smelt and Leistra 1974; Topp et al. 1986; O'Connor et al. 1990; Bell 1992; Schroll and Scheunert 1992). The fat-liking chemicals of lipophilics can be petitioned to leaf lipids; thus, variations in plant foliage lipid contents are partly behind differences in plant ability to absorb organic substances (Buckley 1982). The important role of leaf lipids was confirmed experimentally in studies by Simonich and Hites (1995) on accumulation of pollutants in plant foliage recording acquisition of polycyclic aromatic hydrocarbons (PAHs) by plants. However, Bohme et al. (1999) found that such factor may not have a marked role in the uptake of organic substances by plant. Trapp and Matthies (1995) and Collins et al. (2006b) reported findings affirming the importance of leaf lipids in uptake of pollutants by plants (Schröder and Collins 2002; Bell 1992).

18.2.8 Growth of Plant and Status of Xenobiotics Inside It

Growth of plant and the magnitude it reaches are important in xenobiotic phyto-removal of xenobiotics from soil. The improvement in plant growth leads to the increase in adherence and in leaf uptake of substances, particularly with plants of long growing seasons or perennials (Thorne et al. 2004; Collins and Cunningham 2005; Collins and Finnegan 2010). There may be a case of supplying plants with nutrients in order to acquire the most efficient removal of unwanted soil pollutants. Estimation of uptake extent of volatile substance by plants was done through the mathematical models from practical results of experimental findings done by many researchers (Trapp and Matthies 1995; McLachlan 1999; Collins and Cunningham 2005; Collins and Finnegan 2010).

18.2.9 Mechanisms Performed by Plants to Alleviate or Nullify the Harmful Effects of Xenobiotics

Plants can perform a variety of mechanisms to alleviate or nullify the harmful effect of xenobiotics. Sandermann (1992) reported three-phase metabolism of xenobiotics by plants: (a) by transformation, i.e., changing the structure of the substance into a less harmful one; (b) by confining the substance inside a confined space within the plant tissue; and (c) by conjugation, i.e., joining of two or more compounds. The processes are mostly performed through enzyme actions. One of the mechanisms done by plants to nullify the harmful effect of xenobiotics is oxidative coupling where abiotic catalyst such as ferric chloride links the xenobiotic chemical to organic matter (Bollag 2002). In such mechanism, the mediating catalysts, which include cupric hydroxide, are involved in spontaneous reactions incorporating non-phenolic compounds into humic polymers. Peroxide enzymes contain porphyrin ring and require peroxides for detoxifying polluted liquids. Plant cell walls have vacuoles used for storing foreign materials besides the plant nutrients (Hermam and Larkind 1999; Hames and Hooper 2005). The USDA (2015) concluded that the three-phase action performed by plant for metabolizing xenobiotic substances are (1) initial reactions as oxidation, reduction, or hydrolysis; (2) primary conjugation with endogenous substrates such as sugars, amino acids, and glutathione; and (3) secondary conjugation to form insoluble substances or sequestrate such substances in vacuoles.

18.2.10 Metabolizing Xenobiotics

Metabolizing foreign substances which enter plants is one of the defense mechanisms that plants can use. Komives and Gullner (2005) reported that after xenobiotic entering the plants, their functional groups are subjected to the reactions leading to formation of polar, water-soluble, chemically, and biologically reactive substances. These reactions are important with hydrophobic, chemically stable organic xenobiotics such as polycyclic aromatic hydrocarbons and (poly)chlorinated aliphatic and aromatic hydrocarbons. The reactions involve a number of transformations such as hydrolysis, reduction, and oxidation catalyzed by cytochrome P450 which are heme-containing enzymes, most of which catalyze reactions of nicotinamide-adenine-dinucleotide phosphate (NADP) and O_2-dependent hydroxylation. Transgenic plants with specific enzymatic activities may be used in this respect.

Plants can metabolize substances such as benzo(a)pyrene, trichloroethylene, and benzene through various mechanisms (Ugrekhelidze and Phiriashvili 2000; Shang and Gordon 2002; Brady et al. 2003). Metabolizing xenobiotics would decrease their harmful effect to plant (Schröder and Collins 2002, 2011). Modification of organisms and plants through modern genetic modifications and other techniques could help in making metabolism of such substances more efficient (Rylott and Bruce 2009) as shown by work on endophytic bacteria (Weyens et al. 2009a, b).

Acquisition and the uptake of xenobiotics by plants through their roots and adherence to their bodies is commonplace in many species (Little and Wiffen 1977; Pinder et al. 1991; Schreiber and Schonherr 1992). Leaf surface area and leaf hair characteristics are considered vital importance in the efficiency of leaves to absorb, adsorb, and adhere different substances (McCrady and Maggard 1993; McCrady 1994; Simonich and Hites 1995). Broader leaves of grater fine hairs lead to more of substances getting adsorbed, absorbed, and adhered to foliage of plants (Little and Wiffen 1977). Through air and wind, soil particles, along with substances they adsorb, can adhere to plant foliage (Pinder et al. 1991). Closeness of foliage to soil surface would increase amounts of substances adhered to plant leaves (Pinder et al. 1991; McFarlane 1995). Some soil constituents could volatilize, and consequently the volatilized substances would be taken up by plant leaves (Collins and Bell 1997).

18.3 Chemo-remediation of Xenobiotics from Soil

The technology of chemical remediation depends on using chemical substances in detoxifying or lessening the harmful effect of xenobiotic substances in soil, materials such as solvents, and oxidizing chemicals. On chemical remediation, Abdel-Salam et al. (2015) reviewed the techniques of chemo-remediating heavy-metal contaminated soils by subjecting the soil to addition of some chemicals such as acids, chelators, and immobilizers. Acids and acidic materials used for remediating contaminated soils include synthetic aminopolycarboxylic acids, natural acids, low-molecular-weight organic acids, and acidic humic substances (Evangelou et al. 2007). Examples of such materials are ethylenediaminetetraacetic acid (EDTA), hydroxyethylenediaminetetraacetic acid (HEDTA), diethylenetriaminepentaacetic acid (DTPA), trans-1,2-cyclohexylene dinitrilo-tetraacetic acid (CDTA), ethylene glycol tetra-acetic acid (EGTA), ethylenediamine-N,N′-bis (*o*-hydroxyphenyl) acetic acid (EDDHA), N-(2-hydroxyethyl)-iminodiacetic acid (HEIDA), and N, N0-bis(2-hydroxybenzyl)ethylenediamine-N,N′-di-acetic acid (HBED). Also, solvents, which dissolve non-water-soluble substances such as polychlorinated biphenyl (PCBs), are bellowing to those materials. An organo-mineral complex (OMC) consisting of the zeolite/clinoptilite clay minerals and humic acids was used by Dercova et al. (2006) to adsorb pollutants of pentachlorophenols (PCPs) from contaminated soils. The results indicated high efficiency, in polluted soils, particularly when the technique was combined with the use of bioremediation utilizing special bacterial isolates. Dolomitic pulverized limestones were used to remediate soils contaminated with heavy metals (de-Valle-Zermeño et al. 2015), with successful results proving the effectiveness of this method in stabilization of such pollutants and guaranteeing environmentally friendly alkali reservoirs for long-term stabilization of heavy metals and metalloids. Therefore, the use of Mg-carbonates in the form of dolomitic limestones may be a practical means in remediation of polluted soils.

18.4 Degradation of Xenobiotics in Soil

Many xenobiotics cause a variety of harmful biological effects due to their toxicity to man and environment as well as their persistence in the environment. Some microorganisms can degrade many xenobiotics in soil (Alexander 1999). Microorganisms capable of degrading xenobiotics must live with no restraints in the contaminated soil. The xenobiotic compound must be capable of inducing enzymatic activity (within the microorganism) appropriate for detoxification. Solubility of the xenobiotic compound may cause decreased degradation. The microenvironment conditions (including pH, temperature, and others) must be conducive to degradation. Karpouzas and Singh (2006) followed degradation of organophosphorus pesticides and found that most were degraded in soil by the indigenous soil microorganisms. They concluded that there are a number of genes/enzymes, which enable microorganisms to degrade such compounds, and they may be engineered to enhance the degrading efficiency of microorganisms. Cloransulam-methyl pesticides were reported to degrade relatively quickly in soil (Wolt et al. 1996).

References

Abdel-Salam A, Salem HM, Abdel-Salam MA, Seleiman MF (2015) Phyto and chemical removal of heavy metal-contaminated soils. In: Sherameti I, Varma A (eds) Heavy metal contamination of soils: monitoring and remediation, Series soil biology. Springer, Zürich, pp 299–308

Alexander M (1999) Biodegradation and bioremediation. Elsevier Science, Amsterdam

Alkorta I, Garbisu C (2001) Phytoremediation of organic contaminants in soils. Bioresour Technol 79:273–276

Barac T, Taghavi S, Borremans B, Provoost A, Oeyen L, Colpaert J, Vangronsveld J, van der Lelie D (2004) Engineered endophytic bacteria improve phytoremediation of water-soluble, volatile, organic pollutants. Nat Biotechnol 22:583–588

Barak E, Jacoby B, Dinoor A (1983) Adsorption of systemic pesticides on ground stems and the apoplectic pathway of stems as related to lignification and lipophilicity of the pesticides. Pestic Biochem Physiol 20(2):194–202

Bell RM (1992) Higher plant accumulation of organic pollutants from soil. United States Environmental Protection Agency (USEPA), Washington, DC

Bohme F, Welsch-Paulsch K, McLachlan MS (1999) Upltake of airborne semi-volatile organic compounds in agricultural plants: field measurements of interspecies variability. Environ Sci Technol 33(11):1805–1813

Bollag J (2002) Immobilization of pesticides in soil through enzymatic reactions. In: Agathos SN, Reineke W (eds) Biotechnology for the environments: soil remediation. Kluwer Academic, Dordrecht

Brady CAL, Gill RA, Lynch PT (2003) Preliminary evidence for the metabolism of benzo(a) pyrene by plantagolanceolata. Environ Geochem Health 25(1):131–137

Briggs GG, Bromilow RH, Evans AA (1982) Relationships between lipophilicity and the distribution of non-ionized chemicals in barley shoots following uptake by the roots and translocation of non-ionized chemicals by barley. Pestic Sci 13:495–504

Bromilow RH, Chamberlain K (1995) Principles governing uptake and transport of chemicals. In: Trapp S, McFarlane JC (eds) Plant contamination: modeling and simulation of organic chemical processes. Lewis, London, pp 38–64

Buckley EH (1982) Accumulation of airborne polychlorinated-biphenyls in foliage. Science 216 (4545):520–522
Chamberlain AC (1991) Radioactive aerosols. Cambridge University Press, Cambridge
Chapman PM (2007) Determining when contamination is pollution: weight of evidence determinations for sediments and effluents. Environ Int 33:492–501
Collins CD, Bell JN (1997) Absorption and adsorption of benzene by plants, following exposure during the final part of the growth cycle. Ministry of Agriculture, Fisheries and Food, Contaminants Division, London
Collins CD, Cunningham N (2005) Modelling the fate of sulphur 35 in crops: 2. Development and validation of the CROPS-35 model. Environ Pollut 133(3):439–445
Collins CD, Finnegan E (2010) Modeling the plant uptake of organic chemicals, including the soil-air-plant pathway. Environ Sci Technol 44(3):998–1003
Collins CD, Fryer M, Grosso A (2006a) Plant uptake of non-ionic organic chemicals. Environ Sci Technol 40(1):45–52
Collins CD, Martin I, Fryer ME (2006b) Evaluation of models for predicting plant uptake of chemicals from soil (Science Report SC050021/SR). Environment Agency, Bristol
Collins C, Martin I, Doucette W (2012) Plant uptake of xenobiotics. In: Schroder P, Collins CD (eds) Organic xenobiotics and plants: from mode of action to ecophysiology, Plant ecophysiology. Springer, London
Compton HR, Harosi DM, Hirsch SR, Wrobel JG (1998) Pilot-scale use of trees to address voc contamination. In: Wickramanayake GB, Hinchee RE (eds) Bioremediation and phytoremediation of chlorinated and recalcitrant compounds. Battelle Press, Columbus
Cousins IT, Mackay D (2001) Strategies for including vegetation compartments in multimedia models. Chemosphere 44(4):643–654
Cunningham SD, Ow DW (1996) Promise and prospects of phytoremediation. Plant Physiol 110:715–719
Dercova K, Sejakova Z, Skokanova M, Barancikova G, Mkovnikova J (2006) Potential use of organo-mineral complex (OMC) for bioremediation of pentachlorophenol (PCP) in soil. Int Biodeterior Biodegrad 58:248–253
de-Valle-Zermeño R, Formosa J, Chimenos JM (2015) Low-grade magnesium oxide by-products for environmental solutions: characterization and geochemical performance. J Geochem Explor 152:143–144
Doucette WJ (2003) Quantitative structure-activity relationships for predicting soil-sediment sorption coefficients for organic chemicals. Environ Toxicol Chem 22(8):1771–1788
Dreicer M, Hakonson TE, White GC, Whicker FW (1984) Rainsplash as a mechanism for soil contamination of plant-surfaces. Health Phys 46(1):177–187
Duarte-Davidson R, Jones KC (1996) Screening the environmental fate of organic contaminants in sewage sludge applied to agricultural soils 2. The potential for transfers to plants and grazing animals. Sci Total Environ 185(1–3):59–70
Evangelou MW, Bauer U, Ebel M, Schaeffer A (2007) Evaluation of the effect of small organic acids on phyto-extraction of Cu and Pb from soil with tobacco(Nicotianatabacum). Chemosphere 68:345–353
Fiorenza S, Oubre CL, Ward CH (2000) Phytoremediation of hydrocarbon-contaminated soils. CRC Press, Boca Raton
Haberl R, Grego S, Langergraber G, Kadlec R, Cicalini AR, Dias MS, Novais JM, Aubert S, Gerth A, Thomas H, Hebner A (2003) Constructed wetlands for the treatment of organic pollutants. J Soils Sediment 3:109–124
Hames D, Hooper N (2005) Instant notes in biochemistry. Taylor & Francis Group, New York
Hellstrom A (2004) Uptake of organic pollutants in plants. Department of Environment and Assessments, Swedish University of Agricultural Sciences, Uppsala
Hermam EM, Larkind BA (1999) Protein storage bodies and vacuoles. Plant Cell 11(4):601–613

Hsu FC, Marxmiller RL, Alex YA (1990) Study of root uptake and xylem translocation of cinmethylin and related compounds in de-topped soybean roots using a pressure chamber technique. Plant Physiol 93:1573–1578

Hulster A, Marschner H (1993) Transfer of PCDD/PCDF from contaminated soils to food and fodder crop plants. Chemosphere 27(1–3):439–446

Hulster A, Muller J, Marschne H (1994) Soil-plant transfer of poly-chlorinated dibenzo-P-dioxins and dibenzofurans to vegetables of the cucumber family (Cucurbitaceae). Environ Sci Technol 28(6):1110–1115

Kao AS, Venkataraman C (1995) Estimating the contribution of re-entrainment to the atmospheric deposition of dioxin. Chemosphere 31(10):4317–4331

Karickhoff SW (1981) Semiempirical estimation of sorption of hydrophobic pollutants on natural sediments and soils. Chemosphere 10(8):833–846

Karpouzas DG, Singh BK (2006) Microbial degradation of organo-phosphorus xenobiotics: metabolic pathways and molecular basis. Adv Microb Physiol 51:119–125

Komives T, Gullner G (2005) Phase I xenobiotic metabolic systems in plants. Z-Naturforsch C 60 (3–4):179–185

Little P, Wiffen RD (1977) Emission and deposition of petrol engine exhaust Pb: I. Deposition of exhaust Pb to plant and soil surfaces. Atmos Environ 11:437–447

Mansuy D (2013) Metabolism of xenobiotics: beneficial and adverse effects. Biol Aujourdhui 1 (207):33–37

Mattina MI, Eitzer B, Simon T (2002) Tracking persistent organic pollutants (POPs) through biotic and abiotic processes in the environment: Final Report. United States Environmental Protection Agency (USEPA)

Mattina MI, White JC, Eitzer BD, Iannucci-Berge RW (2003) Uptake and translocation of air-borne and soil-bound persistent organic pollutants by Cucurbita pepo. In: 13th West coast conference contaminated soils, sediments and water, San Diego

McCrady JK (1994) Vapor-phase 2,3,7,8-TCDD sorption to plant foliage: a species comparison. Chemosphere 28(1):207–216

McCrady JK, Maggard SP (1993) Uptake and photo degradation of 2,3,7,8-tetrachlorodibenzo-P-dioxin sorbed to grass foliage. Environ Sci Technol 27(2):343–350

McFarlane JC (1995) Plant contamination: modelling and simulation of organic chemical processes. CRC Press, Boca Raton

McLachlan MS (1999) Framework for the interpretation of measurements of SOCs in plants. Environ Sci Technol 33:1799–1804

O'Connor GA, Kiehl D, Eiceman GA, Ryan JA (1990) Plant uptake of sludge-borne PCBs. J Environ Qual 19:113–118

Parameswaran A, Leitenmaier B, Yang M, Kroneck PM, Welte W, Lutz G, Papoyan A, Kochian LV, Kupper H (2007) A native Zn/Cd pumping P (1B) ATPase from natural overexpression in a hyperaccumulator plant. Biochem Biophys Res Commun 36:51–56

Parrish ZD, White JC, Isleyen M, Gent MP, Iannucci-Berger W, Eitzer BD, Kelsey JW, Mattina MI (2006) Accumulation of weathered polycyclic aromatic hydrocarbons (PAHs) by plant and earthworm species. Chemosphere 64(4):609–618

Pinder JE, McLeod KW, Fide RF, Shenod KC (1991) Mass loading of soil particles on a pasture grass. J Environ Radioactiv 13:341–354

Rikken MG, Lijzen JP, Cornelese AA (2001) Evaluation of model concepts on human exposure. RIVM, Bilthoven

Rylott EL, Bruce NC (2009) Plants disarm soil: engineering plants for the phytoremediation of explosives. Trends Biotechnol 27(2):73–81

Sandermann H (1992) Plant metabolism of xenobiotics. Trends Biochem Sci 17(2):82–84

Schnoor JL (1997) Phytoremediation. Pittsburgh: National Environmental Technology Applications Center. Technol Eval Rep TE-97-01, National Environmental Technology Applications Center and the University of Pittsburgh, Pennsylvania

Schreiber L, Schonherr J (1992) Uptake of organic-chemicals in conifer needles-surface adsorption and permeability of cuticles. Environ Sci Technol 26(1):153–159
Schröder P (2007) Exploiting plant metabolism for the phytoremediation of organic xenobiotics. In: Willey N (ed) Phytoremediation (methods and reviews). Humana, Totowa, pp 251–289
Schröder P, Collins CJ (2002) Conjugating enzymes involved in xenobiotic metabolism of organic xenobiotics in plants. J Phytorem 4(4):247–265
Schröder P, Collins CD (2011) Organic xenobiotics and plants: from mode of action to ecophysiology. Springer, Dordrecht
Schroll R, Scheunert I (1992) A laboratory system to determine separately the uptake of organic chemicals from soil by plant roots and by leaves after vaporization. Chemosphere 24:97–108
Schwitzguébel JP, Vanek T (2003) Fundamental advances in phytoremediation for xenobiotic chemicals. In: McCutcheon SC, Schnoor JL (eds) Phytoremediation: transformation and control of contaminants. Wiley, Hoboken, pp 123–157
Schwitzguébel JP, Page V, Martin-Dias S, Davies LC, Vasilyeva G, Strijakova E (2011) Using plants to remove forign compounds from contaminated water and soil. In: Collins P (ed) Organic xenobiotics and plants. Plant ecophysiology series. Springer, London
Seleiman MF, Santanen A, Jaakkola S, Ekholm P, Hartikainen H, Stoddard FL, Mäkelä PSA (2013) Biomass yield and quality of bioenergy crops grown with synthetic and organic fertilizers. Biomass Bioenerg 59:477–485
Shang TQ, Gordon MP (2002) Transformation of C^{14} Trichloro-ethylene poplar suspension cells. Chemosphere 47(9):957–962
Simonich SL, Hites RA (1995) Organic pollutant accumulation in vegetation. Environ Sci Techol 29(12):2905–2914
Smelt JH, Leistra MA (1974) Hexachlorobenzene in soils and crops after soil treatment with pentachloro-nitrobenzene. Agric Environ 1:65–71
Suresh B, Ravishankar GA (2004) Phytoremediation—a novel and promising approach for environmental clean-up. Crit Rev Biotechnol 24:97–124
Susarla S, Medina V, McCutcheon S (2002) Phytoremediation: an ecological solution to organic chemical contamination. Ecol Eng 18:647–658
Thorne M, Maul P, Robinson P (2004) The PRISM food chain modelling software: model structures for PRISM 2.0. Report QRS-1198A-1. Food Standards Agency, London
Topp E, Scheunert I, Attar A, Kort F (1986) Factors affecting the uptake of C14-labulled organic chemicals by plants from soil. Ecotoxicol Environ Saf 11:219–228
Trapp S, Matthies M (1995) Generic one compartment model for the uptake of organic chemicals by foliar vegetation. Environ Sci Technol 29:2333–2338
Trapp M, McFarlane C (1995) Plant contamination. Lewis, Boca Raton
Trapp S, Pussemeir L (1991) Model calculations and measurements of uptake and translocation of carbamates by bean plants. Chemosphere 22:327–345
Ugrekhelidze V, Phiriashvili V (2000) Uptake and transformation of some water phenolic pollutants by common duckweed (*Lemna minor* L.). Fresenius Environ Bull 9(7–8):483–488
USDA (2015) Metabolism of herbicides or xenobiotics in plants. United States Department of Agriculture (USDA), National Institute of Food and Agriculture (NIFA), University of California, Davis
USEPA (1996) Soil screening guidance: technical background document, EPA/540/R95/128. United States Environmental Protection Agency (USEPA), Washington, DC
Weyens N, van der Lelie AT, Smeets K, Taghavi S, Newman L (2009a) Bio-augmentation with engineered endophytic bacteria improves contaminant fate in phytoremediation. Environ Sci Technol 43(24):9413–9418
Weyens N, Taghavi S, Barac T, van der Lelie D, Boulet J, Artois T, Carleer R, Vangronsveld J (2009b) Bacteria associated with oak and ash on a TCE-contaminated site: characterization of isolates with potential to avoid evapotranspiration of TCE. Environ Sci Pollut Res 16:830–843
White JC (2002) Differential bioavailability of field-weathered p, p′-DDE to plants of the *Cucurbita* and *Cucumis* genera. Chemosphere 49(2):143–152

Wild SR, Berrow ML, Mcgrath S, Jones KC (1992) Polynuclear aromatic-hydrocarbons in crops from long-term field experiments amended with sewage-sludge. Environ Pollut 76(1):25–32
Wolt JD, Smith JK, Sims JK, Duebelbeis DO (1996) Products and kinetics of cloransulam-methyl aerobic soil metabolism. J Agric Food Chem 44:324–332

Chapter 19
Soil–PCB–PGPR Interactions in Changing Climate Scenarios

Saeed Ahmad Asad

19.1 Introduction

The last five decades have witnessed a speedy deterioration of the environment due to significant increase in industrial production of chemicals, and perhaps all sectors including agriculture, chemical industries, military invasions, mining and transportation have contributed one way or the other to pollute the environment (Graham and Ramsden 2008). The concentrations of anthropogenic toxic substances have exceeded the permissible limits due to excessive production of these compounds, and according to some estimates, it has been reported to be in billions of tons (Kvesitadze et al. 2004). A variety of contaminants exist in the environment, but those with high persistency in bioaccumulation and causing toxicity to living biota due to their occurrence in food chain classified as the most dangerous and persistent organic pollutants (POPs) are one of those. These organic compounds are resistant to chemical, biological and photolytic processes due to their persistence and pose serious threats to environment and living beings on the planet. Many POPs in the past were used as pesticide, pharmaceutical and industrial chemicals by the industrialized nations until they were banned in the late 1970s, but in the developing world, this practice is still going on (Ritter et al. 2007). Although some POPs arise after natural processes (e.g. volcanoes and biosynthetic pathways), most are man-made. Amongst synthetic POPs, polychlorinated biphenyls (PCBs) accumulate in various niches of biosphere and are of great concern to human health and cause of ecological imbalance (Graham and Ramsden 2008). PCBs have been reported to cause multiple toxic responses in living organisms (Fitzgerald et al. 2001). For example, Furukawa et al. (2004) reported reduced growth of algal plants

S.A. Asad (✉)
Center for Climate Research and Development (CCRD), COMSATS University, Park Road, Islamabad 45550, Pakistan
e-mail: saeed.asad@comsats.edu.pk

© Springer International Publishing AG 2017

M.Z. Hashmi et al. (eds.), *Xenobiotics in the Soil Environment*, Soil Biology 49,
DOI 10.1007/978-3-319-47744-2_19

when exposed to even very minute concentrations (0.3–10 ppm) of PCBs in aqueous solution.

Another study by Strek and Weber (1982) observed a significant reduction in fresh weight and height of soybean, beets and pigweed growing at PCB-contaminated site. Toxicity caused by PCBs has also been reported in human beings and other members of kingdom Animalia. Once entered in the human body through the skin and/or lungs, they are transported to the blood stream and accumulate in the adipose tissue (Safe et al. 1990). Although toxicity caused by PCBs depends on age and point of accumulation in the body, they have been reported to cause cancer, liver damages, and gall bladder (Borja et al. 2005; Tsai et al. 2007). All of 209 congeners of PCBs are linked to cause diseases in plants and animals one way or the other, but 36 (Table 19.1) are classified as the most dangerous depending on their toxicity, elevated accumulation in the animal tissue and prevalence in the environment (McFarland and Clarke 1989). Despite all these toxicities and diseases caused by these noxious compounds, no serious and sustainable strategies have been planned to remove/degrade these pollutants.

19.2 Role of Rhizosphere Soil in Plant Growth

Rhizosphere, a narrow zone of soil around plant roots, is a sink of root secretions and inhabits an enormous assembly of microbial communities. Plant roots recruit special bacteria from bulk soil in the root surroundings (Pieterse et al. 2014). According to careful analysis, one gramme of rhizosphere may contain up to 10^{11} microbial cells and >30,000 prokaryotic species (Mendes et al. 2011). The number and diversity of microbial communities in the rhizosphere is much larger than plant (Berendsen et al. 2012), and this narrow zone of soil is an important determinant of

Table 19.1 Priority groups of PCB congeners classified as the most dangerous

Group 1 (IUPAC no.)		Group 2 (IUPAC no.)	Group 3 (IUPAC no.)	Group 4 (IUPAC no.)
A	B	87	18	37
77	118	99	44	81
126	128	101	49	105
169	138	153	52	114
	156	180	70	119
	170	183	74	123
		194	151	157
			177	158
			187	167
			201	168
				189

Groups and numbers as assigned by the International Union of Pure and Applied Chemistry (IUPAC) (Source: Anyasi 2012)

the health and growth of plant. Root exudates take part in signalling to mediate the interactions between root-root-microbes present in the rhizosphere. Microbial communities in the rhizosphere are primarily involved to enhance plant growth, either by increasing resistance against diseases and/or augmenting plant nutrition. These microbial compositions are not only influenced by climatic changes but also by soil types and plant species growing in the same soil (Miethling et al. 2000). This is perhaps due to varying composition of root exudates and root cell components which further may vary with plant age, fertilizer applications and root zone. Soils also impact the rhizosphere microbes to the greater extent as soils having varying degrees of organic and inorganic contaminants, physico-chemical properties, pH and carbon availability; mineral nutrition will contain different microbial populations (Degens et al. 2000). Contradictory reports exist in the literature regarding the impact of soil type and plant species on the microbial communities in the rhizosphere. For example, many studies have demonstrated that soil has more impact on the rhizosphere microbes than plant species (Buyer et al. 1999 and the references therein), whilst studies conducted by Berendsen et al. (2012) showed greater impact by plant species than soil type. Haney et al. (2015) investigated the interactions between plant growth-promoting rhizobacteria (PGPR) strain, *Pseudomonas fluorescens* and model plant *Arabidopsis thaliana* and concluded that growth of *A. thaliana* largely depended on the performance of *P. fluorescens*. They further elaborated that rhizosphere changes mediated by these bacterial strains may impart larger effects on plant health. Another group of PGPR bacteria 'phosphate solubilizers' regulate a series of processes ranging from transformation to availability of tightly bound phosphorus, and these microbes are particularly effective to release the sparingly available soil P which along with enhancing the plant growth enriches the rhizosphere (Barea and Richardson 2014). Panke-Buisse et al. (2015) further investigated the role of rhizosphere on growth of development in brassica plants (*A. thaliana*) whereby specific microbiota profiles assembled near plant roots and delayed the flowering time of host plant as compared with control. Based on their observations, these researchers hypothesized that microbiomes may modify plant traits depending on microbial strains present in the rhizosphere at a particular growth stage of the plant.

19.3 Plant Growth-Promoting Rhizobacteria

Plant growth-promoting rhizobacteria (PGPR) are a group of bacteria living in the few millimetre radius of plant roots and are known to enhance the plant growth through a variety of direct and indirect mechanisms. These mechanisms are nitrogen fixation, phosphate solubilization, phytohormone production, resistance against diseases, siderophore production, promoting plant–microbe symbiotic relationships and systemic resistance induction. Since decades, PGPR have been used along with chemical fertilizers to promote plant growth, but more recently their role to remediate the contaminated land has been the topic of many researchers around the globe. The role of PGPR in remediating the heavy metals and metalloids (inorganic contaminants) from the soil is well researched and documented (Nadeem et al. 2014). PGPR are known to improve the metal uptake by plants through changing the metal bioavailability by production of phytohormones, releasing chelating agents and changing pH (Ma et al. 2011). These plant beneficial bacteria produce metabolites responsible to deter pathogenicity or siderophores which limit the Fe availability, a precursor for many pathogens to grow and multiply (Bhattacharyya and Jha 2012). Likewise, PGPR induce resistance in plants against diseases, a mechanism termed as systemic resistance. Along with many direct and indirect mechanisms by which PGPR facilitate plant growth, these microorganisms secrete key enzymes (e.g. ACC-deaminase, chitinase) and rhizobitoxine exopolysaccharides, the key metabolites helping plants to survive under stressful conditions. More recently, involvement of PGPR in the remediation of organic contaminants has been the subject of interest for many researchers, but limited to very basic research and that too under artificial settings which do not reflect their efficiency under natural environmental conditions.

19.4 Role of PGPR in Plant Development in Polluted Soil

PGPR are well researched to alleviate the plants against many stresses originating from the polluted rhizosphere, but certainly not all the PGPR strains are equally effective due to their inability to compete and survive in the polluted environment. Successful microbes enhance plant growth in polluted environment by a variety of mechanisms including induced systemic resistance, production of exopolysaccharides, etc. (Upadhyay et al. 2011). Major mechanism employed by PGPR to enhance growth under polluted environment is ethylene, a phytohormone to function at its lower concentrations (Glick et al. 2007). Polluted rhizosphere, regardless of the nature of pollutants, usually enhances the ethylene levels due to increased concentrations of 1-aminocyclopropane-1-carboxylic acid (ACC), a precursor of ethylene biosynthesis. Enhanced level of ACC is known to reduce the root development, thus affecting both nutritional and physiological functions of plant growth (Alarcon et al. 2012). Many PGPR in the rhizosphere contain ACC deaminase, an enzyme to degrade ACC into ammonia and α-ketobutyrate (Glick et al. 2007). This lowering of ACC results in reduced ethylene concentration and thus promotes root growth.

Pathogens are another problem for plants growing in polluted environments. PGPR induce resistance in plants against pathogens and diseases through various biocontrol mechanisms such as parasitism, competition, antibiosis, etc. (Beneduzi et al. 2012). These organisms also secrete antimicrobial compounds, induce systemic resistance and decrease plant vulnerability by increasing resistance against pathogen (Alizadeh et al. 2013).

PGPR are ubiquitous, as many plant species rely only on these microorganisms because of the lack of immediate effector-triggered immunity to respond to the plant pathogens (Weller 2007). Wang et al. (2015) isolated and characterized microbial strains from the rhizosphere of brassica (rape seed) and wheat plants growing on land contaminated with recalcitrant organic pollutants, hydrocarbon, salinity and heavy metals. The strains isolated and used in this study such as *Pseudomonas brassicacearum*, *P. protegens* and *P. chlororaphis* showed inhibitory effects against pathogens and exhibited growth-promoting effects under controlled as well as field conditions. Genetic studies of these PGPR morphotypes revealed that *phzF*, *acdS*, *prnD*, *pltC* and *phlD* genes were involved to produce cellulase, siderophores, pyoluteorin, pyrrolnitrin, IAA, protease, 2,4-diacetylphloroglucinol and phenazines. Interestingly all microbial population used in this study also proved to be resistant against organic and inorganic pollutants and salinity along with producing growth-promoting hormones.

19.5 Effect of Changing Climate on the Functioning of PGPR

Changing climate has significantly reduced the soil biodiversity and altered the ecosystems dynamics. Soil microbial composition is more sensitive to climate change events like rainfall patterns and CO_2, where some microbial populations were reported to be overgrown whilst others were reduced to extinction influenced by these climatic change indicators (Ng et al. 2015). Variations in precipitation (drought and wetting periods) have been reported to cause toxicity in soil microbial life by hindering the growth of plant growth promoters and stimulating the growth of some mycotoxigenic species particularly under water stress (Magan et al. 2011). Available information suggest that as a result of toxicity induced by moisture stressing conditions, many staple crops (e.g. cereals) could be infected by fungi from *Aspergillus* and *Fusarium* species contaminating the edible parts with mycotoxins like secondary metabolites which being carcinogenic (aflatoxins) may pose serious threats to animal and human health (Wagacha and Muthomi 2008). Plant growth-promoting microbial communities change drastically with varying moisture conditions. Drift in seasonal precipitation directly affects the microbial functional dynamics (Cregger et al. 2012). Forchetti et al. (2007) identified many subpopulations of endophytic bacteria colonizing sunflower under drought or over wetting conditions. Interestingly, Forchetti and co-workers reported only one stain to be able to survive under drought conditions which was more resistant and better able to adapt to varying moisture regimes. To the worst end, these extreme weather events may result in enhanced efflux of CO_2 disturbing the net ecosystem carbon balance and elevated leaching of nitrogen and impacting the C/N ratio (Gordon et al. 2008). Elevated CO_2 may enhance carbon allocation to root zone resulting in different C/N ratio and altered composition of root exudates and hence the microbial population (Haase et al. 2007). Enhanced supply of CO_2 and rainfall due to changing climate scenarios are also believed to regulate available soil C and N (Johnson et al. 2012) and fungal abundance in prairie grass which is most likely due to additional precipitation (Van Gestel et al. 1993).

19.6 Xenobiotics in Soils

The term 'xenobiotic' originates from the Greek words ξένος (xenos) meaning foreigner and βίος (bios) meaning life. Xenobiotic is most often used in the context of pollutants such as polychlorinated biphenyls (PCBs) which are extremely foreign substances in the biological systems and did not exist in nature before they were synthesized by human. PCBs are persistent organic pollutants (POPs) and being one of the toxic groups of pollutants require special attention to be eliminated from the environment as stated in the Stockholm convention for universal elimination. Toxicity and hazardousness of these organic compounds is attributed to their resistance against natural degradation, tendency to accumulate in the lipid and perhaps due to higher degree of halogenation (Younas et al. 2013). Legacy problem is that once entered in the soil; PCBs pave their ways into the whole environment through various biological, chemical and physical means (Abhilash et al. 2013). Although many of the POPs listed under Stockholm convention are already banned in many countries, however before ban release into the environment has serious repercussions on the natural and managed ecosystems and finally on the end user of products, originating from those contaminated habitats. Perugini et al. (2012) observed traces of dioxin-like PCBs in milk of livestock grazing on PCB-contaminated land. Similarly, Leat et al. (2011) observed these compounds in poultry products because of poultry feed originating from PCB-contaminated habitats. Despite the continued efforts by developed nations and significant success towards elimination from the environment, more joint efforts are needed at global scale for gauging the risk factor posed by these contaminants to all living beings including human along with framing suitable methodologies to eliminate these persistent toxicants (Abhilash and Yunus 2011).

19.7 Polychlorinated Biphenyls

Polychlorinated biphenyls (PCBs), persistent organic pollutants, are a group of man-made organic compounds consisting of 1–10 chlorine atoms adhered to two benzene rings (Yang et al. 2014). There are 209 congeners of PCBs with an empirical formulae of $C_{12}H_{10-n}Cl_n$ (n = 1–10; Fig. 19.1) (Larsson et al. 2000), out of which 130 orientations are used commercially. Soil is an important reservoir of PCBs, and their possible entry points into the human beings and food chain are through contaminated water, accidental contacts, landfills, combustion processes, uncontrolled waste dumping and exposure to contaminated atmosphere (Stohs 2014).

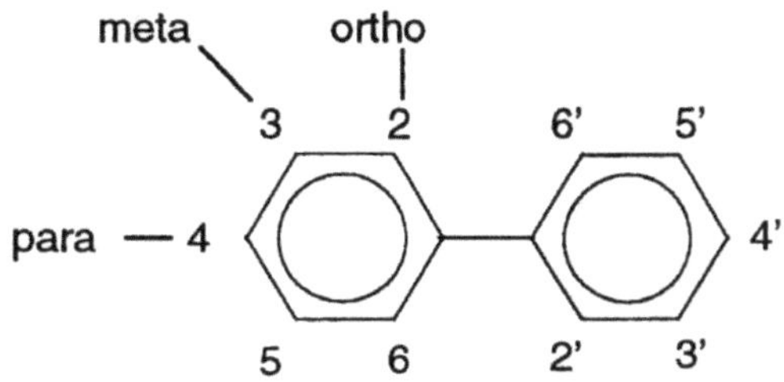

Fig. 19.1 Structural formulae of the PCBs with the spatial terms: orthor, meta and para (Barbalace 2003; Anyasi 2012)

PCBs were first time reported by Jensen in 1966 in a white-tailed sea eagle near Stockholm (Andersson 2000), but today these noxious compounds are found in nearly all ecological niches right from the sea bottom to the polar regions. The United States Environmental Protection Agency (USEPA) has placed these noxious compounds at no. 12 amongst the list of persistent organic pollutants (POPs).

19.8 Environmental Occurrences of PCBs

PCBs being xenobiotic do not exist naturally in the environment but are manufactured and released into the environment by human beings through industrial effluents, volatilization from soil and water bodies, incineration of wastes containing PCBs, landfills containing hazardous wastes, leakages from electrical equipment and uncontrolled disposal of spills (Bremle and Larsson 1998). Most of these organic toxicants find their ways into the environment during manufacturing, transportation and disposal points. Water contains generally higher concentrations of these organic pollutants due to unchecked disposal of PCBs containing effluents into water bodies; thus, contamination of water sources mostly results from environmental cycling (Anyasi and Atagana 2011). Ironically, sediments become the reservoirs of PCBs which continuously release these compounds into the water which accumulate in the sea animals and enter into the food chain (Borja et al. 2005). All PCBs used in the USA were manufactured and marketed by Monsanto, Illinois, USA (IARC 1978), amounting to 571,000 metric tons (Hansen 1999) until their manufacturing and distribution was banned in 1976 in the USA. Although new production of PCBs is banned according to Stockholm convention, current existence in the environment is because of bioaccumulation in the environment which continues from soil to water bodies, air, sediments, etc. (Eisenreich et al. 1999). Surprisingly, PCB concentration is greater in the higher trophic level as compared to lower level in the food chain (Willman et al. 1997) despite they are partitioned between different media and transformed via photolysis, metabolism and microbial activity.

19.9 Remediation of PCBs from Soil

Removal of PCBs from the environment has had been the challenging issue for scientists, as no single technology exists that can guarantee 100 % removal of these toxic compounds from the environment (Chary and Yates 2000). Different techniques for clean-ups of environment from PCBs include (1) thermal (incineration), (2) chemical (oxidation, vapour extraction), (3) mechanical (soil excavation) and (4) biological (use of plants and microbes). The use of biological means to clean up the contaminated environment is regarded as 'bioremediation' which is cost effective and environmentally friendly compared with chemical and thermal techniques and also ensures the complete transformation of pollutants rather than transferring from one phase to another. Bioremediation is not a new technology as evidenced by the compost piles which existed as far back as 6000 BC and the creation of the first biological sewage treatment plant in Sussex, UK. However, the word 'bioremediation' did not appear in peer-reviewed scientific literature until 1987 (Leung 2004). Bacteria transform the contaminants through reactions taking place as a part of their metabolic processes. For decades, PGPR were known to remove inorganic compounds from the rhizosphere, but credible evidences suggest their involvement in degradation of organic compounds primarily PCBs, in association with plants (Teng et al. 2015). These scientists observed that a PGPR bacterium *Mesorhizobium* sp. (ZY1 strain) isolated from the nodule of *Astragalus sinicus* L. remediated the 20.5 % of PCB compounds from the soil in 10 days after application as compared with control. They also observed that consortia of *A. sinicus* and *Mesorhizobium* removed >50 % of this organic pollutant from the soil in the same time frame which suggests that synergistic association between plant and PGPR may enhance the PCB degradation potential of rhizosphere microflora. Literature also suggests the role of PGPR in remediating the soil contaminated with hydrocarbons (Khan et al. 2013) where these bacteria secrete various enzymes capable to reduce the toxicity and evapotranspiration and enhance the degradation of volatile hydrocarbons (Li et al. 2012). But real challenge for remediation of PCBs is climate change affecting microbial biomass, functional diversity of microorganism, bioavailability of organic pollutants and microbe–plant signalling in the soil system.

19.9.1 Use of PGPR for Degradation of Organic Pollutants

PCBs are hydrophobic in nature and thus adhere to soil particles/sediments. The success of PCB remediation depends on the desorption ability of these compounds from soil and sediments (Gomes 2013). The concentration of these organic pollutants is usually low where in most cases is less than 1 % of the contaminated mass (Zhou et al. 2004). Research conducted so far indicates that traditional techniques (e.g. thermal desorption, land farming, solvent extraction, etc.) to clean up

Table 19.2 PGPR bacteria involved in remediation of organic pollutants

PGPR/rhizobacteria	Type of pollutant degraded	References
Sinorhizobium meliloti	Hydrocarbon. Produced auxins in inoculated plants	Golubev et al. (2011)
Rhizobacterium gordonia	Degradation of hydrocarbon. ACC deaminase and siderophore production	Hong et al. (2011)
Azospirillum brasilense	Hydrocarbon degradation and produced auxin	Muratova et al. (2010)
Rhizobium leguminosarum	Hydrocarbon degradation	Johnson et al. (2004)
Pseudomonas putida	Naphthalene	Kuiper et al. (2001)
Enterobacter sp.	Degradation of pyrene production of siderophore/ IAA/phosphate solubilization	Sheng et al. (2008)
Rhizobium meliloti	PCB remediation	Teng et al. (2010)
Pseudomonas putida Flav1-1	PCB degradation	Narasimhan et al. (2003)
Pseudomonas putida PML2	PCB degradation	Narasimhan et al. (2003)
Pseudomonas fluorescens F113	Metabolization of PCBs	Villacieros et al. (2005)
Azospirillum lipoferum strains 15	Crude oil degradation	Muratova et al. (2005)
Azospirillum brasilense	Polycyclic aromatic hydrocarbon (PAH) degradation	Huang et al. (2004)
Pseudomonas putida	PCB degradation	Nabavi et al. (2013)
Sinorhizobium meliloti	PCB (PCB 28), 2,4,4′-TCB degradation	Tu et al. (2011)
Pseudomonas sp.	PCB (PCB 77), degradation	Tao et al. (2012)

PCB-contaminated sites are too expensive. Hence, biological remediation which is low cost and sustainable technology comes to address this issue. Different PGPRs and the respective organic pollutants degraded by these bacterial strains are shown in (Table 19.2). From cost point of view, biological methods range between 1/10 and 1/2 of chemical and physical methods.

Degradation of PCBs via PGPR or other bacteria can broadly be categorized into (1) an aerobic reductive dechlorination—whereby PCBs are converted into less chlorinated compounds by accepting electrons (Vasilyeva and Strijakova 2007)—and (2) aerobic oxidation, which results in breakdown of biphenyl structure leading to complete mineralization (Pieper 2005). Reductive chlorination of commercial mixtures by microbes reduces the dioxin-like toxicity of PCB congeners to convert them into readily degradable moiety (Zhou et al. 2004). Major pathways involved in the biodegradation of PCBs are shown in Fig. 19.2. Selection or adaptation of degradation route depends on the temperature and pH which determine the

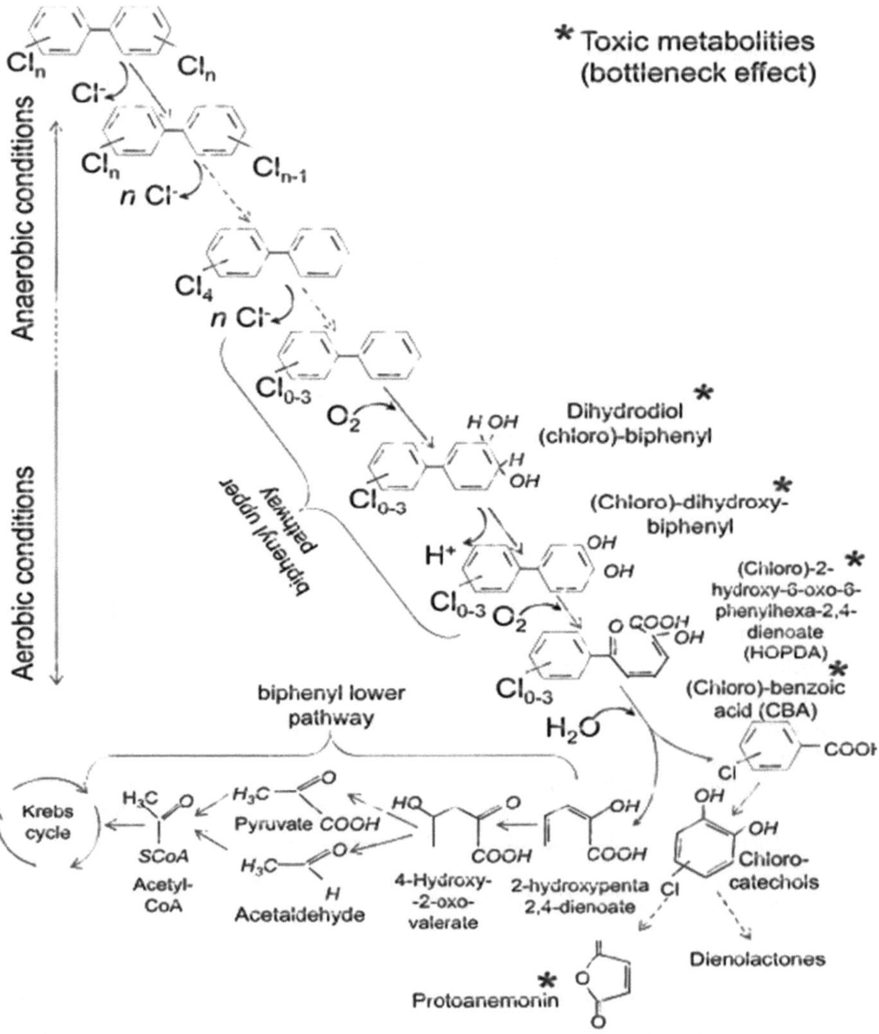

Fig. 19.2 Major pathways in the biodegradation of PCBs. *Asterisks* indicate the very toxic compounds having bottleneck effects in the degradation (Adapted from Passatore et al. 2014)

microbial strains and their degrading abilities at these thermal conditions (Wiegel and Wu 2000). Chemical structure of PCBs determines the rate of their degradation by microorganisms, as higher numbers of chlorine atoms increase their chemical stability, thus lowering their aqueous solubility. PCB congeners with more chlorine atoms have been found to be more resistant for microbial bioremediation (Furukawa and Fujihara 2008). Moreover, PCB congeners which are di-tetra substituted are degraded poorly due to lack of action point by respective enzyme, whilst congeners where chlorine atoms are located on one ring are easily degradable. Chlorination of PCBs depends on the concentration of pollutant. Ideally, the

PCB concentration between 50 and 800 mg kg^{-1} yields better results, because concentrations less than 50 mg is not enough to sustain microbial growth and concentration greater than upper limit may induce toxicity for bacterial strains (Bedard 2008). However, bioremediation either by using plants and/or microorganisms is slower than chemical and physical remediation because of their high tendency to bind with soil organic matter and thus rendering low bioavailability and also because of low expression of catabolic genes. Transformation of PCBs by PGPR unlike many other conversions is enhanced at reduced energy rather than demanding additional source of carbon to facilitate transformation. This is because of secretions of root exudates that serve as energy source for PGPR to enhance degradation of PCBs (Shann 2001). In double or multiprocess remediation, the consortia of PGPR and specific PCB-degrading bacterial strains resulted in successful remediation (Bhandary 2007). *Pseudomonas* bacteria, the prominent group of PGPR, have been reported to degrade the organic contaminants with the aid of several enzymes where the genes coding is often located on large plasmids weighing approximately 50–200 kb (Cork and Krueger 1991). This enzyme-assisted degradation of xenobiotics converts these compounds into either catechol or protocatechuate which can easily be metabolized by all soil biota. It is worth mentioning that halogenated compounds are also degradable by these enzymes but at a slower rate (Glick 2010).

Plants had always been the first choice to remediate the organic pollutants, but it was a relatively slow process (Zeeb et al. 2006). However, a consortia of plants and bacteria yield very promising results for the degradation of organic compounds (Kuiper et al. 2004). Therefore, addition of highly efficient microbial strains to degrade industrial chemicals from the soil is highly desirable. Rhizobium is the best example which is not only nitrogen fixer in legumes but a great selection to remove the organic pollutants, namely, PCBs, TNT and PAHs, from the soil very effectively (Yang and Lee 2008).

19.10 How Climatic Factors Are Affecting the Remediation of PCBs by PGPR

Human economic activity, chemical emissions and persistent organic pollutants including PCBs are likely to be altered by climate change. Under changing climate scenarios, the remediation of PCBs will be highly challenging because increasing temperature will enhance the volatility and dispersion of these pollutants over the long distances (Miraglia et al. 2009). Changing climatic conditions may influence the plant–rhizosphere–pollutant interactions in many ways. For example, the level of root exudates secreted into the rhizosphere significantly affects the microbial biomass, activities and structure of microbes inhabiting the rhizosphere (Walker et al. 2003). Increased root exudation is known to reduce the soil pH and increase organic matter which increases the bioavailability and degradation of PCBs by microorganisms (Kim and Kang 2010). Increasing CO_2 levels may have fertilization effect on plants affecting plant growth which would ultimately alter the root exudates (Kuzyakov et al. 2007), thus changing the rhizospheric microbial community accordingly. Climate change triggered elevated concentration of CO_2 will increase plant biomass (Abhilash et al. 2015) which will enhance root exudates and microbial activity in the rhizosphere (Abhilash and Dubey 2014).

Elevated atmospheric temperature may also affect the behaviour of PCBs in soil causing a challenge for their bioremediation by PGPR. Temperature has been reported to drastically affect the microbial degradation and remediation of PCBs as it may hinder the microbial communities involved and also influence the bioavailability of PCBs by creating imbalance between the ratio of PCBs that are dissolved and the ones adsorbed on organic matter which affects their bioavailability in the environment. Climatic changes may alter degradation/remediation of PCBs by altering the physico-chemical properties of soil, i.e. nutrients, organic matter, particle size and texture. Changing climatic scenarios are also known to greatly influence the soil biochemistry and microbial composition and thus influence the bioremediation of these contaminants (Wu et al. 2009). All these alterations may reduce the bioavailability of pollutants, and ultimately remediation will be very passive and in extreme case may be stopped. Similarly, aerobic respiration of microbes may be hindered due to limited availability of oxygen and thus will result in reduced biotransformation of pollutants (Walton et al. 1994). Nutrient imbalance is another repercussion of seasonal and long-term changes in climate change indicator which will result in extinction of native species of bacteria and introduction of aliens which may not be able to remove these organic pollutants as effectively as native ones. Thus, end result would be partial degradation of contaminant whilst leaving the resistant compounds in the soil. Moreover, climate change may influence the leaching and volatilization of PCBs (Miraglia et al. 2009), but mechanisms of how climate change will affect the microbe–pollutant interaction require further investigation.

19.11 Strategies to Improve PGPR-Assisted Remediation of PCBs in Changing Climate

Legacy pollutants such as PCBs persist and bioaccumulate in the environment, and global climate change may strongly influence their mobility to the remote ecological niches and thus increase the human and ecosystem exposures. PCBs are primarily based in static compartments (soil/sediments) migrating from one ecosystem to distant locations via mobile sources such as water, air, etc. It is well established that stronger winds, river flow and lake and ocean currents influenced by climate change affect the global migration patterns of legacy compounds. Similarly, transformation and hydrolysis of PCBs are greatly influenced by climate change and so will be remediation of these compounds (Balbus et al. 2013).

Bioremediation of PCBs can be enhanced by adopting various strategies. For example, Mohn and Stewart (2000) suggested that nutrient additions (phosphorus and nitrogen) to PCB-contaminated soils may enhance the biodegradation of these compounds. Care should be taken that concentration of nutrients should not reach toxicity level. Optimum concentrations as recommended by USEPA (1995) are C/N ratio of 10:1–100:1 for enhanced removal of hydrocarbons, whilst C/N/P ratio of 120:10:1 are optical concentrations for PCB removal from the soil (Braddock et al. 1997). However, concentration above these limits has inhibitory effects on soil microbes. Temperature is another critical factor to enhance bioremediation in changing climate, and soil heating (15–20 °C) might be considered to speed up the bioremediation (Mohn and Stewart 2000). Similarly, composting could also be used for this purpose since microbes in compost increase the temperature which enhances the bioremediation of PCBs. Bioremediation of these organic pollutants can be accelerated by addition of nonindigenous PGPR which after acclimatization with the environment have proved to be very effective to remediate these compounds. This practice has yielded satisfactory results in cold climate like Antarctic. Removal of other barriers like oxygen deficiency, heavy metals and low water content of soil may accelerate the process of remediation (Beyer et al. 2000).

19.12 Conclusions and Future Recommendations

Bacterial role to remediate the contaminated land with both inorganic (heavy metals) and organic (POPs) contaminants is well established under controlled settings. For the removal of organic compounds such as PCBs, pilot studies under field conditions have proved to be effective but not in diverse environmental and climatic conditions, and research is progressing for the widespread dispersal of this technology despite technical difficulties which are fixable. Plant growth-promoting rhizobacteria (PGPR) along with growth-promoting attributes are ideal organisms to remediate the PCB-contaminated environment for two reasons. (1) They alleviate the plant biota growing on contaminated land against stress imposed by contaminants and increase their biomass/efficiency to take up and store more and more contaminants into their body, and (2) they are very diverse group of microorganisms and capable to survive and multiply at extreme weather and climatic conditions. Before making recommendations to use PGPR as possible remediators of PCB-polluted site, regional trials at different climatic regimes (temperature, CO_2 and moisture level) are highly recommended for future experiments Thorough understanding of PGPR behaviours (survival and functions) in different climate extremes may help to decide the possible combinations of microbe–microbe or microbe–plant to devise effective strategy for removal/degradation of xenobiotics from the environment. Although bioremediation is an effective technology to remediate the PCB-contaminated environment, exploration of other technologies such as rhizoremediation involving the consortia of plants and bacteria, natural attenuation and bioaugmentation may help to address the problem. Moreover, study of molecular interactions between microbes and plants and communication mechanisms may help to achieve the target of contaminant elimination from the environment.

References

Abhilash PC, Dubey RK (2014) Integrating aboveground–belowground responses to climate change. Curr Sci 106:1637–1638

Abhilash PC, Yunus M (2011) Can we use biomass produced from phytoremediation? Biomass Bioenergy 35:1371–1372

Abhilash PC, Dubey RK, Tripathi V, Srivastava P, Verma JP, Singh HB (2013) Adaptive soil management. Curr Sci 104:2–3

Abhilash PC, Tripathi V, Dubey RK, Edrisi SA (2015) Coping with changes: adaptation of trees in a changing environment. Trends Plant Sci 20:137–138

Alarcon MV, Lloret PG, Iglesias DJ, Talon M, Salguero J (2012) Comparison of growth responses to auxin 1-naphthaleneacetic acid and the ethylene precursor 1-aminocyclopropane-1-carboxilic acid in maize seedling root. Acta Biol Cracov Ser Bot 54:16–23

Alizadeh H, Behboudi K, Ahmadzadeh M, Javan-Nikkhah M, Zamioudis C, Pieterse CM, Bakker PA (2013) Induced systemic resistance in cucumber and *Arabidopsis thaliana* by the combination of *Trichoderma harzianum* Tr6 and *Pseudomonas* sp. Ps14. Biol Control 65:14–23

Andersson P (2000) Physico-chemical characteristics and quantitative structure-activity relationships of PCBs. Dissertation, Umea University, Sweden

Anyasi RO (2012) Bioremediation of polychlorinated biphenyls (PCBs)-contaminated soil by phytoremediation with *chromolaena odorata* L. Dissertation, University of South Africa

Anyasi RO, Atagana HI (2011) Biological remediation of polychlorinated biphenyl (PCB) in the environment by microorganisms and plants. Afr J Biotechnol 10:18916–18939

Balbus JM, Boxall AB, Fenske RA (2013) Implications of global climate change for the assessment and management of human health risks of chemicals in the natural environment. Environ Toxicol Chem 32:62–78

Barbalace RC (2003) The chemistry of polychlorinated biphenyls. http://EnvironmentalChemistry.com. Accessed 21 Sept 2013

Barea JM, Richardson AE (2014) Phosphate mobilization by soil microorganisms. In: Lutenberg B (ed) Principles of plant-microbe interactions. Springer, Berlin, pp 225–234

Bedard DL (2008) A case study for microbial biodegradation: anaerobic bacterial reductive dechlorination of polychlorinated biphenyls-from sediment to defined medium. Annu Rev Microbiol 62:253–270

Beneduzi A, Ambrosini A, Passaglia LMP (2012) Plant growth-promoting rhizobacteria (PGPR): their potential as antagonists and biocontrol agents. Genet Mol Biol 35:1044–1051

Berendsen RL, Pieterse CM, Bakker PA (2012) The rhizosphere microbiome and plant health. Trends Plant Sci 17:478–486

Beyer L, Bölter M, Seppelt RD (2000) Nutrient and thermal regime microbial biomass, vegetation of Antarctic soils in the Windmill Islands region of East Antarctica (Wilkes Land). Arct Antarct Alp Res 32:30–39

Bhandary A (2007) Remediation technologies for soil and groundwater. US Environmental Council Science, Reston

Bhattacharyya PN, Jha DK (2012) Plant growth-promoting rhizobacteria (PGPR): emergence in agriculture. World J Microbiol Biotechnol 28:1327–1350

Borja J, Taleon DM, Auresenia J, Gallardo S (2005) Polychlorinated biphenyls and their biodegradation. Process Biochem 40:1999–2013

Braddock JF, Ruth ML, Catterall PH, Walworth JL, McCarthy KA (1997) Enhancement and inhibition of microbial activity in hydrocarbon-contaminated Arctic soils: implications for nutrient-amended bioremediation. Environ Sci Technol 31:2078–2084

Bremle G, Larsson P (1998) PCB in the air during land filling of contaminated lake sediment. Atmos Environ 32:1011–1019

Buyer JS, Roberts DP, Russek-Cohen E (1999) Microbial community structure and function in the spermosphere as affected by soil and seed type. Can J Microbiol 45:138–144

Chary LK, Yates L (2000) Technologies for destruction of PCBs: feasible alternatives to incineration and landfilling www.ipen.org/technolo. Accessed 19 June 2011

Cork DJ, Krueger JP (1991) Microbial transformation of herbicides and pesticides. Adv Appl Microbiol 36:1–66

Cregger MA, Schadt CW, McDowell NG, Pockman WT, Classen AT (2012) Response of the soil microbial community to changes in precipitation in a semiarid ecosystem. Appl Environ Microbiol 78:8587–8594

Degens BP, Schipper LA, Sparling GP, Vojvodic VM (2000) Decrease in organic C reserves in soils can reduce the catabolic diversity of soil microbial communities. Soil Biol Biochem 32:189–196

Eisenreich SJ, Jeremiason JD, Paterson MJ (1999) Biochemical cycling of PCBs in lakes of variable tropical status: a paired-lake experiment. Limnol Oceanogr 44:889–902

Fitzgerald ED, Hwang SA, Deres DA, Bush B, Cook K, Worswick P (2001) The association between local fish consumption and DDE, mirex, and HCB concentration in the breast milk of Mohawk women at Akwesasne. J Expos Anal Environ Epid 11:381–388

Forchetti G, Masciarelli O, Alemano S, Alvarez D, Abdala G (2007) Endophytic bacteria in sunflower (*Helianthus annuus* L.): isolation, characterization, and production of jasmonates and abscisic acid in culture medium. Appl Microbiol Biotechnol 76:1145–1152

Furukawa K, Fujihara H (2008) Microbial degradation of polychlorinated biphenyls: biochemical and molecular features. J Biosci Bioeng 105:433–449

Furukawa K, Suenaga H, Goto M (2004) Biphenyl dioxygenases: functional versatilities and directed evolution. J Bacteriolog 186:5189–5196

Glick BR (2010) Using soil bacterial to facilitate phytoremediation. Biotechnol Adv 28:367–374

Glick BR, Cheng Z, Czarny J, Duan J (2007) Promotion of plant growth by ACC deaminase-producing soil bacteria. Eur J Plant Pathol 119:329–339

Golubev SN, Muratova AY, Wittenmayer L, Bondarenkova AD, Hirche F, Matora LY, Merbach W, Turkovskaya OV (2011) Rhizosphere indole-3-acetic acid as a mediator in the Sorghum bicolor–phenanthrene–*Sinorhizobium meliloti* interactions. Plant Physiol Biochem 49:600–608

Gomes HI (2013) Overview of *in situ* and *ex situ* remediation technologies for PCB-contaminated soils and sediments and obstacles for full-scale application. Sci Total Environ 445:237–260

Gordon H, Haygarth PM, Bardgett RD (2008) Drying and rewetting effects on soil microbial community composition and nutrient leaching. Soil Biol Biochem 40:302–311

Graham C, Ramsden JJ (2008) Introduction to global warming. In: Proceeding of the NATO advanced research workshop on complexity and security Tbilisi, Georgia. IOS, Amsterdam, pp 147–184

Haase S, Neumann G, Kania A, Kuzyakov Y, Romheld V, Kandeler E (2007) Elevation of atmospheric CO_2 and N-nutritional status modify nodulation, nodule-carbon supply, and root exudation of *Phaseolus vulgaris* L. Soil Biol Biochem 39:2208–2221

Haney CA, Samuel BS, Bush J, Ausubel FM (2015) Associations with rhizosphere bacteria can confer an adaptive advantage to plants. Nat Plants. doi:10.1038/nplants.2015.51

Hansen LG (1999) The ortho side of PCBs: occurrence and disposition. Kluwer Academic, Norwell

Hong SH, Ryu H, Kim J, Cho KS (2011) Rhizoremediation of diesel contaminated soil using the plant growth-promoting *Rhizobacterium gordonia* sp. S2RP-17. Biodegradation 22:593–601

Huang XD, El-Alawi Y, Penrose DM, Glick BR, Greenberg BM (2004) A multi-process phytoremediation system for removal of polycyclic aromatic hydrocarbons from contaminated soils. Environ Pollut 130:465–476

IARC (1978) IARC monographs on the evaluation of the carcinogenic risk of chemicals to humans. Lyon. http://monographs.iarc.fr/ENG/Monographs/vol1-42/

Jensen S (1966) Report of new chemical hazards. New Sci 32:612

Johnson DL, Maguire KL, Anderson DR, McGrath SP (2004) Enhanced dissipation of chrysene in planted soil: the impact of a rhizobial inoculum. Soil Biol Biochem 36:33–38

Johnson SL, Kuske CR, Carney TD, Housman DC, Gallegos-Graves LV, Belnap J (2012) Increased temperature and altered summer precipitation have differential effects on biological soil crusts in a dry land ecosystem. Global Change Biol 18:2583–2593

Khan S, Afzal M, Iqbal S, Khan QM (2013) Plant–bacteria partnerships for the remediation of hydrocarbon contaminated soils. Chemosphere 90:1317–1332

Kim S, Kang H (2010) Effects of elevated CO_2 and Pb on phytoextraction and enzyme activity. Water Air Soil Pollut 219:365–375

Kuiper I, Bloemberg GV, Lutenberg BJ (2001) Selection of a plant–bacterium pair as a novel tool for rhizostimulation of polycyclic aromatic hydrocarbon degrading bacteria. Mol Plant Microbe Interact 14:1197–1205

Kuiper I, Lagendijk EL, Bloemberg GV, Lugtenberg BJ (2004) Rhizoremediation: a beneficial plant-microbe interaction. Mol Plant Microbe 17:6–15

Kuzyakov Y, Hill PW, Jones DL (2007) Root exudates components change litter decomposition in a simulated rhizosphere depending on temperature. Plant Soil 290:293–305

Kvesitadze G, Khatisashvili G, Sadunishvili T (2004) Mechanisms to detoxify selected organic contaminants in higher plants and microbes, and their potential use in landscape management. European Research Office, London, pp. 5–6

Larsson P, Andersson A, Broman D, Nordback J, Lundberg E (2000) Persistent organic pollutants (POPs) in pelagic systems. AMBIO 29:202–209

Leat EHK, Bourgeon S, Borga K, Strom H, Hanssen SA, Gabrielsen GW, Petersen E, Olafsdottir K, Magnusdottir E, Fisk AT, Ellis S (2011) Effects of environmental exposure and diet on levels of persistent organic pollutants (POPs) in eggs of a top predator in the North Atlantic in 1980 and 2008. Environ Pollut 159:1222–1228

Leung M (2004) Bioremediation: techniques for cleaning up. Emess J Biotechnol 2:18–22

Li H-Y, Wei D-Q, Shen M, Zhou ZP (2012) Endophytes and their role in phytoremediation. Fungal Divers 54:11–18

Ma Y, Prasad MN, Rajkumar M, Freitas H (2011) Plant growth promoting rhizobacteria and endophytes accelerate phytoremediation of metalliferous soils. Biotechnol Adv 29:248–258

Magan NA, Medina A, Aldred D (2011) Possible climate-change effects on mycotoxins contamination of food crops pre- and postharvest. Plant Pathol 60:150–163

McFarland VA, Clarke JU (1989) Environmental occurrence, abundance and potential toxicity of polychlorinated biphenyl congeners: consideration for a congener-specific analysis. Environ Health Perspect 81:225–239

Mendes R, Kruijt M, Bruijn ID, Dekkers E, van der Voort M, Schneider JH, Piceno YM, DeSantis TZ, Andersen GL, Bakker PA, Raaijmakers JM (2011) Deciphering the rhizosphere microbiome for disease-suppressive bacteria. Science 332:1097–1100

Miethling R, Wieland G, Backhaus H, Tebbe CC (2000) Variation of microbial rhizosphere communities in response to crop species, soil origin, and inoculation with *Sinorhizobium meliloti* L 33. Microb Ecol 40:43–56

Miraglia M, Marvin HJ, Kleter GA, Battilani PA, Brera C, Coni E, Cubadda F, Croci L, De Santis B, Dekkers S, Filippi (2009) Climate change and food safety: an emerging issue with special focus on Europe. Food Chem Toxicol 47: 1009–1021

Mohn WW, Stewart GR (2000) Limiting factors for hydrocarbon biodegradation at low temperatures in Arctic soils. Soil Biol Biochem 32:1161–1172

Muratova AL, Turkovskaia OV, Antoniuk LP, Makarov OE, Pozdnyakova LI, Ignatov VV (2005) Oil-oxidizing potential of associative rhizobacteria of the genus *Azospirillum*. Microbiologiia 74:210–215

Muratova AY, Bondarenkova AD, Panchenko LV, Turkovskaya OV (2010) Use of integrated phytoremediation for cleaning-up of oil-sludge-contaminated soil. Appl Biochem Microbiol 46:789–794

Nabavi BF, Nikaeen M, Amin MM, Hatamzadeh M (2013) Isolation and identification of aerobic polychlorinated biphenyls degrading bacteria. Int J Environ Health Eng 2:47

Nadeem SM, Ahmad M, Zahir ZA, Javaid A, Ashraf M (2014) The role of mycorrhizae and plant growth promoting rhizobacteria (PGPR) in improving crop productivity under stressful environments. Biotechnol Adv 32:429–448

Narasimhan K, Basheer C, Bajic VB, Swarup S (2003) Enhancement of plant-microbe interactions using a rhizosphere metabolomics-driven approach and its application in the removal of polychlorinated biphenyls. Plant Physiol 132:146–153

Ng E-L, Patti AF, Rose MT, Schefe CR, Smernik RJ, Cavagnaro TR (2015) Do organic inputs alter resistance and resilience of soil microbial community to drying? Soil Biol Biochem 81:58–66

Panke- Buisse K, Poole AC, Goodrich JK, Ley RE, Kao-Kniffin J (2015) Selection on soil microbiomes reveals reproducible impacts on plant function. ISME J 9:980–989

Passatore L, Rossetti S, Juwarkar AA, Massacci A (2014) Phytoremediation and bioremediation of polychlorinated biphenyls (PCBs): state of knowledge and research perspectives. J Hazard Mater 278:189–202

Perugini M, Nuñez EGH, Baldi L, Esposito M, Serpe FP, Amorena M (2012) Predicting dioxin-like PCBs soil contamination levels using milk of grazing animal as indicator. Chemosphere 89:964–969

Pieper DH (2005) Aerobic degradation of polychlorinated biphenyls. Appl Microbiol Biotechnol 67:170–191

Pieterse CMJ, Zamioudis C, Berendsen RL, Weller DM, Van Wees SC, Bakker PA (2014) Induced systemic resistance by beneficial microbes. Annu Rev Microbiol 52:347–375

Ritter L, Solomon KR, Forget J, Stemeroff M, O'Leary C (2007) Persistent organic pollutants. United Nations Environment Programme. Accessed 01 July 2016

Safe S, Yao C, Davis D (1990) Development of toxic equivalency factors for polychlorinated biphenyls (PCBs). Organohalogen Compd 2:55–59

Shann JR (2001) Bioremediation through rhizosphere technology coats, vol 563. American Chemical Society, Washington, DC

Sheng X, Chen X, He L (2008) Characteristics of an endophytic pyrene-degrading bacterium of *Enterobacter* sp. 12J1 from Allium macrostemon Bunge. Int Biodeter 62:88–95

Stohs SJ (2014) Polychlorinated biphenyls (PCBs). In: Wexler P (ed) Encyclopaedia of toxicology, 3rd edn. Academic, Oxford, pp 1035–1037

Strek HJ, Weber JB (1982) Adsorption and reduction in bioactivity of polychlorinated biphenyl (Aroclor 1254) to red root Pigweed by soil organic matter and montmorillonite clay. Soil Sci Soc Am J 46:318–322

Tao C, Min-min Y, Xiong-yuan SI, You-bin SI (2012) Isolation, identification and characteristics of PCB77-degrading strain. Environment Science Technology, School of Resource and Environment, Anhui Agricultural University, Hefei

Teng Y, Luo Y, Sun X, Tu C, Xu L, Liu W, Li Z, Christie P (2010) Influence of arbuscular mycorrhiza and *Rhizobium* on phytoremediation by Alfalfa of an agricultural soil contaminated with weathered PCBs: a field study. Intl J Phytorem 12:516–533

Teng Y, Li X, Chen T, Zhang M, Wang X, Li Z, Luo Y (2015) Isolation of the PCB-degrading bacteria *Mesorhizobium* sp. ZY1 and its combined remediation with *Astragalus sinicus* L. for contaminated soil. Intl J Phytorem 18(2):141–149. doi:10.1080/15226514.2015.1073667

Tsai PC, Ko YC, Huang W, Liu HS, Guo YL (2007) Increased liver and lupus mortalities in 24-year follow-up of the Taiwanese people highly exposed to PCBs and PDDF. Sci Total Environ 247:216–232

Tu C, Teng Y, Luo Y, Li X, Sun X, Li Z, Liu W, Christie P (2011) Potential for biodegradation of polychlorinated biphenyls (PCBs) by *Sinorhizobium meliloti*. J Hazard Mater 186:1438–1444

Upadhyay SK, Singh JS, Singh DP (2011) Exopolysaccharide-producing plant growth- promoting rhizobacteria under salinity condition. Pedosphere 21:214–222

USEPA (1995) How to evaluate alternative cleanup technologies for underground storage tank sites: a guide for corrective action plan reviewers. U.S. Government Printing Office, Washington, DC, EPA-510-B-95-007 http://www2.epa.gov/ust

Van Gestel M, Merckx R, Vlassak K (1993) Microbial biomass responses to soil drying and rewetting: the fate of fast- and slow-growing microorganisms in soils from different climates. Soil Biol Biochem 25:109–123

Vasilyeva G, Strijakova E (2007) Bioremediation of soils and sediments contaminated by polychlorinated biphenyls. Microbiology 76:639–653

Villacieros M, Whelan C, Mackova M, Molgaard J, Sánchez-Contreras M, Lloret J, de Cárcer DA, Oruezábal RI, Bolanos L, Macek T, Karlson U (2005) Polychlorinated biphenyl rhizoremediation by *Pseudomonas fluorescens* F113 derivatives, using a *Sinorhizobium meliloti nod* system to drive *bph* gene expression. Appl Environ Microbiol 71:2687–2794

Wagacha JM, Muthomi JW (2008) Mycotoxins in Africa: current status, implications to food safety and health and possible management strategies. Int J Food Microbiol 124:187–201

Walker TS, Bais HP, Grotewold E, Vivanco JM (2003) Root exudation and rhizosphere biology. Plant Physiol 132:44–51

Walton BT, Guthrie EA, Hoylman AM (1994) Toxicant degradation in the rhizosphere. In: Anderson TA, Coats JR (eds) Bioremediation through rhizosphere technology, vol 563. ACS Series, Washington, DC, pp 11–26

Wang X, Mavrodi DV, Linfeng KE, Mavrodi OV, Yang M, Thomashow LS, Zheng N, Weller DM, Zhang J (2015) Biocontrol and plant growth-promoting activity of rhizobacteria from Chinese fields with contaminated soils. Microbial Biotechnol 8:404–418

Weller DM (2007) *Pseudomonas* biocontrol agents of soil borne pathogens: looking back over 30 years. Phytopathology 97:250–256

Wiegel J, Wu Q (2000) Microbial reductive dehalogenation of polychlorinated biphenyls. FEMS Microbiol Ecol 32:1–15

Willman EJ, Manchester-Neesvig JB, Armstrong DE (1997) Influence of ortho-substitution on pattern of PCB accumulation in sediments; plankton and fish in fresh water estuary. Environ Sci Technol 31:3712–3718

Wu H, Tang S, Zhang X, Guo J, Song Z, Tian S, Smith DL (2009) Using elevated CO_2 to increase the biomass of a *Sorghum vulgare* var. sudanense hybrid and *Trifolium pratense* L. and to trigger hyperaccumulation of cesium. J Hazard Mater 170:861–870

Yang CF, Lee CM (2008) Enrichment, isolation, and characterization of 4-chlorophenol-degrading bacterium *Rhizobium* sp 4-CP-20. Biodegradation 19:329–336

Yang GX, Zhuang HS, Chen HY, Ping XY, Bu D (2014) A sensitive immunosorbent bio-barcode assay based on real-time immuno-PCR for detecting 3, 4, 3′, 4′-tetrachlorobiphenyl. Anal Bioanal Chem 406:1693–1700

Younas A, Hilber I, Rehman S, Khawaja M, Bucheli TD (2013) DDT factory in Pakistan revisited for remediation: severe DDT concentrations in soils and plants from within the area. Environ Sci Pollut Res 20:1966–1976. doi:10.1007/s11356-012-1317-y

Zeeb BA, Amphlett JS, Rutter A, Reimer KJ (2006) Potential for phytoremediation of polychlorinated biphenyl-(PCB-) contaminated soil. Int J Phytorem 8:199–221

Zhou W, Anitescu G, Rice PA, Tavlarides LL (2004) Supercritical fluid extraction–oxidation technology to remediate PCB-contaminated soils/sediments: an economic analysis. Environ Prog 23:222–231

Chapter 20
Impact of Biochar on Soil Fertility and Behaviour of Xenobiotics in Soil

Hossain M Anawar, Vladimir Strezov, Farjana Akter, Md Abdul Kader, and Zakaria M. Solaiman

20.1 Introduction

Application of biochar to soil can increase carbon (C) sequestration (Lehmann et al. 2011), improve soil health and fertility, reduce soil greenhouse gas (GHGs) emissions (i.e. N_2O) (Ippolito et al. 2012; Spokas et al. 2009; Kookana et al. 2011) and mitigate climate change impacts (Kookana 2010; Cabrera et al. 2014; Mandal et al. 2015). In addition, biochar increases soil organic matter, nutrient content and availability, pH and soil water retention and aggregation (Piccolo et al. 1996; Glaser et al. 2002), but lowers soil bulk density (Laird 2008; Sohi et al. 2010), erosion potential and leaching of pesticides and nutrients to surface and groundwater (Laird 2008). Biochar application has demonstrated improvement in quality for the soil physical, chemical and biological properties (Glaser et al. 2002; Lehmann and Rondon 2006) which results in favourable impacts on plant growth and development, such as enhanced seed germination, crop yield, shoot height and biomass production of crop plants and trees (Kishimoto and Sugiura 1985;

H.M. Anawar • V. Strezov (✉)
Department of Environmental Sciences, Faculty of Science and Engineering, Macquarie University, 2109 Sydney, NSW, Australia
e-mail: vladimir.strezov@mq.edu.au

F. Akter
Sher-E-Bangla Agricultural University, Sher-E-Bangla Nagar, Dhaka 1207, Bangladesh

M.A. Kader
Department of Soil Science, Bangladesh Agricultural University, Mymensingh 2202, Bangladesh

Z.M. Solaiman
Soil Biology and Molecular Ecology Group, School of Earth and Environment (M087), and The UWA Institute of Agriculture, The University of Western Australia, Crawley, WA 6009, Australia

© Springer International Publishing AG 2017
M.Z. Hashmi et al. (eds.), *Xenobiotics in the Soil Environment*, Soil Biology 49,
DOI 10.1007/978-3-319-47744-2_20

Chidumayo 1994; Mbagwu and Piccolo 1997; Joseph et al. 2015a, b). Due to extreme affinity of biochars for essential plant nutrients, it can provide a slow release mechanism (Sanchez et al. 2009).

In the last decades, high production of agricultural crops by modern agricultural practices led to an intensive use of pesticides. Thus, different xenobiotic compounds, such as insecticides, fungicides, herbicides, chlorinated derivatives, polycyclic aromatic hydrocarbons (PAHs) and others, are introduced inadvertently or deliberately into soil and water ecosystems (Magan et al. 2010). These contaminants can have medium- to long-term stability and persistence in soils causing major environmental concerns and significant toxic impacts on the soil ecosystem and potentially impact human health and environmental quality through impacts on soil, surface water and groundwater resources (Ashman and Puri 2002; Han et al. 2004). An important fraction of pesticides applied to crops enters directly through soil surfaces without interacting with target species, but affecting other organisms or being adsorbed to soil particles (Kohler and Triebskorn 2013). These xenobiotic compounds can develop different secondary effects and adversely affect biodiversity or activity of soil microorganisms (Perucci et al. 2000), biological soil function and soil quality.

In order to mitigate and remediate the soils contaminated by xenobiotic compounds, the use of biochar has been proposed as a potential, inexpensive and natural amendment (Anawar et al. 2015). The greater surface area, high microporosity and chemical functional groups in the biochar contribute to its high sorption capacity (Chun et al. 2004), which enables reduction of the mobility and bioavailability of xenobiotic compounds and other organic contaminants in soil, stabilize soil organic matter levels and block toxic pollutants in the soil (Chen and Yuan 2011). Biochar can be used for remediation of agrochemicals (fungicides, herbicides, insecticides and growth regulators) or pesticide-contaminated soils because it contributes to reduction of the availability of the xenobiotics present in soils or sediments to the organisms and limits the transport into the receiving environments (Burgess et al. 2009). The mechanism of biochar interaction with xenobiotics in soils requires assembly of knowledge and research in this field to improve understanding, research and development in the area. Therefore, the aims of this book chapter are to (1) examine biochar production properties and their effects on soil fertility, physical, chemical and biological properties; (2) consider the fate and behaviour of xenobiotics in soil, illustrating their interaction with soil constituents; and (3) review the potential pathways for remediation through mobility and bioavailability reduction of the xenobiotic compounds using biochar as a management tool.

20.2 Impact of Biochar on Soil Fertility

Biochar application in soils contributes significantly to increase in soil pH; base saturation; electrical conductivity (EC); exchangeable Ca, Mg, K, Na and available P; cation exchange capacity (CEC) in highly weathered soils with low-ion retention capacities (Mbagwu and Piccolo 1997); and grain and biomass yield of crops (Oguntunde et al. 2004; Blackwell et al. 2015). Biochars with high pH (e.g. >9.5) can provide liming capacity (Clay and Malo 2012) to soils. The pH of an eroded Palouse soil increased from 6.0 to 6.8 after amendment of coal ash at the rate of 110 Mg ha^{-1} (Cox et al. 2001). The hardwood and conifer charcoals augmented the exchangeable bases in sandy and loamy soils (Tryon 1948). Biochar application reduced the leaching of NH^{4+} from an unfertilized Ferralsol due to its high C content (Lehmann et al. 2002). Biochars have the greatest positive impacts on degraded and low fertility soils, whereas they have low or minimal impacts on highly productive soils (Woolf et al. 2010). For example, neither biochars from *Pinus taeda* and swine manure solids pyrolysed at 350 °C significantly increased the wetland muck's (organic humus) OC contents (Novak et al. 2015).

Due to its high surface area, negative surface charge and surface charge density, biochar can retain nutrients and act as a fertiliser depending on feedstock type (Liang et al. 2006). The crop yields and productivity are increased when these retained plant nutrients are available to plants (Joseph et al. 2015a, b). However, Spokas et al. (2012) reported some contradictory results. When biochars produced from papermill waste were applied without fertilisers, the radish biomass production was increased in the two types of soils used (calcarosol and ferrosol), but in case of wheat biomass production, it increased in the ferrosol but decreased in the calcarosol soil (Van Zwieten et al. 2010). Therefore, co-application of both biochar and fertiliser is a recommended step for synergetic improvement of the effect on biomass production, evident in most plants studied, except for wheat and radish cultivation in the calcarosol soil (Van Zwieten et al. 2010). Analysing the results of 76 studies, Cabrera et al. (2014) reported influence of both soil and biochar properties on soil system response to biochar.

20.2.1 *Effect of Co-composted Biochar on Soil Fertility*

The pure biochar application has moderately negative to positive yield effects in temperate soils. The co-composted biochars (2 %, w/w) (compost + biochar) have shown positive effects increasing the biomass yield of *Chenopodium quinoa* up to 305 % in a sandy-poor soil due to nutrient-enrichment, particularly with the anions nitrate and phosphate, while addition of 2 % (w/w) untreated biochar decreased the biomass to 60 % of the control (Kammann et al. 2015). These growth-promoting (co-composted biochars) as well as growth-reducing (pure biochars) effects were more pronounced at lower nutrient-supply levels. The ageing and nonconventional ion-water bonding in micro- and nanopores captured nitrate ion in biochar particles that was largely protected against leaching, partly plant available, and did not stimulate N_2O emissions. The results finally suggest that biochar amended with biowaste (N-rich) provides more positive agronomic benefit reducing nutrient losses from biowastes and agricultural soils.

20.2.2 Physicochemical Properties of Biochars from Different Sources and Their Fertility Effect

Table 20.1 shows the physicochemical properties of different biochars and their effects on soil fertility.

Surface areas, micropore surface areas and bulk densities of shoot biochars from 13 species of both native and non-native trees and shrubs showed variable results (Vaughn et al. 2015). All biochars had very strongly basic pH values (>9.0) and EC values varied greatly correlating with levels of potassium found in the biochars,

Table 20.1 Physical and chemical characteristics of different types of biochars and their effects on soil fertility and biomass yield

Biochar types	Physical property	Chemical property	Soil fertility and yield	References
Shoots from 13 native and non-native trees and shrubs	Variable surface, micropore surface, bulk density and moisture sorption	All biochars had basic pH value (>9.0), different EC and similar ash content	Increased soil fertility and yield	Vaughn et al. (2015)
Weed biomass, maize stalk and pine needles	NA	Increased soil pH, SOC, nutrients like N, P and K	Enhanced soil and crop productivity	Mandal et al. (2015)
Poultry litter and tree green waste biochar	NA	Poultry litter biochar had higher available N and P than green waste	Poultry litter did higher soil fertility than green waste	Bai et al. (2015)
Wheat, rice, maize and pearl millet	Maize biochar has stronger surface functional groups and most stable than other biochars	Wheat and rice biochars have higher CEC, K and Si and lower pH, C, N and P than maize and pearl	Maize biochar has higher nutrient values and C stability	Purakayastha et al. (2015)
Carbon nanotube (CNT) hickory, bagasse biochars	CNT-biochar has higher thermal stability, surface area and pore volume	Excellent sorbent material for removing dyes and organic pollutants	Decreased soil contaminant	Inyang et al. (2014)
Eucalyptus wood biochar		Lower-temperature biochars give more benefits than higher temperature	Enhanced soil and crop productivity	Butnan et al. (2015)
Birch (*Betula pendula*) biochar from contaminated and control soil	NA	Increased ryegrass shoot K and Zn	No adverse effect on soil bacterial community	Evangelou et al. (2014)

NA not available

while peak moisture sorption correlated with surface areas. Biochars prepared from seven different weed biomass (viz. *Ageratum conyzoides*, *Lantana camera*, *Gynura* sp., *Setaria* sp., *Avena fatua*, Maize stalk, Pine needles) significantly increased soil pH (from 0.26 to 0.3), SOC (1.62–1.74 %), available nutrients like N (4.5–-21.3 mg kg^{-1}), P (3.32–3.68 mg kg^{-1}) and K (by 20 % above control) and maize biomass yield in subtropical northeast India (Mandal et al. 2015). The poultry litter biochar demonstrated greater positive agronomic effects with significant increase in soil fertility and bioavailability of N and P content in contrast to the green waste biochar. The results indicated that poultry litter is a better feedstock to produce biochar comparing to tree crop green waste (Bai et al. 2015).

The wheat and rice biochar exhibited higher CEC, K and Si content, but lower pH value, C, N and P contents compared to the maize and pearl millet biochars (Purakayastha et al. 2015). The application of maize biochar with higher nutrient content (N and P) and C stability is highly recommended to enhance soil fertility and long-term C sequestration, while rice biochar can increase higher microbial activities in restoring biological fertility of degraded soils. The lower-temperature biochar derived from eucalyptus wood provided higher benefits to two contrasting textured soils (a loamy-sand ultisol and a silty-clay-loam oxisol) compared to the higher temperature biochar. The 1–2 % (w/w) rate of the low-temperature biochar was appropriate for these soils, while the higher rate is more suited for the finer-textured soil (Butnan et al. 2015). Neither the trace element-contaminated biochar (from birch, *Betula pendula*) nor the non-contaminated biochar had adverse effect on the bacterial community of the soil suggesting safe use of biochar prepared from plants grown on trace element-contaminated soils (Evangelou et al. 2014). Hybridised carbon nanotube (CNT)-biochar nanocomposite can efficiently remove dyes and organic pollutants from wastewater (Inyang et al. 2014).

20.2.3 Impact of Biochars on Physical and Mechanical Properties of Clayey Soil

The expansive clayey soils with the characteristics of swell-shrinkage, cracking and stickiness have usually low crop yield. The application of biochars prepared from wheat straw, woodchips and wastewater sludge has positive impact on the swell-shrinkage behaviour, mechanical strength and surface cracking of a clayey soil (Zong et al. 2014). All biochars significantly decreased the coefficient of linear extensibility of the soil, and among them, wheat straw has the greatest effect. Biochar significantly reduced the formation of soil surface cracks and length of the cracks suggesting suitability of biochar as a soil amendment to improve the poor physical properties of the clayey soil, specifically with reduction in swell-shrinkage, tensile strength and surface area density of cracking.

20.2.4 Interactive Effects of Biochar Ageing in Soils Related to Feedstock, Pyrolysis Temperature and Historic Charcoal Production

Biochars are recalcitrant forms of carbon, long-term persistent and can remain in the soil for more than 1000 years (Skjemstad et al. 2002) due to the slow microbial degradation and chemical oxidation rates (Sanchez et al. 2009) and interactions of biochars with soil materials such as ions, organic matter and clays (Clay and Malo 2012). Although the charge characteristics of biochar can undergo changes through ageing processes (with time) in soils, these changes depend on the initial biochar properties affected by feedstock and pyrolysis temperature and soil properties (Heitkotter and Marschner 2015). While the CEC and the amount of acid functional groups of corn digestate-derived biochars markedly increased during incubation, pine chip biochars showed no or only slight increases in CEC and twofold lower increase in acid functional groups.

The pyrolytic process parameters, such as temperature, heating rate and pressure, can change the recovery amounts and the physicochemical properties of biochars (Yaman 2004; Clay and Malo 2012). Increasing pyrolytic temperature from 300 to 800 °C decreased biochar recovery from 67 to 26 %, but increased C concentration of the char from 56 to 93 % compared with char prepared at lower temperatures (Katyal et al. 2003; Okimori et al. 2003). The higher flows of sweep gas in the pyrolytic process can reduce the particle size but increase the heating values (Katyal et al. 2003; Demirbas 2004). The biochar properties depend on the heating conditions, heating rate, residence time, catalytic effect of the biomass mineral matter and reactor designs. Feedstock type, quality and initial physical characteristics of the material (e.g. particle size, shape and structure) can impact the type, physicochemical properties and amounts of biochar formed (Bridgewater et al. 1999). The source materials of biochar and pyrolytic process parameters combined can produce various types of biochars for soil amendment (Lehmann et al. 2009).

20.2.5 Effects of Aged and Fresh Biochars on Soil Acidity

Biochar amendment has positive effects on soil acidity management. Biochar prepared from *Pinus massoniana* bark at 450 °C reduced the soil acidity with fresh biochar exhibiting more remarkable effects, but the efficiency decreased to a certain extent after a short-term ageing before being added to soils (Zhao et al. 2015).

20.2.6 Use of Alkaline Slag and Crop Residue Biochars on Acidic Soil

The alkaline slag, crop straw (peanut, canola) biochars and lime increased soil pH and decreased soil (acidic ultisol) acidity with the higher rates showing greater performance. The levels of one or more soil exchangeable base cations increased in case of all applied amendments (Masud et al. 2014). The soil exchangeable Ca^{2+} were positively affected by lime treatment, while the alkaline slag treatment increased exchangeable-Ca^{2+} and Mg^{2+} levels. The biochars and combined applications of alkaline slag with biochars positively improved soil exchangeable Ca^{2+}, Mg^{2+} and K^{+} and soil available P. The combined application of alkaline slag with biochars demonstrated the most favourable effect on reduction of soil acidity; increased soil Ca, Mg, K and P levels; and enhanced the uptake of Ca, Mg, K and P by soybean plants, albeit all other amendments also showed variable enhanced uptake of one or more nutrients of N, P, K, Ca and Mg by soybean.

20.2.7 Effect of Biochar on Cycling of Organic Carbon and Soil Microbial Community

Biochar application may significantly affect the flow of low molecular weight dissolved organic carbons through the microbial community in soils, most commonly found in the rhizosphere, through a variety of mechanisms (Farrell et al. 2015). Biochar type and soil type affect the soil microbial abundance and community structure that play key roles for the biogeochemical cycling of nutrients and organic matter turnover (Marks et al. 2014; Paz-Ferreiro et al. 2015). For example, sewage sludge biochars, rich in mineral matter and ash, significantly increased the fungi to bacteria ratio, while this ratio was unaffected after addition of a high fixed carbon biochars from Miscanthus and pine wood (Elmer et al. 2015). The DNA analysis shows significantly higher size of the total bacterial community in the soil with aged biochar compared to the control soil without biochar, and higher abundance in the biochar soil was also the case for denitrifiers and ammonia-oxidising bacteria. Interestingly, the relative proportion of nitrous oxide reductases to nitrite reductases was higher in the biochar soil than in the control soil indicating an overall higher potential for reduction of nitrous oxide (N_2O) emission either via complete denitrification or by N_2O reduction by non-denitrifiers (Interreg IVB 2014).

20.2.8 Contributions of Biochar Properties to Enhanced Biological Nitrogen Fixation

Biochar amendment increased the maize biomass and N use efficiency in the red soil ($p < 0.05$), but not in the other four agricultural soils (Zhu et al. 2015). Some studies suggested that biochar application caused increases in biological nitrogen fixation. However, the underlying mechanisms for this positive effect are still unclear. The biochar additions at a rate of 15 t ha^{-1} resulted in an average 262 % increase in shoot biomass of common beans (*Phaseolus vulgaris* L.), 164 % increase in root biomass, 3575 % increase in nodule biomass and a 2126 % increase in N derived from atmosphere (Ndfa) over the control (Guerena et al. 2015).

20.2.9 Effect of Rice Plantation on the Stability of Biochar in Paddy Soil

Wu et al. (2015) showed that although rice plants had no significant influence on the bulk characteristics and decomposition rates of the biochar, they enhanced the surface oxidation of biochar particles. The isotopic study (C-13 labelling) showed that rice plants improve biochar carbon incorporation into soil microbial biomass with approximately 0.05 % of the carbon in biochar incorporated into the rice plants during the whole rice-growing cycle. These results revealed decrease in stability of biochar in paddy soils by the root exudates and transportation of biochar particles into rice plants. Therefore, the impact of plants should be considered when predicting carbon sequestration potential of biochar in soil systems.

20.3 Impact of Biochar on the Xenobiotics in Soil

Widespread use of the pesticides used for improving crop productivity and resistance to disease can lead to environmental contamination of the soils (Ahmed et al. 2015). Studies have found pesticide residue in 60 % of groundwater samples from urban and agricultural areas of the United States (Gilliom et al. 2006). In the second half of the twentieth century, the random disposal of xenobiotic chemicals, such as pesticides, polycyclic aromatic hydrocarbons (PAHs) and dioxin into the environment created the serious environmental contamination (Gianfreda and Rao 2008) with negative and irreversible effects on soil quality and health, soil microbial and biochemical activities, humans, plants and animals. Biochar can significantly contribute to the sorption and sequestration of xenobiotic chemicals in soil due to its highly carbonaceous and aromatic nature, and high specific surface area (Accardi-Dey and Gschwend 2003; Lohmann et al. 2005; Ogbonnaya and Semple 2013), thereby reducing the transport and toxicity of organic contaminants in the

Table 20.2 Role of biochar in regulating the fate of genobiotics in agroecosystem

Function of biochar	Response	Source
Adsorption of pesticide, herbicide and fungicide due to mainly high specific surface area	Improves soil environment and soil biology	Cao et al. (2009), Accardi-Dey and Gschwend (2003)
Decrease in leaching of xenobiotics to waterways	Improves groundwater and surface water quality	Lü et al. (2012), Xu et al. (2012), Jones et al. (2011)
Less available to microorganism thereby slows down the biodegradation	Increases residence time in soil and develop eco-toxicology to soil organisms	Interreg IVB (2014), Yadav and Loper (2000)
Less available to plant and thereby reduces efficacy	Increases the current agronomic doses of pesticides except chiral pesticides and fungicides	Yang et al. (2010), Yu et al. (2009)
Modify the enantioselectivity behaviour of chiral pesticides and fungicides	Reduces the current agronomic doses of chiral pesticides and fungicides	Gamiz et al. (2015)
Enhance microbial catabolic activities to degrade common petroleum contaminants	Cleans the environment	Ogbonnaya et al. (2014)
Graphene-coated biochars can be used to remove organic pollutants	Cleans the aquatic environment from pollution	Ghaffar and Younis (2014)

soil environment, their plant availability and efficacy (Kookana 2010) and their potential risks to human and ecosystem health (Yu et al. 2010; Beesley et al. 2010). The detailed roles of biochar in regulating the fate of genobiotics in agroecosystem have been provided in Table 20.2.

Biochars produced from wheat and rice residues were found to sorb diuron 2500 times more effectively than soil over the concentrations of 0–6 mg l^{-1} (Yang and Sheng 2003a, b). The sorption coefficient of diuron, its isotherm nonlinearity and the extent of sorption-desorption hysteresis increased markedly with increasing contents of biochars in soil (Yu et al. 2006). A large body of publications reported the impact of biochar on sorption and dissipation of pesticides, including atrazine, terbuthylazine, diuron, isoproturon and pyrimethanil (Cao et al. 2009; Martin et al. 2012; Sopeña et al. 2012; Spokas et al. 2009; Wang et al. 2010; Yu et al. 2010, 2006). However, Martin et al. (2012) observed loss in biochar's sorptive capability over time with ageing/weathering in the soil environment. Generally, high temperature produced biochar increases the sorption and persistence of xenobiotics to soils.

Three of the six different biochars increased the leaching of the herbicide MCPA (4-chloro 2-methylphenoxyacetic acid) in a previous study (Cabrera et al. 2011), where the addition of two of these biochars had insignificant effect on mobility of fluometuron (Fig. 20.1).

Lower specific surface area and high DOC content of these biochars was identified as the factor for increase in leaching (Cabrera et al. 2011, 2014). Biochar can directly interact with pesticides either neutralising its efficacy or requiring higher

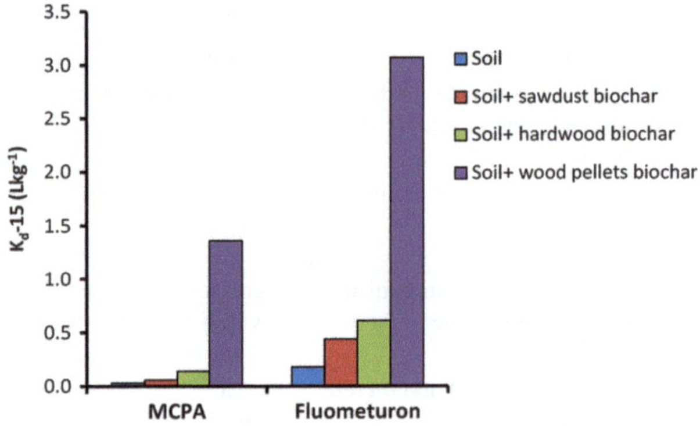

Fig. 20.1 Sorption coefficient (when fitting with Freundlich equation) of MCPA and Fluometuron in a sandy loam soil treated with different biochars (2 % w/w) [adapted from Cabrera et al. (2011)]

application rates, which enhance environmental contamination risks (Yang et al. 2010; Anawar et al. 2015).

20.3.1 Adsorption of Organic Contaminants on Graphene-Coated Biochars

Nanotechnology contributed to incorporation of the engineered nanoparticles in conventional biochar systems and enhanced biochar functions to some extent, such as carbon sequestration, soil fertility, wastewater treatment and sorption of both inorganic and organic pollutants to their surfaces, but reduced the environmental fate and mobility. The graphene coating on biochar surface increases the surface properties of the biochars, exhibiting the greatest surface area, pore volume and carbon content (Ghaffar and Younis 2014). The mechanisms responsible for the sorption of organic contaminants onto the biochars were the electrostatic attraction, pi-pi interaction between graphene sheets on the biochar surface and aromatic domain. The results inferred that graphene-coated biochars can be used as a stable, cheap sorbent and biomaterial to remove organic pollutants from aqueous media (Ghaffar and Younis 2014).

20.3.2 Degradation of Pesticides Using Fresh and Aged Biochar

The application of fresh biochar to contaminated soil profiles would substantially increase sorption of the xenobiotic pollutants, thereby decreasing their transport and leaching, and also affect the degradation rate of these substances (Interreg IVB 2014). However, the presence of aged biochar, e.g. more than 70 years in the soil, had no effect on the adsorption to soil and degradation rates of MCPA and chlorpyrifos compared to their adsorption and degradation rates in the control soil without biochar. There was a remarkable increased degradation rate of glyphosate in the soil with biochar, probably due to the lower adsorption in this soil compared to the control soil. However, there was a significant increase in the adsorption of diuron.

20.3.3 Degradation of Pesticides in Soil with Added Biochar Using Mixing and Layering Contact

Skogens kol biochar (<1 mm) mixed into the soil (1.9 % organic carbon) at 10 t ha^{-1} did not have any significant influence on pesticide degradation (soil moisture at 60 % of water holding capacity and 20 °C). When the same amount of biochar was placed in layer below the pesticide-amended soil, it increased the degradation half-lives from 84 to 156 days for diuron, from 5–6 to 15 days for MCPA and from 24–30 to 74 days for chlorpyrifos (Interreg IVB 2014).

20.3.4 Environmental Factors and Bioremediation of Xenobiotics Using White Rot Fungi

Microbial metabolism is one of the most recognised mechanisms for degradation of xenobiotic compounds in soil (Armstrong et al. 1967; Häggblom 1992). However, it may become compromised under low moisture and nutrient conditions, where the persistence of triazines and other xenobiotic compound may increase (Yadav and Loper 2000). Therefore, the soil type, moisture, pH, organic matter and clay content affect the binding of individual and mixtures of pesticides in soil and finally control the effectiveness of bioremediation strategies. Biochar addition to soils can significantly increase the abundance of mycorrhizal fungi (Solaiman et al. 2010; Blackwell et al. 2015), especially Trametes versicolor and Phanerochaete chrysosporium, that, in turn, show the differential effect on the degradation of single and mixtures of pesticides (Magan et al. 2010).

20.3.5 Role of Biochar to Enhance Microbial Catabolic Activity to Degrade Phenanthrene

Soils contain a wide variety of indigenous microflora that possesses catabolic potential to degrade organic contaminants, such as PAHs (Rhodes et al. 2010). However, the catabolic potential of microbes is influenced by contaminant concentration, its bioavailability, chemical stability (Tian et al. 2002; Semple et al. 2007), the presence of co-contaminants (Couling et al. 2010), nonaqueous phase liquids (NAPLs) (Lee et al. 2003) and soil organic matter (Gourlay et al. 2005). Biochar has demonstrated the potential role to enhance microbial catabolic activity for degradation of common petroleum contaminants (Ogbonnaya et al. 2014) and alkanes in soils (Bushnaf et al. 2011). However, it depends on the contaminant concentration, ageing period and soil properties. The biochar addition to low OM pasture soil resulted in a decrease in the 14C-phenanthrene mineralisation lag phase with larger mineralisation rates following 20 days ageing. Higher extent of 14C-phenanthrene mineralisation was observed in the high OM loam soil, which was more prominent with 0.01 % biochar amendment (Cornelissen and Gustafsson 2004).

20.3.6 Effect of Pyrochars and Hydrochars on Sorption of Herbicide Isoproturon in Soil

Both pyrochars and hydrochars, added biochar content, carbonization type and raw materials have statistically significant effects on the sorption of isoproturon to soils (Eibisch et al. 2015). Pyrochar reduced the concentration of isoproturon by a factor of 10–2283, while hydrochar reduced it by a factor of 3–13 in all treatments. The proportion of non-extractable pesticide residues in pyrochars-amended soil was higher than hydrochar-amended soils due to their higher microporosity, surface area and ash content, lower water extractable carbon contents and O-functional groups.

20.3.7 Influence of Biochar on the Chiral Fungicide in Soil

Chiral pesticides are an emerging class of organic pollutants, currently accounting for more than 25 % of used pesticides. As a result, the contamination problem caused by chiral pesticides is a matter of concern, and factors affecting enantioselective processes of chiral pesticides in soil should be clearly understood. Soil amendment with biochar (2 % w/w) resulted in three times higher sorption of metalaxyl enantiomers compared to unamended soil, but no enantioselectivity in the process was observed (Gamiz et al. 2015). This result demonstrated the biochar ability to alter the enantioselectivity behaviour of soil metalaxyl by its high sorption capacity. Biochar could contribute to reduction of the agronomic doses currently used for chiral pesticides, address the contamination problems associated with their use and also act as an immobilising amendment in soil remediation.

20.3.8 Interactions Between Biochar and Pyrimethanil Fungicide in Soil

Pyrimethanil is an aniline-pyrimidine fungicide commonly used for the control of grey mould and leaf scab on fruit, vegetables and ornamentals (Tomlin 2000). It is a moderately persistent substance in soil with a half-life of 23–54 days (EFSA 2006; Vanni et al. 2003), and it is the most toxic for a nontarget aquatic plant (*Lemnar minor*) among the three fungicides widely used in Champagne's vineyards (Verdisson et al. 2001). Biochar prepared from *Eucalyptus* spp. woodchips at 850 °C had higher surface area and microporosity and showed stronger effect on the sorption of pesticide than biochars prepared at 450 °C. Sorption coefficient and isotherm nonlinearity of the amended soils progressively increased with the content of biochar in the soil (Yu et al. 2010).

20.3.9 Interactions of Biochar with Chlorpyrifos, Fipronil, Carbofuran, Aminocyclopyrachlor, Bentazone and Pyraclostrobin Pesticides in Soil

The loss of the pesticides was higher in the biochar-amended soils with plants (*Allium tuberosum* and *Allium cepa*) than biochar-amended soils without plants due to plant uptake and enhanced microbial degradation. The plant uptake of the pesticides from the amended soils decreased markedly with increasing biochar content in the soil (Yang et al. 2010). High temperature biochar was more effective in reducing the bioavailability of these pesticides from the soil indicating suitability of biochars for sequestering or degradion of pesticide residues (chlorpyrifos, fipronil and carbofuran) in contaminated soils and reduction in plant uptake (Yu et al. 2009). Aminocyclopyrachlor and bentazone herbicides were almost completely sorbed by the silt loam soils amended with biochars produced from wood pellets (Cabrera et al. 2014). However, soils amended with macademia nut biochar exhibited lower herbicide sorption, which was attributed to the competition for sorption sites between the biochar DOC and the herbicides. Therefore, biochars with high surface areas and low DOC contents can increase the sorption of highly mobile pesticides in soils. The pyraclostrobin is highly sorbed to soil and biochar addition did not further increase its sorption.

20.3.10 Remediation and Atrazine Leaching from Biochar-Amended Soils

Atrazine is the most commonly used herbicide and frequently detected in drinking water aquifers and shallow groundwater beneath agricultural areas (Barbash et al. 2001; Delwiche et al. 2014). The pine chip biochar readily sorbs atrazine (Cao et al. 2009; Zheng et al. 2010) and potentially decreases high atrazine leaching to groundwater (Spokas et al. 2009; Zheng et al. 2010; Delwiche et al. 2014), but heterogeneous soil conditions, especially preferential flow paths, may reduce the effect of biochar. In fact, the preferential flow through macropores increases contaminant movement through the soil profile (Kookana et al. 1998; Akhtar et al. 2003). Since biochars contain colloidal-sized particles that move through soil pore water flows (Zhang et al. 2010; Abiven et al. 2011), colloid-facilitated transport could actually enhance atrazine mobility and leaching in the presence of biochar (Cabrera et al. 2011). Biochar reduced the downward movement of simazine (one type of triazine herbicides) in response to artificial rainwater, thus potentially reducing the risk of groundwater contamination (Fig. 20.2).

The effects of biochar on pesticides depended on the type of biochar used and the soil (Jones et al. 2011). For example, more simazine tended to leach from high fertility soil or if the biochar had small particles (diameter of less than 2 mm). Deisopropylatrazine was irreversibly sorbed on biochars, and greater sorption was observed with higher surface area (4.7–2061 mg g^{-1}) of biochar (Uchimiya et al. 2012).

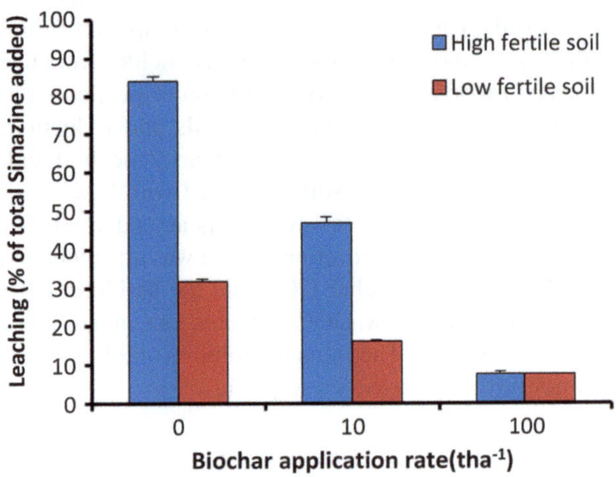

Fig. 20.2 Leaching of simazine from two soils containing different amount of biochar [adapted from Jones et al. (2011)]

20.4 Conclusions

The application of biochars to agricultural and poor quality degraded land can increase soil pH; base saturation; EC; CEC; exchangeable Ca, Mg, K and Na; microbial activity; colonisation of mycorrhizal fungi in rhizosphere around plant roots; nutrients (N, P); and soil fertility, which finally increase grain and biomass yield. The biochar addition to soils can effectively sequester or degrade xenobiotic chemicals and reduce their uptake by plants and food chain contamination. Depending on the type and amount of biochar applied, the changes in soil properties associated with the application (e.g. soil pH, EC) as well as the physicochemical properties of the biochar itself may impact the use, rates, efficacious properties and fates of pesticides used in agronomic management (Clay and Malo 2012). Conversely, if the biochar reduces the efficacy of soil-applied herbicides or other pesticides, it may have negative impacts requiring higher herbicide application rates and monetary implications for growers. Increased sorption to soils and recalcitrance of pesticides leading to longer residence times in the environment is desirable if bioactivity is still acceptable, and it controls the target pest. However, longer residence time may also create some environmental problems, such as greater leaching potential or carry-over problems into the following season. Therefore, prior to its use, the suitability of biochar application and the reasons for application should be defined clearly and the outcomes should be closely monitored to monitor any adverse impact.

References

Abiven S, Hengartner P, Schneider MPW, Singh N, Schmidt MWI (2011) Pyrogenic carbon soluble fraction is larger and more aromatic in aged charcoal than in fresh charcoal. Soil Biol Biochem 43:1615–1617

Accardi-Dey A, Gschwend PM (2003) Reinterpreting literature sorption data considering both absorption into organic carbon and adsorption onto black carbon. Environ Sci Technol 37: 99–106

Ahmed G, Anawar HM, Takuwa DT, Chibua IT, Singh GS, Sichilongo K (2015) Environmental assessment of fate, transport and persistent behavior of dichlorodiphenyltrichloroethanes and hexachloroethanes in land and water ecosystems. Int J Environ Sci Technol 12:2741–2756

Akhtar MS, Steenhuis TS, Richards BK, McBride MB (2003) Chloride and lithium transport in large arrays of undisturbed silt loam and sandy loam soil columns. Vadose Zone J 2:715–727

Anawar HM, Akter F, Solaiman ZM, Strezov V (2015) Biochar is an emerging panacea for remediation of soil contaminants from mining, industry and sewage waste. Pedosphere 25: 654–665

Armstrong DE, Chesters G, Harris RF (1967) Atrazine hydrolysis in soil. Soil Sci Soc Am J 31: 61–66

Ashman MR, Puri G (2002) Essential soil science: a clear and concise introduction to soil science. Blackwell Science, Oxford

Bai SH, Xu CY, Xu ZH, Blumfield TJ, Zhao HT, Wallace H, Reverchon F, Van Zwieten L (2015) Soil and foliar nutrient and nitrogen isotope composition ($\delta^{15}N$) at 5 years after poultry litter and green waste biochar amendment in a macadamia orchard. Environ Sci Pollut Res 22: 3803–3809

Barbash JE, Thelin GP, Kolpin DW, Gilliom RJ (2001) Major herbicides in groundwater: results from the national water-quality assessment. J Environ Qual 30:831–845

Beesley L, Moreno-Jiménez E, Gomez-Eyles JL (2010) Effect of biochar and greenwaste compost amendments on mobility, bioavailability and toxicity of inorganic and organic contaminants in a multi-element polluted soil. Environ Pollut 158:2282–2287

Blackwell P, Storer P, Anawar HM, Joseph S, Gilkes RJ, Solaiman ZM (2015) Influences of biochar and biochar-mineral complex on mycorrhizal colonisation and nutrition of wheat and sorghum. Pedosphere 25:686–695

Bridgewater AV, Meier D, Radlein D (1999) An overview of fast pyrolysis of biomass. Org Geochem 30:1479–1493

Burgess RM, Perron MM, Friedman CL, Suuberg EM, Pennel KG, Cantwell MG, Pelletier MC, Ho KT, Serbst JR, Ryba SA (2009) Evaluation of the effects of coal fly ash amendments on the toxicity of a contaminated marine sediment. Environ Toxicol Chem 28(1):26–35

Bushnaf KM, Puricelli S, Saponaro S, Werner D (2011) Effect of biochar on the fate of volatile petroleum hydrocarbons in an aerobic sandy soil. J Contam Hydrol 126:208–215

Butnan S, Deenik JL, Toomsan B, Antal MJ, Vityakon P (2015) Biochar characteristics and application rates affecting corn growth and properties of soils contrasting in texture and mineralogy. Geoderma 237:105–116

Cabrera A, Cox L, Spokas KA, Celis R, Hermosin MC, Cornejo J et al (2011) Comparative sorption and leaching study of the herbicides fluometuron and 4-chloro-2-methylphenoxyacetic acid (MCPA) in a soil amended with biochars and other sorbents. J Agric Food Chem 59:12550–12560

Cabrera A, Cox L, Spokas K, Hermosin MC, Cornejo J, Koskinen WC (2014) Influence of biochar amendments on the sorption-desorption of aminocyclopyrachlor, bentazone and pyraclostrobin pesticides to an agricultural soil. Sci Total Environ 470–471:438–443

Cao X, Ma L, Gao B, Harris W (2009) Dairy-manure derived biochar effectively sorbs lead and atrazine. Environ Sci Technol 43:3285–3291

Chen B, Yuan M (2011) Enhanced sorption of polycyclic aromatic hydrocarbons by soil amended with biochar. J Soils Sed 11:62–71

Chidumayo EN (1994) Effects of wood carbonization on soil and initial development of seedlings in miombo woodland, Zambia. Forest Ecol Manage 70:353–357

Chun Y, Sheng G, Chiou CT, Xing B (2004) Compositions and sorptive properties of crop residue-derived chars. Environ Sci Technol 38(17):4649–4655

Clay SA, Malo DD (2012) The influence of biochar production on herbicide sorption characteristics. In: Hasaneen MN (ed) Herbicides—properties, synthesis and control of weeds. InTech, Rijeka

Cornelissen G, Gustafsson O (2004) Sorption of phenanthrene to environmental black carbon in sediment with and without organic matter and native sorbates. Environ Sci Technol 38: 148–155

Couling NR, Towell MG, Semple KT (2010) Reprinted from biodegradation of PAHs in soil: influence of chemical structure, concentration and multiple amendment. Environ Pollut 158: 3411–3420

Cox D, Bezdicek D, Fauci M (2001) Effects of compost, coal ash, and straw amendments on restoring the quality of eroded Palouse soil. Biol Fert Soils 33:365–372

Delwiche KB, Lehmann J, Walter MT (2014) Atrazine leaching from biochar-amended soils. Chemosphere 95:346–352

Demirbas A (2004) Effects of temperature and particle size on bio-char yield from pyrolysis of agricultural residues. J Anal Appl Pyrol 72:243–248

EFSA (European Food Safety Authority) (2006) Conclusion regarding the peer review of the pesticide risk assessment of the active substance: pyrimethanil. Summ EFSA Sci Rep 61:1–5

Eibisch N, Schroll R, Fuss R, Mikutta R, Helfrich M, Flessa H (2015) Pyrochars and hydrochars differently alter the sorption of the herbicide isoproturon in an agricultural soil. Chemosphere 119:155–162

Elmer WH, Lattao CV, Pignatello JJ (2015) Active removal of biochar by earthworms (*Lumbricus terrestris*). Pedobiologia 58(1):1–6

Evangelou MWH, Brem A, Ugolini F, Abiven S, Schulin R (2014) Soil application of biochar produced from biomass grown on trace element contaminated land. J Environ Manage 146: 100–106

Farrell M, Macdonald LM, Baldock JA (2015) Biochar differentially affects the cycling and partitioning of low molecular weight carbon in contrasting soils. Soil Biol Biochem 80:79–88

Gamiz B, Pignatello JJ, Hermosín MC, Cox L, Celis R (2015) Influence of biochar on the enantio-selective behavior of the chiral fungicide metalaxyl in soil. vol 17. Geophysical Research Abstracts, EGU2015-15367, 2015, EGU General Assembly 2015

Ghaffar A, Younis MN (2014) Adsorption of organic chemicals on graphene coated biochars and its environmental implications. Green Proces Synth 3(6):479–487

Gianfreda L, Rao MA (2008) Interactions between xenobiotics and microbial and enzymatic soil activity. Critical Rev Environ Sci Technol 38(4):269–310

Gilliom RJ, Barbash JE, Crawford CG, Hamilton PA, Martin JD, Nakagaki N, Nowell LH, Scott JC, Stackelberg PE, Thelin GP, Wolock DM (2006) The quality of our nation's waters—pesticides in the nation's streams and ground water, 1992–2001: US Geological Survey Circular 1291, 172 p

Glaser B, Lehmann J, Zech W (2002) Ameliorating physical and chemical properties of highly weathered soils in the tropics with charcoals: a review. Biol Fertil Soils 35:219–230

Gourlay C, Tusseau-Vuillemin MH, Mouchel JM, Garric J (2005) The ability of dissolved organic matter (DOM) to influence benzo[a]pyrene bioavailability increases with DOM biodegradation. Ecotoxicol Environ Saf 61:74–82

Guerena DT, Lehmann J, Thies JE, Enders A, Karanja N, Neufeldt H (2015) Partitioning the contributions of biochar properties to enhanced biological nitrogen fixation in common bean (*Phaseolus vulgaris*). Biol Fertil Soils 51:479–491

Häggblom MM (1992) Microbial breakdown of halogenated aromatic pesticides and related compounds. FEMS Microbiol Rev 9:29–71

Han MJ, Choi HT, Song HG (2004) Degradation of phenanthrene by trametes versicolor and its laccase. J Microbiol 42:94–98

Heitkotter J, Marschner B (2015) Interactive effects of biochar ageing in soils related to feedstock, pyrolysis temperature, and historic charcoal production. Geoderma 245:56–64

Interreg IVB Project Biochar: Climate Saving Soils (2014) The effect of freshly added and aged biochar on pesticide degradation and adsorption in soil and the effect of long-term exposure to biochar on selected key taxonomic and functional community composition of soil microbial communities. www.biochar-interreg4b.eu. Accessed Feb 2014

Inyang M, Gao B, Zimmerman A, Zhang M, Chen H (2014) Synthesis, characterization, and dye sorption ability of carbon nanotube-biochar nanocomposites. Chem Eng J 236:39–46

Ippolito JA, Laird DA, Busscher WJ (2012) Environmental benefits of biochar. J Environ Qual 41: 967–972

Jones DL, Edward-Jones G, Murphy DV (2011) Biochar mediated alterations in herbicide breakdown and leaching in soil. Soil Biol Biochem 43:804–813

Joseph S, Anawar HM, Storer P, Blackwell P, Chia C, Lin Y, Munroe P, Donne S, Horvat J, Wang J, Solaiman ZM (2015a) Effect of enriched biochars containing nanophase magnetic iron particles on mycorrhizal colonisation, plant growth, nutrient uptake and soil quality improvement. Pedosphere 25:749–760

Joseph S, Pow D, Dawson K, Mitchell DRG, Rawal A, Hook J, Sarasadat Taherymoosavi S, Zwieten LV, Rust J, Donne S, Munroe P, Pace B, Graber E, Thomas T, Nielsen S, Ye J, Lin Y, Pan GX, Liquang L, Solaiman ZM (2015b) Feeding biochar to cows: an innovative solution for improving soil fertility and farm productivity. Pedosphere 25:666–679

Kammann CI, Schmidt HP, Messerschmidt N, Linsel S, Steffens D, Muller C, Koyro HW, Conte P, Stephen J (2015) Plant growth improvement mediated by nitrate capture in co-composted biochar. Sci Rep 5:11080

Katyal S, Thambimuthu K, Valix M (2003) Carbonisation of bagasse in a fixed bed reactor: influence of process variables on char yield and characteristics. Renew Energy 28:713–725

Kishimoto S, Sugiura G (1985) Charcoal as a soil conditioner. In: Symposium on forest products research, International achievements for the future, 5 (April 1985), Republic of South Africa, pp 12–23

Kohler HR, Triebskorn R (2013) Wildlife ecotoxicology of pesticides: can we track effects to the population level and beyond? Science 16:759–765

Kookana RS (2010) The role of biochar in modifying the environmental fate, bioavailability, and efficacy of pesticides in soils: a review. Soil Res 48(7):627–637

Kookana RS, Baskaran S, Naidu R (1998) Pesticide fate and behavior in Australian soils in relation to contamination and management of soil and water: a review. Aust J Soil Res 36:715–764

Kookana RS, Sarmah AK, Van Zwieten L, Krull E, Singh B (2011) Biochar application to soil: agronomic and environmental benefits and unintended consequences. Adv Agron 112:103–143

Laird DA (2008) The charcoal vision: a win-win-win scenario for simultaneously producing bioenergy, permanently sequestering carbon, while improving soil and water quality. Agron J 100:178–181

Lee PH, Doik KJ, Semple KT (2003) The development of phenanthrene catabolism in soil amended with transformer oil. FEMS Microbiol Lett 228:217–223

Lehmann CJ, Rondon M (2006) Bio-char soil management on highly-weathered soils in the tropics. In: Uphoff NT (ed) Biological approaches to sustainable soil systems. CRC Press, Boca Raton, pp 517–530

Lehmann CJ, da Silva Jr JP, Rondon M, da Silva CM, Greenwood J, Nehls T, Stein C, Glaser B (2002) Slash-and-char – a feasible alternative for soil fertility management in the central Amazon? In: Proceedings 17th World Congress Soil Science, Bangkok, August

Lehmann J, Czimczik C, Laird D, Sohi S (2009) Stability of biochar in the soil. In: Lehmann J, Joseph S (eds) Biochar for environmental management. Earthscan, London, pp 183–205

Lehmann J, Rillig MC, Thies J, Masiello CA, Hockaday WC, Crowley D (2011) Biochar effects on soil biota: a review. Soil Biol Biogeochem 43:1812–1836

Liang B, Lehmann J, Solomon D, Kinyangi J, Grossman J, O'Neill B et al (2006) Black carbon increases cation exchange capacity in soil. Soil Sci Soc Am J 70:1719–1730

Lohmann R, MacFarlane JK, Gschwend PM (2005) Importance of black carbon to sorption of native PAHs, PCBs, and PCDDs in Boston and New York harbor sediments. Environ Sci Technol 39:141–148

Lü J, Li J, Li Y, Chan B, Bao Z (2012) Use of rice straw biochar simultaneously as the sustained release carrier of herbicides and soil amendment for their reduced leaching. J Agric Food Chem 60:6463–6470

Magan N, Fragoeiro S, Bastos C (2010) Environmental factors and bioremediation of xenobiotics using white rot fungi. Mycobiology 38(4):238–248

Mandal S, Verma BC, Ramkrushna GI, Singh RK, Rajkhowa DJ (2015) Characterization of biochar obtained from weeds and its effect on soil properties of North Eastern Region of India. J Environ Biol 36:499–505

Marks EAN, Mattana S, Alcaniz JM, Domene X (2014) Biochars provoke diverse soil mesofauna reproductive responses in laboratory bioassays. Europ J Soil Biol 60:104–111

Martin SM, Kookana RS, Van Zwieten L, Krull E (2012) Marked changes in herbicide sorption desorption upon ageing of biochars in soil. J Hazard Mater 231–232:70–78

Masud MM, Li JY, Xu RK (2014) Use of alkaline slag and crop residue biochars to promote base saturation and reduce acidity of an acidic ultisol. Pedosphere 24(6):791–798

Mbagwu JSC, Piccolo A (1997) Effects of humic substances from oxidized coal on soil chemical properties and maize yield. In: Drozd J, Gonet SS, Senesi N, Weber J (eds) The role of humic substances in the ecosystems and in environmental protection. IHSS, Polish Society of Humic Substances, Wroclaw, pp 921–925

Novak JM, Sigua GC, Spokas KA, Busscher WJ, Cantrell KB, Watts DW, Glaz B, Hunt PG (2015) Plant macro- and micronutrient dynamics in a biochar-amended wetland muck. Water Air Soil Pollut 226:2228

Ogbonnaya U, Semple KT (2013) Impact of biochar on organic contaminants in soil: a tool for mitigating risk? Agronomy 3:349–375

Ogbonnaya U, Oyelami A, Matthews J, Adebisi O, Semple KT (2014) Influence of wood bochar on phenanthrene catabolism in soils. Environments 1:60–74

Oguntunde PG, Fosu M, Ajayi AE, van de Giesen N (2004) Effects of charcoal production on maize yield, chemical properties and texture of soil. Biol Fertil Soils 39:295–299

Okimori Y, Ogawa M, Takahashi F (2003) Potential of CO_2 emission reductions by carbonizing biomass waste from industrial tree plantation in Sumatra, Indonesia. Mitig Adapt Strat Glob Change 8(3):261–280

Paz-Ferreiro J, Liang C, Fu S, Mendez A, Gasco G (2015) The effect of biochar and its interaction with the earthworm Pontoscolex corethrurus on soil microbial community structure in tropical soils. PLoS One 10(4):e0124891

Perucci P, Dumontet S, Bufo SA, Mazzatura A, Casucci C (2000) Effects of organic amendment and herbicide treatment on soil microbial biomass. Biol Fertil Soils 32:17–23

Piccolo A, Pietramellara G, Mbagwu JSC (1996) Effects of coal derived humic substances on water retention and structural stability of Mediterranean soils. Soil Use Manage 12:209–213

Purakayastha TJ, Kumari S, Pathak H (2015) Characterisation, stability, and microbial effects of four biochars produced from crop residues. Geoderma 239:293–303

Rhodes AH, McAllister LE, Chen R, Semple KT (2010) Impact of activated charcoal on the mineralisation of 14C-phenanthrene in soils. Chemosphere 79:463–469

Sanchez ME, Lindao E, Margaleff D, Martinez O, Moran A (2009) Pyrolysis of agricultural residues from rape and sunflower: production and characterization of biofuels and biochar soil management. J Anal Appl Pyrol 85:142–144

Semple KT, Doick KJ, Wick LY, Harms H (2007) Microbial interactions with organic contaminants in soil: definitions, processes and measurements. Environ Pollut 150:166–176

Skjemstad JO, Reicosky DC, Wilts AR, McGowan JA (2002) Charcoal carbon in U.S. agricultural soils. Soil Sci Soc Am J 66:1249–1255

Sohi SP, Krull E, Lopez-Capel E, Bol R (2010) A review of biochar and its use and function in soil. In: Sparks DL (ed) Advances in agronomy. Elsevier Academic, San Diego, pp 47–82

Solaiman ZM, Blackwell P, Abbott LK, Storer P (2010) Direct and residual effect of biochar application on mycorrhizal root colonisation, growth and nutrition of wheat. Aust J Soil Res 48: 546–554

Sopeña F, Semple K, Sohi S, Bending G (2012) Assessing the chemical and biological accessibility of the herbicide isoproturon in soil amended with biochar. Chemosphere 88:77–83

Spokas K, Koskinen WC, Baker JM, Reicosky DC (2009) Impacts of wood biochar additions on greenhouse gas production and sorption/degradation of two herbicides in a Minnesota soil. Chemosphere 77:574–581

Spokas KA, Cantrell KB, Novak JM, Archer DW, Ippolito JA, Collins HP et al (2012) Biochar: a synthesis of its agronomic impact beyond carbon sequestration. J Environ Qual 41:973–989

Tian Y, Zheng T, Wang X (2002) PAHs contamination and PAH-degrading bacteria in Xiamen Western Sea. Chem Speciat Bioavailab 14:25–33

Tomlin CDS (2000) The pesticide manual, 12th edn. British Crop Protection Council, Alton, p. 1250

Tryon EH (1948) Effect of charcoal on certain physical, chemical, and biological properties of forest soils. Ecol Monogr 18:81–115

Uchimiya M, Wartelle LH, Boddu VM (2012) Sorption of triazine and organophosphorus pesticides on soil and biochar. J Agric Food Chem 60(12):2989–2997

Van Zwieten L, Kimber S, Morris S, Chan KY, Downie A, Rust J et al (2010) Effects of biochar from slow pyrolysis of papermill waste on agronomic performance and soil fertility. Plant Soil 327:235–246

Vanni A, Fontana F, Cignetti A, Gennari M (2003) Behavior of pyrimethanil in soil: abiotic and biotic processes. In: Proceedings of the XII Symposium Pesticide Chemistry, Piacenza, pp 233–238

Vaughn SF, Kenar JA, Eller FJ, Moser BR, Jackson MA, Peterson SC (2015) Physical and chemical characterization of biochars produced from coppiced wood of thirteen tree species for use in horticultural substrates. Ind Crop Prod 66:44–51

Verdisson S, Couderchet M, Vernet G (2001) Effect of procymidone, fludioxonil and pyrimethanil on two non-target aquatic plants. Chemosphere 44(3):467–474

Wang H, Lin K, Hou Z, Richardson B, Gan J (2010) Sorption of the herbicide terbuthylazine in two New Zealand forest soils amended with biosolids and biochars. J Soil Sediment 10: 283–289

Woolf D, Amonette JE, Street-Perrott FA, Lehmann J, Joseph S (2010) Sustainable biochar to mitigate global climate change. Nat Commun 1(56):1–9

Wu MX, Feng QB, Sun X, Wang HL, Gielen G, Wu WX (2015) Rice (*Oryza sativa* L) plantation affects the stability of biochar in paddy soil. Sci Rep 5

Xu T, Lou L, Luo L, Cao R, Duan D, Chen Y (2012) Effect of bamboo biochar on pentachlorophenol leachability and bioavailability in agricultural soil. Sci Total Environ 414:727–731

Yadav JS, Loper JC (2000) Cytochrome P450 oxidoreductase gene and its differentially terminated cDNAs from the white rot fungus Phanerochaete chrysosporium. Curr Genet 37:65–73

Yaman S (2004) Pyrolysis of biomass to produce fuels and chemical feedstocks. Energ Conver Manage 45(5):651–671

Yang YN, Sheng GY (2003a) Enhanced pesticide sorption by soils containing particulate matter from crop residue burns. Environ Sci Technol 37(16):3635–3639

Yang YN, Sheng GY (2003b) Pesticide adsorptivity of aged particulate matter arising from crop residue burns. J Agric Food Chem 51(17):5047–5051

Yang XB, Ying GG, Peng PA, Wang L, Zhao JL, Zhang LJ, Yuan P, He HP (2010) Influence of biochars on plant uptake and dissipation of two pesticides in an agricultural soil. J Agric Food Chem 58(13):7915–7921

Yu XY, Ying GG, Kookana RS (2006) Sorption and desorption behaviours of diuron in soils amended with charcoal. J Agric Food Chem 54:8545–8550

Yu X, Ying G, Kookana RS (2009) Reduced plant uptake of pesticides with biochar additions to soil. Chemosphere 76:665–671

Yu XY, Pan LG, Ying GG, Kookana RS (2010) Enhanced and irreversible sorption of pesticide pyrimethanil by soil amended with biochars. J Environ Sci 22:615–620

Zhang W, Niu J, Morales VL, Chen X, Hay AG, Lehmann J et al (2010) Transport and retention of biochar particles in porous media: effect of pH, ionic strength, and particle size. Ecohydrology 3:497–508

Zhao RD, Coles N, Kong Z, Wu JP (2015) Effects of aged and fresh biochars on soil acidity under different incubation conditions. Soil Tillage Res 146:133–138

Zheng W, Guo M, Chow T, Bennett DN, Rajagopalan N (2010) Sorption properties of greenwaste biochar for two triazine pesticides. J Hazard Mater 181:121–126

Zhu QH, Peng XH, Huang TQ (2015) Contrasted effects of biochar on maize growth and N use efficiency depending on soil conditions. Int Agrophys 29(20):257–266

Zong YT, Chen DP, Lu SG (2014) Impact of biochars on swell-shrinkage behavior, mechanical strength, and surface cracking of clayey soil. J Plant Nutr Soil Sci 177(6):920–926

Chapter 21
Environmental Biodegradation of Xenobiotics: Role of Potential Microflora

Manoj Kumar, Ram Prasad, Pankaj Goyal, Priyanku Teotia, Narendra Tuteja, Ajit Varma, and Vivek Kumar

21.1 Introduction

The term xenobiotic (Greek xenos + bioticos, which means "strange" and "life-related," respectively) means a chemical substance that is not a natural component of a living organism exposed to it, i.e., a strange, exogenous substance or anthropogenic material. This definition also covers the substances strange to the target organisms; hence it is used for most poisons and drugs. An important group of xenobiotics are chemical compounds produced by humans, with artificial chemical structure, to which organisms have not adjusted through prior evolution. Both natural and anthropogenic activities result in accumulation of wide ranges of toxic xenobiotic compounds in the environment and thus cause a global concern (Gren 2012). Primarily, xenobiotics are those compounds that are alien to a living individual and have a propensity to accumulate in the environment. Principal xenobiotics include pesticides, fuels, solvents, alkanes, polycyclic hydrocarbons (PAHs), antibiotics, synthetic azo dyes, pollutants (dioxins and polychlorinated biphenyls), and polyaromatic, chlorinated, and nitroaromatic compounds (Guermouche M'rassi et al. 2015). The main concern with xenobiotic compounds is the toxicity threat they pose to public health. It is quite shocking that some xenobiotic compounds (phenols, biphenyl compounds, phthalates, etc.) act as endocrine disruptors (Itoh et al. 2000).

In early times, we had an unlimited abundance of land and resources; today, due to our carelessness and negligence in using them, however, the resources in the world show, in lesser degree (Vidali 2001). The quick growth of various industries

M. Kumar • R. Prasad • P. Goyal • N. Tuteja • A. Varma • V. Kumar (✉)
Amity Institute of Microbial Technology, Amity University, Noida, Uttar Pradesh, India
e-mail: vivekbps@gmail.com

P. Teotia
Department of Botany, CCS University, Meerut, India

© Springer International Publishing AG 2017
M.Z. Hashmi et al. (eds.), *Xenobiotics in the Soil Environment*, Soil Biology 49,
DOI 10.1007/978-3-319-47744-2_21

in the past century has extremely increased the release of toxic waste effluents into water bodies along with groundwater (Sethy et al. 2011). Environmental pollution caused by the release of these wide range of compounds (i.e., persistent organic pollutants (POPs)) from industries is creating a disturbance to the ecosystem (Gursahani and Gupta 2011), causing climatic changes, reduction of water levels in the ground as well as oceans, melting of icecaps, global warming, ozone layer depletion due to photochemical oxidation, etc. (Sharma et al. 2011), and this made ecologists to focus more on impacts of pollution and its reduction.

Biodegradation is a microorganism-mediated transformation of contaminants into nonhazardous or less hazardous substances (Lin et al. 2010, 2014a). Microorganisms are nature's recyclers, converting toxic organic compounds to innocuous compounds, often carbon dioxide and water (Leys et al. 2005). The appropriate use of various organisms like bacteria, fungi, and algae for efficient bioremediation of pollutants has been reported by Vidali (2001). As per the opinion of Hamzah et al. (2010), most of the organisms, predominantly bacteria, are known for their detoxifying abilities. They mineralize, transform, or immobilize the pollutants. Bacteria play a crucial role in biogeochemical cycles for sustainable development of the biosphere.

The enormous genetic diversity of microorganisms, their metabolic plasticity and high reproduction rates, and the capacity for horizontal gene transfer ensure the development and adaptation of microorganisms to rapidly changing conditions of the environment (Kumar et al. 2016). Bioremediation can be effective only when environmental conditions permit microbial growth and activity. Bioremediation involves the manipulation of environmental parameters (pH, temperature, moisture, and oxygen) to allow microbial growth and degradation procedure at a faster rate (Selvam and Vishnupriya 2013). The development of recombinant genetically modified organisms (GMOs) is very significant for the bioremediation of complex waste; through this we can identify the gene responsible for specific compound degradation (Karpouzas and Singh 2006).

21.2 Xenobiotics Sources

Basically there are two types of sources for occurrence of xenobiotics in environment, direct and indirect.

21.2.1 Straight Sources

The prime direct source of xenobiotics is wastewater and solid residual releases from the industries like chemical and pharma, plastics, paper and pulp mills, textile mills, and agricultural (enhancement products like pesticides, herbicides, etc.). Some of the common residual compounds in the wastewater and other effluents

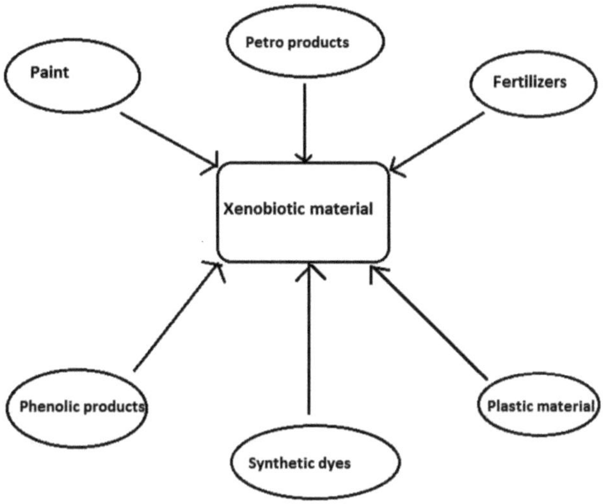

Fig. 21.1 Various sources of xenobiotic compounds in nature

are phenol, hydrocarbons, different dyes, paint effluents, pesticides and insecticides, etc. (Fig. 21.1).

A. Plastic Material "Plastic materials are huge molecules which are held together by strong forces which can be broken only by breaking the energy forces such as light" (Kathiresan 2003). Plastic material is tough, is sturdy, and degrades sluggishly owing to the molecular bonds and interactions. Plastics are made of polystyrene and polyvinyl chloride and polyethylene and its derivatives. Nowadays plastics (from crude oil) are used as fuels in industries since they break down into liquid hydrocarbons (Raaman et al. 2012). Microbial degradation of plastics gained importance in the last few years, but the fragmented compounds released by these also lead to further environmental issues.

B. Paint Material Paints: Volatile organic compounds and additives like emulsifiers and texturizers in paint are considered harmful which can be degraded by different means like chemicals (water as solvent), hygroscopic stresses, and microbial sources (Dixit et al. 2015).

C. Phenolic Components Phenol is one among the most prevalent chemical and pharma pollutants due to its toxicity even at lower concentrations and formation of substituted compounds during oxidation and disinfection processes. Its direct effects on the environment include depletion of ozone layer, effect on the earth's heat balance, reduced visibility, and adding acidic air pollutants to the atmosphere (Yeom et al. 2010). Phenol removal from the industrial wastewaters is very much necessary, prior to the wastewater discharge, so as to decrease all these effects. Phenol being a carcinogenic compound requires biodegradation method which results in minimum secondary metabolites and harmless end products (Prpich et al. 2006).

D. Petro-products Petroleum effluents mainly contain polycyclic (polynuclear) aromatic hydrocarbons, saturated hydrocarbons, and many nitrogen-, sulfur-, and oxygen-containing organic compounds (Gojgic-Cvijovic et al. 2012). Remediation of such petro-compounds using physicochemical treatments is not cost effective and may lead to further instabilities in environment, thus giving importance to biotreatments, which had an impact on reduction of these recalcitrants. Microorganisms that biodegrade these components are isolated from various environments, particularly from petroleum-contaminated sites (Prakash et al. 2014). Saturated hydrocarbons having the straight chain (*n*-alkanes) are most susceptible to microbial attack than branched alkanes. The aromatic fraction is more difficult to degrade, and susceptibility of biodegradation decreases as the aromacity increases in the compound (Milić et al. 2009).

E. Dyes and Pigments Dye agglomeration is the major cause for the persistence of xenobiotics, and their presence in aquatic bodies will affect photosynthetic activity in aquatic life due to reduced light penetration even at low concentrations (Kumari et al. 2014). A number of industrial processes, such as textile industries, paper printing, and photography, use synthetic dyes extensively, which usually have complex aromatic molecular structures. Azo (Black B, Turq Blue GN, Yellow HEM, Red HEFB, and Navy HER), anthraquinone, and phthalocyanine dyes are commonly used dyes in these industries (Vigneeswaran et al. 2012; Shahid et al. 2013). The degradation of these dyes produces aromatic amines, which may be carcinogenic and mutagenic. Microorganism (living or dead biomass) has the ability not only to decolorize dyes but also detoxify them (Hemapriya and Vijayanand 2014) by adsorption of dyes on microbial surfaces because of the presence of negatively charged ligands in cell wall components.

21.2.2 Subsidiary Sources

Subsidiary or indirect sources of xenobiotics include nonsteroidal anti-inflammatory drugs, pharmaceutical products, chemical fertilizers, pesticide residues, etc. Pharmaceutically active compounds, being an indirect source of xenobiotics, are discharged directly by manufacturers of the pharmaceuticals or effluents from hospitals which have performed their biologically intended effect and are passed onto the environment in either their complete or fragmented state. These mainly include hormones, anesthetics, and antibiotics which bioaccumulate in an organism and passed on the other through the common food chain (Iovdijová and Bencko 2010). Biomaterials developed from the synthetic polymers have the biocompatibility, but their degradation into toxic substances in the body is a cause for concern (Baun et al. 2008). Even though they are the indirect sources, they cause adverse effect on the ecological cycle.

Pollution of aquatic bodies and soil is a worldwide problem that can result in uptake and accumulation of toxic chemicals in food chains and also harm to the flora and fauna of affected habitats. Studies of bioaccumulation characteristics of various ecosystems are essential for long-term planning of industrial waste disposal in ecosystem (Iyovo et al. 2010). Bioaccumulation of pesticides and biomagnification processes lead to toxic behavioral effect on animals and mankind. DDT, having a half-life of 10 years, and BHC are chemicals used in pesticides that accumulate in the plant or in plant parts like fruits and vegetables, though both the pesticides are banned now globally.

Nonsteroidal anti-inflammatory drugs (NSAIDs) are the large miscellaneous chemicals used as drugs in human beings and animals for curing the inflammation, body pain, and fever (analgesic aspects). Diclofenac salt uses in animals has been informed to have led to a sharp fall in the vulture population in Pakistan, 95 % decline in 2003 and 99.9 % decline as of 2008 (Oaks et al. 2004).

21.3 Microbial Role in Bioremediation

Microbes epitomize half of biomass of our globe, and the human activity disturbs the environment and initiated the xenobiotic chemicals on the earth. The microbes exhibit capability to biodegrade xenobiotic compounds using their endo- and exoenzymes and by using their metabolic pathways for exploiting them as novel carbon sources and to cleanse toxic compounds (Singh et al. 2014). Microbes show eco-friendly behavior to overcome environmental pollution and to help in biodegradation of xenobiotic compounds. Microorganisms apply two modes of action for degradation of xenobiotics compound: (a) aerobic biodegradation and (b) anaerobic biodegradation. Aerobic biodegradation processes require excess oxygen delivery systems, because it is necessary to supply continuous oxygen due to biofouling in subsurface remedial applications (Sharma and Fulekar 2009); when bioreactors are applied, its energy costs and sludge production are high (Kumar et al. 1994a, b). Anaerobic habitats, including sludge digesters, groundwater, sediments, waterladen soils, gastrointestinal contents, feedlot wastes, and landfill sites (Kumar and Singh 1998), and some xenobiotic compounds (e.g., tetrachloroethylene, polychlorinated biphenyls (PCBs), and nitro-substituted aromatics) can be effectively transformed or mineralized by anaerobic bacteria (Zhang and Bennett 2005).

In situ bioremediation procedure consists of basically three vital steps:

1. Bioattenuation: It is related to monitoring of natural progress of biodegradation to guarantee that contaminant declines with sampling time.
2. Biostimulation: The intentional stimulation of natural xenobiotic remediating microbes by electron acceptors, water molecule, nutrient addition, and/or electron donors.
3. Bioaugmentation: It is the addition of laboratory-grown potential bacteria that have suitable and biodegradative abilities.

Normally, the microbes use two pathways for biodegradation of xenobiotics, aerobic and anaerobic conditions.

In aerobic bioremediation, the basic equation will be

$$\text{Xenobiotics} + \text{oxygen} \overset{\text{Microbe}}{\rightarrow} CO_2 + H_2O + \text{left over residue}$$

In the case of anaerobic bioremediation, it is

$$\text{Xenobiotics} \overset{\text{Microbe}}{\rightarrow} CO_2 + CH_4 + H_2O + \text{left over residue}$$

In aerobic biodegradation, CO_2 is produced along with some amount of water. In absence of oxygen, anaerobic biodegradation process starts and methane gas is generated instead of CO_2. The conversion of biodegradable materials to gases like carbon dioxide, methane, and nitrogen compounds is called mineralization. The mineralization process is completed when all the biodegradable biomass is consumed and all the carbon is converted into carbon dioxide (Kyrikou and Briassoulis 2007). Alkanes consisting long carbon chains and straight structures considered to be more prone to aerobic biodegradation. Aerobic degradation pathway of alkane degradation is the oxidation of the terminal methyl group into a carboxylic acid through an alcohol intermediate and after all completes mineralization through β-oxidation (Le and Coleman 2011). Aerobic biodegradation process of aromatic compounds comprises of their oxidation by molecular oxygen; after oxidation steps, intermediates are the outcome, and then they enter into central metabolic pathways, including the Krebs cycle and β-oxidation.

Some xenobiotic pollutants are not mineralized by an aerobic degradation system because they are greatly recalcitrant owing to increase in halogenations in their structures. Replacement of halogen, nitro-, and sulfo-groups on the aromatic ring increases the electrophilicity of the target molecule. These xenobiotic compounds resist the electrophilic attack by enzyme oxygenases in aerobic degradation process. Some of the recalcitrants that persist under aerobic condition are the polychlorinated biphenyls (PCBs), chlorinated dioxins, and some complex and banned pesticides like DDT (Lin et al. 2014). It is essential to overawe the high stubbornness of halogenated xenobiotic compounds from biosphere; in achieving these, the reductive attacks by anaerobic microorganisms are of boundless worth. On the other hand, anaerobic bacteria carried out reductive dehalogenation either by the complimentary reaction or by using a new type of anaerobic respiration. This procedure decreases the degree of chlorination and makes the product more available and manageable for mineralization process by aerobic bacteria (Ferguson and Pietari 2000). During anaerobic degradation process, the reductive dehalogenation is the first step of biodegradation of PCBs (polychlorinated biphenyls); dehalogenation process is carried out under anaerobic conditions where organic substrates act as electron donors.

There are vast numbers of potential microbes, especially the bacteria, which carry out the bioremediation of xenobiotics. The common major groups of anaerobic bacteria that have capability of biodegrading xenobiotic compounds are *Acidovorax* spp., *Bordetella* spp., *Pseudomonas* spp., *Sphingomonas* spp., *Variovorax* spp., *Veillonella alkalescens*, *Desulfovibrio* spp., *Desulfuromonas michiganensis*, *Desulfitobacterium dehalogenans*, *D. oleovorans*, *G. metallireducens*, and *D. acetonicum*. Anaerobic sulfate-reducing bacteria and methanogenic bacterial conditions can be useful to isolate pure culture of anaerobic bacteria to carry out xenobiotic degradation research work (Jiang and Fan 2008; Zhang and Bennet 2005). Anaerobic microbes can also use and exploit substituted and intricate aromatic compounds in a way that do not disturb the benzene nucleus in the ring. On the other hand, sulfate-reducing bacteria (SRB) represent a huge group of anaerobic microorganisms that play a crucial role in numerous biogeochemical cyclic processes and also able to biodegrade the crude oil (Ferradji et al. 2014; Liu et al. 2014a). The sulfate-reducing bacteria are obligated anaerobic bacteria, which utilize sulfate as final electron acceptor during the process of anaerobic respiration and, therefore, generate hydrogen sulfide (H_2S gas) by sulfate reduction. Anaerobic degradation process is also a renewable energy source; here the biogas is generated from the anaerobic digestion. It mainly consists of methane (CH_4) that can be collected easily and applied for eco-friendly power generation or as a fuel, which has been proved on a greater scale (Boetius et al. 2000). Different xenobiotic compounds biodegraded by various microbes are mentioned in Table 21.1.

21.4 Role of Microbial Enzymes in Bioremediation

Bioremediation is a microbial secreted enzymatic process which transforms a xenobiotic pollutant to innocuous products, which blends naturally with the environment; therefore, the toxicity is removed or reduced to a greater extent.

21.4.1 Oxidoreductases

These enzymes slice the chemical bonds and reposition the electrons from a reduced organic compound (called as donor) to another chemical substrate (known as acceptor). During this oxidation reduction process, the chemical pollutants or contaminants are oxidized to inoffensive and harmless compounds (Karigar and Rao 2011). The oxidoreductases cleanse toxic xenobiotic products like phenolic or anilinic compounds, either by the process of polymerization, or copolymerization with other substrates, or binding with the humic substances. The microbial enzymes have also been used in decolorization and bioremediation of azo dyes (Husain 2006; Rani et al. 2014).

Table 21.1 Xenobiotics biodegraded by the microbes

S. no.	Microbe targeting the xenobiotic	Target xenobiotic	Place	References
1	Phenanthrene	*Pseudomonas* sp. Ph6	China	Sun et al. (2014)
	Phenanthrene	*Massilia sp.* strain Pn2	China	Liu et al. (2014b)
2	Anthracene	*Microbacterium* sp. strain SL10	Lagos, Nigeria	Salam et al. (2014)
3	Naphthalene	*Streptomyces* spp.	Algeria	Ferradji et al. (2014)
4	Pentachlorophenol	*Kocuria* sp. CL2	India	Karn et al. (2011)
5	Chloroaniline	*Acinetobacter baylyi* strain GFJ2	Thailand	Hongsawat and Vangnai (2011)
6	Fluoranthene	*Herbaspirillum chlorophenolicum*	China	Xu et al. (2011)
8	1,2,4-Trichlorobenzene (1,2,4-TCB)	*Bordetella* sp.	Germany	Wang et al. (2007)
9	2-Chlorobenzoic acid	*Pseudomonas* sp., *Enterobacter* sp., *Acinetobacter* sp., and *Corynebacterium* sp.	Iran	Kafilzadeh et al. (2012)
10	Pyrene	*Klebsiella oxytoca* PYR-1	China	Zhang and Zhu (2012)
11	HCH/lindane (1,2,3,4,5,6-hexachlorocyclohexane)	*Sphingobium czechense* LL01	India	Niharika et al. (2013)
12	DDT (dichlorodiphenyltrichloroethane)	*Pseudoxanthobacter liyangensis* sp. nov.	China	Liu et al. (2014b)
13	DDT (dichlorodiphenyltrichloroethane)	*Serratia marcescens* DT-1P	India	Bidlan and Manonmani (2002)
14	DDT (dichlorodiphenyltrichloroethane)	*Novosphingobium arabidopsis* sp. nov.	Taiwan	Lin et al. (2014)
13	Phthalate	*Achromobacter denitrificans* strain SP1	India	Pradeep et al. (2015)
14	Phthalate	*Arthrobacter* sp.C21	China	Wen et al. (2014)
15	Phthalate	Wen et al. (2014)	China	Wu et al. (2010)
16	Endosulfan compounds	*Paenibacillus* sp. ISTP10	India	Kumari et al. (2014)
17	Endosulfan compounds	*Stenotrophomonas maltophilia* and *Rhodococcus erythropolis*	India	Kumar et al. (2007)
18	Endosulfan compounds	*Klebsiella pneumonia*	South Korea	Kwon et al. (2002)

(continued)

Table 21.1 (continued)

S. no.	Microbe targeting the xenobiotic	Target xenobiotic	Place	References
19	Vinyl chloride	*Micrococcus* species	India	Patil and Bagde 2012)
20	Vinyl chloride	*Sphingopyxis* sp. PVA3	Japan	Yamatsu et al. (2006)
20	Diuron DCMU (3-(3,4-dichlorophenyl)-1,1-dimethylurea)	*Arthrobacter* sp. BS2 and *Achromobacter* sp. SP1	France	Devers-Lamrani et al. (2014)
	Diuron DCMU (3-(3,4-dichlorophenyl)-1,1-dimethylurea)	*Pseudomonas* sp. and *Stenotrophomonas* sp.	France	Batisson et al. (2007)
	Atrazine (2-chloro-4-ethylamino-6-isopropylamino-1,3-5-triazine)	*Raoultella planticola*	Israel	Swissa et al. (2014)
	Atrazine (2-chloro-4-ethylamino-6-isopropylamino-1,3-5-triazine)	*Bacillus subtilis* strain HB-6	China	Wang et al. (2014)
	Atrazine (2-chloro-4-ethylamino-6-isopropylamino-1,3-5-triazine)	*Arthrobacter* sp. HB-5	China	Wang et al. (2011)
	Propanil	*Xanthomonas* sp.	Mexico	Herrera-Gonzalez et al. (2013)
	Propanil	*Catellibacterium nanjingense* sp. nov.	China	Zhang et al. (2012)
	PCE (tetrachloroethylene or perchloroethylene)	*Dehalococcoides* spp.	Germany	Kranzioch et al. (2014)
	PCE (tetrachloroethylene or perchloroethylene)	*Propionibacterium* sp. HK-1and *Propionibacterium* sp. HK-3	South Korea and Japan	Chang et al. (2011)

21.4.2 Monooxygenases

These enzymes transfer one atom of molecular oxygen to the organic compound (Karigar and Rao 2011). Monooxygenases can be categorized into two subclasses based on the presence of cofactors, flavin-dependent monooxygenases and P450 monooxygenases. Flavin-dependent monooxygenases contain flavin as prosthetic group and NADP or NADPH as coenzyme. P450 monooxygenases are heme-containing oxygenases that persist in both eukaryotes and prokaryotes. Monooxygenases act as biocatalysts in the bioremediation process and synthetic chemistry because they are highly regionselective and stereoselective on a wide range of substrates (Karigar and Rao 2011). Monooxygenases catalyze enormous reactions such as desulfurization, dehalogenation, denitrification, ammonification, hydroxylation, biotransformation, and biodegradation of various aromatic and aliphatic compounds.

21.4.3 Dehalogenases

Dehalogenase plays an important role in the degradation of chlorinated pollutant (Copley 1998). Some anaerobic microorganisms exploit dehalorespiration and use halogenated compounds as terminal electron acceptors (Le and Coleman 2011). An example of this process is the conversion of either perchloroethylene(PCE), dichloroethylene (DCE) (Schumacher and Holliger 1996), ethylene, or ethane depending on the conditions. Magnuson et al. (1998) reported the partial purification of two reductive dehalogenases from *Dehalococcoides ethenogenes* strain 195; both enzymes are membrane proteins. The first enzyme PCE reductive dehalogenase reduces PCE to TCE, and the second enzyme TCE-reductive dehalogenase reduces TCE, *trans*-DCE, *cis*-DCE, 1,1-dichloroethene, and vinyl chloride (Patil and Bagde 2012).

21.4.4 Phosphotriesterases

Phosphotriesterases (PTEs) are microbial isolated enzyme which hydrolyze and detoxify organophosphate pesticides (OPs). This reduces OP toxicity, and it decreases the ability of OPs to inactivate AchE (Shen et al. 2010; Theriot and Grunden 2010). These enzymes mainly hydrolyze phosphoester bonds like P–O, P–F, P–NC, and P–S, and these hydrolysis mechanisms include water molecule in the phosphorus center (Ortiz-Hernandez et al. 2003).

21.4.5 Dioxygenases

These are multicomponent enzyme systems that incorporate molecular oxygen to the substrate. On the basis of the complexity of the degradation pathways, the biodegradation phenomenon can be categorized into two types: (1) convergent mode and (2) divergent mode of degradation (Eltis and Bolin 1996). In the convergent mode, structurally varied aromatic compounds are converted to aromatic ring cleavage substrates catechol, gentisate, protocatechuate, and their derivatives (Meer et al. 1992). In divergent mode, metal-dependent dioxygenase channels operate, and dihydroxylated intermediates are formed by one of the two possible pathways: the meta-cleavage pathway or the ortho-cleavage pathway (Takami et al. 1997).

21.4.6 Oxygenases

These are classified under the oxidoreductase group of enzymes (E.C. Class 1) (Karigar and Rao 2011). Oxidation reaction is the major enzymatic reaction of aerobic biodegradation and is catalyzed by oxygenases. Oxygenases oxidize the substrates by transferring oxygen from molecular oxygen (O_2) and utilize FAD/NADH/NADPH as the co-substrate. Oxygenases metabolize organic compounds; they increase their reactivity and water solubility and cleave the aromatic ring (Arora et al. 2010). On the basis of the number of oxygen atoms used for oxidation, oxygenases can be further categorized into two groups, (1) monooxygenases and (2) dioxygenases, which have been discussed earlier.

21.5 Future Facets

During the past many years, there has been a boundless work of development in the field of the bioremediation of xenobiotic compounds. Numerous novel microbes bearing bioremediation prospective have been isolated from various ecological niches, and several new remediation pathways have been explicated. However, this information and data is far from complete knowledge. Biotransformation of organosulfide compounds is yet to be explored owing to its complex nature. Efficiency of xenobiotic compound biodegradation can be meaningfully enhanced by addressing vital issues such as tolerance to various xenobiotic compounds, the constitutive expression of catabolic genes and their raw substrate specificity, and the kinetics and stability of the enzyme which has been encoded. Though, the usefulness and efficacy of the constructed organisms in relation to the environmental pollution problem in the ecosystem is yet to be explored and tested.

Most of the microbes, which biodegrade xenobiotic compounds, bear plasmids which encode for the catabolic genes. To depict and describe the suitable genes and to augment the process of biodegradation through improved constructed potential strains, a proper, well-designed management is prerequisite. And due to the same reasons, the microbial degradation machinery is a spanning spectrum from the environmental monitoring point of view which ultimately leads to biodegradation as well. In bioremediation process, presently, molecular techniques and approaches are being applied to characterize the genetic material of numerous bacteria from the several ecological samples. Comparing with the standard and prevalent microbiological techniques and approaches, the molecular procedures provide us with more complete and inclusive interpretation of in situ microorganism population and its response to concocted bioremediation and normal lessening processes. Additional dominant molecular procedure known as metagenomic libraries has been thrived for identification of the desired catabolic genes. Fundamentally, metagenomic technique is a culture-dependent microbial genomic analysis; this technique is either a function-driven tactic or sequence-driven method, of entire microbial communities, which provides the access to recover unknown sequences. The regular and constant contact with the contaminants and prolonged exposure to their presence is the basics of struggle against xenobiotic compounds, since such processes enable the evolution and progression of new, more or less safe processes of xenobiotic remediation by microorganisms.

21.6 Conclusions

Microbial diversity, the richness of species in environmental sites, provides a huge reservoir of resources which we can utilize for our benefit. However, little is known about the true diversity of bacterial life. Despite the acknowledged value of microorganisms, our understanding of their diversity and many of their key roles in sustaining global life support systems is still very scarce. This is because the vast majority of bacteria are non-culturable by standard methods and we have only recently acquired the skills to explore this aspect of microbial biodiversity. Exploring the range of microbial biodiversity is the key to developing effective and environment-friendly "green" technologies. Bioremediation is one such process that exploits the catabolic abilities of microorganisms to degrade harmful and toxic xenobiotics. We have been able to restore what once were irreversibly polluted sites in some cases, attesting to the usefulness of this clean-up process. However, to maximize the potential benefits of the microbial community in combating pollution problems, it is vital that we have fundamental understanding of a microbe's degradative potential under various conditions, its biochemical systems, and its molecular biology.

Environmental problems caused by the industrial effluents is mainly due to accumulation of pollutants and other fragmented compounds, which in turn form into other substitutes (natural or manmade), finally forming a xenobiont. There is a quick need to degrade these xenobiotic compounds in an eco-friendly way. Various techniques like microbial remediation, phytoremediation, and photoremediation and their subtypes have been discussed. Each having their own ways of degrading, these xenobionts also have negative impact on the environment (side effects due to fragmentations and bioaccumulations). Photoremediation is a novel equipment-based technique which is rapid but also has a negative impact on the environment. Being a solar-driven technique, phytoremediation is restricted to particular sites containing contaminants. Although slow, on the whole, microbial bioremediation was found to cover a wide range of recalcitrant degradation and is known to be a better choice because of its nature of degradation.

References

Arora PK, Srivastava A, Singh VP (2010) Application of monooxygenases in dehalogenation, desulphurization, denitrification and hydroxylation of aromatic compounds. J Bioremed Biodegrad 1:1–8

Batisson I, Pesce S, Hoggan PB, Sancelme M, Bohatier J (2007) Isolation and characterization of diuron degrading bacteria from lotic surface water. Microbial Ecol 54(4):761–770

Baun A, Sørensen SN, Rasmussen RF, Hartmann NB, Koch CB (2008) Toxicity and bioaccumulation of xenobiotic organic compounds in the presence of aqueous suspensions of aggregates of nano-C(60). Aquat Toxicol 86(3):379–387

Bidlan R, Manonmani HK (2002) Aerobic degradation of dichlorodiphenyl trichloroethane (DDT) by *Serratia marcescens* DT-1P. Process Biochem 38:49–56

Boetius A, Ravenschlag K, Schubert CJ, Rickert D, Widdel F, Gieseke A, Amann R, Jorgense BB, Witte U, Pfannkuche O (2000) A marine microbial consortium apparently mediating anaerobic oxidation of methane. Nature 407:623–626

Chang YC, Ikeutsu K, Toyama T, Choi DB, Kikuchi S (2011) Isolation and characterization of tetrachloroethylene- and cis-1, 2-dichloroethylene-dechlorinating *Propionibacteria*. J Ind Microbiol Biotechnol 38:1667–1677

Copley SD (1998) Microbial dehalogenases: enzymes recruited to convert xenobiotic substrates. Curr Opin Chem Biol 2:613–617

Devers-Lamrani M, Pesce S, Rouard N, Martin-Laurent F (2014) Evidence for cooperative mineralization of diuron by *Arthrobacter* sp. BS2 and *Achromobacter* sp. SP1 isolated from a mixed culture enriched from diuron exposed environments. Chemosphere 117:208–215

Dixit R, Wasiullah MD, Pandiyan K, Singh UB, Sahu A, Shukla R, Singh BP, Rai JP, Sharma PK, Lade H, Paul D (2015) Bioremediation of heavy metals from soil and aquatic environment: an overview of principles and criteria of fundamental processes. Sustainability 7:2189–2212

Eltis LD, Bolin JT (1996) Evolutionary relationships among extradioldioxygenases. J Bacteriol 178:5930–5937

Ferguson JF, Pietari JM (2000) Anaerobic transformations and bioremediation of chlorinated solvents. Environ Pollut 107(2):209–215

Ferradji FZ, Mnif S, Badis A, Rebbani S, Fodil D, Eddouaouda K, Sayadi S (2014) Naphthalene and crude oil degradation by biosurfactant producing *Streptomyces* spp. isolated from Mitidja plain soil (North of Algeria). Int Biodeterior Biodegrad 86(C):300–308

Gojgic-Cvijovic GD, Milic JS, Solevic TM, Beskoski VP, Ilic MV, Djokic LS, Narancic TM, Vrvic MM (2012) Biodegradation of petroleum sludge and petroleum polluted soil by a bacterial consortium: a laboratory study. Biodegradation 23:1–14

Gren I (2012) Microbial transformation of xenobiotics. Chemik 66(8):835–842

Guermouche M'rassi A, Bensalah F, Gury J, Duran R (2015) Isolation and characterization of different bacterial strains for bioremediation of n-alkanes and polycyclic aromatic hydrocarbons. Environ Sci Pollut Res Int 22(20):15332–15346

Gursahani YH, Gupta SG (2011) Decolourization of textile effluent by a thermophilic bacteria *Anoxybacillus rupiensis*. J Pet Environ Biotechnol 2:111–120

Hamzah A, Rabu A, Farzarul R, Yussoff NA (2010) Isolation and characterization of bacteria degrading Sumandak and South Angsi oils. Sains Malaysiana 39(2):161–168

Hemapriya J, Vijayanand S (2014) Ecofriendly bioremediation of a triphenylmethane dye by textile effluent adapted bacterial strain vp-64. Int J Curr Microbiol Appl Sci 3(9):983–992

Herrera-Gonzalez VE, Ruiz-Ordaz N, Galindez-Mayer J, Juarez-Ramirez C, Santoyo-Tepole F, Montiel EM (2013) Biodegradation of the herbicide propanil, and its 3, 4-dichloroaniline by product in a continuously operated biofilm reactor. World J Microbiol Biotechnol 29:467–474

Hongsawat P, Vangnai AS (2011) Biodegradation pathways of chloroanilines by *Acinetobacter baylyi* strain GFJ2. J Hazard Mater 186(2/3):1300–1307

Husain Q (2006) Potential applications of the oxidoreductive enzymes in the decolorization and detoxification of textile and other synthetic dyes from polluted water: a review. Crit Rev Biotechnol 26(4):201–221

Iovdijová A, Bencko V (2010) Potential risk of exposure to selected xenobiotic residues and their fate in the food chain-part I: classification of xenobiotics. Ann Agric Environ Med 17(2): 183–192

Itoh K, Fujita M, Kumano K, Suyama K, Yamamoto H (2000) Phenolic acids affect transformations of chlorophenols by a *Coriolus versicolor laccase*. Soil Biol Biochem 32:85–91

Iyovo GD, Du G, Chen J (2010) Sustainable bioenergy bioprocessing: biomethane production, digestate as biofertilizer and as supplemental feed in algae cultivation to promote algae biofuel commercialization. J Microb Biochem Technol 2(4):100–106

Jiang W, Fan W (2008) Bioremediation of heavy metal contaminated soils by sulfate reducing bacteria. Ann NY Acad Sci 1140:446–454

Kafilzadeh F, Nikvarz M, Jabbari S, Tahery Y (2012) Evaluation of biodegradation of 2-chlorobenzoic acid by isolated bacteria from landfill soils in Shiraz, Iran. Afr J Microbiol 6(27):5708–5714

Karigar CS, Rao SS (2011) Role of microbial enzymes in the bioremediation of pollutants: a review. Enzym Res 2011:11, 805187

Karn SK, Chakrabarti SK, Reddy MS (2011) Degradation of pentachlorophenol by *Kocuria* sp. CL2 isolated from secondary sludge of pulp and paper mill. Biodegradation 22:63–69

Karpouzas DG, Singh BK (2006) Microbial degradation of organophosphorus xenobiotics: metabolic pathways and molecular basis. Adv Microb Physiol 51:119–185

Kathiresan K (2003) Polythene and plastic degrading microbes in an Indian mangrove soil. Rev Biol Trop 51(3–4):629–633

Kranzioch I, Ganz S, Tiehm A (2014) Chloroethene degradation and expression of *Dehalococcoides dehalogenase* genes in cultures originating from Yangtze sediments. Environ Sci Pollut Res 22(4):3138–3148

Kumar ARC, Singh R (1998) Methanogenesis by enrichment cultures from distillery effluent. Indian J Microbiol 38:109–110

Kumar V, Anand RC, Singh R (1994a) Enrichment and isolation of acetoclastic methanogens from distillery effluent. Ann Biol 10(2):253–256

Kumar V, Anand RC, Singh R (1994b) Methanogenesis by pure isolates from distillery effluent digester. Ann Biol 10(2):257–260

Kumar K, Devi SS, Krishnamurthi K, Kanade GS, Chakrabarti T (2007) Enrichment and isolation of endosulfan degrading and detoxifying bacteria. Chemosphere 68(2):317–322

Kumar M, Varma A, Kumar V (2016) Ecogenomics based microbial enzyme for biofuel industry. Sci Int 4:1–11

Kumari M, Ghosh P, Swati, Thakur IS (2014) Microcosmic study of endosulfan degradation by *Paenibacillus* sp. ISTP10 and its toxicological evaluation using mammalian cell line. Int Biodeter Biodegrad 96:33–40

Kwon GS, Kim JK, Kim TK, Sohn HY, Koh SC, Shin KS, Kim DG (2002) *Klebsiella pneumoniae* KE-1 degrades endosulfan without formation of the toxic metabolite, endosulfan sulphate. Microbiol Lett 215:255–259

Kyrikou J, Briassoulis D (2007) Biodegradation of agricultural plastic films: a critical review. J Polym Environ 15:125–150

Le NB, Coleman NV (2011) Biodegradation of vinyl chloride, cis-dichloroethene and 1,2-dichloroethane in the alkene/alkane oxidising *Mycobacterium* strain NBB4. Biodegradation 22:1095–1108

Leys NM, Bastiaens L, Verstraete W, Springael D (2005) Influence of the carbon/nitrogen/phosphorus ratio on polycyclic aromatic hydrocarbons degradation by *Mycobacterium* and *Sphingomonas* in soil. Appl Microbiol Biotechnol 66:726–736

Lin C, Gan L, Chen ZL (2010) Biodegradation of naphthalene by strain *Bacillus fusiformis* (BFN). J Hazard Mater 182(1/3):771–777

Lin SY, Hameed A, Liu YC, Hsu YH, Lai WA, Huang HI, Young CC (2014) *Novosphingobium arabidopsis* sp. nov., a DDT resistant bacterium isolated from the rhizosphere of *Arabidopsis thaliana*. Int J Syst Evol Microbiol 64(2):594–598

Liu H, Yao J, Yuan Z, Shang Y, Chen H, Wang F, Masakorala K, Yu C, Cai M, Blake RE, Choi MMF (2014a) Isolation and characterization of crude oil degrading bacteria from oil water mixture in Dagang oil field, China. Int Biodet Biodegrad 87:52–59

Liu XM, Chen K, Meng C, Zhang C, Zhu JC, Huang X, Li SP, Jiang JD (2014b) *Pseudoxanthobacter liyangensis* sp. nov., isolated from dichlorodiphenyl trichloroethane contaminated soil. Int J Syst Evol Microbiol 64:3390–3394

Magnuson JK, Stern RV, Gossett JM, Zinder SH, Burris DR (1998) Reductive dechlorination of tetrachloroethene to ethene by a two component enzyme pathway. Appl Environ Microbiol 64: 1270–1275

Meer JR, van der de Vos WM, Harayama S, Zehnder AJB (1992) Molecular mechanisms of genetic adaptation to xenobiotic compounds. Microbiol Rev 56:677–694

Milić JS, Beškoski VP, Iilic MV, Ali SAM, Gojgic-Cvijovic GD, Vric MM (2009) Bioremediation of soil heavily contaminated with crude oil and its products: composition of the microbial consortium. J Serb Chem Soc 74(4):455–460

Niharika N, Moskalikova H, Kaur J, Khan F, Sedlackova M, Hampl A, Damborsky J, Prokop Z, Lal R (2013) *Sphingobium czechense* sp. nov. isolated from a hexachlorocyclohexane dump site. Int J Syst Evol Microbiol 63:723–728

Oaks JL, Gilbert M, Virani MZ, Watson RT, Meteyer CU, Rideout BA, Shivaprasad HL, Ahmed S, Chaudhry MJ, Arshad M, Mahmood S, Ali A, Khan AA (2004) Diclofenac residues as the cause of vulture population decline in Pakistan. Nature 427:630–633

Ortiz-Hernandez ML, Quintero-Ramirez R, Nava-Ocampo AA, Bello-Ramirez AM (2003) Study of the mechanism of *Flavobacterium* sp. for hydrolyzing organophosphate pesticides. Fundam Clin Pharmacol 17(6):717–723

Patil R, Bagde US (2012) Isolation of polyvinyl chloride degrading bacterial strains from environmental samples using enrichment culture technique. Afr J Biotechnol 11(31):7947–7956

Pradeep S, Josh MKS, Binod P, Devi RS, Balachandran S, Anderson RC, Benjamin S (2015) *Achromobacter denitrificans* strain SP1 efficiently remediates di (2-ethylhexyl) phthalate. Ecotoxicol Environ Saf 112:114–121

Prakash A, Bisht S, Singh J, Teotia P, Kela R, Kumar V (2014) Biodegradation potential of petroleum hydrocarbons (PHCs) by bacteria and mixed bacterial consortium isolated from contaminated sites. Turk J Eng Environ Sci 38(1):31–40

Prpich GP, Adams RL, Daugulis AJ (2006) Ex situ bioremediation of phenol contaminated soil using polymer beads. Biotechnol Lett 28(24):2027–2031

Raaman N, Rajitha N, Jayshree A, Jegadeesh R (2012) Biodegradation of plastic by *Aspergillus* spp. isolated from polythene polluted sites around Chennai. J Acad Indus Res 1(6):313–317

Rani B, Kumar V, Singh J, Bisht S, Teotia P, Sharma S, Kela R (2014) Bioremediation of dyes by fungi isolated from contaminated dye effluent sites for bio-usability. Braz J Microbiol 45(3): 1055–1063

Salam LB, Obayori OS, Olatoye NO (2014) Biodegradation of anthracene by a novel actinomycete, microbacterium sp. isolated from tropical hydrocarbon contaminated soil. World J Microbiol Biotechnol 30(1):335–341

Schumacher W, Holliger C (1996) The proton/electron ration of themenaquinone dependent electron transport from dihydrogen to tetrachloroethene in *Dehalobacter restrictu*. J Bacteriol 178: 2328–2333

Selvam K, Vishnupriya B (2013) Bioremediation of xenobiotic compounds and heavy metals by the novel actinobacteria. Int Pharm Chem Sci 2(3):1589–1597

Sethy NK, Jha VN, Sahoo SK, Shukla AK, Tripathi RM (2011) Ground water ingestion dose due to intake of radionuclide (Natural U and 226Ra) to population around uranium mining complex at Jaduguda. J Ecosyst Ecograph 1:104

Shahid A, Singh J, Bisht S, Teotia P, Kumar V (2013) Biodegradation of textile dyes by fungi isolated from North Indian field soil. Env Asia 6(2):51–57

Sharma J, Fulekar MH (2009) Potential of *Citrobacter freundii* for bioaccumulation of heavy metal copper. Biol Med 1(3):7–14

Sharma SK, Saxena M, Mandal TK, Ahammed YN, Pathak H (2011) Variations in mixing ratios of ambient ammonia, nitric oxide and nitrogen dioxide in different environments of India. J Food Process Technol 1:101

Shen YJ, Lu P, Mei H, Yu HJ, Hong Q, Li SP (2010) Isolation of a methyl parathion degrading strain *Stenotrophomonas* sp. SMSP-1 and cloning of the ophc2 gene. Biodegradation 21:785–792

Singh R, Singh P, Sharma R (2014) Microorganism as a tool of bioremediation technology for cleaning environment: a review. Proc Int Acad Ecol Environ Sci 4(1):1–6

Sun K, Liu J, Gao Y, Jin L, GuY, Wang, W (2014) Isolation, plant colonization potential and phenanthrene degradation performance of the endophytic bacterium *Pseudomonas* sp. Ph6-gfp. Sci Rep 4(5462): 1–11

Swissa N, Nitzan Y, Langzam Y, Cahan R (2014) Atrazine biodegradation by a monoculture of *Raoultella planticola* isolated from a herbicides wastewater treatment facility. Int Biodeterior Biodegrad 92:6–11

Takami H, Kudo T, Horikoshi K (1997) Isolation of extradiol dioxygenasegenes that is phylogenetically distant from other meta cleavage dioxygenase genes. Biosci Biotechnol Biochem 61:530–532

Theriot CM, Grunden AM (2010) Hydrolysis of organophosphorus compounds by microbial enzymes. Appl Microbiol Biotechnol 89:35–43

Vidali M (2001) Bioremediation. An overview. Pure Appl Chem 73:1163–1172

Vigneeswaran M, Prabakaran V, Arthysurendr S (2012) Biodegradation and bioremediation of azo dye entrenched soil by *Pseudomonas* sp. Int J Chem Anal Sci 3(5):1381–1384

Wang F, Grundmanna S, Schmid M, Dorflera U, Rohererb S, Muncha JC, Hartmann A, Jiang X, Schroll R (2007) Isolation and characterization of 1,2,4-trichlorobenzene mineralizing *Bordetella* sp. and its bioremediation potential in soil. Chemosphere 67(5):896–902

Wang J, Zhu L, Liu A, Ma T, Wang Q, Xie H, Wang J, Jiang T, Zhao R (2011) Isolation and characterization of an *Arthrobacter* sp. Strain HB-5 that transforms atrazine. Environ Geochem Health 33:259–266

Wang J, Zhu L, Wang Q, Wang J, Xie H (2014) Isolation and characterization of atrazine mineralizing *Bacillus subtilis* Strain HB-6. PLoS One 9(9):1–8

Wen ZD, Gao DW, Wu WM (2014) Biodegradation and kinetic analysis of phthalates by an arthrobacter strain isolated from constructed wetland soil. Appl Microbiol Biotechnol 98: 4683–4690

Wu XL, Wang YY, Liang RX, Dai QY, Chao WL (2010) Degradation of Di-n-butyl phthalate by newly isolated *Ochrobactrum* sp. Bull Environ Contamin Toxicol 85:235–237

Xu HX, Wu HY, Qiu YP, Shi XQ, He GH, Zhang JF, Wu JC (2011) Degradation of fluoranthene by a newly isolated strain of *Herbaspirillum chlorophenolicum* from activated sludge. Biodegradation 22:335–345

Yamatsu A, Matsumi R, Atomi H, Imanaka T (2006) Isolation and characterization of a novel poly (vinyl alcohol) degrading bacterium, *Sphingopyxis* sp. PVA3. Appl Microbiol Biotechnol 72 (4):804–811

Yeom SH, Daugulis AJ, Lee SH (2010) Bioremediation of phenol contaminated water and soil using magnetic polymer beads. Process Biochem 45(9):1582–1586

Zhang C, Bennett GN (2005) Biodegradation of xenobiotics by anaerobic bacteria. Appl Microbiol Biotechnol 67:600–618

Zhang D, Zhu L (2012) Effects of Tween 80 on the removal, sorption and biodegradation of pyrene by *Klebsiella oxytoca* PYR-1. Environ Pollut 164:169–174

Zhang J, Chen SA, Zheng JW, Cai S, Hang BJ, He J, Li SP (2012) *Catellibacterium nanjingense* sp. nov., a propanil degrading bacterium isolated from activated sludge, and emended description of the genus *Catellibacterium*. Int J Syst Evol Microbiol 62:495–499

Chapter 22
Biotransformation of Xenobiotic Compounds: Microbial Approach

Rupesh Kumar Basniwal, Nim J. Singh, Manoj Kumar, Vivek Kumar, Narendra Tuteja, Ajit Varma, and Pankaj Goyal

22.1 Introduction

The processes of biodegradation, mineralization, and transformation conducted by microorganisms are the foundation of contemporary scuffle against the rising volumes of xenobiotics released into the environment. Xenobiotic refers to a strange, exogenous substance or anthropogenic material substance that is not a natural component of a living organism exposed to it. Some of xenobiotics are chemical compounds which are produced by humans and are having artificial chemical structure, to which organisms have not acclimatized through prior evolution (Gren 2012). Other important groups of xenobiotics are aromatic compounds with chlorine substituent. Major types of xenobiotics include pesticides, fuels, solvents, alkanes, polycyclic hydrocarbons (PAHs), antibiotics, synthetic azo dyes, pollutants (dioxins and polychlorinated biphenyls), and polyaromatic, chlorinated, and nitro-aromatic compounds.

Natural and anthropogenic activities along with continuous efflux from population and industrial drainages result in the accumulation of wide ranges of toxic xenobiotic compounds and created a thoughtful impact on the pristine nature of our environment which is of a serious concern (Gienfrada and Rao 2008; Sinha et al. 2009).

R.K. Basniwal
Amity Institute for Advanced Research and Studies, Amity University, Noida, Uttar Pradesh, India

N.J. Singh
School of Biotechnology, Gautam Buddha University, Noida, Uttar Pradesh 201308, India

M. Kumar • V. Kumar • N. Tuteja • A. Varma • P. Goyal (✉)
Amity Institute of Microbial Technology, Amity University, Noida, Uttar Pradesh, India
e-mail: pankaj.goyals@gmail.com

© Springer International Publishing AG 2017

M.Z. Hashmi et al. (eds.), *Xenobiotics in the Soil Environment*, Soil Biology 49,
DOI 10.1007/978-3-319-47744-2_22

It has also reported that the *Amblyomma americanum* tick produces 2,6-dichlorophenol as pheromone and the 2,5-dichlorophenol in the same function was also found in grasshoppers (Gribble 2004). Therefore, as far as microorganisms are concerned, xenobiotics may be defined as the substances with the structure with which the microorganism has had no direct contact during their evolution. It is also noteworthy to point out that all xenobiotics are not necessarily toxic. If it had not been for the presence of the sets of genes encoding the enzymes of xenobiotic decomposition pathways, both aerobic and anaerobic, which are located on the mobile genetic components of the cell, such as plasmids or transposons, life on Earth could not have survived.

Xenobiotic compounds, due to their structure and strange nature to the natural microorganisms, are released into the environment and can linger there for months or years. However, this does not mean that these compounds are never degraded, since some of the anthropogenic substances are sensitive to microbial degradation. Moreover, adaptive processes such as horizontal gene transfer allow the microbes to participate in the biotransformation of xenobiotics processes. Constant contact with xenobiotic contaminants and prolonged exposure to their presence enable the evolution of new, more or less secure processes of xenobiotic biotransformations by microorganisms (Kim and Crowley 2007).

Although most of the organisms have detoxifying abilities such as mineralization, transformation, and/or immobilization of pollutants, microorganisms, particularly bacteria (Table 22.1), play a crucial role in biogeochemical cycles for sustainable development of the biosphere (Tropel and Van der Meer 2004).

A number of microbial communities have been identified as potential degrader of pollutants that can adapt to use these chemicals as their novel growth and energy substrates (Janssen et al. 1985).

Evaluation of the xenobiotic contaminated areas is quite essential with respect to enumeration and detection of the xenobiotic degrader; however, enumeration and monitoring of xenobiotic-degrading bacterial populations in contaminated environment with traditional microbiological methods can acquire an unreasonable length of time (Lloyd-Jones et al. 1999). To conquer this challenge, modern molecular approaches are proficient to epitomize the nucleic acids of microorganisms contained in the microbial community from environmental samples (Widada et al. 2002).

Biotransformation process involves the change of the structure of the original chemical compounds to such degree that its original characteristic properties change as well. During this process, physicochemical properties of compounds, such as solubility or bioavailability, as well as the toxicity level of the given xenobiotic are modified. If the compound undergoes biodegradation, it is then used as a growth substrate, viz., the source of carbon, nitrogen, phosphorus, or sulfur and energy by the microorganism conducting the process. It often happens in the natural environment and accompanies by transformations of other compounds, a phenomenon known as co-metabolism, co-oxidation, or accidental or free metabolism (Horvath 1972).

Table 22.1 Xenobiotics and their degrading bacterial genera (Source: Sinha et al. 2009)

Xenobiotic compounds	Bacteria degrading the compounds
Pesticides	
Endosulfan compounds	*Mycobacterium* sp.
Endosulphate compounds	*Arthrobacter* sp.
HCH	*Pseudomonas putida*
2,4-D	*Alcaligenes eutrophus*
DDT	*Dehalospirilum multivorans*
Halogenated organic compounds	
Vinyl chloride	*Dehalococcoides* sp.
Atrazine	*Pseudomonas* sp.
PCE	*Dehalococcoides ethenogenes*195
PAH compounds	
Napthalene	*Pseudomonas putida*
PCP	*Psedomonas* sp.
3CBA	*Arthrobacter* sp.
1,4DCB	*Alcaligenes* sp.
2,3,4-chloroaniline	*Pseudomonas* sp.
2,4,5-T	*Pseudomonas* sp.
Fluoranthene	*Pseudomonas cepacia*
Pyrene	*Mycobacterium*
	Sphingomonas paucimobilis
Phthalate compounds	
Phthalate	*Burkholderia cepacia*
PCB	*Rhodococcus*RHA1
Dioxins	*Dehalococcoides* sp.
RDX	*Desulfovibrio* sp.
Benzene	*Dechloromonas* sp.
Petroleum products	
	Achromobacter sp.
	Acinetobacter sp.
	Micrococcus sp.
	Nocardia sp.
	Bacillus sp.
	Flavobacterium sp.
Azo dyes	
	Bacillus sp.
	Pseudomonas sp.
	Sphingomonas sp.
	Xanthomonas sp.

With a number of subsequent co-metabolic reactions, the xenobiotic, which in the presence of a single microbial strain would only undergo minor transformation, might become completely mineralized (Knackmuss 1996). Aranda et al. (2003) described two strains of *Sphingopyxis chilensis* which, if both present in the culture at the same time, conducted complete mineralization of the nongrowth substrate of 2,4,6-trichlorophenol. The strains did not degrade 2,4,6-trichlorophenol on their own. Because no accumulation of intermediate products of this xenobiotic was observed during co-metabolism, a new definition was proposed for this type of complete degradation process, i.e., secondary utilization. However, the processes of co-metabolism of xenobiotics do not benefit the conducting microorganisms in any measurable way, with the exception of eliminating a potentially toxic substance from their environment or at least reducing its toxicity due to the change of its molecular structure (Gren 2012). It is also possible that the toxicity of the co-metabolic transformation product might turn out to be higher than that of the original compound. For example, cleavage product of the aromatic ring of 3-chlorocatechol by catechol 2,3-dioxygenase is, acyl chloride. If the ring-cleaving dioxygenase does not have the capability to remove the chlorine substituent, then the chlorine substituent in acyl chloride is substituted by an enzyme. In consequence, the enzyme undergoes acylation and becomes inactive (Van Hylckama et al. 2000). Another example is the production of ethanol during vinyl acetate transformations (Nieder et al. 1990). As a result of hydrolysis of vinyl acetate, the produced vinyl alcohol undergoes spontaneous isomerization to ethanol, which, however, has strong toxic effect on microorganism cells. The condition for defense against its toxicity is the efficient system of dehydrogenases oxidizing ethanol to acetic acid or temporarily reducing it to ethanol (Gren et al. 2009).

Biotransformation of xenobiotics leads to the production of compounds that contribute to the increase of humification processes and then either return temporarily to the environment, or are stored in the soil, the sludge; dissolve in surface and/or groundwater; are released into the atmosphere; or aggregate in the food chain. It is also to be noticed that the described biotic transformations of xenobiotics in the environment may be accompanied by abiotic chemical transformations in the soil and sediments and abiotic photochemical transformations in the water and in the air (Gren 2012).

22.2 Bioavailability of Xenobiotics

Biotransformation of xenobiotic compounds directly depends upon the chemical structure and concentration of the respective xenobiotic, type and number of microorganisms capable of degrading or transforming the xenobiotic, as well as the physicochemical properties of the environment to which the xenobiotic is released or in which it accumulates (Oleszczuk 2007). Bioavailability, thus, may be defined as the total volume of the contaminant found in the soil or bottoms in free state, i.e., not permanently bound to the matrix and can be absorbed by the organism. Soil comprising solid, liquid, and gas phases is among the most diverse system to which xenobiotics may be released. The solid phase includes fragments of rocks and minerals, humus, animal and plant remnants, and mineral-organic particles. The liquid phase is water with dissolved mineral and organic substances, as well as gases, retained by capillary forces between soil aggregates and lumps. The mineral and organic compounds dissolved in the water constitute the soil water retention. The soil air is saturated by vapor and contains approximately ten times more carbon dioxide than atmospheric air and fills the soil spaces between solid particles that have not been taken by water.

Bioavailability of a xenobiotic depends on its state, i.e., solid, liquid, or gas; water solubility; and capability of adsorbing and adhering to solid particles of soil or sediment; however, only water-dissolved fraction of the xenobiotic is available to the microorganisms, and the direct contact of the xenobiotic with the microorganism's cell is the condition of the biological transformation of the xenobiotic (Gren 2012). At the same time it should be pointed out that the vast majority of xenobiotics exhibit significant hydrophobicity, and thus, after release into the environment, they are immobilized on the solid particles of the matrix by the process of sorption or in the structures of the organic matter by occlusion.

The process releasing the contamination (desorption) is the result of a collaboration between the physicochemical factors, such as a change of humidity, reaction or surface properties of the sorbent, and biological factors including microorganisms, plants, and animals. The released contaminants are transported by the way of diffusion and dispersion, which may lead to the xenobiotic coming into direct contact with the microorganism's surface. Passing the physiological barrier of cellular membranes of microorganisms is the key stage in the process of transformation of xenobiotics, taking place with the participation of more or less specialized enzymes of xenobiotic decomposition pathways (Oleszczuk 2007). The process of transporting the xenobiotic into the cell is not always crucial to the process of its transformation because of the fact that this may occur with the participation of extracellular enzymes, for example, transformation of xenobiotic esters that occur with the participation of extracellular lipases and esterases (Ozcan et al. 2009; Smacchi et al. 2000) or the transformation of chlorophenols with the participation of extracellular laccase (*p*-diphenol oxidase), isolated from the fungus *Coriolus versicolor* (Itoh et al. 2000). This enzyme transfers electrons and protons from ortho- or para-diphenols to oxygen.

The transport of xenobiotics into the microbial cells relatively involves the participation of ATPase-dependent carriers during the transfer of xenobiotics into the microorganism cells such as the StyE protein which is responsible for the active transport of styrene in *Pseudomonas putida* CA-3 (Mooney et al. 2006), or the TfdK protein, participating in the transfer of 2,4-dichlorophenoxyacetic acid into the cells of *Ralstonia eutropha* JMP134 strain (Leveau et al. 1998). Porin proteins located in the outer cellular membrane, such as the XylN protein, also participate in the transport of *m*-Xylene and its analogues through the outer membrane (Kasai et al. 2001) or the TbuX protein which takes part in the transport of toluene in the *Pseudomonas pickettii* PKO1 strain (Kahng et al. 2000).

22.3 Biochemical Decomposition of Xenobiotics

Biotransformation of xenobiotics through microbes can take place either in aerobic or anaerobic conditions which are described in detail as follows.

22.3.1 Aerobic Conditions

Microorganisms participating in the aerobic transformations of xenobiotics include several species including *Pseudomonas*, *Escherichia*, *Acinetobacter*, *Alcaligenes*, *Rhodococcus*, *Micrococcus*, *Streptomyces*, *Sphingobium*, *Pandoraea*, *Gordonia*, *Bacillus*, and *Moraxella*. During aerobic conditions, biotransformation of xenobiotics involves the major role of molecular oxygen, regardless whether their chemical structure is aliphatic or aromatic (Cao et al. 2009; Sinha et al. 2009). Hydroxylation reactions involving oxygenases (usually mono- or dioxygenases) of xenobiotics' biotransformation are of high importance and are often the limiting stage for their metabolism by microbes (Ullrich and Hofrichter 2007). Resulting products of aliphatic oxidation are carboxylic acid intermediates that are also important intermediates involved in the biotransformation of fatty acids in the cell (Van Hamme et al. 2003).

During the biotransformation of xenobiotics with aromatic structure, activity of specific oxygenases depends on the original xenobiotic structure, with the possible production of one of the key intermediate metabolites, i.e., catechol, protocatechuic acid, gentisic acid, or hydroquinone, having characteristic feature of the presence of two hydroxyl groups located either in ortho- or para-position. If the structure of the compound undergoing microbial transformation already contains a hydroxyl group, the transformation involves monooxygenase which transports one of the molecular oxygen atoms to the aromatic rings and reduces the other atom to water (Ullrich and Hofrichter 2007).

The abovementioned situation could be reversed in case the original structure of the aromatic compound does not have hydroxyl substituents. Further transformation of xenobiotics requires introducing two hydroxyl groups to the ring, and the transformation is catalyzed by ring-hydroxylating dioxygenase (Vaillancourt et al. 2006). Upper pathways of decomposition of xenobiotics also include the transformation of additional substituents, such as halogens, nitro group, sulfo group, or azo group. The presence of these substituents contributes to the increase of xenobiotics' "strangeness" in the environment as well as to increase their resistance to microbial decomposition (Jindrova et al. 2002; Ye et al. 2004; Kulkarni and Chaudhari 2007).

The second significant stage in the decomposition of aromatic structure is the destruction of this structure by the activity of aromatic ring-cleaving dioxygenases and the production of unsaturated aliphatic acids. The type of dioxygenase enzyme involved in the ring cleavage depends on the structure of the central intermediate, whereby ring-cleaving dioxygenases are divided into two main groups:

(a) Intradiol dioxygenases cleave the carbon–carbon bond of the aromatic ring, provided that both carbon atoms have a hydroxyl substituent and the product of this reaction is the *cis,cis*-muconic acid or its derivative.
(b) Extradiol dioxygenases cleave the carbon–carbon bond of the aromatic ring if only one of the carbons in question is hydroxylated and the product of the reaction is the hydroxymuconic semialdehyde or its appropriate derivative. Therefore, the substrates of their activity include not only catechol but also

compounds with two hydroxyl groups located in the para position and/or having a carboxyl or amine group in the position of the second hydroxyl substituent. Greater diversity of substrates for extradiol dioxygenases makes this enzyme group more versatile and useful in comparison to intradiol dioxygenases.

Constitution of two main pathways of decomposition of aromatic rings in xenobiotic biotransformation involves aromatic ring-cleaving dioxygenases along with enzymes responsible for further transformation of cleavage products of this stable structure (Van Pee and Unversucht 2003).

If the microbial cells exhibit activity of intradiol dioxygenase, then the xenobiotic decomposition pathway is described as ortho, while the presence of extradiol dioxygenase labels the pathway of further reactions as metatype. Both of these pathways involve mainly hydrolysis, (cyclo)isomerization, and reduction of compounds, producing carboxyl acids which, upon further processing, become the Krebs cycle intermediates (Harwood and Parales 1996). If the additional substituent has not been eliminated prior to the cleavage of the ring, then the xenobiotic transformations on ortho- or meta-pathways trigger appropriate reactions, for example, the removal of halogen substituent during decomposition of mono- and dichlorophenols with the participation of appropriate dehydrogenases (Janssen et al. 2001).

22.3.2 Anaerobic Conditions

Xenobiotic biotransformations under conditions of oxygen deficiency are currently less known. Microbial strains belong mainly to bacteria reducing nitrate(V), sulfate (VI), ferrum(III), vanadium(V), and chromium(VI) ions, as well as photosynthesizing purple bacteria and fermentation bacteria such as *Desulfobacterium*, *Clostridium*, *Methanococcus*, *Thauera*, *Pelotomaculum*, *Desulfotomaculum*, *Syntrophobacter*, *Syntrophus*, *Desulfovibrio*, *Methanospirillum*, *Methanosaeta*, *Azoarcus*, or *Geobacter* (Weelink et al. 2010).

Under anaerobic conditions, benzoyl-CoA is mainly produced by numerous transformations. In the presence of a carbon substituent in the original structure, biotransformations are directed such a way so that it becomes a carboxyl group, which is then bonded to coenzyme A. However, if the compound does not have an alkyl substituent, its transformation process comprises carboxylation of the ring in the presence of carbon dioxide. Consequently, 4-hydroxybenzoic acid is produced from phenol and naphthoic acid from naphthalene.

Benzoyl-CoA is the final aromatic product of transformation, since further stages of its conversion include reduction of the aromatic ring with gradual saturation of unsaturated bonds and the creation of unsaturated cyclic systems. The aromatic ring opening is hydrolytic and results in the production of 3-hydroxypimelic-CoA which then, through oxidation and decarboxylation, can be finally conversed into acetyl-CoA molecules (Weelink et al. 2010).

Condensation with fumarate is also a transformation process during anaerobic transformations of alkanes and produces alkyl derivative of succinate which, after condensation with coenzyme A, enters the β-oxidation pathway.

22.4 Xenobiotics and Catabolic Gene Organization

Biotransformation of xenobiotics involves a network of enzymes present in the bacterial system. Manipulation of the catabolic genes from degradative enzymes could boost up the process. Recalcitrance of several toxic pollutants may decrease with the mobilization of silent sequences into the functional catabolic routes and advancement of substrate range by gradual or spontaneous mutations. Approaches have also been made for adaptation in bacterial populations to specific xenobiotic compounds by gene transfer and to characterize and compare the genes involved in degradation of identical or similar xenobiotic compounds in nearly diverse or more isolated bacterial genera from diverse topologies. Sinha et al. (2009) illustrated the following:

(a) Evolutionarily related catabolic genes and their clusters have been derived from very distant locations in bacterial genera.
(b) The phylogeny of the catabolic genes is not compatible with that of the 16S rRNA genes of the related hosts.
(c) Genes for the degradation of synthetic pollutants are often associated with plasmids and transposons.
(d) Evolutionary related catabolic genes and entire gene modules are involved in the degradation of structurally similar but different xenobiotic compounds.

22.4.1 Mechanism of Gene Action

It varies among microorganisms and is described as follows:

(a) Genes are organized on single operon system either on plasmid or main chromosome, e.g., phenol (Khan et al. 2001).
(b) Genes are organized on two operon systems, e.g., polychlorinated biphenyls, and are characterized by plasmid genome (Shimizu et al. 2001).
(c) Genes are organized on more than two operon systems, e.g., 2,4-dichlorophenoxyaceticacid, etc. Genes are characterized by plasmid genome, and the capability or incapability of bacteria to do the entire degradation depends on the existence of the complementary enzymes encoded by genes of the chromosome (Don and Pemberton 1981).
(d) Genes are organized on transposons, e.g., 2,4,5-trichlorophenoxyacetate (2,4,5-T is a herbicide). *Pseudomonas* AC1100 contains two insertion elements,

RS110 selected as IS931 and IS932; they play a significant role in degradation of 2,4,5-T (Chaudhry and Chapalamadugu 1991).

22.4.2 Biotransformation and Molecular Approaches

Xenobiotic-transforming bacteria may harbor plasmids which code for catabolic genes. A proper management is required in order to characterize the appropriate genes and to enhance the process of transformation through improved constructed strains (Jain et al. 2005; Sinha et al. 2009).

This is achieved by utilization of modern molecular approaches which provide a more comprehensive interpretation of the in situ microbial community and its response to both engineered bioremediation and natural attenuation processes (Sinha et al. 2009). These approaches include PCR amplification and subsequent analysis of bacterial rRNA genes by sequencing, preparing metagenomics libraries, RFLP, dot blot, Southern blot, denaturing gradient gel electrophoresis (DGGE), and microarrays. Polychlorinated biphenyl (PCB) catabolic genes have been used to measure the level of PCB-degrading organisms in soil microbial communities with the help of dot blot technique (Walia et al. 1990). To quantify the degradation of 2,4-dichlorophenoxyacetic acid (2,4-D), tfdA and tfdB gene probes have been used and identified with the help of Southern hybridization technique (Holben et al. 1992). Metagenomics, either with function-driven or sequence-driven approach, has been proved to be very powerful tool for the identification of the desired catabolic genes and also provides access to retrieve unknown sequences (Schloss and Handelsman 2003).

22.5 Conclusion

Xenobiotics are of major concern and pose the greatest threat to the current world. For the past few decades, a lot of emphasis has been given for the scientific study for the biodegradation of xenobiotic compounds, which is one such process that exploits the catabolic abilities of microorganisms to degrade harmful and toxic xenobiotics. With this respect, microbial diversity proves to be foremost resource, and numerous novel strains of microbes with bioremediation potential have been isolated. Elucidation of new biochemical degradation pathways has also been given with novel ideas for potential biotransformation of xenobiotic compounds. However, despite the accredited worth of microorganisms, consideration of microbial diversity in sustaining comprehensive life-support systems is still very limited. This could be attributed to the viable but non-culturable (VBNC) nature of microorganisms. The efficiency of xenobiotic degradation can also be meaningfully improved by addressing key issues as tolerance to various xenobiotics, constitutive expression of the catabolic genes, and the substrate specificity, kinetics, and the stability of the encoded enzyme. Therefore, it is essential to explore the microbial biodiversity in order to develop effective and environment-friendly "green" technologies.

References

Aranda C, Godoy F, Becerra J, Barra R, Martinez M (2003) Aerobic secondary utilization of a non-growth and inhibitory substrate 2,4,6-trichlorophenol by *Sphingopyxis chilensis* S37 and *Sphingopyxis*-like strain S32. Biodegradation 14:265–270

Cao B, Nagarajan K, Loh KC (2009) Biodegradation of aromatic compounds: current status and opportunities for biomolecular approaches. Appl Microbiol Biotechnol 85:207–213

Chaudhry GR, Chapalamadugu S (1991) Biodegradation of halogenated organic compounds. Microbiol Rev 55:59–79

Don RH, Pemberton JM (1981) Properties of six pesticide degradation plasmids isolated from *Alcaligenes paradoxus* and *Alcaligenes eutrophus*. J Bacteriol 145:681–686

Gienfrada L, Rao MA (2008) Interactions between xenobiotics and microbial and enzymatic soil activity. Crit Rev Environ Sci Technol 38:269–310

Gren I (2012) Microbial transformation of xenobiotics. Chemik 66(8):835–842

Greń I, Gąszczak A, Szczyrba E, Łabużek S (2009) Enrichment, isolation and susceptibility profile to the growth substrate of bacterial strains able to degrade vinyl acetate. Pol J Environ Stud 18:383–387

Gribble GW (2004) Amazing organohalogens. Am Sci 92:342–346

Harwood CS, Parales EP (1996) The β-ketoadipate pathway and the biology of self-identity. Ann Rev Microbiol 50:553–558

Holben WE, Schroeter BM, Calabrese VGM, Olsen RH, Kukor JK, Biederbeck UD, Smith AE, Tiedje JM (1992) Gene probe analysis of soil microbial populations selected by amendment with 2,4- dichlorophenoxyacetic acid. Appl Environ Microbiol 58:3941–3948

Horvath RS (1972) Microbial co-metabolism and the degradation of organic compounds in nature. Bacteriol Rev 36(2):146–161

Itoh K, Fujita M, Kumano K, Suyama K, Yamamoto H (2000) Phenolic acids affect transformations of chlorophenols by a *Coriolus versicolor* laccase. Soil Biol Biochem 32:85–90

Jain RK, Kapur M, Labana S, La B, Sarma PM, Bhattacharya D, Thakur IS (2005) Microbial diversity: application of microorganisms for the biodegradation of xenobiotics. Curr Sci 89:101–112

Janssen DB, Scheper A, Dijkhuizen L, Witholt B (1985) Degradation of halogenated aliphatic compounds by *Xanthobacter autotrophicus* GJ10. Appl Environ Microbiol 49:673–677

Janssen DB, Oppentocht JE, Poelarends GJ (2001) Microbial dehalogenation. Curr Opin Biotechnol 12:254–260

Jindrova E, Chocova M, Demnerova K, Brenner V (2002) Bacterial aerobic degradation of benzene, toluene, ethylbenzene and xylene. Folia Microbiol 47(2):83–90

Kahng HY, Byrne AM, Olsen RH, Kukor JJ (2000) Characterization and role of tbuX in utilization of toluene by *Ralstonia pickettii* PKO1. J Bacteriol 182(5):1232–1238

Kasai Y, Inoue J, Harayama S (2001) The TOL plasmid pWWO xylN gene product from *Pseudomonas putida* is involved in m-xylene uptake. J Bacteriol 183:6662–6670

Khan AA, Wang RF, Cao WW, Doerge DR, Wennerstrom D, Cerniglia CE (2001) Molecular cloning, nucleotide sequence and expression of genes encoding a polycyclic ring dioxygenase from *Mycobacterium* sp. strain PYR-1. Appl Environ Microbiol 67:3577–3585

Kim JS, Crowley DE (2007) Microbial diversity in natural asphalts of the Rancho La Brea Tar Pits. Appl Environ Microbiol 73:4579–4591

Knackmuss HJ (1996) Basic knowledge and perspectives of bioelimination of xenobiotic compounds. J Biotechnol 51:287293

Kulkarni M, Chaudhari A (2007) Microbial remediation of nitro-aromatic compounds: an review. J Environ Manage 85:496–501

Leveau JHJ, Zehnder AJB, van der Meer JR (1998) The tfdK gene product facilitates uptake of 2,4-dichlorophenoxyacetate by *Ralstonia eutropha* JMP134 (pJP4). J Bacteriol 180 (8):2237–2241

Lloyd-Jones G, Laurie AD, Hunter DWF, Fraser R (1999) Analysis of catabolic genes for naphthalene and phenanthrene degradation in contaminated New Zealand soils. FEMS Microbiol Ecol 29:69–79

Mooney A, O'Leary ND, Dobson ADW (2006) Cloning and functional characterization of the styE gene, involved in styrene transport in *Pseudomonas putida* CA-3. Appl Environ Microbiol 72 (2):1302–1309

Nieder M, Sunarko B, Meyer O (1990) Degradation of vinyl acetate by soil, sewage, sludge, and the newly isolated aerobic bacterium V2. Appl Environ Microbiol 56:3023–3031

Oleszczuk P (2007) Organic pollutants in sewage sludge-amended soil. Part II fate of contaminants in soils. Ecol Chem Eng 14(2):185–191

Ozcan B, Ozyilmaz G, Cokmus C, Caliskan M (2009) Characterization of extracellular esterase and lipase activities from five halophilic archeal strains. J Ind Microbiol Biotechnol 36:105–111

Schloss PD, Handelsman J (2003) Biotechnological prospects from metagenomics. Curr Opin Microbiol 14:303–310

Shimizu S, Kobayashi H, Masai E, Fuduka M (2001) Characterization of the 450-kb linear plasmid in a polychlorinated biphenyl degrader, *Rhodococcus* sp. strain RHA1. Appl Environ Microbiol 67:2021–2022

Sinha S, Chattopadhyay P, Pan I, Chatterjee S, Chanda P, Bandyopadhyay D, Das K, Sen SK (2009) Microbial transformation of xenobiotics for environmental bioremediation. Afr J Biotechnol 8(22):6016–6020

Smacchi E, Gobbetti M, Rossi J, Fox PF (2000) Purification and characterization of a extracellular esterase from *Arthrobacter nicotianae* 9458. Lait 80:255–260

Tropel D, Van der Meer JR (2004) Bacterial transcriptional regulators for degradation pathways of aromatic compounds. Microbiol Mol Biol Rev 68:474–500

Ullrich R, Hofrichter M (2007) Enzymatic hydroxylation of aromatic compounds. Cell Mol Live Sci 64:271–277

Vaillancourt FH, Bolin JT, Eltis LD (2006) The ins and outs of ring-cleaving dioxygenases. Crit Rev Biochem Mol Biol 41:241–246

Van Hamme JD, Singh A, Ward OP (2003) Recent advances in petroleum microbiology. Microbiol Mol Biol Rev 67(4):503–510

Van Hylckama Vlieg JET, Poelarends GJ, Mars AE (2000) Detoxification of reactive intermediates during microbial metabolism of halogenated compounds. Curr Opin Microbiol 3:257–262

Van Pee KH, Unversucht S (2003) Biological dehalogenation and halogenation reactions. Chemosphere 52:299–309

Walia S, Khan A, Rosenthal N (1990) Construction and applications of DNA probes for detection of polychlorinated biphenyl degrading genotypes in toxic organic-contaminated soil environments. Appl Environ Microbiol 56:254–259

Weelink SAB, van Eekert MHA, Stams AJM (2010) Degradation of BTEX by anaerobic bacteria: physiology and application. Rev Environ Sci Biotechnol 9:359–365

Widada J, Nojiri H, Omori T (2002) Recent developments in molecular techniques for identification and monitoring of xenobiotic-degrading bacteria and their catabolic genes in bioremediation. Appl Microbiol Biotechnol 60:45–59

Ye J, Singh A, Ward OP (2004) Biodegradation of nitroaromatics and other nitrogencontaining xenobiotics. World J Microbiol Biotechnol 20:117–121

Chapter 23
Environmental Bioremediation: Biodegradation of Xenobiotic Compounds

Pankaj Goyal and Rupesh Kumar Basniwal

23.1 Introduction

During the past few years, xenobiotics in the environment have increased at alarming level and pose threat to flora and fauna. This is because of malpractices in handling wastes. According to Ashman and Puri (2002), people have used soil as a quick and convenient disposal method for xenobiotic compounds, but these contaminants in the soil can find their own way in different environmental sectors like groundwater, river, sea, and other atmosphere. The major classes of xenobiotic compounds are polycyclic aromatic hydrocarbons, halogenated aromatic or aliphatic hydrocarbons, azo or nitro compounds, organic sulfonic acids, and other synthetic polymers. According to Rao and Hornsby (2001), the use of xenobiotics in the form of pesticides constitutes an important aspect of modern agriculture, as they are necessary for economical pest management because they improve quality of life, chiefly in the area of public health by controlling the spread of various diseases like yellow fever, malaria, etc.

The conventional techniques like low temperature thermal desorption, chemical treatment, and incineration used for remediation of xenobiotic-contaminated sites have certain limitations and disadvantages (Frazar 2000). Low temperature and thermal desorption technology are an ex situ cleaning technology for removing volatile organic compounds from soils and sediments, but their higher running cost limits their use (Gavrilescu 2005). Incineration is also a good technology for destruction of xenobiotic compounds, but it emits potential toxic gases during the

P. Goyal
Amity Institute of Microbial Technology, Amity University, Noida, Uttar Pradesh, India

R.K. Basniwal (✉)
Amity Institute of Advanced Research and Studies (M&D), Amity University, Noida, Uttar Pradesh, India
e-mail: rup4kumar@gmail.com; rkbasniwal@amity.edu

© Springer International Publishing AG 2017

M.Z. Hashmi et al. (eds.), *Xenobiotics in the Soil Environment*, Soil Biology 49,
DOI 10.1007/978-3-319-47744-2_23

process and further removing them adds up additional cost to the method (Kearney and Wauchope 1998; Zhang and Quiao 2002). According to Schoefs et al. (2004), biological techniques like bioremediation and phytoremediation are safe, feasible, and economical because they offer cost-effective detoxification of xenobiotic compounds (Zhang and Quiao 2002). Biodegradation or detoxification performance of xenobiotics depends on various parameters like physicochemical environment, types of microorganism (aerobic or anaerobic bacteria and fungi), and the metabolic fate pathway. Biotransformations basically proceed due to the body requirement of nutrients like sulfur, carbon, and nitrogen components for its own growth and fulfillment of energy demand. It means microbial ecology, physiology, and evolution also play major roles in the biodegradability and biotransformation of xenobiotic compounds.

Recent advances in the molecular techniques including DNA fingerprinting, microarrays, and metagenomics have also played a great role in biotransformation of these compounds. Hence, broad knowledge about xenobiotic compounds and in-depth research about microorganism's biodegradability potential enable scientists and researchers to establish the rules for prediction of biodegradability of xenobiotic compounds. To date, several reviews have turned up for highlighting the xenobiotic pollution problem, but comprehensive understanding of these microbial transformations is still in its infancy. Therefore, this attempt is to understand the diversified microbial species role in the biodegradation of xenobiotic compounds in the context of available tools and techniques in the field of molecular microbiology and biotechnology.

23.2 What Is Bioremediation?

Bioremediation is a process of detoxification of contaminants in the environment through biological means like microorganisms, plants, or microbial or plant enzymes. Complete bioconversion of an organic pollutant to its inorganic constituents by a single species or a consortium of microorganisms is called mineralization. According to Pointing (2001), bioremediation is a process for reducing contaminant levels upto untraceable amount or within levels as set by regulatory organizations or become nontoxic or completely mineralized or transformed into carbon dioxide. Co-metabolism is another type of bioremediation process in which pollutant substrate takes place without the provision of carbon or energy for the microorganisms (Skipper 1998).

23.3 Benefits of Bioremediation

Bioremediation is a natural, less expensive, equivalent to any physicochemical method and has a reduced impact on natural ecosystems. According to Kearney and Wauchope (1998) and Vidali (2001), a complete biodestruction of target pollutants is possible on the site without the need of excavation or transporting waste from one place to another place, and finally, it requires small energy input and preserves the soil fertility or structure (Hohener et al. 1998).

23.4 Drawbacks of Bioremediation

- This process requires involvement of microorganism and so limited to only biodegradable compounds.
- Moreover, there are some doubts that the resultant products of bioremediation may be more toxic or persistent compared to parent compounds.
- It is difficult to scale up at industrial level due to many environmental influencing factors.
- This is complex and more time taking process compared to any other available physicochemical process like incineration (Vidali 2001).
- It has the capability to alter the toxicity of individual compounds in a mixture of organic pollutants (Hernando et al. 2003), but in case of pesticide mixtures, a few species of bacteria or white rot fungi are capable of degrading the individual components (Gadd 2001).
- In most studies, individual pesticide or herbicides were used for the study of biodegradation, but according to Grigg et al. (1997) and Memic et al. (2005) for practical results, a combination of pesticide-contaminated agrochemical soil with herbicides and their remediation strategies must take into account the presence of multiple contaminants.
- One type of microbial species can't degrade all types of pollutants. Henceforth, during the development of research and engineering tools for bioremediation process, we have to consider all types of associated environmental factors along with individual degradation behavior and their behavior in a complex mixture of pollutants (Vidali 2001).

23.5 Impacts of Environmental Factors on Bioremediation

The factors which can influence the biodegradation of pollutants are type of available microbial species for biodegradation, availability of contaminants to the microbial population, and environmental factors like temperature, pH, and presence of O_2 or other electron acceptors and nutrients. That's why control and optimization of bioremediation processes is a composite system of many factors (Vidali 2001). According to Vidali (2001), following optimum conditions are required for the degradation of any pollutant, i.e., soil moisture (25–28 % of water holding capacity), soil pH (5.5–8.8), oxygen content (aerobic, minimum air-filled pore space of 10 %), temperature (15–45 °C), contaminants not too toxic, heavy metals (Total content 2000 ppm), and the type of soil (low clay or silt content). According to Singleton (2001), solubility of contaminants, pH of soil, availability of nutrients, and level of oxygen and temperature are not always optimal for the growth of microorganism on contaminant for biotransformation and that's the possible reason behind the low biodegradability of highly biodegradable compounds (Romantschuk et al. 2000). Strong absorption of pesticide to soil reduces the bioavailability to microorganism and thus turns into low biodegradability of the pesticide. Henceforth, absorption of chemicals or pollutants at larger scale may reduce the biodegradation level (Singleton 2001). Lower temperature in China, Russia, Europe, and Northern parts prohibits the efficient microbial biodegradation of pollutants, and the same is also true for the deeper soil layers (Romantschuk et al. 2000). Further as mentioned by Romantschuk et al. (2000), aerobic and anaerobic environment also play a critical role in biodegradation of pollutants through fungal microorganism because in the presence of lower oxygen they show lower rate of biodegradation and therefore result in toxic intermediates. Osmolality environment has also a role in maintaining the integrity of microorganism and in biodegradability of contaminants. Exposure of microorganisms toward higher salty environment results in loss of cell water along with their important ingredients, and to counteract this, cells start to build up the organic osmolyte concentration in the cells and maintain their integrity and internal cell water (Kempf and Bremer 1998). According to Jennings and Burke (1990), these organic osmolyte compounds are able to resist water potential change and thus maintain turgidity while having no significant effect on enzyme activity (Ramirez et al. 2004).

23.6 Xenobiotic, Biodegradation, Biotransformation, Co-metabolism, Biomagnification

23.6.1 Xenobiotic

"Xenobiotic" is a term derived from Greek word "xenos" (foreign or forgiener). It means unnatural or foreign compound found in biosphere. Xenobiotic compound resists complete biodegradation or biotransformation and that's why they persist in the environment for long times like months or even years. Recalcitrant nature of xenobiotic molecules is due to its non-physiological chemical bonds or substituents in their structure which blocks the catabolic microbial enzymes' attack. Type, number, relative position of bond, and substituent affect the xenobiotic character (Fetzner 1999).

23.6.2 Biodegradation

Pollutants are transformed by microorganism through the biodegradation process for energy, nutrient elements like nitrogen and carbon, or as final acceptor for respiratory system for their growth. Biodegradation involves the breakdown of organic compounds into smaller or simple organic or inorganic compounds. Complete biodegradation process involves the complete oxidation of organic compounds into inorganic compounds, i.e., CO_2 and H_2O, and the process is known as "mineralization."

23.6.3 Biotransformation

It involves the partial or complete alteration in the original molecular structure of compounds through the microorganism which affects the toxicity, solubility, and mobility of the compounds in the environment.

23.6.4 Co-metabolism

Microbial population transforms the compound without fulfilling their carbon or energy requirements from the compound which is termed as co-metabolism. It means the organism does not have benefit from co-metabolism. The mechanism behind this is that primary substrate works as an inducer able to secrete microbial enzymes which further alter the molecular structure of other compounds.

23.6.5 Bio-magnification

Metabolic dead-end products and persistent xenobiotic compounds accumulate in the environment due to their non-biodegradability nature and become part of food chain leading to bio-amplification of the compound concentration, and the process is termed as bio-magnification.

23.7 Bioavailability and the Rate of Biodegradation

Bioavailability of pollutants for microorganisms depends on their half-life, solubility in water, structure and concentration of the compound, availability of the type of microbial population, and their surrounding physicochemical environment. Long half-life of pollutant increases the possibilities of their less availability for biodegradation due to entrapment into micro pores of soil. More water-soluble contaminants easily biodegrade compared to less soluble ones. Concentration of the pollutants also plays a critical role in the biodegradation process because their low concentration is not able to induce secretion of enzymes in the microorganisms and their higher concentration may become toxic to microorganisms which ultimately turned into lower biodegradation rate. More persistent nature of pollutants shows more resistance toward any physical, structural, or chemical changes. Other environmental factors which affect the biodegradability of compound are salinity, pH, temperature, availability of nutrients, water, and terminal electron acceptor species. Moreover, the presence of other type or new competent microorganism species for the same pollutant can affect the biodegradation and biodegradability of pollutants.

23.8 Role of Chemotaxis in Biodegradations

Chemotaxis is a process of microorganism movement toward chemical gradient and this enhances the bioavailability of pollutants to microorganism. Ralstonia sp. is chemotactic toward different types of chemical compounds like 3-methyl 4-nitrophenol, 4-nitrocatechol, and ortho- and para-nitrobenzoate (PNB), and the compounds can easily and completely be degraded by this organism (Bhushan et al. 2000; Samanta et al. 2000). On the other hand, this Ralstonia strain does not degrade *o*-nitrophenol, *p*-nitroaniline, 2,3-dinitrotoluene, naphthalene, phenanthrene, or salicylic acid, and this is because of not showing chemotaxis toward abovementioned compounds. This is clear indication of some relationship between chemotaxis and biodegradation.

23.9 Biological Mechanisms of Transformation

Microorganisms start to secrete enzymes (which were coded by their own genetic material) when in contact with contaminant either through chemotaxis process or inoculated manually. They produce degradative enzymes that can oxidize, reduce, dehalogenate, dealkylate, deaminate, and hydrolyze hazardous chemicals such as organic nitro or chloro derivatives and pesticide compounds in the soil environment. The enzymes can be intracellular or extracellular, and each type of enzyme has specific conditions for optimum activity (Skipper 1998). Once a contaminant has been enzymatically transformed into a less toxic compound, it can often be metabolized by various microorganisms through various pathways.

23.10 Resources of Xenobiotic Compounds

Resources of xenobiotic compounds are plastics, effluents of paint, paper, pharma, textile mills, hospitals, and agricultural soil pollutants like pesticides and herbicides. Phenol is a derivative of benzene (Gayathri and Vasudevan 2010), which contains carcinogenic activity and has many adverse effects on environment like harming the ozone layer, affecting visibility, and adding acidic air pollutants to the atmosphere (Jame et al. 2010). Xenobiotic plastics are made of polystyrene and polyvinyl chloride, polyethylene, and their derivatives. Recently, plastics from crude oil are source of fuels for the industries because of their convertibility into liquid hydrocarbons. Petroleum derivative polyolefin is inert and hydrophobic in nature and chain has high molecular weight, so its biodegradation is quite difficult (Santhoskumar et al. 2010a, b). Aromacity nature of compound also has a role in biodegradation rate because their increasing concentration results in less biodegradability of the compound (Mirdamadian et al. 2010). Reason for resistance of dye toward degradation is because of their agglomeration nature, and this has adverse effect on the aquatic species of plant in terms of photosynthesis rate (Abdelkader et al. 2011; Elaziouti et al. 2011). A large number of pesticides and insecticides like organo-phosphorous, benzimidazoles, methyl parathion, morpholine, and toxic paper mill effluent like di-, tri-, tetra-, and pentachlorophenols, tetrachloroguaiacols, and tetrachlorocatechols are extensively used in the field of environment, and they all biodegrade slowly (Lalithakumari 2011). Many pharmaceutically derived less biodegradable compounds like hormones, anesthetics, and antibiotics are bioaccumulated through various routes of food chain (Heberer 2002). Although biomaterials developed from the artificial polymers show biocompatibility, but their further degradation into toxic substances is a serious concern of body (Reddy and Yang 2011). Other common xenobiotic pesticide compounds DDT (having half-life of 10 years) and BHC can bioaccumulate in plants, fruits, vegetables, and even in animal's milk (Karanth 2000).

23.11 Xenobiotic Degradation Through Bacteria

Bacteria which contain large capacity to degrade a wide range of xenobiotic chemicals are aerobic bacteria (*Pseudomonas, Escherichia, Sphingobium, Pandoraea, Rhodococcus, Gordonia, Bacillus, Moraxella, Micrococcus*), *anaerobic bacteria* (*Pelatomaculum, Desulfotomaculum, Syntrophobacter, Syntrophus, Desulphovibrio, Methanospirillum, Methanosaeta*), methanotrophic bacteria, methanogenic bacteria, cyanobacteria, and sphingomonads (Sinha et al. 2009). Due to plasmid-borne mechanism, sphingomonads are highly efficient in degrading a wide range of xenobiotics like synthetic polymers and aromatic compounds. The cyanobacterial mats are used for bioremediation of oil derivatives (Raeid 2011; Bordenave et al. 2009). *Pseudomonas* species of bacteria has been reported for complete and partial mineralization of organo-phosphorous pesticides, fungicides, aromatic or aliphatic hydrocarbons, phenols, hexavalent heavy metals, and dye (Lalithakumari 2011; Poornima et al. 2010; Wasi et al. 2010; Joe et al. 2011). *Bacillus* sp. of bacteria has been also reported for their ability to degrade benzimidazole compounds and oil spills (Amin 2010; Owolabi et al. 2011). *Azotobacter* sp. can remove the less toxic trivalent Cr through biosorption. Synthetic dye can be degraded by thermophilic bacteria like *Anoxybacillus pushchinoensis, Anoxibacillus kamchatkensis*, and *Anoxybacillus flavithermus* (Gursahani and Gupta 2011). Azo degradation can be done by using lactic acid bacteria which is efficient under both anaerobic and aerobic conditions (Elbanna et al. 2010). Polyethylene xenobiotic compounds are degraded by *Brevibacillus borstelensis* and *Rhodococcus ruber* for the sole source of carbon (Hadad et al. 2005). *Proteus mirabilis*, *Pseudomonas aeruginosa*, and *Micrococcus* are urease-producing bacteria which produce a novel metabolic by-product, i.e., "Microbial Concrete" which can be further used for remediating and restoring the building structures (Reddy and Yang 2011). Bio-desulfurization of crude oil can be done with the help of *Rhodococcus erythropolis* bacteria (Amin 2011). The involvement of plasmid-encoded catabolic sequences in pesticide degradation has been widely documented (Somasundaram and Coats 1990). Bacterial species of *Nocardia* and *Pseudomonas* utilize atrazine aerobically for the sole source of carbon, nitrogen, and energy (Mandelbaum et al. 1995). *Achromobacter*, *Pseudomonas*, and *Flavobacterium*are able to utilize carbofuran, a carbamate pesticide, as a growth substrate (Aislabie and Jones 1995). Recently, "Super Bugs" concept has been developed to degrade a wide range of xenobiotic compounds (Kensuke Furukawa 2003).

23.12 Xenobiotic Degradation Through Fungi (Myco-remediation)

Myco-remediation plays a pivotal role in breaking down numerous toxic substances like petroleum hydrocarbons, polychlorinated biphenyls, heavy metals, phenolic derivatives, persistent pesticides, etc. Many fungal species like basidiomycetes and ascomycetes have the potential to degrade lignocellulose materials present in dead wood, paper, and pulp effluents. Natural oil component, i.e., polycyclic aromatic hydrocarbons, is degraded by laccases (a copper containing enzyme) found in basidiomycetes (Cho et al. 2009). An edible rot fungus, *Pleurotus pulmonarius*, is known for its ability to degrade crude oil (Olusola and Anslem 2010). *Trichoderma harzianum* (bio-fungicide) produces cellulolytic enzymes which are extensively used by textile and paper industries for the degradation of cellulose (El-Bondkly et al. 2010). According to Mascoma, conversion of paper sludge to ethanol by *Saccharomyces cerevisiae* is 85 % without the addition of commercial enzymes. Fungal biomasses can uptake easily considerable quantities of organic pollutants from aqueous solution by adsorption (Kurnaz and Buyukgungor 2009). *Mucor*,

Rhizopus, Aspergillus carbonaruius, Aspergillus niger, Saccharomyces cerevisiae, Botrytis cinerea, Neurospora crossa, Phanerochaete chrysosporium, and Lentinus sajor-caju fungal species have been extensively studied for heavy metal biosorption (Kumar et al. 2009). For heavy metals' (Pb, Au, Co, and Cu) bio-absorption, *Saccharomyces cerevisiae* has been found to be a suitable and efficient biosorber (Rajesh et al. 2011). Under the new technology, development by implementing fungal symbiont into plant works efficiently for bio-adsorption (Kumar et al. 2010). Mycorrhizal fungi (a symbiotic association of actinomycetes fungus in the root zone of vascular plant) like *Morchella conica* and *Tylospcno fibrilnsa* naturally biodegrade the organic pollutants (Bennet et al. 2002). Fungal species were used for cleaning environment after the discovery of *Phanerochaete chrysosporium* which was able to metabolize a number of pollutants (Sasek 2003). Lignin/persistent toxic environmental pollutants can be degraded by extracellular lignin-degrading enzymes (Barr and Aust 1994; Cameron et al. 2000). Mycelial nature of fungi helps in rapid colonization of substrates and enables them for deep penetration into the substrate or pollutants (Reddy and Mathew 2001; Fragoeiro and Magan 2008) which can further maximize by physical, mechanical, and enzymatic contact with the surrounding environment (Maloney 2001). These fungi survive on chief material, i.e., lignocellulose which is abundant in nature, and they can tolerate a wide range of environmental conditions, such as temperature, pH, and moisture levels (Maloney 2001). White rot fungi mostly belonging to basidiomycetes are those fungi that are capable of extensively degrading lignin (a heterogeneous polyphenolic polymer) within lignocellulosic substrates (Pointing 2001). Lignin, which provides strength and structure to the plant, is extremely recalcitrant. The three major families of lignin-modifying enzymes believed to be involved in lignin degradation are laccases, lignin peroxidases, and manganese peroxidases. The key step in lignin degradation by laccase or the ligninolytic peroxidases involves the formation of free radical intermediates, which are formed when one electron is removed or added to the ground state of a chemical (Reddy and Mathew 2001). Nonspecific degradation function of the enzymes makes them suitable for the action on many susceptible xenobiotic compounds (Field et al. 1993; Barr and Aust 1994). The fungus can effectively degrade very low concentrations of a pollutant to non-detectable or nearly non-detectable levels (Bumpus and Aust 1986). *P. chrysosporium* (thermophilic) was the first fungus to be associated with degradation of organopollutants, in 1985, because it has been extensively studied as a model microorganism in research on the mechanism of lignin degradation (Sasek 2003). In straw cultures, *P. chrysosporium* was able to degrade 91 % of the herbicide in 14 days of incubation. *Pleurotus ostreatus* is a saprophytic basidiomycete and a natural decomposer because it secretes enzymes and acids that degrade organic polymers. *Pycnoporus coccineus* has been used in bioremediation research mainly for its effective extracellular laccase production (Alves et al. 2004; Pointing et al. 2000). *Trametes versicolor* most versatile in white-rotters has many research applications in bioremediation, effluent treatment, the pulp and paper industry, the food industry, synthetic chemistry, biofuels, cosmetics, biosensors, and the textile industry, among others (Nyanhongo et al. 2007). *Phlebiopsis gigantea* is used as a biological control of annosum root rot, caused by *Heterobasidion annosum*, in Western Europe.

23.13 Xenobiotic Degradation Through Algae (Phycoremediation)

Phycoremediation is defined as the use of macro/microalgae to remove or biodegrade pollutants from the environment. Algae possess the ability to take up toxic heavy metals from the environment and this is because they have potential metal ion binding sites in their cell wall (Imamul et al. 2007). Algal species which play role in degradation are *Chlorococcum* sp., *Chroococcus* sp., *Desmococcus* sp, *Dactylococcopsis* sp, and *Chlamydomonas* sp. (Sivasubramanian et al. 2010). Biosorption through blue green algae has distinct advantages over the conventional methods because they are easy to operate, cost-effective for the treatment of large volumes of wastewaters, selective, and more efficient (Parameswari et al. 2010).

23.14 Xenobiotic Degradation Through Plants (Phytoremediation)

Phytoremediation: Phytoremediation (also called as green remediation, botanoremediation, agro remediation, and vegetative remediation) is a method that uses green or higher terrestrial plants for treating chemically polluted soils (Wenzel et al. 1999), reducing the amount of hazardous compounds (Maria and Walter 2006). Phytoremediation is one of most eco-friendly technique to target the organic and inorganic pollutants in the water, soil, and air simultaneously. Plants have the capacity to withstand relatively high concentrations of organic xenobiotic chemicals without toxic effects (Briggs et al. 1982), and they able to take up and convert these chemicals quickly into less toxic compounds (Kolb and Harms 2002). Rhizodegradation is the process of biodegradation of petroleum hydrocarbons, pesticides, polychlorinated biphenyls (PCBs), surfactants, and chlorinated solvents in soil through the influence of plant roots, finally leading to destruction or detoxification of an organic contaminant. Chlorinated solvents, some organic herbicides, and trinitrotoluene can be degraded through secreted enzymes within plant, and this method is called phytodegradation or phytotransformation (Shukla et al. 2010). Phytomining is another important process for removing metals from soil, and, in some cases, incorporation of plant incinerations will help metal reuse (Shukla et al. 2010). For example, *Brassica juncea*, *Berkheya coddii*, *Alyssum bertolonii*, *Thlaspi caerulescens*, and *Thlaspi goesingense* are some of the plants involved in Phytomining. Phytohydraulics is the use of deep-rooted plants to degrade groundwater contaminants like methyl-tert-butyl-ether (MTBE) that come into contact with their roots (Pilon-Smits et al. 1999). Indian mustard appeared to have potential for phytostabilization; in this process, soil and water contaminants are immobilized through the plant.

Recent trends in phytoremediation use bacteria like *Pseudomonas, Burkholderia, Bacillus,* and *Azospirillum* an endophyte (Ram and Srivastava 2008). Transgenic or engineered plants also help to increase tolerance and metabolism of xenobiotic chemicals for the phytoremediation (Jaanis 2010; Annette and Jerald 2001; Sonoki et al. 2011). Pros of phytoremediation are clean, cost-effective, and environment-friendly technology, as opposed to mechanical cleanup methods such as soil excavation or pumping polluted groundwater (Wei et al. 2004), and major cons of this technology are longer time requirement because this process depends on plant growth. Although phytoremediation is a promising technique for removing pollutants from the environment, we just need to find the right plant for the right pollutant (Zhai 2011).

23.15 Photodegradation

Photodegradation is degradation of a molecule which has the capability to absorb photons, particularly those wavelengths found in sunlight, such as infrared radiation, visible light, and ultraviolet light. Congo red dye used in the cellulose industries (cotton textile, wood pulp, and paper) has long been abandoned, primarily because of its tendency to change color and its toxicity. Recent advances to degrade Congo red include photocatalytic degradation using ZnO/UV-A (Elaziouti et al. 2011; Santhoskumar et al. 2010a, b). Nowadays, photodegraded films are used to evaluate biodegradation using microorganisms such as *Aspergillus niger* and *Penicillium funiculosum* for degradation of both natural and synthetic plastics.

23.16 Biodegradation Pathways of Xenobiotic Compounds

Microorganisms are able to degrade various xenobiotics like pyrene, chloroanilines, pentachlorophenol, petroleum hydrocarbons, chlorinated aliphatics, benzene, toluene, phenol, naphthalene, fluorine, dichlorobenzenes, etc., into nontoxic compounds. The degradation pathway can be divided into (1) aerobic and (2) anaerobic pathways. Final products of aerobic and anaerobic degradation are carbon dioxide and methane (Swift 1998; Grima et al. 2002; Kyrikou and Briassoulis 2007), and the process is termed as mineralization. Mineralization process is completed when all the biodegradable biomass carbon is converted into carbon dioxide or methane (Kyrikou and Briassoulis 2007).

23.16.1 Aerobic Biodegradation Pathway

In response to immense turnover of the aromatic compounds in the carbon cycle, well-organized channels of aerobic catabolism have been evolved separately with evolution (Chauhan et al. 2008). Thus, it appears that aerobic culture techniques are relatively simple, as well as efficient and generally applicable through oxidative degradation (Adriaens and Vogel 1995). The aerobic degradation pathway starts by inserting one or two oxygen molecules into xenobiotic compound with the help of monooxygenases or hydroxylating dioxygenase enzymes and generates di-hydroxy aromatic compounds which further process central intermediates like protocatechuates, catechols, gentisates, homoprotocatechuates, homogentisates, hydroquinones, or hydroxyquinols, and finally they transform into tricarboxylic acid cycle intermediates and are channelized into the Kreb's cycle for energy generation (Harayama and Timmis 1992; Dagley 1975; Wilson and Bouwer 1997; Sims and Overcash 1983). According to Hayaishi and Nozaki (1969), major reactions catalyzed by di-oxygenases enzymes for aerobic biodegradation of aromatic compounds are either through ortho, meta, or indole ring cleavage pathways. According to Mishra et al. (2001), mode of aerobic ring cleavage is different for dichlorodiphenyltrichloroethane (DDT), pentachlorophenol (PCP), and 1,2,3,4,5,6-hexachlorocyclohexane (HCH) because initially they are reduced into less chlorinated intermediates and finally transformed by microbes. Aromatic ring-degrading enzymes evolve in such a way that they can act on diversified range of xenobiotic compounds due to their less specificity toward particular substrate (Copley 2000). According to Janssen et al. (2005), a key step in aerobic mineralization pathways of several halogenated pollutants is cleavage of carbon–halogen bond through bacterial dehalogenase enzymes. In case of aerobic alkane degradation, the terminal methyl group transforms into carboxylic acid and their complete mineralization takes place after the beta-oxidation (Leahy and Colwell 1990). Aerobic degradation pathway of alkane degradation is the oxidation of the terminal methyl group into a carboxylic acid through an alcohol intermediate and after complete mineralization takes place through β-oxidation (Leahy and Colwell 1990; Cookson 1995; Vander et al. 1992; Zhang and Bennet 2005). Esterases, permeases, and dioxygenases enzymes are involved in aerobic degradation of phthalate isomers (Vamsee-Krishna and Phale 2008). In most cases, the generated intermediate 3,4-dihydroxybenzoate was further degraded by bacteria into acetyl-CoA and succinyl-CoA as end products (Hara et al. 2007). Aerobic decolorization and reduction of azo dyes is carried out in the presence of additional carbon sources by various strains of microorganisms like *Pseudomonas, Sphingomonas, Xanthomonas, Aeromonas, and Bacillus* during aerobic conditions (Stolz 2001). In a co-metabolic pathway, bacteria are able to use reduced azo dyes as a sole source of carbon and energy (Yatome et al. 1993; Dykes et al. 1994; Stolz 2001).

23.16.2 Anaerobic Biodegradation Pathway

Pollutants which are not biodegradable in aerobic condition due to persistent nature of attached groups like halogen, nitro, and sulfo groups can be degraded by substituting them with degradable groups in the presence of the anaerobic condition. Xenobiotic compounds like phenols, phthalates, and hydrocarbons including benzene-toluene-ethyl benzene-xylene (BTEX) are oxidized and mineralized under anaerobic conditions (Reuter et al. 1994). Halogenated pesticides work as terminal electron accepter for reduction of chlorine component as mentioned by Cutter et al. (2001), and the process is termed as halorespiration or dehalorespiration, e.g., complete dechlorination of tetrachloroethylene or perchloroethylene (PCE) by bacteria *Dehalococcoides ethenogenes* (Magnuson et al. 2000). The use of trinitrotoluene (TNT) as a terminal electron acceptor has been reported by Esteve-Nunez et al. (2000), and according to Janssen et al. (2005), cyclo-trimethylene-trinitramine (RDX), azo dyes, and carbon tetrachloride are biotransformed by co-metabolic reduction processes. A key enzyme benzoate coenzyme-A ligase plays a major role in the degradation of monocyclic aromatic compounds through benzoyl coenzyme-A (CoA) pathway and formed acetyl-CoA (Schink et al. 2000). Anaerobic degradation of phthalate compound starts initially by decarboxylation, then reduction followed by ring cleavage, and ultimately the beta-oxidation pathway (Qiu et al. 2004; Zhang and Bennet 2005). Generally, xenobiotics like pesticides, DDT, polychlorinated biphenyls, chlorinated dioxins, etc., show resistance toward electrophilic attack by oxygenases in aerobic degradation, but as suggested by Van Agteren et al. (1998), the anaerobic reductive dehalogenation of persistent xenobiotics makes them more susceptible toward aerobic mineralization (Kazumi et al. 1995; Song et al. 2000 and Vargas et al. 2000). Biodegradation through anaerobic bacteria can be seen on various places like gastrointestinal contents, sludge digesters, groundwater, sediments, and landfill sites that depend on their availability and type of xenobiotic compounds (Williams 1977). According to Zhang and Bennet (2005), methogenic and sulfate-reducing environments are favorable for culturing and isolation of pure anaerobic bacteria like *Veillonella alkalescens, Clostridia, Acidovorax, Bordetella Desulfobacterium, Variovorax Desulfovibrio, Methanococcus, Pseudomonas, Methanosarcina, Sphingomonas*, and *Desulfitobacterium halogenans* dehalogenating bacteria. As suggested, aromatic compounds can serve as electron shuttles by accepting electrons for changing their ring substituent (Gibson and Harwood 2002). In case of sulfate-reducing bacteria, sulfate works as final electron acceptor to degrade crude oil and generates hydrogen sulfide (Barton and Hamilton 2007; Boetius et al. 2000; Sahrani et al. 2008). The biogas (methane) production through anaerobic degradation process is economical and eco-friendly, and this can be scaled up at larger scale for applications in power sectors (Lier et al. 2001; Angelidaki and Sanders 2004; Holm-Nielsen et al. 2009). So, we can generate power and manage waste simultaneously while reducing the emission of landfill gas into the atmosphere (Dolfing and Bloemen 1985; Angelidaki and Ahring 1993; Soto et al. 1993).

23.16.3 Co-metabolic Pathway

A growth of *Pseudomonas methanica* on methane creates oxidation of ethane to acetic acid, propane to propanoic acid, and butane to butanoic acid and methyl ethyl ketone which generate the concept of co-metabolism (Jensen 1963). Consequently, co-metabolism refers to oxidation of substances without utilization of the energy derived from the oxidation to support microbial growth and does not infer the presence or absence of growth substrate during the oxidation, and this technique is generally employed for the biochemical study of microbial aromatic metabolism (Horvath 1972). Example of co-metabolism is generation of dalapon during rapid decomposition of the herbicide without increase in bacterial numbers in soil (Burge 1969). Expression of the action of catechol-1,6-dioxygenase (meta-cleaving enzyme) by *Achromobacter* bacteria has been accomplished by using the technique of co-metabolism (Horvath and Alexander 1970), and this technique is found to be very useful for the degradation of mixtures of polycyclic aromatic hydrocarbons (Chauhan et al. 2008). *Nocardia, Pseudomonas, Xanthomonas, Bacillus, Brevibacterium, Flavobacterium, Aspergillus, Azotobacter, Trichoderma, Vibrio, Achromobacter, Arthrobacter, Hydrogenomonas, Microbacterium, Micrococcus*, and *Streptomyces* species of bacteria show the phenomena of co-metabolism in a very well-organized manner (Beam and Perry 1973).

23.17 Microbial Enzymes Involved in Bioremediation

As mentioned below, various microbial enzymes are involved in the biotransformation of toxic pollutants.

23.17.1 Microbial Oxidoreductases

These enzymes cleave chemical bonds and transfer the electrons from a reduced organic substrate (donor) to another chemical compound (acceptor) (Karigar and Rao 2011). Detoxification, decolorization, and degradation of phenolic and azo dyes compounds are the examples of microbial oxidoreductases' enzyme activity (Park et al. 2006; Williams 1977; Vidali 2001; Husain 2006).

23.17.2 Microbial Oxygenases

This oxidizes the substrates by transferring oxygen from molecular oxygen (O_2) to xenobiotic compound in the presence of co-substrates like FAD or NADH or

NADPH (Karigar and Rao 2011). These enzymes enhance the substrate reactivity and water solubility and cleave the structure of compound and metabolize into nontoxic compounds (Arora et al. 2009). On the basis of the number of oxygen atoms used for oxidation, oxygenase enzymes can be further divided into two groups, i.e., monooxygenases and dioxygenases. Monooxygenases act as biocatalysts in the bioremediation process and transfer one atom of molecular oxygen to the organic compound and show broad stereo selectivity of substrates (Arora et al. 2009, 2010; Cirino and Arnold 2002; Karigar and Rao 2011). Biodegradation or biotransformation of various aromatic and aliphatic compounds through desulfurization, dehalogenation, denitrification, ammonification, and hydroxylation are the examples of monooxygenase enzyme activity (Arora et al. 2010). In the convergent mode of dioxygenases (which incorporate double atomic oxygen to the substrate), structurally different aromatic compounds are converted into catechol, gentsate, protocatechuate, and derivatives of ring cleavage substrates (Meer et al. 1992). In other divergent modes of dioxygenases, metal-dependent dioxygenase channels operate and dihydroxylated intermediates are formed by *meta* (m) or *ortho* (o) cleavage pathway (Harayama and Rekik 1989; Eltis and Bolin 1996; Takami et al. 1997).

23.17.3 Microbial Dehalogenases

Dehalogenase plays an important role in the degradation of chlorinated pollutant (Copley 1998) which works as terminal electron acceptors for anaerobic microorganisms (Wohlfarth and Diekert 1997), e.g., bioconversion of per-chloro-ethylene (PCE) to either dichloroethylene (DCE) or ethylene or ethane (Scholz-Muramatsu et al. 1995).

Phosphotriesterases (PTEs) Organophosphate pesticides are hydrolyzed and degraded by this enzyme which mainly hydrolyzes phosphoester bonds like P–O, P–F, P–NC, and P–S (Ortiz-Hernandez et al. 2003).

23.18 Evolution of Molecular Techniques to Study Catabolic Gene

The environment contains several xenobiotic-degrading bacteria, and their identification, culturing, and monitoring through traditional microbiological methods are very time-consuming (Lloyd-Jones et al. 1999). According to Widada et al. (2002), the diversity of catabolic genes in bacteria can be investigated by two different approaches from environmental samples: first is *culture-dependent* and second is *culture-independent* method. In culture-dependent method, nucleic acid is extracted from isolated bacterial culture from environmental samples, and in culture-independent method, nucleic acid is directly extracted from environmental samples (Okuta et al. 1998; Watanabe et al. 1998; Lloyd-Jones et al. 1999). Recently, many molecular approaches have been used to characterize the DNA of different bacteria from the environment (Hurt et al. 2001). For in situ microbial community, the molecular techniques give us more in-depth elucidation in comparison with standard microbiological methods. According to Sinha et al. (2009), many molecular techniques are used to identify and analyze the biodegrading gene in the microorganism like PCR, RFLP, Southern blotting, Microarrays, Meta genomic libraries, etc. To analyze the degradation pathways of toluene and napthalene, DNA hybridization techniques can be used to study their associated degrading plasmid TOL and NAH (Sayler and Layton 1990). According to Hosein et al. (1997), these techniques are very useful for monitoring the xylE and ndoB genes involved in creosote degradation in soil communities. In these studies, colonies were hybridized by entire plasmids as probes and the relationship between plasmid concentrations and the rates of mineralization was computed. Mills et al. (2003) have explained about nutrient alterations on microbial community during bioremediation of petroleum-contaminated soils through amplicon length heterogeneity PCR (LH-PCR) and terminal restriction fragment length polymorphism (TRFLP) molecular techniques. Fulthorpe and Wyndham (1989) have mentioned about the study of chlorobenzoate-degrading genes of *Alcaligenes* strain through combination with most-probable-number (MPN) and colony hybridization technique. RT-PCR molecular technique is very useful for the study of metabolically active structural genes and their identification and expression in the group of microorganisms like the study of highly PCB-polluted soil for active bacterial community (Weller and Ward 1989; Nogales et al. 1999; Widala et al. 2002a, b). The genetic fingerprinting technique gives us a profile of the genetic diversity in a microbial community. MALDI-TOF-MS study for restriction fragments of PCR-amplified product analysis (Taranenko et al. 2002), T-RFLP (terminal restriction fragment length polymorphism) for polymorphism study, and denaturing high-performance liquid chromatography (DHPLC) for detecting single base-pair mutations in a specific sequence (Taliani et al. 2001) and metagenomic libraries for finding information about unknown sequences (Schloss and Handelsman 2003) are the techniques which are used for the study of genes for recalcitrant degradation.

In the recent past, many techniques have been evolved for the study of biodegradation of xenobiotic compounds, but through the molecular techniques we were able to identify responsible catabolic genes and capable of exploring the unknown community of microorganism for particular recalcitrant degradation in a very short time period. Many novel genes responsible for xenobiotic degradation were patented by many scientists through the molecular study approach.

23.19 Screening Methods for Xenobiotic Biodegraders

Each microbial species differs in its biodegradation competencies; therefore, screening of microorganisms for potential hydrocarbon is an essential tool. For example, if we want to screen the bacteria for utilizing 2,6-dichlorophenol indophenol (DCPIP) hydrocarbon, then microbial oxidation of this compound results in color change which can be monitored through UV-Visible spectroscopy by measuring their wavelength at 600 nm. Crude oil-degrading bacteria can be also screened through redox indicator, i.e., 2,6 dichlorophenol indophenol method and total plate count method. The bacteria which show high potential to degrade the oil turn up the medium into colorless medium. The efficiency was measured by measuring the optical density (OD) and total cell count which were further screened by gravimetric analysis. For enumerating the phenanthrene-degrading bacteria from petroleum-contaminated sites, an over-layer technique was employed. In this technique, a mixture of contaminated soil and water was prepared along with fine particles of phenanthrene (aromatic polycyclic hydrocarbon). Phenanthrene-degrading bacteria embedded in the over-layer were recognized by a clear halo zone appearing on the opaque phenanthrene layer. The biodegadation of spent engine oil by bacteria was explained by Ismaila et al. (2014) and Sunday et al. (2006), and according to them, these bacteria were isolated from the rhizosphere of *Cajan cajan* and *Lablab purpureus* (legumes) which were grown in the contaminated soil. Further GC–MS analysis confirmed the presence of lower molecular weight hydrocarbon in the residual oil which was indication of biodegradation. In turbidometric analysis, if more viable cells are present in contaminated sites, then it shows indirectly more biodegradability capacities of the bacterial cells (Mandri and Lin 2007; Kumari et al. 2013; Okoh 2003; Olajide and Ogbeifun 2010). Gravimetric analysis (Mbachu et al. 2014) of hydrocarbon degradation can also be done; in this method, the remaining percentage of hydrocarbon was calculated and compared with control. In the other method, we can observe the width of the oil layer after several days of incubation; decreasing width of oil layer also indicates hydrocarbon degradation (Khan and Rizvi 2011).

23.20 Future Perspective of Bioremediation

In the past few years, there has been a great deal of progress in the study and research of the biodegradation of xenobiotic compounds. Recently, many microbes with incredible potential of biodegradation of xenobiotic compounds have been isolated, and their new degradation pathways have been elucidated. Molecular approach for biodegradation study is still in its infancy. Nevertheless, the knowledge about biotransformation of organo-sulfide compounds is yet to be explored. To analyze the maximum efficiency of xenobiotics, degradation of microorganisms shows a mirror to us time to time and this is because of its dependency on place to place environmental conditions. Lignolytic and many other xenobiotic-degrading enzymes secreted by particular microorganism show less specificity toward the substrate which prohibits their use at large industrial scale. The use of genetically modified organism for biodegradation of oil spill and petroleum derivatives compounds has a bright future. The utility of constructed organisms in dealing with problems related to environmental pollution in nature is yet to be tested on practical grounds. Many factors like tolerance to xenobiotics, constitutive expression of the catabolic genes, substrate specificity, kinetics, and the stability of the encoded enzyme, etc., can have influence on the efficiency of constructed organisms. Although bioremediation has many drawbacks, but its future is still bright, because of having green and environmental friendly technique to clean up anthropogenically polluted environment and having many great advantages over existing physical and chemical methods for remediation. Phytoremediation is also another attractive area for the pollutants' bioremediation, but many untapped wild species and the limited knowledge about various phytoremediation mechanisms, including the regulation of enzyme systems that degrade pollutants, prohibit us to fully utilize their extreme potential for bioremediation process. Therefore, based on this review, it may be concluded that economical and green cleaning of environment without microorganism (bacteria, algae, and fungi) and plants is not possible.

References

Abdelkader E, Nadjia L, Ahmed B (2011) Degradation study of phenazin neutral red from aqueous suspension by paper sludge. J Chem Eng Process Technol 2:109–114

Adriaens P, Vogel TM (1995) Biological treatment of chlorinated organics. In: Young LY, Cerniglia CE (eds) Microbial transformation and degradation of toxic organic chemicals. Wiley, New York, pp 435–486

Aislabie J, Jones G (1995) A review of bacterial-degradation of pesticides. Aust J Soil Res 33(6): 925–942

Alves AMCR, Record E, Lomascolo A, Scholtmeijer K, Asther M, Wessels JGH, Wösten HAB (2004) High efficient production of laccase by the basidiomycete *Pycnoporus cinnabarinus*. Appl Environ Microbiol 70:6379–6384

Amin GA (2010) A potent biosurfactant producing bacterial strain for application in enhanced oil recovery applications. J Petrol Environ Biotechnol 1:104–110

Amin GA (2011) Integrated two-stage process for biodesulfurization of model oil by vertical rotating immobilized cell reactor with the bacterium rhodococcus erythropolis. J Petrol Environ Biotechnol 2:107

Angelidaki I, Ahring BK (1993) Thermophilic anaerobic digestion of livestock waste: the effect of ammonia. Appl Microbio Biotech 38:560–564

Angelidaki I, Sanders W (2004) Assessment of the anaerobic biodegradability of macropollutants. Rev Environ Sci Biotechnol 3:117–129

Annette CD, Jerald LS (2001) Advances in phytoremediation. Environ Health Perspect 109: 163–168

Arora PK, Kumar M, Chauhan A, Raghava GP, Jain RK (2009) OxDBase: a database of oxygenases involved in biodegradation. BMC Res Notes 2:1–9

Arora PK, Srivastava A, Singh VP (2010) Application of monooxygenases in dehalogenation, desulphurization, denitrification and hydroxylation of aromatic compounds. J Bioremed Biodegrad 1:1–8

Ashman M, Puri G (2002) Essential soil science a clear and concise introduction to soil science. Blackwell Science, Oxford

Barr D, Aust S (1994) Mechanisms white rot fungi use to degrade pollutants. Environ Sci Technol 28(2):78–87

Barton LL, Hamilton WA (2007) Sulphate reducing bacteria: environmental and engineered system. Cambridge University, Cambridge, p. 558

Beam HW, Perry JJ (1973) Co-metabolism as a factor in microbial degradation of cyclo-paraffinic hydrocarbons. Arch Microbiol 91:87–90

Bennet JW, Wunch KG, Faison BD (2002) Use of fungi biodegradation manual of environmental microbiology. ASM Press, Washington, DC

Bhushan B, Samanta SK, Chauhan A, Chakraborti AK, Jain RK (2000) Chemotaxis and biodegradation of 3-methyl-4-nitrophenol by Ralstonia sp. SJ98. Biochem Biophys Res Commun 275:129–133

Boetius A, Ravenschlag K, Schubert CJ, Rickert D, Widdel F, Gieseke A, Amann R, Jorgense BB, Witte U, Pfannkuche O (2000) A marine microbial consortium apparently mediating anaerobic oxidation of methane. Nature 407:623–626

Bordenave S, Goni-Urriza M, Caumette P, Duran R (2009) Differential display analysis of cDNA involved in microbial mats response after heavy fuel oil contamination. J Microbial Biochem Technol 1:001–004

Briggs GG, Bromilow RH, Evans AA (1982) Relationships between lipophilicity and root uptake and translocation of non-ionised chemicals by barley. Pestic Sci 13:495–504

Bumpus J, Aust S (1986) Biodegradation of environmental pollutants by the white rot fungus Phanerochaete chrysosporium: involvement of the lignin degrading system. Bioessays 6: 166–170

Burge WD (1969) Populations of dalapon-decomposing bacteria in soil as influenced by additions of dalapon or other carbon sources. Appl Microbiol 17:545–550

Cameron MD, Timofeevski S, Aust SD (2000) Enzymology of P Phanerochaete chrysosporium with the respect to the degradation of recalcitrant compounds and xenobiotics. Appl Microbiol Biotechnol 54:751–758

Chauhan A, Faziurrahman OJG, Jain RK (2008) Bacterial metabolism of polycyclic aromatic hydrocarbons: strategies for bioremediation. Indian J Microbiol 48:95–113

Cho NS, Wilkolazka AJ, Staszczak M, Cho HY, Ohga S (2009) The role of laccase from white rot fungi to stress conditions. J Fac Agric Kyushu Univ 54:81–83

Cirino PC, Arnold FH (2002) Protein engineering of oxygenases for biocatalysis. Curr Opin Chem Biol 6(2):130–135

Cookson JT (1995) Journal of bioremediation engineering: design and application. McGraw-Hill, New York

Copley SD (1998) Microbial dehalogenases: enzymes recruited to convert xenobiotic substrates. Curr Opin Chem Biol 2:613–617

Copley SD (2000) Evolution of a metabolic pathway for degradation of a toxic xenobiotic: the patchwork approach. Trends Biochem Sci 25:261–265

Cutter LA, Watts JEM, Sowers KR, May HD (2001) Identification of a microorganism that links its growth to the reductive dechlorination of 2,3,5,6-chlorobiphenyl. Environ Microbiol 3: 699–709

Dagley S (1975) Microbial degradation of organic compounds in the biosphere. Am Sci 63(6): 681–688

Dolfing J, Bloemen GBM (1985) Activity measurement as a tool to characterize the microbial composition of methanogenic environments. J Microbiol Methods 4:1–12

Dykes GA, Timm RG, Von HA (1994) Azoreductase activity in bacteria associated with the greening of instant chocolate puddings. Appl Environ Microbiol 60:3027–3029

Elaziouti A, Laouedj N, Ahmed B (2011) Effect of pH solution on the optical properties of cationic dyes in dye/maghnia montmorillonite suspensions. J Chem Eng Process Technol 2: 113–117
Elbanna K, Hassan G, Khider M, Mandour R (2010) Safe biodegradation of textile azo dyes by newly isolated lactic acid bacteria and detection of plasmids associated with degradation. J Bioremed Biodegrad 1:110–118
El-Bondkly AM, Aboshosha AAM, Radwan NH, Dora SA (2010) Successive construction of ß-glucosidase hyperproducers of *Trichoderma Harzianum* using microbial biotechnology techniques. J Microbial Biochem Technol 2:070–073
Eltis LD, Bolin JT (1996) Evolutionary relationships among extradioldioxygenases. J Bacteriol 178:5930–5937
Esteve-Nunez A, Lucchesi G, Philipp B, Schink B, Ramos JL (2000) Respiration of 2,4,6-trinitrotoluene by *Pseudomonas* sp strain JLR11. J Bacteriol 182(5):1352–1355
Fetzner S (1999) Bacterial dehalogenation. Appl Microbiol Biotechnol 50:633–657
Field J, Jong E, Feijo-Costa G, Bont J (1993) Screening for ligninolytic fungi applicable to the biodegradation of xenobiotics. Trends Biotechnol 11:44–49
Fragoeiro S, Magan N (2008) Impact of *Trametes versicolor* and *Phanerochaete chrysosporium* on differential breakdown of pesticide mixtures in soil microcosms at two water potentials and associated respiration and enzyme activity. Int Biodeterior Biodegrad 62(4):376–383
Frazar C (2000) The bioremediation and phytoremediation of pesticide contaminated sites, national network of environmental studies. National Network of Environmental Studies Fellow, Washington, DC
Fulthorpe RR, Wyndham RC (1989) Survival and activity of a 3-chlorobenzoate catabolic genotype in a natural system. Appl Environ Microbiol 55:1584–1590
Furukawa K (2003) 'Super bugs' for bioremediation. Trends Biotechnol 21:187–189
Gadd G (ed) (2001) Fungi in bioremediation. Cambridge University, Cambridge
Gavrilescu M (2005) Fate of pesticides in the environment and its bioremediation. Eng Life Sci 5(6):497–526
Gayathri KV, Vasudevan N (2010) Enrichment of phenol degrading moderately halophilic bacterial consortium from saline environment. J Bioremed Biodegrad 1:104–111
Gibson J, Harwood CS (2002) Metabolic diversity in aromatic compound utilization by anaerobic microbes. Ann Rev Microbiol 56:345–369
Grigg BC, Bischoff M, Turco RF (1997) Cocontaminant effects on degradation of triazine herbicides by a mixed microbial culture. J Agric Food Chem 45:995–1000
Grima S, Bellon-Maurel V, Feuilloley P, Silvestre F (2002) Aerobic biodegradation of polymers in solid-state conditions: a review of environmental and physicochemical parameter settings in laboratory simulation. J Polym Environ 8(4):183–195
Gursahani YH, Gupta SG (2011) Decolourization of textile effluent by a thermophilic bacteria Anoxybacillus rupiensis. J Petrol Environ Biotechnol 2:111–117
Hadad D, Geresh S, Sivan A (2005) Biodegradation of polyethylene by the thermophilic bacterium *Brevibacillus borstelensis*. J Appl Microbiol 98:1093–1100
Hara H, Eltis LD, Davies JE, Mohn WW (2007) Transcriptomic analysis reveals a bifurcated terepthalate degradation pathway in *Rhodococcus* sp. strain RHA1. J Bacteriol 189:1641–1647
Harayama S, Rekik M (1989) Bacterial aromatic ring cleavage is classified into two different families. J Biol Chem 264:15328–15333
Harayama S, Timmis KN (1992) Aerobic biodegradation of aromatic hydrocarbons by bacteria. In: Sigel H, Sigel A (eds) Metal ions in biological systems, vol 28. Marcel Dekker, New York, pp 99–156
Hayaishi O, Nozaki M (1969) Nature and mechanism of oxygenases. Science 164(3878):389–396
Heberer T (2002) Occurrence, fate and removal of pharmaceutical residues in the aquatic environment – a review of recent research data. Toxicol Lett 131:5–17
Hernando MD, Ejerhoon M, Fernández-Alba AR, Chisti Y (2003) Combined toxicity effects of MTBE and pesticides measured with *Vibrio fischeri* and *Daphnia magna* bioassays. Water Res 37:4091–4098

Hohener P, Hunkeler D, Hess A, Bregnard T, Zeyer J (1998) Methodology for the evaluation of engineered in situ bioremediation: lessons from a case study. J Microbiol Methods 32(2): 179–192
Holm-Nielsen JB, T-Al S, Oleskowicz-Popiel P (2009) The future of anaerobic digestionand biogas utilization. Bioresour Technol 100:5478–5484
Horvath RS (1972) Microbial co-metabolism and the degradation of organic compounds in nature. Bacteriol Rev 36(2):146–155
Horvath RS, Alexander M (1970) Co-metabolism: a technique for the accumulation of biochemical products. Can J Microbiol 16:1131–1132
Hosein SG, Millette D, Butler BJ, Greer CW (1997) Catabolic gene probe analysis of an aquifer microbial community degrading creosote-related polycyclic aromatic and heterocyclic compounds. Microbial Ecol 34:81–89
Hurt RA, Qiu X, Wu L, Roh Y, Palumbo AV, Tiedje JM, Zhou J (2001) Simultaneous recovery of RNA and DNA from soils and sediments. Appl Environ Microbiol 67:4495–4503
Husain Q (2006) Potential applications of the oxidoreductive enzymes in the decolorization and detoxification of textile and other synthetic dyes from polluted water: a review. Crit Rev Biotechnol 26(4):201–221
Imamul Huq SM, Abdullah MB, Joardar JC (2007) Bioremediation of arsenic toxicity by algae in rice culture. Land Contam Reclamat 15:327–334
Ismaila HY, Ijahb UJJ, Riskuwac ML, Allamina IA, Isaha MA (2014) Assessment of phytoremediation potentials of legumes in spent engine oil contaminated soil. Eur J Environ Saf Sci 2(2):59–64
Jaanis J (2010) Impact of phytoremediation and bioaugmentation on the microbial community in oil shale chemical industry solid waste. Tarutu University, Tarutu
Jame SA, Rashidul Alam AKM, Fakhruddin ANM, Alam MK (2010) Degradation of phenol by mixed culture of locally isolated *Pseudomonas* species. J Bioremed Biodegrad 1:102–112
Janssen DB, Dinkla IJT, Poelarends GJ, Terpstra P (2005) Bacterial degradation of xenobiotic compounds: evolution and distribution of novel enzyme activities. Environ Microbiol 7:1868–1882
Jennings DH, Burke RM (1990) Compatible solutes-the mycological dimension and their role as physiological buffering agents. New Phytol 116:277–283
Jensen HL (1963) Carbon nutrition of some microorganisms decomposing halogen substituted aliphatic acids. Acta Agric Scand 13:404–412
Joe J, Kothari RK, Raval CM, Kothari CR, Akbari VG et al (2011) Decolorization of textile dye remazol black b by *Pseudomonas aeruginosa* CR-25 isolated from the common effluent treatment plant. J Bioremed Biodegrad 2:118–125
Karanth NGK (2000) Challenges of limiting pesticide residues in fresh vegetables: the Indian experience food safety management in developing countries. CIRAD-FAO:11–13
Karigar CS, Rao SS (2011) Role of microbial enzymes in the bioremediation of pollutants. Enzyme Res 76:1–11
Kazumi J, Haggblom MM, Young LY (1995) Diversity of anaerobic microbial processes in chlorobenzoate degradation: nitrate, iron, sulfate and carbonate as electron acceptors. Appl Microbiol Biotechnol 43:929–936
Kearney P, Wauchope R (1998) Disposal options based on properties of pesticides in soil and water. In: Kearney P, Roberts T (eds) Pesticide remediation in soils and water. Wiley, Chichester
Kempf B, Bremer E (1998) Uptake and synthesis of compatible solutes as microbial stress responses to high-osmolality environments. Arch Microbiol 170:319–330
Khan JA, Rizvi SHA (2011) Isolation and characterization of micro-organism from oil contaminated sites. Advan Appl Sci Res 2(3):455–460
Kolb M, Harms H (2002) Metabolism of fluoranthne in different plant cell cultures and intact plants. Environ Toxicol Chem 19:1304–1310

Kumar KK, Prasad MK, Sarma GVS, Murthy CVR (2009) Removal of Cd (II) from aqueous solution using immobilized *Rhizomucor tauricus*. J Microbial Biochem Technol 1:015–021

Kumar NK, Reddy DSR, Venkateswarlu P (2010) Application of response surface methodology for optimization of chromium biosorption from an aqueous solution onto *Syzigium cumini* (java) seed powder. J Microbial Biochem Technol 2:020–027

Kumari N, Vashishtha A, Saini P, Menghani E (2013) Isolation, identification and characterization of oil degrading bacteria isolated from the contaminated sites of Barmer, Rajasthan. Int J Biotechnol Bioeng Res 4(5):429–436

Kurnaz SU, Buyukgungor H (2009) Assessment of various biomasses in the removal of phenol from aqueous solutions. J Microbial Biochem Technol 1:047–050

Kyrikou J, Briassoulis D (2007) Biodegradation of agricultural plastic films: a critical review. J Polym Environ 15:125–150

Lalithakumari D (2011) Microbes: "a tribute" to clean environment. University of Madras, Chennai

Leahy JG, Colwell RR (1990) Microbial degradation of hydrocarbons in the environment. Microbiol Rev 54:305–315

Lier JBV, Tilche A, Ahring BK, Macarie H, Moletta R, Dohanyos M, Hulshoff Pol LW, Lens P, Verstraete W (2001) New perspectives in anaerobic digestion. Water Sci Technol 43(1):1–18

Lloyd-Jones G, Laurie AD, Hunter DWF, Fraser R (1999) Analysis of catabolic genes for naphthalene and phenanthrene degradation in contaminated New Zealand soils. FEMS Microbiol Ecol 29:69–79

Magnuson JK, Romine MF, Burris DR, Kingsley MT (2000) Trichloroethene reductive dehalogenase from *Dehalococcoides ethenogenes*: sequence of tceA and substrate range characterization. Appl Environ Microbiol 66:5141–5147

Maloney S (2001) Pesticide degradation. In: Gadd G (ed) Fungi in bioremediation. Cambridge University, Cambridge

Mandellbaum RT, Allan DL, Wackett LP (1995) Isolation and characterisation of a *Pseudomonas* sp that mineralises the s-triazine herbicide atrazine. Appl Environ Microbiol 61:1451–1457

Mandri T, Lin J (2007) Isolation and characterization of engine oil degrading indigenous microrganisms in Kwazulu-Natal, South Africa. Afr J Biotechnol 6(1):23–27

Maria NDSU, Walter WW (2006) Phytoextraction of metal polluted soils in Latin America. Environmental Applications of Poplar and Willow Working Party

Mbachu AE, Onochie CC, Agu KC, Okafor OI, Awah NS (2014) Hydrocarbon degrading potentials of indigenous bacteria isolated from auto-mechanic workshops at mgbuka-nkpor, Nigeria. J Global Biosci 3(1):321–326

Meer JR, van der de Vos WM, Harayama S, Zehnder AJB (1992) Molecular mechanisms of genetic adaptation to xenobiotic compounds. Microbiol Rev 56:677–694

Memic M, Vrtacnik M, Vatrenjak-Velagic V, Wissiak Grm KS (2005) Comparative biodegradation studies of pre-emergence broadleaf and grass herbicides in aqueous medium. Int Biodeterior Biodegrad 55:109–113

Mills DK, Fitzgerald K, Litchfield CD, Gillevet PM (2003) A comparison of DNA profiling techniques for monitoring nutrient impact on microbial community composition during bioremediation of petroleum-contaminated soils. J Microbiol Methods 54(1):57–74

Mirdamadian SH, Emtiazi G, Golabi MH, Ghanavati H (2010) Biodegradation of petroleum and aromatic hydrocarbons by bacteria isolated from petroleum-contaminated soil. J Petrol Environ Biotechnol 1:102–109

Mishra V, Lal R, Srinivasan (2001) Enzymes and operons mediating xenobiotic degradation in bacteria. Crit Rev Microbiol 27:133–166

Nogales B, Moore ERB, Abraham WR, Timmis KN (1999) Identification of the metabolically active members of a bacterial community in a polychlorinated biphenyl polluted moorland soil. Environ Microbiol 1:199–212

Nyanhongo G, Gübitz G, Prakit S, Leitner C, Haltrich D, Ludwig P (2007) Oxidoreductases from *Trametes* spp. in biotechnol: a wealth of catalytic activity. Food Technol Biotechnol 45(3): 250–268

Okoh AI (2003) Biodegradation of bonny light crude oil in soil microcosm by some bacterial strains isolated from crude oil flow stations saver pits in Nigeria. Afr J Biotechnol 2(5): 104–108

Okuta A, Ohnishi K, Harayama S (1998) PCR isolation of catechol 2, 3-dioxygenase gene fragments from environmental samples and their assembly into functional genes. Gene 212: 221–228

Olajide PO, Ogbeifun LB (2010) Hydrocarbon biodegrading potentials of a *Proteus vulgaris* strain isolated from fish samples. Am J Appl Sci 7(7):922–928

Olusola SA, Anslem EE (2010) Bioremediation of a crude oil polluted soil with *Pleurotus pulmonarius* and *Glomus mosseae* using *Amaranthus hybridus* as a test plant. J Bioremed Biodegrad 1:111–118

Ortiz-Hernandez ML, Quintero-Ramirez R, Nava-Ocampo AA, Bello-Ramirez AM (2003) Study of the mechanism of *Flavobacterium* sp. for hydrolyzing organophosphate pesticides. Fundam Clin Pharmacol 17(6):717–723

Owolabi RU, Osiyemi NA, Amosa MK, Ojewumi ME (2011) Biodiesel from household/restaurant waste cooking oil (WCO). J Chem Eng Process Technol 2:112–118

Parameswari E, Lakshmanan A, Thilagavathi T (2010) Phycoremediation of heavy metals in polluted water bodies. Electron J Environ Agric Food 9:808–814

Park JW, Park BK, Kim JE (2006) Remediation of soil contaminated with 2, 4-dichlorophenol by treatment of minced shepherd's purse roots. Arch Environ Contam Toxicol 50(2):191–195

Pilon-Smits EAH, de Souza MP, Hong G, Amini A, Bravo RC et al (1999) Selenium volatilization and accumulation by twenty aquatic plant species. J Environ Qual 28:1011–1017

Pointing S (2001) Feasibility of bioremediation by white rot fungi. Appl Microbiol Biotechnol 57:20–33

Pointing S, Jones E, Vrijmoed L (2000) Optimization of laccase production by *Pycnoporus sanguineus* in submerged liquid culture. Mycologia 92:139–144

Poornima K, Karthik L, Swadhini SP, Mythili S, Sathiavelu A (2010) Degradation of chromium by using a novel strains of *Pseudomonas* species. J Microbial Biochem Technol 2:095–099

Qiu YL, Sekiguchi Y, Imachi H, Kamagata Y, Tseng IC, Cheng SS, Ohashi A, Harada H (2004) Identification and isolation of anaerobic, syntropic phthalate isomer degrading microbes from methanogenic sludges treating wastewater from terepthalate manufacturing. Appl Environ Microbiol 70:1617–1626

Raeid MMA (2011) Unraveling the role of cyanobacterial mats in the cleanup of oil pollutants using modern molecular and microsensor tools. World Congress on Biotechnology, New Delhi

Rajesh D, Anju H, Radha S, Poonam AS (2011) Saccharomyces cerevisiae: a potential biosorbent for biosorption of uranium. Int J Eng Sci Technol 3:5397–5407

Ram SV, Srivastava PN (2008) Phytoremediation-green for environmental clean. In: The 12th World lake conference, Jaipur, pp 1016–1021

Ramirez M, Chulze S, Magan N (2004) Impact of osmotic and matric water stress on germination, growth, mycelial water potentials and endogenous accumulation of sugars and sugar alcohols in *Fusarium graminearum*. Mycologia 96(3):470–478

Rao P, Hornsby A (2001) Behaviour of pesticides in soils and water. Fact Sheet SL40. University of Florida, Institute of Food and Agricultural Sciences, Gainesville

Reddy C, Mathew Z (2001) Bioremediation potential of white rot fungi. In: Gadd G (ed) Fungi in bioremediation. Cambridge University, Cambridge

Reddy N, Yang Y (2011) Plant proteins for medical applications. Trends Biotechnol 29(10): 490–498

Reuter P, Rabus R, Wilkes H, Aeckersberg F, Rainey FA, Jannasch HW, Widdel F (1994) Anaerobic oxidation of hydrocarbons in crude oil by new types of sulphate-reducing bacteria. Nature 372:455–458

Romantschuk M, Sarand I, Petanene R, Peltola R, Jonsson-Vihanne M, Koivula T, Yrjala K, Haahtela K (2000) Means to improve the effect of in situ bioremediation of contaminated soil: an overview of novel approaches. Environ Pollut 107(2):179–185

Sahrani FK, Ibrahim Z, Yahya A, Aziz M (2008) Isolation and identification of marine sulphate reducing bacteria, *Desulfovibrio* sp. and *Citrobacter freundii* from Pasir Gudang, Malaysia. Science 47:365–371

Samanta SK, Bhushan B, Chauhan A, Jain RK (2000) Chemotaxis of a *Ralstonia* sp. SJ98 toward different nitroaromatic compounds and their degradation. Biochem Biophys Res Commun 269: 117–123

Santhoskumar AU, Palanivelu K, Sharma SK, Nayak SK (2010a) A new synthesis of nickel 12-- hydroxy oleate formulation to improve polyolefi n's degradation. J Bioremed Biodegrad 1: 108–115

Santhoskumar AU, Palanivelu K, Sharma SK, Nayak SK (2010b) Comparison of biological activity transistion metal 12 hydroxy oleate on photodegradation of plastics. J Bioremed Biodegrad 1: 109–116

Sasek V (2003) Why mycoremediations have not yet come to practice. In: Sasek V et al (eds) The utilization of bioremediation to reduce soil contamination: problems and solutions. Kluwer Academic, Amsterdam, pp 247–276

Sayler GS, Layton AC (1990) Environmental application of nucleic acid hybridization. Ann Rev Microbiol 44:625–648

Schink B, Philipp B, Muller J (2000) Anaerobic degradation of phenolic compounds. Naturwissenschaften 87:12–23

Schloss PD, Handelsman J (2003) Biotechnological prospects from metagenomics. Curr Opin Microbiol 14:303–310

Schoefs O, Perrier M, Samson R (2004) Estimation of contaminant depletion in unsaturated soils using a reduced-order biodegradation modeland carbon dioxide measurement. Appl Microbiol Biotechnol 64:256–261

Scholz-Muramatsu H, Neumann A, Mebmer M, Moore E, Diekert G (1995) Isolation and characterization of *Dehalospirillum multivorans* gen. nov., sp. nov., a tetrachloroethene utilizing, strictly anaerobic bacterium. Arch Microbiol 163:48–56

Shukla KP, Singh NK, Sharma S (2010) Bioremediation: developments, current practices and perspectives. Genet Eng Biotechnol J 3:1–20

Sims RC, Overcash MR (1983) Fate of polynuclear aromatic compounds (PNAs) in soil plant system. Residue Rev 88:1–68

Singleton I (2001) Fungal remediation of soils contaminated with persistent organic pollutants. In: Gadd G (ed) Fungi in bioremediation. Cambridge University, Cambridge

Sinha S, Chattopadhyay P, Pan I, Chatterjee S, Chanda P (2009) Microbial transformation of xenobiotics for environmental bioremediation. Afr J Biotechnol 8:6016–6027

Sivasubramanian V, Subramanian VV, Muthukumaran M (2010) Bioremediation of chrome-sludge from an electroplating industry using the micro alga *Desmococcus olivaceus*—a pilot study. J Algal Biomass Utln 3:104–128

Skipper H (1998) Bioremediation of contaminated soils. In: Sylvia D et al (eds) Principles and applications of soil microbiology. Prentice Hall, Upper Saddle River

Somasundaram L, Coats R (1990) Influence of pesticide metabolites on the development of enhanced biodegradation. In: Racke KD, Coats JR (eds) Enhanced biodegradation of pesticides in the environment. American Chemical Society, Washington, DC

Song B, Palleroni NJ, HaGgblom MM (2000) Isolation and characterization of diverse halobenzoate degrading denitrifying bacteria from soils and sediments. Appl Environ Microbiol 66(8): 3446–3453

Sonoki T, Kajita S, Uesugi M, Katayama Y, Iimura Y (2011) Effective removal of bisphenol a from contaminated areas by recombinant plant producing lignin peroxidase. J Petrol Environ Biotechnol 2:105–111

Soto M, Mendez R, Lema JM (1993) Methanogenic activity tests theoretical basis and experimental setup. Water Res 27:850–857

Stolz A (2001) Basic and applied aspects in the microbial degradation of azo dyes. Appl Microbiol Biotechnol 56:69–80

Sunday AA, Matthew OI, Olukayode OA, Olakunle DT, Olatope SO (2006) Microbial degradation of petroleum hydrocarbons in a polluted tropical stream. J Am Sci 2(3):48–47
Swift G (1998) Requirements for biodegradable water soluble polymers. Polym Degrad Stab 59: 19–24
Takami H, Kudo T, Horikoshi K (1997) Isolation of extradiol dioxygenasegenes that is phylogenetically distant from other meta cleavage dioxygenase genes. Biosci Biotechnol Biochem 61:530–532
Taliani MR, Roberts SC, Dukek BA, Pruthi RK, Nichols WL, Heit JA (2001) Sensitivity and specificity of denaturing high pressure liquid chromatography for unknown protein C gene mutations. Genet Test 5:39–44
Taranenko NI, Hurt R, Zhou J, Isola NR, Huang H, Lee SH, Chen CH (2002) Laser desorption mass spectrometry for microbial DNA analysis. J Microbiol Method 48:101–106
Vamsee-Krishna C, Phale PS (2008) Bacterial degradation of phthalate isomers and their esters. Indian J Micriobiol 48:19–34
Van Agteren MH, Keuning S, Janssen DB (1998) Handbook on biodegradation and biological treatment of hazardous organic compounds. Kluwer Academic, Dordrecht
Vander JRM, Vos WMD, Harayama S, Zehnder AJB (1992) Molecular mechanisms of genetic adaptation to xenobiotic compounds. Microbiol Rev 56:677–694
Vargas C, Song B, Camps M, Haggblom MM (2000) Anaerobic degradation of fluorinated aromatic compounds. Appl Microbiol Biotechnol 53:342–347
Vidali M (2001) Bioremediation an overview. Pure Appl Chem 73(7):1163–1172
Wasi S, Tabrez S, Ahmad M (2010) Isolation and characterization of a *Pseudomonas fluorescens* strain tolerant to major indian water pollutants. J Bioremed Biodegrad 1:101–109
Watanabe K, Teramoto M, Futamata H, Harayama S (1998) Molecular detection, isolation and physiological characterization of functionally dominant phenol degrading bacteria in activated sludge. Appl Environ Microbiol 64:4396–4402
Wei SH, Zhou QX, Wang X, Cao W, Ren LP (2004) Potential of weed species applied to remediation of soils contaminated with heavy metals. J Environ Sci 16:868–873
Weller R, Ward DM (1989) Selective recovery of 16S ribosomal RNA sequences from natural microbial communities in the form of complementary DNA. Appl Environ Microbiol 55:1818–1822
Wenzel WW, Lombi E, Adriano D (1999) Biogeochemical processes in the rhizosphere: role in phytoremediation of metal-polluted soils. In: Prasad MNV, Hagemeyer J (eds) Heavy metal stress in plants—from molecules to ecosystems. Springer, Heidelberg, pp 273–303
Widada J, Nojiri H, Omori T (2002) Recent developments in molecular techniques for identification and monitoring of xenobiotic-degrading bacteria and their catabolic genes in bioremediation. Appl Microbiol Biotechnol 60:45–59
Williams PP (1977) Metabolism of synthetic organic pesticides by anaerobic microorganisms. Residue Rev 66:63–135
Wilson L, Bouwer E (1997) Biodegradation of aromatic compounds under mixed oxygen/denitrifying conditions: a review. J Ind Microbiol Biotechnol 18:116–130
Wohlfarth G, Diekert G (1997) Anaerobic dehalogenases. Curr Opin Biotechnol 8:290–295
Yatome C, Matsufuru H, Taguchi T, Ogawa T (1993) Degradation of 4- dimethylaminoazobenzene-2-carboxylic acid by *Pseudomonas stutzeri*. Appl Microbiol Biotechnol 39:778–781
Zhai G (2011) Phytoremediation: right plants for right pollutants. J Bioremed Biodegrad 2:102
Zhang C, Bennet GN (2005) Biodegradation of xenobiotics by anaerobic bacteria. Appl Microbiol Biotechnol 67:600–618
Zhang J, Quiao C (2002) Novel Approaches for remediation of pesticide pollutants. Int J Environ Pollut 18(5):423–433

[illegible]

Tandy S, [illegible] (20[illegible]) [illegible] in contaminated soil [illegible] Environ [illegible]

Smith C (1996) [illegible] biodegradability of water-soluble polymers [illegible]

[illegible] (1994) [illegible] chemistry [illegible]

[illegible] (20[illegible]) [illegible] liquid [illegible] Env Sci Pollut Res [illegible]

[illegible] NL, Hua L, Zhou J, [illegible], Chen CH (20[illegible]) [illegible] analysis [illegible]

[illegible]

[illegible] DB [illegible] (2010) [illegible] remediation [illegible]

[illegible] of [illegible] compounds and [illegible]

Vogel [illegible] WM, [illegible] S, [illegible] (20[illegible]) Microbial [illegible] Microbiol Rev [illegible]

[illegible] (2000) [illegible] communities. Appl Microbiol Biotechnol [illegible]

[illegible] M (20[illegible]) Bioremediation [illegible]

[illegible] (2010) Isolation and characterization of [illegible] from oil [illegible] water pollutants. J Hazard Mater [illegible]

[illegible] K, [illegible] M, [illegible] (20[illegible]) [illegible] effect [illegible] in contaminated soils. Appl [illegible] Microbiol [illegible]

[illegible] SH, Zhou [illegible], Wang [illegible] (20[illegible]) [illegible] contaminated with heavy metals. J Environ [illegible]

Wagner R, Ward DM (1990) Selective recovery of 16S ribosomal RNA sequences from natural microbial communities in the form of complementary DNA. Appl Environ Microbiol 56:1818–1822

Wenzel WW, [illegible] (1999) Biogeochemical processes in the rhizosphere: role in phytoremediation of metal-polluted soils. In: Prasad MNV, Hagemeyer J (eds) Heavy metal stress in plants—from molecules to ecosystems. Springer, Heidelberg, pp 273–303

[illegible] L, [illegible] (2003) Recent developments of molecular techniques for identification and monitoring of [illegible] and their metabolic genes in bioremediation. Appl Microbiol Biotechnol 60:[illegible]

Williams PT (20[illegible]) [illegible] review [illegible]

[illegible] (20[illegible]) [illegible] J Microbiol Biotechnol [illegible]

[illegible]

[illegible] Appl Microbiol Biotechnol [illegible]

[illegible] (20[illegible]) [illegible]

Zhang [illegible] (20[illegible]) [illegible] Appl Microbiol Biotechnol [illegible]

Zhang [illegible] (20[illegible]) [illegible] for transformation of [illegible] Int J Environ [illegible]

Chapter 24
Prominences on Xenobiotic Degradation Underneath of Ecological Sanitary

Vivek Kumar, Ram Prasad, Pankaj Goyal, Narendra Tuteja, Joginder Singh, Ajit Varma, and Manoj Kumar

24.1 Introduction

A xenobiotic is defined as foreign synthetic substrate or compound that occurs within an organism through unnatural resources, also synthesized by or arrived within the same organism. It may include ingredients which exist in abundant concentrations that are expected in nature either in organism or ecological level.

Toxic Concern on Atmosphere In recent past, human population had an unrestricted wealth in terms of land and forest resources. Presently, due to our sloppiness, greed, and compromising approach in occupying them, the ecological assets in the world show, in lesser degree (Vidali 2001). The quick growth of various industries in the past century has extremely increased the release of toxic waste effluents into water bodies along with groundwater (Sethy et al. 2011). Environmental pollution caused by the release of these wide range of compounds [i.e., persistent organic pollutants (POPs)] from industries is creating disturbance to the ecosystem (Gursahani and Gupta 2011), causing climatic changes, reduction of water levels in the ground as well as oceans, melting of ice caps, global warming, ozone layer depletion due to photochemical oxidation, etc. (Sharma et al. 2011; le Mellec et al. 2010), and this made ecologists to focus more on impacts of pollution and its reduction.

It has already been noticed that industrial effluent is not properly regulated; rather, they become accidental (e.g., chemical or oil spills) which can form toxic and persistent material in native environments. Materials in the ecosystem are

V. Kumar • R. Prasad • P. Goyal • N. Tuteja • A. Varma • M. Kumar (✉)
Amity Institute of Microbial Technology, Amity University, Noida, Uttar Pradesh, India
e-mail: manojjnu@gmail.com

J. Singh
School of Biotechnology, Lovely Professional University, Phagwara, Punjab, India

© Springer International Publishing AG 2017

M.Z. Hashmi et al. (eds.), *Xenobiotics in the Soil Environment*, Soil Biology 49,
DOI 10.1007/978-3-319-47744-2_24

derivative of either biogenic or anthropogenic sources. Anthropogenic compounds are synthetic in nature and play a key role in contaminating the ecosystem (Indu 2006). Xenobiotics are in fact anthropogenic compounds that existed in biological chains or in the defined ecological pyramid which are unnatural and are present in decimal concentration. The latent health risk of a xenobiotic compound is a purpose of its tenacity in the environment as well as the lethality of the synthetic type (Wilson et al. 1985). The matter of xenobiotics in the ecological system has been an advancement for investigation in ecological chemistry for years.

24.2 Genesis of Xenobiotic

The major straight source of xenobiotics is wastewater and solid residual releases from the industries like chemical and pharma, plastics, paper and pulp mills, textile mills, and agriculture (enhancement products like pesticides, herbicides, etc.) (Fig. 24.1a–e). Some of the common residual factors in the wastewater and other effluents are phenol, hydrocarbons, different dyes, paint effluents, pesticides and insecticides, etc. (Gayathri and Vasudevan 2010; Jame et al. 2010; Nagamani et al. 2011; Sridevi et al. 2011).

24.3 Removal of Xenobiotic

Numerous approaches like physicochemical and biological ways have been implemented in the degradation of xenobiotics. There is a proven fact that physicochemical approaches are costly and usually give undesirable yields which are latent and toxic in nature; eventually they require further treatment stages (Sridevi et al. 2011). Such methodologies habitually add fragmented compounds which are

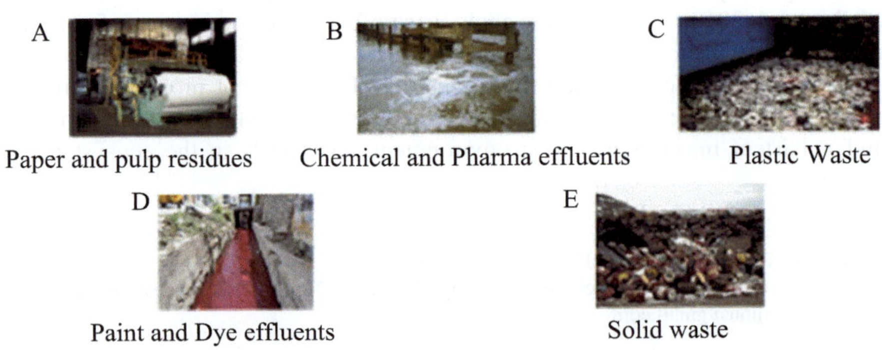

Fig. 24.1 (**a–e**) Different sources of industrial effluents. (**a**) Paper and pulp residues. (**b**) Chemical and pharma effluents. (**c**) Plastic waste. (**d**) Paint and dye effluents. (**e**) Solid waste

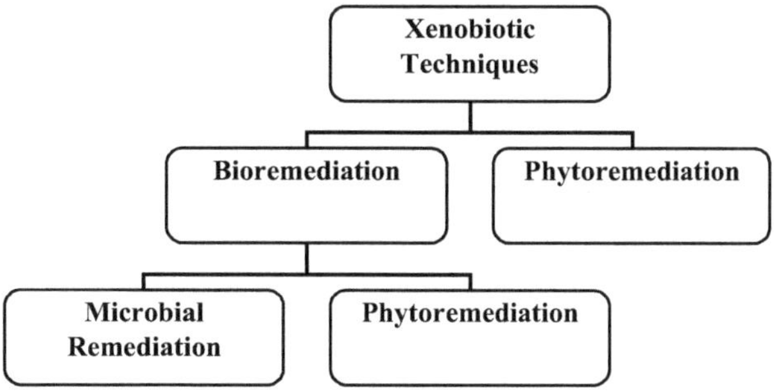

Fig. 24.2 Classification showing different xenobiotic techniques

difficult for further degradation and henceforth create the long-term damage in ecology. In order to get rid of these problems, several other sustainable and eco-friendly techniques have already been evolved, i.e., bioremediation, phytoremediation, etc. (Fig. 24.2).

24.4 What Is Bioremediation?

Microbial degradation of xenobiotics is one of the important ways to remove the environmentally harmful compounds. The potential of microorganisms to metabolize xenobiotic compounds has been recognized as an effective means of toxic and hazardous waste removal (Sridevi et al. 2011; Oaks et al. 2004).

Researchers have used biological system in bioremediation approach and defined as a process that includes microorganisms or their enzymes to retrieve the environment altered by pollutants to its original condition (Oaks et al. 2004). In another diaphragm, bioremediation is being termed as "a treatability technology that exploits biological activity to decrease the toxicity level" (King et al. 1997). Detoxification and mineralization are the vital happenings which occur simultaneously under the effect of biological waylays, where the excess waste material is supposed to be transformed into inorganic substrates, i.e., carbon dioxide, water, and methane (Reshma et al. 2011). Such substrates get accumulated stubbornly in the environment, though nature allows the biodegradation through multiple steps utilizing different biocatalytic systems or in benign presence of microbiota. The best examples are contaminated wastewater, ground- or surface waters, soils, sediments, and air where there has been either accidental or intentional release of pollutants or chemicals which are the sites where bioremediation occurs (Ali Elredaisy 2010; Aghamiri et al. 2011).

24.5 Significant Microbiota for Remediation

Microbiome symbolizes as a predominant biomass of our globe; on the contrary, the human population disturbs the ecology and triggers the xenobiotic influx on the planet. The significant microbes exhibit potential to degrade such lethal chemical, those that are listed as xenobiotic compounds. Microbial system uses their endo- and exo-secreted enzymes under the regime of preset metabolic pathways, exploiting them as novel carbon sources and henceforth cleaning the toxic substances (Singh et al. 2014). It is a proven fact that microbial populations show their eco-friendly behavior to overcome environmental pollution and to help in biodegradation of xenobiotic compounds. To meet such sustainable approach, microorganisms use dual modes of action for dilapidation of xenobiotics compound: (1) aerobic biodegradation and (2) anaerobic biodegradation. Aerobic biodegradation includes rich oxygen delivery systems; it is necessary to stock continuous oxygen due to biofouling at the level of remediation (Sharma and Fulekar 2009); moreover, bioreactor-based application is cost ineffective at the same time produces sludge which is highly expensive for further removal (Kumar et al. 1994a, b). Anaerobic habitats, including sludge digesters, groundwater, sediments, water-laden soils, gastrointestinal contents, feedlot wastes and landfill sites, and some xenobiotic compounds [e.g., tetrachloroethylene, polychlorinated biphenyls (PCBs), and nitro-substituted aromatics], can be effectively transformed or mineralized by anaerobic bacteria (Zhang and Bennett 2005).

In situ bioremediation procedure consists of basically three vital steps:

1. Bioattenuation: It is related to monitoring of natural progress of biodegradation to guarantee that contaminant declines with sampling time.
2. Biostimulation: The intentional stimulation of natural xenobiotic remediating microbes by electron acceptors, water molecule, nutrient addition, and/or electron donors.
3. Bioaugmentation: It is the addition of laboratory-grown potential bacteria that have suitable and biodegradative abilities.

Normally, the microbes use two pathways for biodegradation of xenobiotics, aerobic and anaerobic conditions.

In aerobic bioremediation, the basic equation will be:

$$\text{Xenobiotics} + \text{oxygen} \overset{\text{Microbe}}{\rightarrow} CO_2 + H_2O + \text{left over residue}$$

In case of anaerobic bioremediation:

$$\text{Xenobiotics} \overset{\text{Microbe}}{\rightarrow} CO_2 + CH_4 + H_2O + \text{left over residue}$$

In aerobic biodegradation, CO_2 is produced along with some amount of water. In the absence of oxygen, anaerobic biodegradation process starts and methane gas is generated instead of CO_2. The conversion of biodegradable materials to gases like carbon dioxide, methane, and nitrogen compounds is called mineralization. Mineralization process is completed, when all the biodegradable biomass is consumed and all the carbon is converted into carbon dioxide (Kyrikou and Briassoulis 2007). Alkanes consisting long carbon chains and straight structures considered to be more prone to aerobic biodegradation. Aerobic degradation pathway of alkane degradation is the oxidation of the terminal methyl group into a carboxylic acid through an alcohol intermediate and after all completes mineralization through β-oxidation (Le and Coleman 2011). Aerobic biodegradation process of aromatic compound comprises of their oxidation by molecular oxygen; after oxidation steps, intermediates are the outcome, and then it enters into central metabolic pathways, including the Krebs cycle and β-oxidation.

Some xenobiotic pollutants are not mineralized by an aerobic degradation system because they are greatly recalcitrant owing to increase in halogenations in their structures. Replacement of halogen and nitro and sulfo groups on the aromatic ring increases the electrophilicity of the target molecule. These xenobiotic compounds resist the electrophilic attack by enzyme oxygenases in aerobic degradation process. Some of the recalcitrant that persists under aerobic condition are the polychlorinated biphenyls (PCBs), chlorinated dioxins, and some complex and banned pesticides like DDT (Alcock and Jones 1996). It is essential to overawe the high stubbornness of halogenated xenobiotic compounds from biosphere; in achieving these, the reductive attacks by anaerobic microorganisms are of boundless worth. On the other hand, anaerobic bacteria carried out reductive dehalogenation either by the complimentary reaction or by using a new type of anaerobic respiration. This procedure decreases the degree of chlorination and makes the product more available and manageable for mineralization process by aerobic bacteria (Ferguson and Pietari 2000). During anaerobic degradation process, the reductive dehalogenation is the first step of biodegradation of polychlorinated biphenyls (PCBs); dehalogenation process is carried out under anaerobic conditions where organic substrates act as electron donors.

There are a vast number of potential microbes, especially the bacteria, which carry out the bioremediation of xenobiotics. The common major groups of anaerobic bacteria that have the capability of biodegrading xenobiotic compounds are *Acidovorax* spp., *Bordetella* spp., *Pseudomonas* spp., *Sphingomonas* spp., *Variovorax* spp., *Veillonella alcalescens*, *Desulfovibrio* spp., *Desulfuromonas michiganensis*, and *Desulfitobacterium dehalogenans*, *D. oleovorans*, *G. metallireducens*, and *D. acetonicum*. Anaerobic sulfate-reducing bacteria (SRB) and methanogenic bacterial conditions can be useful to isolate pure culture of anaerobic bacteria to carry out xenobiotic degradation research work (Jiang et al. 2009; Zhang and Bennet 2005). Anaerobic microbes can also use and exploit substituted and intricate aromatic compounds in a way that do not disturb the benzene nucleus in the ring. On the other hand, sulfate-reducing bacteria (SRB) represent a huge group of anaerobic microorganisms that play a crucial role in numerous biogeochemical cyclic processes and also are able to biodegrade the crude oil (Ferradji et al. 2014). The sulfate-reducing bacteria are an obligated anaerobic bacteria, which utilize sulfate as final electron acceptor during the process of anaerobic respiration and, therefore, generate hydrogen sulfide (H_2S gas) by sulfate reduction. Anaerobic degradation process is also a renewable energy source; here the biogas is generated from the anaerobic digestion. It mainly consists of methane (CH_4) that can be collected easily and applied for eco-friendly power generation or as a fuel, which has been proved on a greater scale (Boetius et al. 2000).

24.6 Role of Microbial Enzymes in Bioremediation

Bioremediation is a microbial secreted enzymatic process which transforms a xenobiotic pollutant to innocuous products, which blends naturally with the environment; therefore, the toxicity is removed or reduced to a greater extent.

24.6.1 Enzymatic Activity

Potential enzymes like oxidoreductases, dehalogenases, monooxygenases, phosphotriesterases (PTEs), dioxygenases, oxygenases, etc. are reported from commercial fungal traits and exploited for bioremediation purpose.

These enzymes separate the chemical bonds and repost the electrons from reduced organic compound (called as donor) to another chemical substrate (known as acceptor). In due course of oxidation–reduction reactions, the chemical impurities or pollutants are supposed to be oxidized toward innocuous and harmless compounds (Karigar and Rao 2011). The oxidoreductases rinse the toxic xenobiotics products, i.e., phenolic or aniline compounds, either by the process of polymerization, or copolymerization with other substrates, or binding with the humic substances. The microbial enzymes have already been reported for bioremediation of azo dyes (Kumar et al. 2016; Husain 2006; Rani et al. 2014).

Enzyme-mediated transfer is a well-known phenomenon where one atom of molecular oxygen to the organic compound (Karigar and Rao 2011) in reversible manner has metabolic persuasion at cellular level in prokaryotes. Monooxygenases can be categorized into two subclasses based on the presence cofactor, flavin-dependent monooxygenases and P450 monooxygenases. Flavin-dependent monooxygenases contain flavin as prosthetic group and NADP or NADPH as coenzyme. P450 monooxygenases are heme-containing oxygenases that persist in both eukaryotes and prokaryotes. Monooxygenases act as biocatalysts in the bioremediation process and synthetic chemistry because they are highly regioselectivity and stereoselectivity on a wide range of substrates (Karigar and Rao 2011). Monooxygenases catalyzes enormous reactions such as desulfurization, dehalogenation, denitrification, ammonification, hydroxylation, biotransformation, and biodegradation of various aromatic and aliphatic compounds.

Dehalogenase plays an important role in the degradation of chlorinated pollutant. Some anaerobic microorganisms exploit dehalorespiration and use halogenated compounds as terminal electron acceptors (Le and Coleman 2011). An example of this process is the conversion of perchloroethylene (PCE), either dichloroethylene (DCE) (Schumacher and Holliger 1996), ethylene, or ethane which depends on the conditions. Researchers have reported the partial purification of two reductive dehalogenases from *Dehalococcoides ethenogenes* strain 195; both enzymes are membrane proteins. The first enzyme PCE-reductive dehalogenase reduces PCE to TCE and the second enzyme TCE-reductive dehalogenase reduces TCE, *trans*-DCE, *cis*-DCE, 1,1-dichloroethene, and vinyl chloride (Patil and Bagde 2012).

Phosphotriesterases are microbial isolated enzyme which hydrolyze and detoxify organophosphate pesticides (OPs). This reduces OP toxicity and decreases the ability of OPs to inactivate AchE (Shen et al. 2010; Theriot and Grunden 2010). These enzymes mainly hydrolyze phosphoester bonds like P–O, P–F, P–NC, and P–S, and these hydrolysis mechanisms include water molecule in the phosphorus center (Ortiz-Hernandez et al. 2003).

These are multicomponent enzyme systems that incorporate molecular oxygen to the substrate. On the basis of the complexity of the degradation pathways, the biodegradation phenomenon can be categorized into two types: (1) convergent mode and (2) divergent modes of degradation (Eltis and Bolin 1996). In the convergent mode, structurally varied aromatic compounds are converted to aromatic ring cleavage substrates catechol, gent sate, protocatechuate, and their derivatives. In divergent mode, a metal-dependent dioxygenase channels operate, and dihydroxylated intermediates are formed by one of the two possible pathways: the meta-cleavage pathway or the ortho-cleavage pathway (Takami et al. 1997).

These are classified under the oxidoreductase group of enzymes (EC Class 1) (Karigar and Rao 2011). Oxidation reaction is the major enzymatic reaction of aerobic biodegradation and is catalyzed by oxygenases. Oxygenases oxidize the substrates by transferring oxygen from molecular oxygen (O_2) and utilize FAD/NADH/NADPH as the co-substrate. Oxygenases metabolize organic compounds; they increase their reactivity and water solubility and cleave the aromatic ring (Arora et al. 2010). On the basis of the number of oxygen atoms used for oxidation, oxygenases can be further categorized into two groups: (1) monooxygenases and (2) dioxygenases, which have been discussed earlier.

24.7 Impending Aspects

Since last few decades, there has been a huge development in the field of the bioremediation of xenobiotic compounds. Several novel microbe-mediated bioremediation approaches have been reported targeting rare ecological niches and given interesting remediation pathways. Moreover, existing information on microbial exploitation is still under the scan. Biotransformation of xenobiotic compounds is yet to be reconnoitered owing to its multifarious nature. Efficacy of such xenobiotic compound for its biodegradation can be enhanced by addressing some relevant issues: (1) adapting tolerance mechanism to various xenobiotic materials, (2) the constitutive expression of catabolic candidate genes against the specific raw substrates, and (3) the kinetics and stability of the candidate enzymes which have recently been encoded. Even though usefulness and efficacy of the constructed organisms in terms of environmental pollution problem in ecosystem are yet to be uncovered and tested.

Numerous microbes biodegrade xenobiotic compounds through plasmids which encode for the catabolic genes and further transmit under specific conditions. To elucidate and develop the construct for such candidate genes and its further mobilization through the recombinant process, an improved construct of potential strains is required, and a proper, well-designed management is also a prerequisite. Due to the existing facts, the microbial degradation machinery is a bridging scope from the environmental monitoring point of view and ultimately leads to biodegradation of lethal compounds. In bioremediation process, presently, molecular techniques and approaches are being applied to characterize the genetic material of numerous bacteria from the several ecological samples. Existing data helps the researchers to compare with the standard and prevalent microbiological technique approaches; the molecular procedures facilitate us with more compatible and comprehensive explanation of in situ microorganism population and its response to concocted bioremediation and normal lessening processes. In addition to that, a dominant molecular procedure is required as metagenomic library which has been thrived for identification of the specific catabolic gene pool. In principle, the metagenomics approach is a culture-independent microbial genomic analysis; this approach is in function mode due to the existence of another powerful sequence-driven method. For entire microbial communities, modern tool and techniques which include the direct genetic investigation approach provide the access to recover unknown sequences from rare microbiota. The consistent and persistent contact with the pollutants and long exposure to their occurrence are the fundamentals of fight in contradiction of xenobiotic compounds; meanwhile such developments allow the advancement and evolution of new, more or less safe processes of xenobiotic remediation by microbiomes.

24.8 Conclusions

Microbial assortment, the abundance of species in ecological sites, delivers an enormous pool of resources which we can exploit for our benefit. Though, little is known about the true diversity of bacterial life. Despite the acknowledged value of microorganisms, our understanding of their diversity and many of their key roles in sustaining global life-support systems is still very scarce. This is because the vast majority of bacteria are non-culturable by standard methods, and we have only recently acquired the skills to explore this aspect of microbial biodiversity. Exploring the range of microbial biodiversity is the key to developing effective and environment-friendly "green" technologies. Bioremediation is one such process that exploits the catabolic abilities of microorganisms to degrade harmful and toxic xenobiotics. We have been able to restore what once were irreversibly polluted sites in some cases, attesting to the usefulness of this cleanup process. However, to maximize the potential benefits of microbial community in combating pollution problems, it is vital that we have fundamental understanding of a microbe's degradative potential under various conditions, its biochemical systems, and its molecular biology.

Environmental problems caused by the industrial effluents are mainly due to accumulation of pollutants and other fragmented compounds, which in turn form into other substitutes (natural or manmade), finally forming a xenobiont. There is a quick need to degrade these xenobiotic compounds in an eco-friendly way. Various techniques like microbial remediation, phytoremediation, and photoremediation and their subtypes have been discussed. Each having their own ways of degrading these xenobionts also has a negative impact on the environment (side effects due to fragmentations and bioaccumulations). Photoremediation, a novel equipment-based technique, is rapid but also has a negative impact on the environment. Being a solar-driven technique, phytoremediation is restricted to particular sites containing contaminants. Although slow, on the whole, microbial bioremediation was found to cover a wide range of recalcitrant degradation and is known to be a better choice because of its nature of degradation.

References

Aghamiri SF, Kabiri K, Emtiazi G (2011) A novel approach for optimization of crude oil bioremediation in soil by the Taguchi method. J Pet Environ Biotechnol 2:108

Alcock RE, Jones KC (1996) Dioxins in the environment: a review of trend data. Environ Sci Technol 30:3133–3143

Ali Elredaisy SM (2010) Ecological benefits of bioremediation of oil contaminated water in rich savannah of palogue, Upper Nile Area-Southern Sudan. J Bioremed Biodegrad 1:103

Arora PK, Srivastava A, Singh VP (2010) Application of monooxygenases in dehalogenation, desulphurization, denitrification and hydroxylation of aromatic compounds. J Biorem Biodegrad 1:1–8

Boetius A, Ravenschlag K, Schubert CJ, Rickert D, Widdel F, Gieseke A, Amann R, Jorgense BB, Witte U, Pfannkuche O (2000) A marine microbial consortium apparently mediating anaerobic oxidation of methane. Nature 407:623–626

Eltis LD, Bolin JT (1996) Evolutionary relationships among extradiol dioxygenases. J Bacteriol 178:5930–5937

Ferguson JF, Pietari JM (2000) Anaerobic transformations and bioremediation of chlorinated solvents. Environ Pollut 107:209–215

Ferradji FZ, Mnif S, Badis A, Rebbani S, Fodil D, Eddouaouda K, Sayadi S (2014) Naphthalene and crude oil degradation by biosurfactant producing Streptomyces spp. isolated from Mitidja plain soil (North of Algeria). Int Biodeterior Biodegrad 86(C):300–308

Gayathri KV, Vasudevan N (2010) Enrichment of phenol degrading moderately halophilic bacterial consortium from saline environment. J Bioremed Biodegrad 1:104

Gursahani YH, Gupta SG (2011) Decolourization of textile effluent by a thermophilic bacteria *Anoxybacillus rupiensis*. J Pet Environ Biotechnol 2:111

Husain Q (2006) Potential applications of the oxidoreductive enzymes in the decolorization and detoxification of textile and other synthetic dyes from polluted water: a review. Critical Rev Biotechnol 26:201–221

Indu ST (2006) Xenobiotics: pollutants and their degradation-methane, benzene, pesticides, bioabsorption of metals: http://nsdl.niscair.res.in/jspui/bitstream/123456789/664/1/Xenobiotics.pdf

Jame SA, Rashidul Alam AKM, Fakhruddin ANM, Alam MK (2010) Degradation of phenol by mixed culture of locally isolated pseudomonas species. J Bioremed Biodegrad 1:102

Jiang XW, Liu XH, Wang Y, Leak SJ, Zhou DJ (2009) Genetic and biochemical analyses of chlorobenzene degradation gene clusters in Pandoraea sp. strain MCB032. Arch Microbiol 191:485–492

Karigar CS, Rao, SS (2011) Role of microbial enzymes in the bioremediation of pollutants: a review. Enzym Res 2011:11, 805187

King RB, Long GM, Sheldon JK (1997) Practical environmental bioremediation: the field guide, 2nd edn. Lewis, Boca Raton

Kumar V, Anand RC, Singh R (1994a) Enrichment and isolation of acetoclastic methanogens from distillery effluent. Ann Biol 10:253–256

Kumar V, Anand RC, Singh R (1994b) Methanogenesis by pure isolates from distillery effluent digester. Ann Biol 10:257–260

Kumar M, Kumar V, Varma A, Pal A, Arshi A, Sharma A, Singh J (2016) An efficient approach towards the bio-remediation of copper, cobalt and nickel contamination from environmental field samples. J Soils Sediment 16:1–10

Kyrikou J, Briassoulis D (2007) Biodegradation of agricultural plastic films: a critical review. J Polym Environ 15:125–150

Le NB, Coleman NV (2011) Biodegradation of vinyl chloride, cis-dichloroethane and 1,2-dichloroethane in the alkene/alkane oxidising mycobacterium strain NBB4. Biodegradation 22:1095–1108

le Mellec A, Karg J, Bernacki Z, Slowik J, Korczynski I et al (2010) Effects of insect mass outbreaks on throughfall composition in even aged European Pine Stands—implications for the C and N cycling. J Earth Sci Climat Change 1:101–110

Nagamani B, Chandana Lakshmi MVV, Sridevi V (2011) Enhanced biodegradation of phenol by Pseudomonas pseudomallei with additional carbon sources. World Congress of Biotechnology, New Delhi

Oaks JL, Gilbert M, Virani MZ, Watson RT, Meteyer CU et al (2004) Diclofenac residues as the cause of vulture population decline in Pakistan. Nature 427:630–633

Ortiz-Hernandez ML, Quintero-Ramirez R, Nava-Ocampo AA, Bello-Ramirez AM (2003) Study of the mechanism of *Flavobacterium* sp. for hydrolyzing organophosphate pesticides. Fundam Clin Pharmacol 17:717–723

Patil R, Bagde US (2012) Isolation of polyvinyl chloride degrading bacterial strains from environmental samples using enrichment culture technique. Afr J Biotechnol 11:7947–7956

Rani B, Kumar V, Singh J, Bisht S, Teotia P, Sharma S, Kela R (2014) Bioremediation of dyes by fungi isolated from contaminated dye effluent sites for bio-usability. Braz J Microbiol 45:1055–1063

Reshma SV, Spandana S, Sowmya M (2011) Bioremediation technologies. World Congress of Biotechnology, New Delhi

Schumacher W, Holliger C (1996) The proton/electron ration of themenaquinone dependent electron transport from dihydrogen to tetrachloroethene in Dehalobacter restrictu. J Bacteriol 178:2328–2333

Sethy NK, Jha VN, Sahoo SK, Shukla AK, Tripathi RM et al (2011) Ground water ingestion dose due to intake of radionuclide (natural U and 226Ra) to population around uranium mining complex at Jaduguda. J Ecosyst Ecograph 1:104

Sharma J, Fulekar MH (2009) Potential of Citrobacter freundii for bioaccumulation of heavy metal copper. Biol Med 1:7–14

Sharma SK, Saxena M, Mandal TK, Ahammed YN, Pathak H et al (2011) Variations in mixing ratios of ambient ammonia, nitric oxide and nitrogen dioxide in different environments of India. J Food Process Technol 1:101

Shen YJ, Lu P, Mei H, Yu HJ, Hong Q, Li SP (2010) Isolation of a methyl parathion degrading strain Stenotrophomonas sp. SMSP-1 and cloning of the ophc2 gene. Biodegradation 21:785–792

Singh R, Singh P, Sharma R (2014) Microorganism as a tool of bioremediation technology for cleaning environment: a review. Proc Int Acad Ecol Environ Sci 4:1–6

Sridevi V, Lakshmi MVVC, Swamy AVN, Rao MN (2011) Implementation of response surface methodology for phenol degradation using *Pseudomonas putida* (NCIM 2102). J Bioremed Biodegrad 2:121

Takami H, Kudo T, Horikoshi K (1997) Isolation of extradiol dioxygenase genes that is phylogenetically distant from other meta cleavage dioxygenase genes. Biosci Biotechnol Biochem 61:530–532

Theriot CM, Grunden AM (2010) Hydrolysis of organophosphorus compounds by microbial enzymes. Appl Microbiol Biotechnol 89:35–43

Vidali M (2001) Bioremediation. An overview. Pure Appl Chem 73:1163–1172

Wilson JT, McNabb JF, Cochran JW, Wang TH, Tomson MB et al (1985) Influence of microbial adaptation on the fate of organic pollutants in ground water. Environ Toxicol Chem 4:721–726

Zhang C, Bennett N (2005) Biodegradation of xenobiotics by anaerobic bacteria. Appl Microbiol Biotechnol 67:600–618

Index

© Springer International Publishing AG 2017

M.Z. Hashmi et al. (eds.), *Xenobiotics in the Soil Environment*, Soil Biology 49,
DOI 10.1007/978-3-319-47744-2

B

F

MIX
Papier aus verantwortungsvollen Quellen
Paper from responsible sources
FSC® C105338

If you have any concerns about our products,
you can contact us on
ProductSafety@springernature.com

In case Publisher is established outside the EU,
the EU authorized representative is:
Springer Nature Customer Service Center GmbH
Europaplatz 3, 69115 Heidelberg, Germany

Printed by Libri Plureos GmbH
in Hamburg, Germany